In-Situ Burning for Oil Spill Countermeasures

In-Situ Burning for Oil Spill Countermeasures

Edited by
Merv Fingas

CRC Press
Taylor & Francis Group
Boca Raton London New York

CRC Press is an imprint of the
Taylor & Francis Group, an **informa** business

CRC Press
Taylor & Francis Group
6000 Broken Sound Parkway NW, Suite 300
Boca Raton, FL 33487-2742

© 2018 by Taylor & Francis Group, LLC
CRC Press is an imprint of Taylor & Francis Group, an Informa business

International Standard Book Number-13: 978-1-138-73525-5 (Hardback)

Visit the Taylor & Francis Web site at
http://www.taylorandfrancis.com

and the CRC Press Web site at
http://www.crcpress.com

Contents

Preface

This book is a first in the topic of in-situ burning for oil spill countermeasures. This represents a significant update to the papers and guides on this topic. Much of this material has not been published before and represents a significant advance in providing information on oil spill burning, especially in the field of emissions from oil burning. All chapters in this collection have been reviewed by two reviewers. The peer-reviewers are acknowledged here (in alphabetical order): Alan Allen, Jeff Cantrell, Hamed Farahani, Ben Fieldhouse, Ala Hamdan, Kurt Hansen, Peter Lane, Steve Lehmann, Brian Mitchell, Shon Mosier, Robert Perkins, Hayri Sezer, Xiaochuan (Lydia) Shi, Gary Shigenaka, Chun Yang and Zeyu Yang.

Special thanks go to the authors, many of whom put in their own time to complete their chapters. Many of them were working on spills or other urgent matters during the preparation of this book. This double-duty is greatly appreciated. The authors' names appear throughout the text.

I would also like to thank many other people who provided support and encouragement throughout this project, including my colleagues and the staff at Taylor & Francis. Special thanks also go to Environment Canada, my employer for many years, during which time I completed much of the research summarised here.

Editor

Dr. Merv Fingas is a scientist working on oil and chemical spills. His specialities include oil chemistry, spill dynamics and behaviour, spill treating agents, remote sensing, spill tracking, in-situ burning, and the fate and behaviour of oil spills. He is currently working on oil spill research in Western Canada. Dr. Fingas has a PhD in environmental physics from McGill University, and three master's degrees (chemistry, business and mathematics), all from University of Ottawa. He also has a bachelor of science in chemistry from the University of Alberta and a bachelor of arts from Indiana University. He has more than 900 papers and publications in the field. Fingas has prepared nine books on spill topics; *In-Situ Burning for Oil Spill Countermeasures* is his tenth. Dr. Fingas is also editor of some journals and serves on several editorial boards. In the past he served as an editor of the *Journal of Hazardous Materials*. He has served on four committees for the National Academy of Sciences of the United States on oil spills including the 'Oil in the Sea'. He is chair of several ASTM International and inter-governmental committees on spill matters. He was the founding chairman of the ASTM sub-committee on in-situ burning and chairman of oil spill treating agents and another on oil spill detection and remote sensing, positions he holds today. He was one of three scientists working with the U.S. National Oceanic and Atmospheric Administration (NOAA) to examine the mass balances of the Gulf spill.

Contributors

Merv Fingas
Spill Science
Edmonton, Alberta, Canada

Patrick Lambert
Emergencies Science and Technology
    Section
Environment and Climate Change
    Canada
Ottawa, Ontario, Canada

Neré Mabile
Private Consultant
Houston, Texas

Jacqueline Michel
Research Planning, Inc.
Columbia, South Carolina

1 In-Situ Burning
An Update

Merv Fingas

CONTENTS

1.1 INTRODUCTION

In-situ burning (ISB) is the oldest technique applied to oil spills, especially on land, but burning has not been used extensively on water until recently. This is because burning, although easy to apply, is not well understood. Further, burning oil on water is not intuitive and thus historically, many people did not pursue this course of action.

ISB has been used to deal with land spills since the first land spill. There is little documentation on burning on land and this trend still continues. Of the few documented cases, most were successful and resulted in obvious environmental benefits.

Several reviews of burning have been prepared in the past (API 2015a, b, 2016a; Fingas 2011, 2016a, 2017). These reviews show a steady progression in the science and technology of oil spill burning.

1.2 AN OVERVIEW OF IN-SITU BURNING

It is important to establish in the reader's mind the basic principles of the science of burning. This will establish the basis for the remainder of the discussion.

1.2.1 THE SCIENCE OF BURNING

The fundamentals of ISB are similar to that of any fire, namely, that fuel, oxygen and an ignition source are required. Fuel is provided by the vaporisation of oil. The vaporisation of the oil must be sufficient to yield a steady-state burning, that is, one in which the amount of vaporisation is about the same as that consumed by the fire (Fay 2006). Once an oil slick is burning, it burns at a rate of about 1–4 mm/min. This rate is limited by the amount of oxygen available and the heat radiated back to the oil. The oil burn rate is a function of the oil type as well as conditions such as the presence of ice. If insufficient vapours are produced, the fire will not start or will be quickly extinguished if it does start. The quantity of vapours produced depends on the amount of heat radiated back to the oil. The heat radiated back to the pool has been estimated to be about 0.5%–1% of the heat from a fire for a pool fire. If the oil slick is too thin, some of this heat is conducted to the water layer below it. Since most oils have the same insulation factor, most slicks must be about 0.5–2 mm thick to yield a quantitative burn. Once burning, the heat radiated back to the slick and the insulation is usually sufficient to allow combustion down to about 0.5 mm of oil. The basic concepts of burning oil on water are illustrated in Figure 1.1. Several scientists have noted that burning is a vaporisation process in that, as the oil is burned, the molecular weight and boiling point of the remaining molecules are higher and higher (Van Gelderen et al. 2015b, 2017c). This in fact continues until there are insufficient vapours being boiled off and the fire ceases.

If greater amounts of fuel are vaporised than can be burned, more soot is produced as a result of incomplete combustion, and fuel droplets are released downwind or, more typically, small explosions or fireballs occur. The latter phenomenon is often observed when gasoline or light crudes are burning (Zheng et al. 2014). It has been shown that diesel fuel burns differently than other fuels, with a tendency to atomise rather than vaporise. This results in a heavier soot formation. Soot formation occurs by several processes. One common process is the aggregation of molecular species into larger compounds; another process is the partial combustion of fuels such as diesel fuels (Fingas 2017). Most other fuels evaporate under the influence of heat and do not form droplets as diesel, kerosene or jet fuels do.

The amount of oil that can be removed in a given time depends on the fuel and on the area covered by the oil. Most oil pools burn at a rate of about 1–4 mm/min, which means that the depth of oil is reduced by that value of millimetres per minute. As a rule, oil burn rate is about 2,000–5,000 L/m^2.day (Fingas 2017). Several tests have shown that this does not vary significantly with oil weathering but varies with oil type. Emulsified oil may burn slower as its water content increases the heat requirement. The burn rates for many crude oils in ice are between 1 and 3 mm/min, typically half the rate when ice was not present. This is because of the heat loss to the ice.

The type of oil is relatively unimportant in determining how an oil ignites and burns, except for heavier or emulsified oils. However, heavy oils require longer heating times and a hotter flame to ignite than lighter oils and may often require a primer such as kerosene or diesel fuel.

Oil that is emulsified with water can be ignited. Once started, it is believed that most emulsions burn. Boom tows with un-emulsified oil and emulsified oil can be

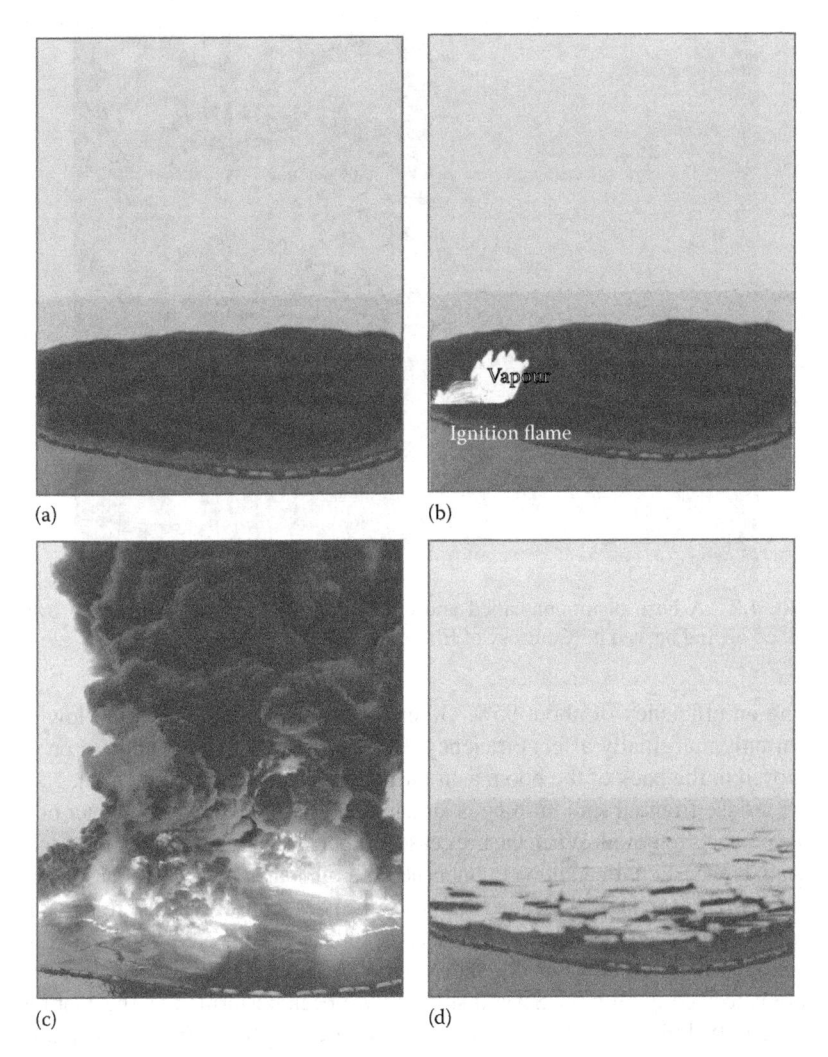

(a) (b)

(c) (d)

FIGURE 1.1 The basic concepts of burning oil on water. Burning on land is similar except that collection of the oil before burning is not needed: (a) oil collected into fire-resistant boom or segregated area, (b) ignition – flame applied to oil creates vapour which ignites, (c) flame spreads over slick and burns oil down to about 1 mm and (d) after vapour is insufficient to maintain combustion, fire ends.

burned by igniting the un-emulsified oil. The heat from burning the un-emulsified oil starts the emulsified oil. Figure 1.2 shows a fire started on an un-emulsified oil that later spread to the emulsified oil and burned it quantitatively.

Burn efficiency is the initial volume of oil before burning, less the volume remaining as residue, divided by the initial volume of the oil. Efficiency is largely a function of oil thickness. For example, a slick of 2 mm burning down to 1 mm yields a maximum efficiency of 50%. A pool of oil 20 mm thick burns to approximately 1 mm,

FIGURE 1.2 A burn of un-emulsified and emulsified oil. The fire later spread over the emulsified oil and burned it. (Courtesy of Elastec, Carmi, IL.)

yielding an efficiency of about 95%. Other factors such as oil type and low water content only marginally affect efficiency. However, if the oil is contained on water and moved to the back of the boom as it burns, efficiency can be very high.

The residue from oil spill burning is largely unburned oil with some lighter or more volatile products removed. When the fire ceases, unburned oil is left that is simply too thin to sustain combustion. In addition to unburned oil, oil is also present that has been subjected to high heat and is thus weathered. Finally, heavier soot particles are re-precipitated into the fire. Highly efficient burns of some types of heavy crude oil may result in oil residue that sinks in seawater. Figure 1.3 shows typical residue after an at-sea burn.

Soot is formed in all fires. Several studies have been performed on the soot formation process (Schnitzler et al. 2014). Soot particles are largely graphitic carbon in layers with particles ranging from 5 to 60 nm. These particles then aggregate to form larger particles. The amount of soot produced is not precisely known because there is no direct means of measuring soot from large fires. It is believed that the amount of soot is about 0.1%–3% for crude oil fires and about 1%–5% for diesel fires. An additional consideration is that the soot precipitates out at a rate approximating the square of the distance from the fire. Thus, a constant percentage of soot for a whole fire may be irrelevant.

The total heat radiated by a given burn has been measured as 1.1 MW/m^2 (Fingas 2017). Calculation reveals that the heat required to vaporise the oil was 6.7 KW/m^2 and the heat lost from conduction through the slick to the underlying water was 2.5 KW/m^2. Thermal radiation is important. One test revealed burning Alaska North Slope oil showed a heat release rate of 176 KW/m^2, diesel fuel 230 KW/m^2 and propane 70 KW/m^2. The heat radiated by a liquid propane fire enhanced by air flow and increased pressures was 180 KW/m^2. The heat flux on booms as a result of these fires was reported as 140–250 KW/m^2 for crude oils, 120–160 KW/m^2 for diesel fuel,

FIGURE 1.3 Some of the residue left over from a multi-hour burn at sea. This amount of residue represents less than 0.1% of the oil that was burned. (Courtesy of Elastec/American Marine, Carmi, IL.)

60–100 KW/m^2 for propane and 100–160 KW/m^2 for enhanced propane burning (Muñoz et al. 2004, 2007).

Flame spreading rates have been measured at several fires. Flame spreading rates do not vary much with fuel type, but vary significantly with wind, especially as this relates wind direction (Fu et al. 2017; Li et al. 2015, 2016). Flame spreading rates range from 0.01 to 0.02 m/s (0.02–0.04 knots). Downwind flame spreading rates range from 0.02 to 0.04 m/s (0.04–0.08 knots), and up to 0.16 m/s (0.3 knots) for high winds. Flame spread through vapour clouds can be as fast as 100 km/h, such as would occur with spills of gasoline or similar volatile fuels.

The flame height of a small fire less than 10 m in diameter is about twice that of the diameter of the fire. The flame height approaches the diameter of the pool up to about 100 m in diameter. Thus, an estimate of flame height for a fire in a boom with a radius of about 10–20 m is about 1.5 times the diameter or 15–30 m.

Several workers reported on findings that there is a vigorous burn phase near the end of a burn on water. Significant amounts of heat are transferred to water near the end of a burn, when slick thickness approaches 1 mm and this heat ultimately causes the water to boil. The boiling injects steam and oil into the flame giving rise to a 'vigorous' burn with the production of steam. This phenomenon occurs only in shallow test tanks because there is little movement of water under the slick to carry the heat away. During burns at sea, no vigorous burning or water heating is observed.

Several studies on burn research were carried out using small-scale test apparatuses (Bellino et al. 2013; Brandvik et al. 2010a–c; Jézéquel et al. 2014; van Gelderen et al. 2015a, b, 2017a, c; van Gelderen and Jomaas 2017). This has resulted in extensive data; however, several parameters such as efficiency and burning rate have been shown to be skewed due to edge effects below about 1 m dimensions (Fingas et al. 2004;

van Gelderen and Jomaas 2017). Burn rates on small-scale burns are typically one-quarter to as much as one-tenth that of larger-scale burns. Efficiencies are often reduced 10%–50% from larger burns. This will be discussed in greater detail later in this chapter.

1.2.2 What Burns and Doesn't Burn

Basically, most oils burn on water, and they burn quantitatively if over about 2 mm thick. On land or wetlands, the situation is similar, although 1 mm thick oils on dry grassland can be burned quantitatively. Light and fresh oils burn readily and can be easily ignited. Heavy oils require a small amount of primer, such as diesel fuel, to start ignition. Once burning, heavy oils burn well, and even emulsified oil breaks down and burns. Table 1.1 shows the ignition characteristics of various oils. This is independent of whether the oil is on land or on water.

1.2.3 Summary of In-Situ Burning Research and Trials

The first reference in the literature to the burning of oil on water was the use of a log boom to burn oil on the Mackenzie River in 1958 (McLeod and McLeod 1972). Failed attempts to ignite the oil spilled from the *Torrey Canyon* in 1968 were widely known (Swift et al. 1968). Extensive research on ISB of oil spills began in the late 1970s and was carried out in North America by Environment Canada, the U.S. Coast Guard (USCG), the U.S. Minerals Management Service (USMMS) and the U.S. National Institute of Standards and Technology (NIST).

TABLE 1.1
Burning Properties of Various Fuels

Fuel	Burnability	Ease of Ignition	Flame Spread	Burning Rate[a] (mm/min)	Efficiency Range (%)
Gasoline	Very high	Very easy	Very rapid – through vapours	3.5–4	95–99
Diesel fuel	High	Easy	Moderate	3–4.0	90–98
Light crude	High	Easy	Moderate	3–3.7	85–98
Medium crude	Moderate	Easy	Moderate	3–3.7	80–95
Heavy crude	Moderate	Medium	Moderate	3–3.5	75–90
Light fuel oil	High	Medium	Moderate	2.5–3	80–90
Weathered crude	Low	Difficult, add primer	Slow	2.8–3.5	50–90
Crude oil with ice	Low	Difficult, maintain heat	Slow	2–2.5	60–90
Heavy fuel oil	Very low	Difficult, add primer	Slow	2.5–2.8	40–70
Waste oil	Very low	Difficult, add primer	Slow	2–2.5	15–50
Emulsified oil	Low	Difficult, primer, high heat	Slow	1–2	40–80

[a] Typical rates only – to get the rate in $L/m^2/h$, multiply by 60.

Over the years, research into ISB has included laboratory-, tank- and full-scale test burns. In the late 1970s several burn tests and studies were carried out in Canada by a consortium of government and industry agencies. Figure 1.4 shows oil resurfaced on ice; Figure 1.5 shows the Beaufort Sea Burn carried out in 1975. Some tests in the early 1980s were performed by ABSORB (now Alaska Clean Seas) and USMMS to evaluate the burning of oil in ice-covered areas. This research covered environmental and oil conditions such as sea state, wind

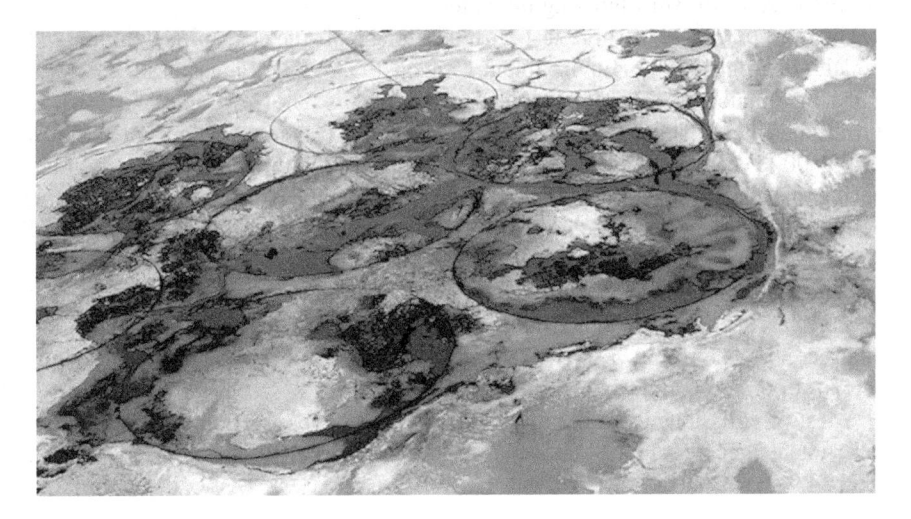

FIGURE 1.4 Oil that has resurfaced from an under-ice experiment in Balaena Bay, Beaufort Sea, Canada. The oil was released under the ice in 1974 and re-surfaced through first-year ice during the sea ice melt in 1975.

FIGURE 1.5 The oil shown in Figure 1.4 being burned.

velocities, air and water temperatures, ice coverage, oil type, slick thickness and degree of oil weathering and emulsification (Tennyson 1994). Several tests have also been performed in an oil spill test tank at the Oil and Hazardous Materials Simulated Environmental Test Tank (OHMSETT) facility in New Jersey. In the early 1990s, several meso-scale burns were performed at the USCG Joint Maritime Test Facility (JMTF) in Mobile, Alabama. Figure 1.6 shows one of the burns at Mobile. Table 1.2 lists some of the tests and burns since the first recorded use of oil spill burning on water.

The most extensive offshore test burn took place off the coast of Newfoundland, Canada, in August 1993 (Environment Canada 1993, 1997; Fingas et al. 1994a, b, 1995a, b). Chapter 5 in this book provides further details. The Newfoundland Offshore Burn Experiment (NOBE) involved 25 agencies from Canada and the United States. Two 50,000 L batches of oil were released and burned within a fire-resistant boom. During this test, more than 2,000 parameters were evaluated using various sampling methods. Figures 1.7 and 1.8 show the NOBE burn. The major findings were that all emission and pollutant levels measured 150 m away from the burn were below health concern levels and that at 500 m from the burn, these levels were difficult to detect. In many cases, pollutants in the smoke plume were less than detected in the original unburned oil. The results also showed that the emission levels from this large burn were lower than found during the meso-scale burns.

A test of emissions from fires was carried out by a consortium of industry and government agencies at a test facility in Calgary, Alberta (Booher and Janke 1997). Figure 1.9 shows one of these burns.

FIGURE 1.6 A view of one of the test burns at the USCG facility in Mobile, Alabama, in 1998. Diesel fuel is being burned here as witnessed by the heavy, black smoke plume.

TABLE 1.2
Summary of Burns or Tests

Year	Country	Location/Incident	Description
1958	Canada	Mackenzie River, North West Territories	First recorded use of in-situ burning, on river using log booms
1967	Britain	Torrey Canyon	Cargo tanks difficult to ignite with military devices
1969	Netherlands series of experiments		Igniter KONTAX tested, many slicks burned
1970	Canada	Arrow	Limited success burning in confined pools
1970	Sweden	Othello/Katelysia	Oil burned among ice and in pools
1970	Canada	Deception Bay	Oil burned among ice and in pools
1973	Canada	Rimouski – experiment	Several burns of various oils on mud flats
1975	Canada	Balaena Bay – experiment	Multiple slicks from under ice oil ignited
1976	United States	Argo Merchant	Tried to ignite thin slicks at sea
1976	Canada	Yellowknife – experiment	Parameters controlling burning not oil type alone
1978–1982	Canada	Series of experiments	Studied many parameters of burning
1979	Mid-Atlantic	Atlantic Empress Aegean Captain	Uncontained oil burned at sea after accident
1979	Canada	Imperial St. Clair	Burned oil in ice conditions
1980	Canada	McKinley Bay – experiment	Several tests involving igniters, different thicknesses
1981	Canada	McKinley Bay – experiment	Tried to ignite emulsions
1983	Canada	Edgar Jourdain	Vessel containing fuels and nearby fuel ignited
1983	United States	Beaufort Sea – experiment	Oil burned in frazil ice
1984	Canada	Series of experiments	Tested the burning of uncontained slicks
1984–1985	United States	Beaufort Sea – experiment	Burning with various ice coverages tested
1984–1986	United States	OHMSETT – experiments	Oil burned among ice but not with high water content; ice concentration not important; emulsions don't burn
1985	Canada	Offshore Atlantic – experiment	Oil among ice burned after physical experiment
1985	Canada	Esso–Calgary – experiments	Several slicks in ice leads burned
1986	Canada	Ottawa – experiments/ analysis	Analysed residue and soot from several burns
1986	United States	Seattle and Deadhorse – experiments; test of the Helitorch and other igniters	
1986–1991	United States	NIST – experiments	Many lab-scale experiments
1986–1991	Canada	Ottawa – analysis on above	Analysed residue and soot from several burns

(*Continued*)

TABLE 1.2 (*Continued*)
Summary of Burns or Tests

Year	Country	Location/Incident	Description
1989	United States	*Exxon Valdez*	Test burn performed using a fireproof boom
1991	United States	First set of Mobile experiments	Several test burns in newly-constructed pan
1992	United States	Second set of Mobile burns	Several test burns in pan
1992	Canada	Several test burns in Calgary	Emissions measured and Ferrocene tested
1993	Canada	Newfoundland, offshore burn	Successful burn on full scale off shore
1994	United States	Third set of mobile burns	Large-scale diesel burns to test sampler
1994	United States	North Slope burns	Large-scale burn to measure smoke
1994	Norway	Series of Spitzbergen burns	Large-scale burns of crude and emulsions
1994	Norway	Series of Spitzbergen burns	Try of uncontained burn
1996	Britain	Burn test	First containment burn test in Britain
1996	United States	Test burns in Alaska	Igniters and boom tested
1997	United States	Fourth set of Mobile burns	Small-scale diesel burns to test booms
1997	United States	North Slope tank tests	Conducted several tests on waves/burning
1998	United States	Fifth set of mobile burns	Small-scale diesel burns to test booms
2001	United States	Boom tests in OHMSETT	Small-scale propane tests of test booms
2002	United States	Small-scale tests in Alaska	Tested burning in frazil and brash ice
2003	Canada	Small-scale tests on heavy oils	Tested procedures to burn heavy and emulsified fuels
2004	Canada	Small-scale tests on heavy oils	Tested procedures to burn heavy and emulsified fuels
2008	Svalbard, Norway	Burns in ice	Tested burning in frazil and brash ice
2010	Gulf of Mexico	411 actual burns	At-sea burn lessons

Tests of various aspects of burning were conducted at the USCG facility in Mobile Bay, Alabama, in 1991, 1992, 1994, 1997 and 1998. More than 50 burns were conducted using crude oil and diesel fuel. Physical parameters were measured as well as emission data. Fireboom test evaluations using diesel fuel were conducted in 1997 and 1998 by the National Institute of Standards and Technology (NIST) and sponsored by the U.S. Coast Guard Research and Development Center and USMMS (Walton et al. 1998, 1999). Five booms were evaluated in 1997 and six in 1998. The test evaluations were conducted in a wave tank designed specifically for evaluating fire-resistant containment booms located at the U.S. Coast Guard Fire and Safety Test Detachment facility in Mobile Bay, Alabama. The wave tank was designed to accommodate a nominal 15-m boom section, forming a circle approximately 5 m in diameter. Figure 1.10 shows a boom undergoing a fire test and Figure 1.11 shows a boom under the influence of waves, without fire present. The test cycle consisted of

FIGURE 1.7 A distant view of the Newfoundland Oil Burn Experiment (NOBE) in 1993. Most of the vessels behind the burn are associated with emission measurements.

FIGURE 1.8 A closer view of the burn at NOBE. The three-point boom tow is partly visible. The helicopter was used for observation and filming.

FIGURE 1.9 One of the burns conducted in Calgary, Alberta, Canada, to study burns and smoke suppression using ferrocene. This particular burn had no smoke suppression.

FIGURE 1.10 A stainless steel boom undergoing fire tests in Mobile, Alabama. Note that this boom is leaking oil from its joints.

FIGURE 1.11 A fire-resistant boom being tested in Mobile, Alabama. The boom is shown undergoing a cool-down period under waves with no fire present.

three 1-h burning periods with two 1-h cool-down periods between the burning periods, in accordance with the American Society of Testing and Materials (ASTM) F-20 Committee standard (ASTM 2152 2013). Four of the six booms evaluated in 1998 were shipped to the OHMSETT facility for post-burn oil containment and tow tests based on ASTM suggestions. In general, there was some degradation of materials in all of the booms.

More tests were conducted in 1996 and 1997 by S.L. Ross Environmental Research Ltd., sponsored by USMMS and the Canadian Coast Guard (Buist et al. 2013; McCourt et al. 2000, 2005). These tests evaluated firebooms using propane rather than the smoke-producing fuels such as diesel or crude oil. The propane test evaluations were conducted in a wave tank located at the Canadian Hydraulic Centre, National Research Council of Canada in Ottawa. The heat flux measured in the 1997 tests with air-enhanced propane was comparable to those measured in the diesel fuel fires.

Two separate fireboom test evaluations using air-enhanced propane were conducted in the fall of 1998 by MAR, Inc. and S.L. Ross Environmental Research Ltd (McCourt et al. 1998, 1999). Both tests were conducted at the OHMSETT facility in Leonardo, New Jersey. The first test was sponsored by USMMS and the U.S. Navy Supervisor of Salvage (SUPSALV). Three candidate fire protection systems were tested and evaluated. Each consisted of a water-cooled blanket designed to be draped over existing oil boom to protect its exposure to an in-situ oil fire. In the second fireboom evaluation, a prototype stainless steel PocketBoom was tested and evaluated using an air-enhanced propane system. The PocketBoom was a redesign of the Dome boom originally developed for use in Arctic seas. Liquid propane from a storage tank was heated to create gaseous propane and piped to an underwater bubbling system.

The test protocol was similar to the ASTM 2152 method. The booms generally survived the tests and showed less degradation than previous models of the same booms.

Studies on ISB were carried out at sea in May 2009, in Svalbard, Norway, with various levels of ice. These tests included studies on weathering of oil, the utility of fire-resistant booms and the use of herding agents for thickening the oil (Brandvik et al. 2010a; Buist 2010; Potter 2010).

1.2.4 THE DEEPWATER HORIZON BURN

The burning that took place at Deepwater Horizon certainly changed the history of in-situ burning. For 35 years the history of ISB had largely been that of small tests, some small burns, many land burns and a few larger tests. What was needed was a few large actual and successful burns at sea to prove that the technique was viable. Indeed, there were about 400 successful burns carried out during the Deepwater Horizon spill and this removed a large part of the oil on the water. This spill is dealt with in detail in Chapter 4 of this book. Table 1.3 summarises some aspects of these burns (Allen et al. 2011; Mabile 2012a, b). The basic technique was to collect oil in a fire-resistant boom (hereinafter called fireboom) and then ignite the oil and slowly pull the fireboom forward to push the oil to the rear or wait if the winds and currents were doing this (Allen et al. 2011; Mabile 2012a, b). The oil was spotted using a fixed-wing aircraft. Two shrimp boats

TABLE 1.3
Summary of the Deepwater Horizon Burns

Amount Burned	35,000–50,000 m^3 (220,000–310,000 Barrels)
Number of fires	411 (396 effective ones)
Time of fires (range)	10 min–12 h
Dates	28 April to 19 August 2010 (83 days)
Location	Approximately 5–25 km (3–15 mi) from source – about approximately 60 km (40 mi) from shore
Average burned/fire	110 m^3 (700 barrels)
Average burn time	Approximately 2 h
Most oil burned in one day	Approximately 9,600 m^3 (approximately 60,000 barrels) (18 June)
Burn teams	8–12
People per burn team	7 or 8
Total people involved	Less than 100
Spotting aircraft	2 King Airs
Spotters	10
Fireboom used	7,000 m (23,000 ft)
Types used	4 types, mostly Elastec/American Marine, then Applied Fabric Technologies
Boom used per burn	Approximately 150 m (approximately 500 ft)
Fireboom lifetimes	Ranging from one to typically 12–14 burns
Large vessels	Approximately 10 supply boats and large shrimp boats
Small vessels	Approximately 20 rigid hull inflatable or aluminium skiffs
Igniters	1,700 handheld with gelled diesel and marine flares

(about 100-ft-long) towed about 150 m (500 ft) of fireboom at about ½–¾ knot to avoid loss of the oil through entrainment under the boom. The tow lines were about 100 m (about 300 ft) for the safety of the tow crews. Once sufficient oil had been collected for a burn and marine and air monitoring approved, ignition was requested. A small boat carrying two persons approached from upwind, and an igniter was dropped over the edge of the boom. The igniters were made from a plastic jar (about 1 L) of gelled diesel fuel, a marine flare and some Styrofoam floats. The flare, once activated, burned down to the bottle of gelled diesel fuel, which started burning and acted as a primer to ignite the oil. Figures 1.12 and 1.13 show some aspects of the burn.

Once lit the heavy, weathered oil burned until most oil was removed. The burn was monitored from the air by trained observers and from larger vessels in the area. The amount burned was gauged by measuring the burning area in the boom and multiplying by the burning time for that area (Allen et al. 2011; Mabile 2012a, b).

Many precautions were taken during the burn. Extensive training was given to the crews and several practice sessions were undertaken. Particulate emissions from the burns were monitored.

1.2.5 BURNING IN DIFFERENT SITUATIONS

No two oil spill situations are the same, so it is helpful to look at several possible scenarios when developing response techniques for spill situations. The following specific spill scenarios include: burning on land, burning an oiled marsh, burning at sea, burning in a protected bay, burning on a river, burning on melt pools in the Arctic and burning in an intertidal zone.

FIGURE 1.12 Initiating an igniter during the Deepwater Horizon spill. The device will then be thrown onto the oil in the boom.

FIGURE 1.13 A surveillance crew monitors a burn during the Deepwater Horizon clean-up.

Burning on land or oiled marshes has been carried out extensively in the past and will, no doubt, serve well as a countermeasure in the future. There is an extensive body of knowledge on this application. This is the subject of Chapter 2 of this book.

The well-known tactic of using towed fireboom at sea to collect and burn oil directly in the boom, has been implemented in practice. As with all booms, this technique has a relative current limitation of 0.4 m/s (0.7 knots) before oil is lost under or over the boom. This can be overcome on the open ocean by towing at the relative velocity, despite the surface current. This means that if the actual current exceeds 0.4 m/s (0.7 knots), the boom tow could be slipping down current. Another limitation of this method is that the fire could propagate to the source of the oil or endanger the tow boats and their crews.

Collecting the oil separately, towing the boom away from a non-burning source and then burning the oil is another technique. This approach prevents the fire from spreading to the oil source. Another advantage is that the oil can be collected using a conventional boom and then transferred to a fire-resistant boom for actual burning. Since fire-resistant boom is more expensive and harder to deploy than conventional boom, this option has some practical and economic benefits.

Burning oil in ice is a standard method for dealing with oil in ice. The natural containment of ice can serve to thicken oil sufficiently for ignition and burning to take place. This technique has often been used to burn oil spills in the Arctic.

1.2.6 Advantages and Disadvantages

ISB has some distinct advantages over other spill clean-up methods, including:

- Rapid removal of large amounts of oil from the water surface
- Significantly reduced volume of oil requiring disposal
- High efficiency rates
- Less equipment and labour required
- May be only clean-up option in some situations, such as oil-in-ice situations

The most significant of these advantages is the capacity to rapidly remove large amounts of oil. When used at the right time, that is, early in the spill before the oil weathers and loses its more flammable components, and under the right conditions, ISB can be very effective at rapidly eliminating large amounts of spilled oil, especially from water. This can prevent oil from spreading to other areas and contaminating shorelines and biota. Compared with mechanical skimming of oil, which generates a large quantity of oil and water that must be dealt with, burning generates a small amount of burn residue. This residue is relatively easy to recover or can be further reduced by repeated burns.

While the efficiency of a burn varies with a number of physical factors, removal efficiencies are generally much greater than those for other response methods such as skimming and the use of chemical dispersants.

In ideal circumstances, ISB requires less equipment and labour than other techniques. It can be applied in remote areas where other methods cannot be used because of distances and lack of infrastructure. Often not enough of these resources are available when large spills occur. Burning is relatively inexpensive in terms of equipment needed and actually conducting the burn operations.

ISB also has disadvantages, including:

- The large black smoke plume created and public fears about toxic emissions to the air and water
- Difficulty in igniting the oil in some circumstances
- Oil on water must be a thicker than about 2–3 mm to burn quantitatively and must usually be contained to achieve this thickness
- Risk of fire spreading to other combustible materials
- Burn residue sometimes requires collection and disposal

The most obvious disadvantage of burning oil is the large black smoke plume that is produced and public concern about emissions. Extensive studies have recently been conducted to measure and analyse these emissions. The second disadvantage is that the oil on water will not burn quantitatively or even ignite if it is not thick enough. Most oils spread rapidly on water and the slick quickly becomes too thin for burning to be feasible. Fire-resistant booms can be used to concentrate the oil into thicker slicks so that the oil can be burned effectively. While this obviously requires equipment, personnel and time, concentrating oil for burning requires less equipment than collecting oil with skimmers.

Finally, burning oil is sometimes not viewed as an appealing alternative to collecting the oil and reprocessing it for reuse. It must be pointed out; however, that recovered oil is usually incinerated as it often contains too many contaminants to be economically reused.

1.2.7 COMPARISON OF BURNING TO OTHER RESPONSE MEASURES

Burning has long been compared to other countermeasures (Dave and Ghaly 2011; Prendergast and Gschwend 2014). ISB is most often compared with the use of dispersants as a countermeasure. Dispersants are chemical spill-treating agents that promote the formation of small droplets of oil that 'disperse' throughout the water column. Dispersants contain surfactants, chemicals like those in soaps and detergents that have both a water-soluble and an oil-soluble component. Surfactants or surfactant mixtures used in dispersants have approximately the same solubility in oil and water, which stabilises oil droplets in water so that the oil disperses into the water column. The comparison between dispersion and other countermeasures is poor because the time and area scales differ widely.

ISB can also be compared to mechanical recovery of oil spills. Mechanical recovery includes the use of booms and skimmers to physically contain the oil and remove it from the water. Booms are limited to waters where the currents, relative to the boom, are less than 0.4 m/s or they must be used in diversionary mode. On the other hand, while recovery using booms and skimmers is slower than removal by ISB or dispersants, the oil is recovered without the potential for air and water pollution. Mechanical recovery works well in sheltered waters such as harbours and marinas where burning should not be conducted, but is impossible in high currents and waves over 2 m.

On-land burning has significant advantages over most techniques. Unless the oil is very thick, the ability to pump the oil is very limited. Any recovery process that takes a long time will allow oil to penetrate the soil.

In some marine spill situations, the best clean-up strategy involves a combination of mechanical recovery techniques and burning for various portions of a spill. For example, burning can be applied in open water, and oil that has already moved closer to shore can be recovered with booms and skimmers. Burning could also be used on open water after the window of opportunity closes for effective use of dispersants. Burning does not preclude the use of other countermeasures on other parts of the slick. When combining different clean-up techniques, the objective should be to find the optimal mix of equipment, personnel and techniques that results in the least environmental impact of the spill.

Given the major differences between time scales and between the methods, Table 1.4 shows the estimated differences between among various countermeasures. The assumptions in this table attempt to equalise the differences between the methods. The comparison in Table 1.4 shows that ISB is favourable for most conditions, especially for the case of heavy oils.

1.2.8 SAFETY

Many points of safety must be considered when using ISB. These include, but are not limited to:

TABLE 1.4

Approximate Comparison of On-Water Countermeasures

	Light Crude		Heavy Crude		Bunker C	
	Hours to Clean	Tons/ Hour	Hours to Clean	Tons/ Hour	Hours to Clean	Tons/ Hour
Brush drum skimmer	7.5	8	30	2	75	1
Large weir skimmer	1.5	40	0.9	71	18	4
Dispersants	0.2	75	0.2	47	0.2	10
In-situ burning	0.2	356	0.3	238	0.3	238

Note: There are many assumptions in the table, including capacities of two average skimmers, dispersant effectiveness, but the burn rate is actual. This comparison is for a 150 m boom filled over time with 75 tons of oil.

- Flashback of burning vapour during ignition
- Spread of fire beyond the control points
- Fumigation by smoke of nearby areas during burning
- Loss of containment of the fire-resistant boom
- Exposure of people to emissions from the fire
- Exposure of biota to emissions or fire
- Issues of safety in marine and airborne operations
- Issues in handling the fire-resistant boom

These issues will be covered in this chapter. Precautions have been shown to eliminate injuries during actual burns. An important facet of ISB is to prepare a comprehensive health and safety plan (HASP).

1.3 DETAILED SCIENCE

1.3.1 PROCESS OF BURNING

The fundamentals of ISB are similar to that of any fire, namely, that fuel, oxygen and an ignition source are required (Crawley 1982; Evans et al. 1991). Fuel is provided by the vaporisation of oil. The vaporisation of the oil must be sufficient to yield a steady-state burning, that is, one in which the amount of vaporisation is about the same as that consumed by the fire. Once an oil slick is burning, it burns at a rate of about 0.5–4 mm/min. This rate is limited by the amount of oxygen available and the heat radiated back to the oil. The oil burn rate is a function of the oil type as well as environmental conditions such as the presence of ice. The 'steady-state' burning implies that the conditions noted above are met (Thompson et al. 1979). If not enough vapours are produced, the fire will either not start or will be quickly extinguished. The amount of vapours produced depends on the amount of heat radiated back to the oil. This has been estimated to be about 1%–2% of the heat from a fire for a pool fire (Buist et al. 1994; Nakakuki 2002). If the oil slick is too thin, some of this heat is conducted to the water layer below it. Since most oils have the same insulation factor, most slicks must be about 0.5–2 mm thick to yield a quantitative burn. Once burning, the heat radiated back to the slick and the insulation is usually sufficient to allow combustion down to about 1 mm of oil. Some have noted that burns of diesel fuel and gasoline go through a three-phase process (Chatris et al. 2001). The first phase occurs when the burn spreads over the whole pool and then increases to the maximum value. The second step is when the fire burns at a relatively constant rate over the entire pool. In the third step, the fire falls back and the rate decreases until the fire is extinguished.

1.3.2 Soot Formation

If greater amounts of fuel are vaporised than can be burned, more soot is produced as a result of incomplete combustion, and fuel droplets are released downwind or, more typically, small explosions or fireballs occur (Xu et al. 2003a, b). The latter phenomenon is often observed when gasoline or light crudes are burning. It has been shown that diesel fuel burns differently than other fuels, with a tendency to atomise rather than vaporise (Frank et al. 2013). This results in an obviously heavier soot formation (Fingas et al. 1996b). Soot formation is an issue that has been studied by several scientists over many years (Apicella et al., 2002; Balthasar and Frenklach 2005; Balthasar et al. 2002; Fingas 2010; Gruenberger et al. 2002; Law 2005; McEnally and Pfefferle, 2009; McEnally et al. 2006; Middlebrook et al. 2012; Murphy and Shaddix 2006; Perring et al. 2011; Smooke et al. 2005; Xiao et al. 2005). Soot formation occurs by several processes. One common process is the aggregation of molecular species into larger compounds; another process is the partial combustion of fuels such as diesel fuels (Wilson et al. 2013). Diesel fuels and kerosene are known to burn with more soot than most other fuels (Dagaut and Cathonnet 2006; Dagaut et al. 2006; Eddings et al. 2005; Morandini et al. 2005; Sazhin 2017). This occurs for several reasons. Diesel fuel and kerosene can form droplets under heat and these droplets often burn only partially, leaving carbonaceous material on the inside or even whole fuel with carbonaceous material or soot on the outside. Most other fuels evaporate under the influence of heat and do not form significant amounts of droplets such as diesel, kerosene or jet fuel.

Soot formation has been the topic of much discussion over the years (Barakat et al. 1998; Zheng et al. 2014). Studies have been carried out on soot values (Fingas 2010; Fingas et al. 1996b). These studies concluded that the value of soot produced varies significantly down-plume because of precipitation of material. Downwind concentrations are likely to be as low as 0.5% for crude oil and about 3% near the fire (Fingas 2010; Johnson et al. 2011). The results of one study indicate an average of 1.8% (Fingas 2010). For diesel, the downwind concentrations may be about 1% and about 7% near the source. Concentrations in the plume downwind and far downwind are likely to be fractions of a percent. There is significant variation, however, for the type of material burned and for environmental conditions.

1.3.3 SLICK THICKNESS

Most, if not all, oils burn on water if sufficient vapours can be produced by the ignition and subsequent fire. Except for light refined products, different types of oils
have not shown significant differences in burning behaviour. Weathered oil requires
a longer ignition time and somewhat higher ignition temperature (Twardus 1980).
This is to produce sufficient vapours to sustain combustion. Alternatively, weathered or heavy oils can be ignited with the addition of a primer (Fingas 2002). At the
time of the *Torrey Canyon* spill (1967), it was not known that the thickness of the
oil would be a limitation. Studies conducted shortly after this incident concluded
that the slicks that did not ignite were below minimum thickness (Glassman and
Hansel 1968). Another group studied oil ignition thicknesses and found that slicks
that were 3 and 6 mm thick burned (Maybourn 1971). Preliminary tests of minimum
burning thicknesses were conducted and it was proposed that all fuels burned at
the 5-mm thickness initially tried (Twardus 1980; Twardus and Bruzustowski 1981).
Bunker C required longer heating times and the addition of a primer. Further testing
on light crudes showed that the minimum thickness for ignition was 0.58–0.62 mm
and the residues varied between 0.35 and 0.58 mm (Twardus and Bruzustowski
1981). This was compared to unconfined fresh oil thicknesses of 0.5–0.6 mm at 0°C,
0.2–0.25 mm at 5°C and 0.5 mm at 10°C. Aged oil showed limiting spreading thicknesses of 1.90–3.0 mm at 0°C, 1.2–2 mm at 5°C and 1.2–1.3 mm at 10°C. Later tests
showed that thicknesses greater than about 0.5 mm burned for all types tested (Fingas
and MacKay 2003; Fingas et al. 2003). Indeed, it is the amount of vapours present
and not the oil thickness that is important.

Burn rates of various crudes were found to decrease at thicknesses from 18 to
1 mm in early studies, but most oils could be ignited at 1–2 mm (Arai et al. 1993). It
was thought that the initial burn thickness depended on variances in the thermal conductivity of the starting oil. In one study, the thermal conductivity of three crude oils
were measured as 130 mW/m K over a 50 K temperature range (Elam et al. 1989).
Little difference was found for oil type or temperature. Overall, many workers have
concluded that the rule of thumb is that the minimum ignitable thickness of oil is
1–3 mm; however, most did not test thin layers nor establish minimums. One study
showed that even heavy oils at thicknesses of 0.5 mm and above could be ignited,
sometimes with the aid of diesel as a primer (Fingas 2002). This again confirms that
it is the amount of vapours present, not the thickness of the oil layer.

Some studies have been conducted of the final thickness of burning oil on water
before it is extinguished. A large number of cases were reviewed in which oil burn
residue, or the thickness of the oil at the end of the burn, was measured (Buist et al.
1994). They found that the average final thickness was 1 mm and the residue ranged
in thickness from about 0.5–2 mm. Thus, it was proposed that 1 mm be adopted as the
rule of thumb for final burn thickness. This has not been substantiated in later studies.

1.3.4 OIL WEATHERING/VOLATILE CONTENT

As a rule, the greater the percentage of volatile compounds in an oil, the more easily
it will ignite and continue to burn. It can therefore be difficult to ignite weathered

oils and heavy crude oils (No. 5 and above) and higher ignition temperatures, primers and/or longer ignition exposure times may be required (Fingas and Punt 2000; Fingas et al. 2005; Maki and Miura 1997). During one burn test, it was found that weathered oils actually burned with an average 7% greater efficiency than fresh oils (Johnson et al. 2011).

1.3.5 HEAVY OILS

Heavy oils were thought to burn poorly if at all; however, results in recent years show that these burn quite well under most circumstances (Fingas et al. 2004). Studies in the past decade have shown much more potential for burning these oils than was previously thought (Fingas and Punt 2000). Burning tests of bitumen, a very heavy oil, along with water have been conducted and shown useful removal potentials. The burning of heavy oils has been studied by Environment Canada over a period of 5 years (Fingas 2002; Fingas et al. 2003, 2004, 2005).

Figure 1.14 shows one such burn, and Figure 1.15 shows the remaining residue. Figure 1.16 illustrates the ignition of a heavy oil.

Heavy oils such as Bunker C burn quite well but yield a highly viscous residue. Figures 1.17 and 1.18 illustrate heavy oil burn residues. This high-viscosity residue has a high asphaltene and resin content. There is no evidence of the presence of soluble components; thus, the residue would exhibit low aquatic toxicity. Examination of the Saturates, Resins, Aromatics, Asphaltenes (SARA) content shows that the values of SARA for the residue can be used to predict burn efficiency. There appears to be a consistent reduction

FIGURE 1.14 A burn of Bunker C. The efficiency of this burn was about 60%. The horizontal objects are thermal probes.

FIGURE 1.15 The residue from the burn in Figure 1.14. Note that the residue is so solid that it can be picked up as a single sheet.

FIGURE 1.16 The ignition of a heavy oil using a barbecue lighter. This is carried out most easily by adding a small amount of primer such as diesel fuel (about 20 mL), and adding a small wick such as cardboard or paper towel.

FIGURE 1.17   The residue of a heavy oil burn. The residue has solidified into chunks with the brittleness of hard candy.

FIGURE 1.18   Another residue of a heavy oil burn. This residue may be moved as a sheet. Parts of this sheet are elastic and parts are brittle.

of saturate and aromatic content in an oil with increasing burn efficiency. This is based on values from 10 burn experiments and 4 oil types. The prediction equation is:

$$\text{Burn Efficiency}\,(\%) = -23000 + 230 * \text{Aromatic}\,\% + 227 * \text{Saturate}\,\%$$
$$+ 254 * \text{Resin}\,\% + 218 * \text{Asphaltene}\,\% \tag{1.1}$$

Several types of oils were used in heavy oil burning tests, as shown in Table 1.5. The results of two series of burning are shown in this table as well. It is interesting

TABLE 1.5
Heavy Oil Burning

Pan Size	Oil Type	Oil Viscosity (mPa.s)	Burn Conditions	Efficiency (%)	Burn Rate (mm/min)	Flame Height (m)	Peak Height (m)
1 m	ESTD bunker	1.5E + 04	Normal	64.8	1.2	1.5	2
1.5 m	ESTD bunker	1.5E + 04	Normal	63.8	1	1	1.5
1 m	Test oil	1.6E + 04	Would not burn	2	No volatiles remaining		
1.5 m	Test oil	1.6E + 04	Would not burn	2	No volatiles remaining		
1 m	Orimulsion[a]	2.5E + 02	Normal	65.6	1.7	3	8
1.5 m	Orimulsion[a]	2.5E + 02	Normal	61.3	2.3	1.5	4
1 m	ESTD bunker	1.5E + 04	Normal	65.9	1.1	3	4
1.5 m	ESTD bunker	1.5E + 04	Normal	70	1	1.5	3
1 m	Bitumen	4.0E + 06	Normal	12.3	1	1.5	2
1.5 m	Bitumen	4.0E + 06	Normal	12.9	0.9	1	1.5
			Overall average	39.3	1.5	1.7	3.3

to note that Orimulsion burning efficiency averages about 40%–60% (excluding the water content of 30%), Bunker C burning averages about 65% and burning bitumen averages about 12%. Orimulsion has certain peculiar burning characteristics such as popping when the water is explosively released (Fingas et al. 2004, 2005; Kadota and Yamasaki 2002). It is suggested that burning of Orimulsion actually takes place as a two-step process: first, vaporisation and water release, and second, the actual combustion. Extremely weathered oils such as the heavily used Bunker test oil did not burn, and analysis of this showed that its calculated burn efficiency per Equation 1.1 was about zero. The burn rate for Orimulsion was found to be between 0.5 and 2 mm/min (Fingas et al. 2003, 2004). This latter study showed that the burn rates for heavy oils varied from 1 to 2 mm/min (Fingas et al. 2004).

Emissions from these heavy oil burns showed very low emissions compared to crude oils and, in particular, there were few volatiles and few polycyclic aromatic hydrocarbons (PAHs) measured in the air.

The residues from all the burns were highly viscous. When cooled, all residues were solid and even 'glassy' in some cases, as shown in Figures 1.15, 1.17 and 1.18. Analysis of the residues showed some concentration of higher molecular weight pyrogenic PAHs.

1.3.6 OIL EMULSIFICATION

In general, unstable oil emulsions can be ignited and will sustain burning because the emulsion is quickly broken down during the burning process (Fingas and Fieldhouse 2009). By contrast, stable oil emulsions are difficult to ignite because a large amount of energy is required to heat the water. Thus, additional energy is required to vaporise the oil in the emulsion before the burning is sustained. Test burns have shown that once an emulsified oil is ignited and has burned long enough, the heat from the burn sometimes breaks down the emulsion and allows the slick to continue to burn (Bech et al. 1992).

All unstable emulsions can be broken down by mechanical means or they break down on their own over time. Based on the commonly accepted definition of stable emulsions (an emulsion that persists for at least 5 days at 15°C), studies have shown that stable and unstable emulsions have different characteristics (Fingas and Fieldhouse 2011). The two most obvious characteristics relate to colour and viscosity. Stable emulsions are reddish brown whereas unstable emulsions are black. The viscosity of stable emulsions is usually more than three orders of magnitude greater than the oil from which the emulsion was made, whereas the viscosity of an unstable emulsion is less than one order of magnitude greater than the original oil. There is also a middle form or mesostable emulsion that usually is brownish in colour and has a viscosity of about 50 times that of the starting oil. Some typical properties of water-in-oil states are given in Table 1.6. The literature has shown that the stability of an emulsion depends on the concentration of asphaltenes and, to a lesser extent, resins in the oil. These compounds form a viscoelastic film at the oil water interface. As well, oil will not create a stable emulsion with a very low (<30%) or very high (>90%) water content. In general, the water content of stable emulsions ranges from 60% to 75%, although there is no correlation between water content and stability of an emulsion within this range (Fingas and Fieldhouse 2011).

TABLE 1.6

Average Properties of the Four Water-in-Oil Types

Averages	Appearance	Day Water Content (%w/w)	Week Water Content (%w/w)	**Rheological Properties on Day of Formation**				
				Viscosity Increase (From Starting)	Complex Modulus (mPa)	Elasticity Modulus (mPa)	Tan Delta (V/E)	Complex Viscosity (mPa.s)
Entrained	Viscous black	44.5	27.5[a]	1.9	8.30E + 05	5.14E + 05	1.73	1.3E + 05
Mesostable	Viscous reddish	64.3	29.6[a]	7.2	1.33E + 05	1.07E + 05	1.7	2.1E + 04
Stable	Solid reddish	80.7	77.4	405	7.50E + 05	7.10E + 05	0.7	1.2E + 05
Unstable	Oil-like	6.1	6.85[a]	0.0	1.10E + 07	3.37E + 06	2.4	1.8E + 06

	Appearance	**Rheological Properties 1 Week after Formation**					
		Viscosity Increase (From Starting)	Complex Modulus Change	Elasticity Modulus Change	Tan Delta Change	Complex Viscosity Change	(increase is shown as number larger than 1)
Entrained	Viscous black	1.9	0.8	0.6	1.7	1.3	
Mesostable	Broken	5.4	0.6	0.5	2.2	0.5	
Stable	Solid reddish	859	1.1	1.1	1.1	1.2	
Unstable	Oil-like	0.0	3.8	2.9	1.7	3.6	

[a] These water content values are high as most were not measured as they were obviously low.

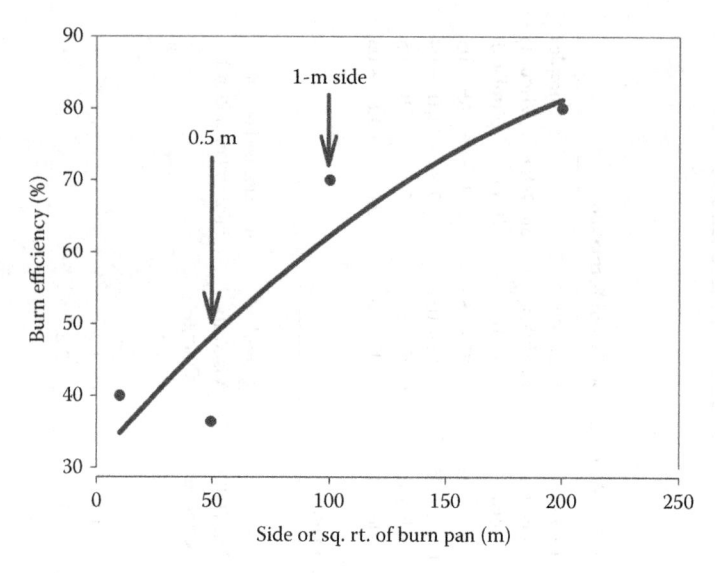

FIGURE 1.19 The efficiency of burning versus pan size with data accumulated over 3 separate studies. The curve was fit using regression. This figure shows that burns of less than about 1-m² in area do not produce representative efficiencies.

Test burns have shown that once an emulsified oil is burned alongside un-emulsified oil, the heat from the burn breaks down the emulsion and allows the slick to continue to burn. The method that works is to burn emulsified oil alongside the emulsified oil and the heat from burning the un-emulsified oil will breakdown the emulsion and enable burning. This is shown in Figure 1.2.

1.3.7 SCALE OF BURNING

Scale of the burn also has an influence on efficiency. Several studies have noted that efficiency increases as burn pan size increases (Bullock et al. 2016, 2017; Fingas 2002; Fingas et al. 2003). This is shown in Figure 1.19. In sum, any burn less than about 1-m squared results in lesser burn efficiencies. This should give pause to several groups who use small-scale tests to measure burn efficiencies. This information is shown in Figure 1.19, which shows an aggregate of data from several tests (Bullock et al. 2016, 2017; Fingas et al. 2003, 2004).

1.4 BURN EMISSIONS

The primary environmental and health concerns related to ISB are the emissions produced by the fire. The measurement of emissions has revealed several facts about the quantity, fate and behaviour of emissions from burning. Overall, emissions are now understood to the extent that emission levels and safe distances downwind can be calculated for fires of various sizes and types. A typical crude oil burn (500 m²) would not exceed health limits for emissions beyond about 500 m from the fire. People and the environment can be protected by ensuring that the burn is kept the minimum distances

CO$_2$ + H$_2$O + By-products
Particulates
Organic compounds
Gases

Oil

Residue

FIGURE 1.20 The basic emissions from the burning of oil.

away from populated and sensitive areas. Figure 1.20 illustrates the emissions from fires. The important point is that the basic products of oil combustion are carbon dioxide and water. The remaining emissions are by-products. Figure 1.21 shows several sets of emission measuring stations which were implemented at a test burn.

FIGURE 1.21 A collection of instruments put out by Environment Canada and the U.S. Environmental Protection Agency to collect emissions and emission data from test burns in Mobile, Alabama.

1.4.1 Particulate Matter

All burns, especially those of diesel fuel, produce an abundance of particulate matter, which is the primary emission from an oil fire that exceeds recommended human health concern levels. Forest and grass fires also produce significant amounts of particulate matter (Myers-Pigg et al. 2016). Concentrations of particulates in emissions from burning diesel are approximately four times that from similar sized crude oil burns at the same distance from the fire. Particulate matter is distributed exponentially downwind from the fire. Concentrations at ground level (1 m) can still be above normal health concern levels (35 $\mu g/m^3$) as far downwind as 500 m from a small crude oil fire. The greatest concern is the smaller or respirable particulates. The PM-10 fraction, or particulates less than 10 μm, are generally about 0.7 of the total particulate concentration (TSP) of all particulates measured. The PM-2.5 fraction is currently the subject of particular concern at this time (Jaligama et al. 2015; Russel and Brunerkreef 2009). It is important to note that currently the fine particles are coming under increasing scrutiny as health concerns. Figure 1.22 shows particulate emission captured on a sampling device at a test burn.

Forest fires yield similar emissions such as particulate matter and PAHs; however, forest fires can yield a number of alkaline materials, which affect water quality (Dahm et al. 2015). Further, the atmospheric degradation by particulate matter is never as severe with oil burns as it is with forest fires simply because of the scale size.

FIGURE 1.22 Soot particles collected on a normally white filter. This measuring device was placed under a smoke plume from a test burn.

1.4.2   Organic Compounds

1.4.2.1   Polycyclic Aromatic Hydrocarbons

Polycyclic aromatic hydrocarbons (PAHs) are ubiquitous in the environment and are considered to be of concern for both humans and the environment. The sources of PAHs are many, but they come primarily from combustion sources (Fingas 2016b). Natural sources such as forest fires are contributors. Generally, PAHs in urban environments are largely assigned to vehicle emissions and then to other combustion sources (Ho et al. 2016). Crude oil contains PAHs; however, oil spills themselves are not a significant contributor of PAHs to the environment. Burning of petroleum could constitute a significant input of PAHs into the environment. In particular, it has been pointed out that the two most significant contributors are diesel engines and home heating using heating fuel similar to diesel fuel. Heating with wood also contributes some PAHs. Sootier flames appear to be associated with more PAHs. The quantity of PAHs in the starting fuel also influences the quantity of PAHs emitted by a burn. Generally low-temperature sources such as diesel engines are larger emitters of PAHs than high-temperature sources such as fires.

Crude oil burns result in PAHs downwind of the fire, but the concentration on the particulate matter, both in the plume and the particulate precipitation at ground level, is often an order of magnitude less than the concentration of PAHs in the starting oil. This includes the concentration of multi-ringed PAHs, which are often created in other combustion processes such as low-temperature incinerators and diesel engines. There is a slight increase in the concentration of multi-ringed PAHs in the burn residue (Fingas 2016b; Topal et al. 2004). When considering the mass balance of the burn, however, most of the five- and six-ringed PAHs are destroyed by the fire. When diesel fuel is burned, the emissions show an increase in the concentration of multi-ringed PAHs in the smoke plume and residue, but a net destruction of PAHs is still found. There may be persistence of PAHs in land after a land burn (Pereira Netto et al. 2004).

The PAH burn data are summarised in Table 1.7 and show that, for the most part, PAHs are destroyed in fires (Fingas 2016b). Updated information on PAHs is given in Chapter 6 of this book. This analysis also shows that there are slight differences between crude oil and diesel burns in that crude oil burns appear to be slightly more efficient in terms of destroying PAHs. Further, the analysis of the PAHs going to various compartments clearly shows that most PAHs are destroyed in the fire, with some remaining with the residue. Further burn efficiency does not appear to change the PAH distribution to any extent.

The main points on PAHs in burning oil can be summarised as follows:

1. Some diesel burns ended up with minor production of higher molecular weight PAHs. This is consistent with the literature; however, minor production did not occur in all cases. This was not found to be significant in any of the crude oil burns; however, the concentration of the U.S. Environmental Protection Agency (EPA) priority 16 compounds can be used as an indicator of burning. The concentration of the EPA priority 16 compounds is higher in the burn residue than the starting concentrations and the concentrations of the alkylated PAHs are lower in the residue than that of the starting oil.

TABLE 1.7
Summary of Burn Results and Assessment of PAH Destruction

Oil Type	Burn Type	Location	Year	% Burn Efficiency	Type (Measured or Estimated)	Measurements	Total PAHs Starting µg/g	Total PAHs Residue µg/g	Overall PAH Destruction %
Alberta medium	Field burn	Offshore	1993	99.9	Measured	Soot, gas, residue	8,607	4,671	99.95
Alberta medium	Field burn 1	Offshore	1993	99.9	Measured	Residue	8,607	5,330	99.94
Alberta medium	Field burn 1	Offshore	1993	99.9	Measured	Residue	8,607	4,011	99.95
Alberta medium	Field burn 2	Offshore	1993	99.9	Measured	Residue	8,607	3,943	99.95
Alberta medium	Field burn 2	Offshore	1993	99.9	Measured	Residue	8,607	4,494	99.95
Alberta medium	Field burn 2	Offshore	1993	99.9	Measured	Residue	8,607	4,017	99.95
Diesel fuel	Test burn 1	Meso tank	1992	98	Measured	Residue	5,920	439	99.85
Diesel fuel	Test burn 2	Meso tank	1992	98	Measured	Residue	5,825	551	99.81
Diesel fuel	Test burn 3	Meso tank	1992	98	Measured	Residue	5,806	635	99.78
Diesel fuel	Test burn 4	Meso tank	1992	98	Measured	Residue	5,697	519	99.82
Diesel fuel	Test burn 5	Meso tank	1992	98	Measured	Residue	5,589	353	99.87
Diesel fuel	Tank burn	Meso tank	1994	99	Measured	Soot, gas, residue	27,510	22,360	99.19
Diesel fuel	Test burn 1	Meso tank	1994	98	Measured	Residue	5,060	1,994	99.21
Diesel fuel	Test burn 2	Meso tank	1994	98	Measured	Residue	4,966	2,464	99.01

(Continued)

TABLE 1.7 (*Continued*)
Summary of Burn Results and Assessment of PAH Destruction

Oil Type	Burn Type	Location	Year	% Burn Efficiency	Type (Measured or Estimated)	Measurements	Total PAHs Starting µg/g	Total PAHs Residue µg/g	Overall PAH Destruction %
Diesel fuel	Test burn 3	Meso tank	1994	98	Measured	Residue	5,243	2,991	98.86
Diesel fuel	Test burn 1	Meso tank	1997	99	Measured	Soot, gas, residue	26,713	19,949	99.25
Diesel fuel	Test burn 2	Meso tank	1997	99	Measured	Soot, gas, residue	26,713	20,966	99.22
Diesel fuel	Test burn 3	Meso tank	1997	99	Measured	Soot, gas, residue	26,713	26,252	99.02
Diesel fuel	Test burn 4	Meso tank	1997	99	Measured	Soot, gas, residue	26,713	29,366	98.9
Diesel fuel	Tank burn	Meso tank	1998	99	Measured	Soot, gas, residue	25,906	25,810	99
Statfjord crude	Tank burn	Small tank	1999	85	Measured	Residue			85 + 40%
Bunker C	Tank burn 2	Small tank	2003	63.8	Measured	Soot, residue	29,419	31	99.96
Test oil – residual	Tank burn 3	Small tank	2003	2	Measured	Soot, residue	849	305	64.81
Test oil – residual	Tank burn 4	Small tank	2003	2	Measured	Soot, residue	849	424	51.1
Orimulsion	Tank burn 5	Small tank	2003	65.6	Measured	Soot, residue	1,459	1,275	69.95
Orimulsion	Tank burn 6	Small tank	2003	61.3	Measured	Soot, residue	1,459	1,322	64.94
Bunker C	Tank burn 7	Small tank	2003	65.9	Measured	Soot, residue	29,419	1,034	98.8
Bunker C	Tank burn 8	Small tank	2003	70	Measured	Soot, residue	29,419	39	99.96
Bitumen	Tank burn 9	Small tank	2003	12.3	Measured	Soot, residue	1,459	1,471	87.60
Bitumen	Tank burn 10	Small tank	2003	12.5	Measured	Soot, residue	1,459	1,020	91.27
Diesel fuel	Test burn	Small tank	2004			Residue			99.40

(*Continued*)

TABLE 1.7 (Continued)
Summary of Burn Results and Assessment of PAH Destruction

Oil Type	Burn Type	Location	Year	% Burn Efficiency	Type (Measured or Estimated)	Measurements	Total PAHs Starting μg/g	Total PAHs Residue μg/g	Overall PAH Destruction %
Louisiana Crude	Test burn	Small tank	2004			Residue			98.90
Macondo	Field burn 1	Offshore	2010	98	Estimated	Residue	6,166	2,271	99.26
Macondo	Field burn 2	Offshore	2010	98	Estimated	Residue	5,363	5,972	97.77
Macondo	Field burn 3	Offshore	2010	98	Estimated	Residue	4,841	651	99.73
Macondo	Field burn 4	Offshore	2010	98	Estimated	Residue	4,893	1,959	99.2
Macondo	Field burn 5	Offshore	2010	98	Estimated	Residue	4,893	1,830	99.25
Macondo	Field burn 6	Offshore	2010	98	Estimated	Residue	4,893	962	99.61
Macondo	Small 5 g	Lab	2012	98	Estimated	Residue	11,459	1,163	99.8
South Louisiana	Test 500 mL	Lab	2012	98	Estimated	Residue	8,091	3,616	99.11
Weathered Iou	Test 500 mL	Lab	2012	98	Estimated	Residue	9,362	3,246	99.31
Emulsified Mac	Test 500 mL	Lab	2012	80	Estimated	Residue	9,362	3,246	93.07
Emulsified Mac	Test 500 mL	Lab	2012	80	Estimated	Residue	2,660	2,643	80.13

Source: Fingas (2016b).

2. Diesel burning is somewhat less efficient than crude oil burning in the destruction of PAHs and does result in more soot with its incumbent PAH load.
3. The quantity of PAHs emitted as vapour is negligible.
4. The quantity of PAHs on the soot is variable, although in most cases it is negligible. The quantity of priority PAHs on the soot maybe somewhat elevated.

1.4.2.2 Carbonyls

Oil burns produce low amounts of partially oxidised material, sometimes referred to as carbonyls or by their main constituents, aldehydes (formaldehyde, acetaldehyde etc) or ketones (acetone etc) (Fingas 2017). Carbonyls from crude oil fires are at very low concentrations and are well below health concern levels even close to the fire. Carbonyls from diesel fires are somewhat higher but also below concern levels. Burning of alcohol-containing fuels results in the release of more carbonyls. Those carbonyls emitted as volatile materials are detailed further in Appendix A.

1.4.2.3 Dioxins and Dibenzofurans

Dioxins and dibenzofurans are highly toxic compounds often produced by burning chlorine-containing organic material. Particulates precipitated downwind and residue produced from several fires have been analysed for dioxins and dibenzofurans. These toxic compounds were at background levels at many test fires, indicating no production by either crude or diesel fires. Generally, Polychlorinated dibenzodioxins (PCDDs) and Polychlorinated dibenzofurans (PCDFs) in urban environments are largely assigned to vehicle emissions and then to other combustion sources (Ho et al. 2016).

A cancer risk estimate based on dioxin exposure was performed at the Deepwater Horizon burns and showed that the risk to workers and residents was quite low (Schaum et al. 2010). Aurell and Gullett measured dioxins and dibenzofurans at the Gulf 2010 burns and noted that the values of dioxins and dibenzofurans were often at background levels (Aurell and Gullett 2010). Comparatively the values measured were higher than those observed for the combustion of waste oil, within the range for biomass burning and very much lower than for the burning of residential waste. The same group also measured emissions from forest burns and found that these emission factors were higher (Aurell and Gullett 2013). Measurements were carried out using an aerostat in both cases.

1.4.2.4 Benzene, Toluene, Ethylbenzene and Xylenes

Benzene, toluene, ethylbenzene and xylenes (BTEX) are often released by oil spills (Avens et al. 2011). Analysis of emissions from several oil burns shows that the BTEX concentration is lower during the burn than from an evaporating oil slick (Fingas et al. 2001a, b). Details of BTEX emissions from historical test burns are given in Appendix A.

1.4.2.5 Cloud Condensing Nuclei

A spill releases cloud condensing nuclei (CNN), whether or not it is burned (Moore et al. 2012). Tests of CNN during burning have shown no increase in CNN.

1.4.3 GASES

1.4.3.1 Volatile Organic Compounds

Volatile organic compounds (VOCs) are organic compounds such as butane, hexane and so on, that have high enough vapour pressures to be gaseous at normal temperatures. When oil is burned, these compounds evaporate and are released. The emission of volatile compounds was measured at several test burns. One-hundred and forty-eight VOCs have been measured from fires and evaporating slicks (Fingas et al. 2001b). The concentrations of VOCs are relatively low or similar in burns compared or similar to an evaporating slick. Concentrations are below human health levels of concern even very close to the fire. Concentrations are highest at the ground (1 m [3.3 ft]) and are distributed exponentially downwind from the fire source. VOCs, although present, do not constitute a major human or environmental threat. Table 1.8 shows the specific emissions of VOCs from a crude oil burn at a large offshore test burn (Fingas et al. 2001a,b). Analysis of these data show that compounds, typically smaller than butane, are in greater abundance from an evaporating slick than from a

---

TABLE 1.8
Volatile Organic Compounds from Crude Oil Spills and Fires Concentration at 125–600 m Downwind of NOBE (600 m² Fire Area) (Concentrations in µg/m³)

Alkanes and Alkenes	Burn	Evap.	Background
2,2,3-Trimethylbutane	3–10	0–2	0–1
2,2,4-Trimethylpentane	2–8	0–11	0–6
2,2,5-Trimethylhexane	0–1	0–1	0–1
2,2-Dimethylbutane	3–16	1–3	1–4
2,2-Dimethylpentane	1–15	1–3	0–1
2,2-Dimethylpropane	0–2	0–0.5	0–0.4
2,3-Dimethylbutane	2–16	1–3	1–4
2,3-Dimethylpentane	3–7	0–13	0–3
2,4-Dimethylhexane	1–65	1–6	0–1
2,4-Dimethylpentane	2–33	0–8	0–2
2,5-Dimethylhexane	1–15	0–4	0–1
2-Methylbutane	1–32	2–180	0–150
2-Methylhexane	6–88	0–20	0–5
2-Methylpentane	16–240	0–110	0–55
3,6-Dimethyloctane	1–4	0–4	0–3
3-Methylheptane	4–74	0–18	0–6
3-Methylpentane	6–280	0–98	0–21
4-Methylheptane	1–17	0–7	0–3
Butane	10–420	1–150	0–110
c-1,3-Dimethylcyclohexane	1–84	0	0
c-1,4/t-1,3-Dimethylcyclohexane	2–10	0–6	0–1
c-2-Butene	0–1	0–4	0–25

(Continued)

TABLE 1.8 (*Continued*)
Volatile Organic Compounds from Crude Oil Spills and Fires Concentration at 125–600 m Downwind of NOBE (600 m^2 Fire Area) (Concentrations in µg/m^3)

Alkanes and Alkenes	Burn	Evap.	Background
Cyclohexane	3–290	0–68	0–5
Cyclopentane	2–102	0–22	0–5
Decane	1–72	0–73	0–15
Dodecane	1–18	0–86	0–25
Heptane	6–1,200	0–97	0–8
Hexane	103–570	0–1,020	0–180
Isobutane (2-Methylpropane)	5–100	0–53	0–110
Isoprene (2-Methyl-1,3-Butadiene)	0–1	0–2	0–2
Methylcyclohexane	3–1,700	0–140	0–5
Methylcyclopentane	0–49	0–210	0–23
Nonane	1–12	0–45	0–6
Octane	0–2	0–38	0–4
Pentane	17–640	0–200	0–66
Propane	11–71	0–67	0–19
Propene	4–15	0–116	0–27
t-1,4-Dimethylcyclohexane	6–15	0–11	0–1
Undecane	1–19	0–87	0–30
Total	485–3,320	0–3,700	0–1,200
Aromatic VOCs			
1,2,3-Trimethylbenzene	2–4	0–16	0–6
1,3,5-Trimethylbenzene	3–4	0–18	0–6
1,4-Diethylbenzene	0–2	0–11	0–6
2-Ethyltoluene	2–6	0–10	0–4
3-Ethyltoluene	0–6	0–25	0–9
4-Ethyltoluene	2–3	0–14	0–5
Benzene	6–8	0–7	0–11
Ethylbenzene	7–14	0–7	0–16
Indan	0–1	0–3	0–2
iso-Butylbenzene	1–5	0–1	0–0.5
Isopropylbenzene	0–5	0–2	0–1
m,p-Xylene	0–21	0–39	0–59
Naphthalene	0–1	0–9	0–8
n-Butylbenzene	0–1	0–3	0–1
n-Propylbenzene	0–2	0–12	0–3
ortho-Xylene	1–12	0–11	0–18
p-Cymene	0–9	0–5	0–2
Toluene	1–20	0–27	0–31

burning slick. Larger VOCs, larger than butanes, are typically in greater abundance from a burning slick than an evaporating slick. Aromatic VOCs such as benzene are also typically greater from an evaporating slick than a burning slick. In any case the ranges of concentrations as shown in Table 1.8 are not that much different between evaporating, background and during burn conditions. Studies of the VOC emissions from diesel burns show similar conclusions (Fingas et al. 2001b). Further details on VOCs are given in Appendix A.

1.4.3.2 Carbon Dioxide

Carbon dioxide is the end result of combustion and is found in increased concentrations around a burn (Fingas et al. 2001b). Normal atmospheric levels are about 300 ppm and levels near a burn can be around 500 ppm, which presents no danger to humans. The three-dimensional distributions of carbon dioxide around a burn have been measured. Concentrations of carbon dioxide are highest at the 1-m level and fall to background levels at the 4-m level. Concentrations at ground level are as high as 10 times that in the plume, and distribution along the ground is broader than for particulates.

1.4.3.3 Carbon Monoxide

Carbon monoxide levels are usually at or below the lowest detection levels of the instruments and thus do not pose any hazard to humans. The gas has only been measured when the burn appears to be inefficient, such as when water is sprayed into the fire. Carbon monoxide appears to be distributed in the same way as carbon dioxide.

1.4.3.4 Sulphur Dioxide

Sulphur dioxide, per se, is usually not detected at significant levels or sometimes not even at measurable levels in the area of an in-situ oil burn. Sulphuric acid, or sulphur dioxide that has reacted with water, is detected at fires and levels, although not of concern, appear to correspond to the sulphur content of the oil.

1.4.3.5 Other Gases

Attempts were made to measure oxides of nitrogen and other fixed gases. None was measured in about 10 experiments. Sartz and Aggarwal (2016, 2017) attempted to measure the emissions from a burn of oil contained using chemical herding agents. It was found that measured gaseous emissions were similar to that noted above. The group was unable to measure gaseous components of the herding agent.

1.4.4 Other Compounds and Emission Factors

There is a concern when burning crude oil about any 'hidden' compounds that might be produced. In one study conducted several years ago, soot and residue samples were extracted and 'totally' analysed in various ways (Fingas et al. 2001b). While the study was not conclusive, no compounds of the several hundred identified were of serious environmental concern. The soot analysis revealed that the bulk of the material was carbon and that all other detectable compounds were present on this carbon matrix in abundances of ppm or less. The most frequent compounds identified were aldehydes, ketones, esters, acetates and acids, which are formed by incomplete oxygenation of the oil. Similar analysis of the residue shows that the same minority compounds are present at about the same levels. The bulk of the residue is unburned oil without some of the volatile components.

Lemieux et al. (2004) used some of the data from the burns referenced in this chapter to calculate emission factors for various compounds. Data were calculated from Fingas et al. (1996a, 1998). These are summarised in Table 1.9. These authors noted that emissions of PAHs were much higher when polymers were burned rather than oils.

Gullett et al. (2017) carried out small-scale burns to measure emission factors, as noted in Tables 1.10 through 1.12. It should be noted that, because of the small scale (0.5 and 1 m^2), these emission factors may be high (see Section 1.3.7).

1.4.5 The Behaviour and Distribution of Emissions

The behaviour of emissions is an important facet of oil spill burning. The overall emission behaviour is illustrated in Figure 1.23. The most important are the soot particles. Many of these initially rise and then most are precipitated back to the ground. It is estimated that before the plume travels 1 km, half of the particles are precipitated downward. Some particles may stay in the direction of the plume for a long time. The plume itself is not dangerous to humans as long it rises and as long as it does not impact human settlements directly such as happens in an inversion. Particulate matter also has some organic compounds such as PAHs adsorbed.

Water vapour, one of the main products of combustion, and light gases are distributed widely and soon reach background levels. Carbon dioxide and other heavier-than-air gases rise somewhat and then slowly sink to the ground. Carbon dioxide is one of the major products of oil combustion and is not dangerous to humans. VOCs, including carbonyls, are transported out of the fire and soon dilute to background levels.

1.4.6 Residue

The residue from burning oil is largely unburned oil with some lighter or more volatile products removed. When the fire ceases, unburned oil is left that is simply too thin to sustain combustion. In addition to unburned oil, oil is also present that has been subjected to high heat and is thus weathered. Highly efficient burns of some types of heavy crude oil may result in an oil residue that sinks in seawater after cooling.

There is concern about the contact of burn residue with wildlife such as birds and mammals. Test of adhesiveness of original oil (a crude and an intermediate fuel oil) and their burn residues shows that their adhesiveness to feathers is similar, although the burn residue is slightly more adhesive (Fritt-Rasmussen et al. 2016).

1.4.7 Safe Distances

Sufficient data are now available to assemble emission data and correlate the results with spatial and burn parameters. The correlations are summarised in a reference

TABLE 1.9
Emissions of Organic Compounds Calculated[a]

Class of Compound	Compound	Emissions from Pool Fires (mg/kg Burned)	
		Diesel Fuel	Crude Oil
VOCs	Benzene	1,020	250
	Toluene	40	
	Ethylbenzene	10	
	Xylenes	25	
	Nonane	13	
	Ethyltoluenes	22	
Carbonyls	Formaldehyde	300	140
	Acetaldehyde	63	32
	Acrolein	39	11
	Acetone	35	20
	Methylethylketone	13	7
	Benzaldehyde	104	44
	Isovaleraldehyde	17	5
	Methylisobutylketone	11	
	2,5-dimethlybenzaldehyde	13	
PAHs	Naphthalene	160	44
	Acenaphthylene	99	4
	Acenaphthene	10	
	Fluorene	1	0.5
	1-Methylfluorene	26	0.2
	Phenanthrene	13	6
	Anthracene	15	1
	Fluoranthene	20	4
	Pyrene	2	5
	Benzo[a,b]fluorine	4	0.3
	Benzo[a]anthracene	5	1
	Chrysene	9	1
	Benzo[b&k]fluoranthene	7	2
	Benzo[a]pyrene	5	1
	Indeno[1,2,3-cd]pyrene		
PCDDs/Fs	HpCDD		7 e-5
	OCDD		1.3 e-4
	TCDF		2.1 e-4
	HxCDF		1.9 e-5
	Total PCDD/F		4.3 e-4

Source: Lemieux, P.M. et al., *Prog. Energ. Comb.*, 30, 1–32, 2004.

[a] Values have generally been rounded off to 2 or 3 significant figures.

TABLE 1.10

Calculated General Emission Factors

Measurement	Units (Mass/Mass Oil)	Emission Factor
CO_2	g/kg	3,023
CO	g/kg	58
CH_4	g/kg	1
$PM_{2.5}$	g/kg	58
DustTrak	PM 1	56
	PM 2.5	57
	PM 4	58
	PM 10	63
	Total	70
Particle number	#/kg	5.5×10^{15}
PCDD/PCDF TEQ	ng TEQ/kg	1.2
PCDD/PCDF	ng Total/kg	35
$PAHs_{16}$	mg/kg	980
$PAHs_{16}$ TEQ	mg BaP eq./kg	29
$PAHs_{16\ particle\ phase}$	mg/kg	288
$PAHs_{16}$ TEQ $_{particle\ phase}$	mg/kg	22
VOCs	g/kg	3.4
Black carbon	g/kg	53
Elemental carbon	g/kg	49
Organic carbon	g/kg	4.1
Total carbon	g/kg	53
$EC/PM_{2.5}$	Ratio	0.82
$BC/PM_{2.5}$	Ratio	0.88
MCE	Ratio	0.978
Mass absorption efficiency	m^2/g	5.45
Mass scattering efficiency	m^2/g	3.57
Single scattering Albedo	Ratio	0.4
Absorption angstrom component	Ratio	0.68

Source: Gullett, B.K. et al., *Mar. Poll. Bull.*, 117, 392–405, 2017.

(Fingas and Punt 2000). Although many correlations were tried, it was found that atmospheric emissions correlated relatively well with distance from the fire and the area covered by the fire. This information was used to develop prediction equations for many pollutants, using the data gathered from the 30 test burns conducted. Sufficient data were available to calculate equations for over 150 individual compounds and for all the major groups. In some cases, however, the data are insufficient to yield high correlation coefficients and low errors. Details of VOC correlations are given in Appendix A.

These correlations significantly increase understanding of ISB in the areas of assessing the importance of specific emissions and classes, predicting a 'safe' distance for burning and predicting concentrations at a given point from the fire. These

TABLE 1.11
Table on VOC Emission Factors

Compound	Emission Factor (mg/kg Oil)
Benzene	1,574
Formaldehyde	311
Acetonitrile	212
Toluene	199
Propene	192
Styrene	145
1,3-Butadiene	107
Acrolein	82
Acetaldehyde	71
Crotonaldehyde	62
m,p-Xylenes	61
Acetone	58
Benzaldehyde	52
n-Nonane	46
n-Octane	45
n-Heptane	36
Butyraldehyde	34
Cyclohexane	27
1,2,4-Trimethylbenzene	25
o-Xylene	23
n-Hexane	22
Ethylbenzene	22

Source: Gullett, B.K. et al., *Mar. Poll. Bull.*, 117, 392–405, 2017.

predictions are based solely on actual data and therefore may be more accurate than theoretical-based predictions. This increased accuracy applies to situations where the conditions are the same as those under which the emissions data were collected. The data were collected with winds between 2 and 5 m/s (4–10 knots) and with no inversions present. The prediction equations for several common emission groupings and specific compounds are given in Table 1.13 for crude oil and in Table 1.14 for diesel fuel. Detailed correlations for VOCs are given in Appendix A.

These data were then used to calculate the difference between the regulated level (typically the time-weighted average recommended exposure to a substance) and the calculated amount of the substance for several burns. Results of a simple exercise of this type are shown in Tables 1.13 and 1.14. These tables show that emissions, especially of particulate matter, are significantly higher from a diesel fire than from a crude oil fire, as had been noted in several studies of particulate emissions (Fingas et al. 1996b; Fingas 2010). Other emissions of concern are similar for diesel and crude oil, although the PAHs are somewhat higher when diesel burns. This calculation confirms that particulate matter is the greatest concern, followed by the PAHs on the particulate matter, and the total VOCs.

TABLE 1.12
Table on PAHs Emission Factors

Compound	Emission Factor (mg/kg Oil)
Naphthalene	352
Acenaphthylene	174
Acenaphthene	2.7
Fluorene	38
Phenanthrene	126
Anthracene	30
Fluoranthene	67
Pyrene	75
Benzo(a)anthracene	15
Chrysene	21
Benzo(b)fluoranthene	11
Benzo(k)fluoranthene	19
Benzo(a)pyrene	19
Indeno(1,2,3-cd)pyrene	12
Dibenz(a,h)anthracene	2
Benzo(ghi)perylene	15
SUM PAH16	980

Source: Gullett, B.K. et al., *Mar. Poll. Bull.*, 117, 392–405, 2017.

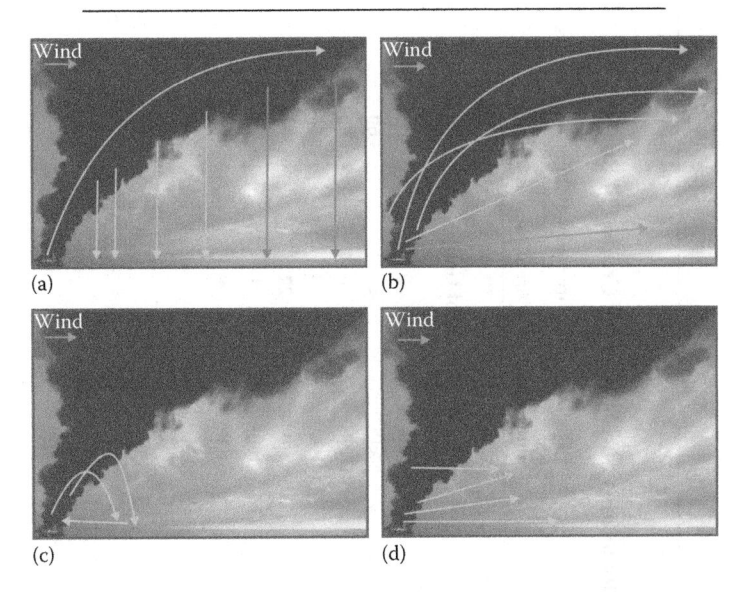

(a) (b)

(c) (d)

FIGURE 1.23 Illustration of the behaviour of emissions. (a) Particulates along with adsorbed organics, for example, PAHs, rise then are precipitated downwind; (b) water vapour and light gases rise and are widely transported and diffused; (c) carbon dioxide and other heavy gases rise then slowly sink and may cycle through the fire; (d) organic gases such as VOCs and carbonyls are widely transported and diffused.

Table 1.13

Emission Calculations and Values for Crude Oil Burns

Substance	Distance from Fire Taken at 300 m			Normal Threshold Limit (µg/m³)	Equations to Calculate		
	Values When Burn Area Taken as 200 m²				Equation Parameters		
	Concentration (mg/m3)	% of Limit	Safe Distance (m)		a	b	c
Total particulates	<0	<0	60	0.2	12.7	0.0347	4.79
PM-10	<0	<0	60	0.15	12.7	0.0347	4.79
PM-2.5	<0	<0	60	0.035	12.7	0.0347	4.79
Total VOCs	6	3	210	161990.6			
PAHs	0	22	90	188.4			
Fixed gases	160	1	0	10,120			
Carbonyls	0	0	0	630			

Equations

$\text{Concentration} = a + b * \text{size} - c * \ln(\text{distance})$

Concentration is mg/m³ for particulates

Concentration is µg/m³ for gases and organics

Size is area of fire in m²

Distance is distance from edge of fire in m

(Continued)

Table 1.13 (*Continued*)
Emission Calculations and Values for Crude Oil Burns

Calculated Values

Summations of Organic Chemicals and Gases

| | ($\mu g/m^3$) | | | ($\mu g/m^3$) | Equations to Calculate | | |
| | Burn Area Taken as | | | Normal Threshold | Equation Parameters | | |
	200 m^2	400 m^2	800 m^2	Limit ($\mu g/m^3$)	a	b	c
Fixed Gases							
Sulphur dioxide	0	0	11	20	19.4	0.0266	5.29
Carbon dioxide	160	264	474	10,000	520	0.523	81.5
Carbon monoxide	0	0	0	100	7.72	0.00124	1.56
Carbonyls							
Acetaldehyde	0	0	42	200	23.3	0.115	12.9
Acetone	0	0	18	170	11.3	0.0445	5.11
Formaldehyde	0	0	26	260	58.4	0.103	20.1
PAHs							
1-Methylnaphthalene	0	1	2	20	1.01	0.00424	0.381
1-Methylphenanthrene	0	0	0	0.3	0.115	4.83E-06	0.0192
2,3,5-Trimethylnaphthalene	0	0	0	10	0.286	0.00053	0.08
2,6-Dimethylnaphthalene	0	0	1	30	0.614	0.0025	0.249
2-Methylnaphthalene	0	0	2	20	1.4	0.00397	0.462
Acenaphthene	0	0	0	5	0.0673	2.13E-05	0.00989
Acenaphthylene	0	0	0	2	0.0673	2.13E-05	0.00989
Anthracene	0	0	0	0.2	0.32	0.000189	0.0653
Benz(a)anthracene	0	0	0	0.2	0.14	1.43E-09	0.398

(*Continued*)

Table 1.13 (*Continued*)
Emission Calculations and Values for Crude Oil Burns

Calculated Values

Summations of Organic Chemicals and Gases

	(µg/m³)			(µg/m³)	Equations to Calculate		
	Burn Area Taken as			Normal Threshold	Equation Parameters		
	200 m²	400 m²	800 m²	Limit (µg/m³)	a	b	c
PAHs							
Benzo(a)pyrene	0	0	0	0.2	0.617	0.000361	0.145
Benzo(b)fluoranthene	0	0	0	0.2	0.108	9.98E-06	0.0229
Benzo(e)pyrene	0	0	0	0.2	0.108	9.98E-05	0.0229
Benzo(g,h,i)perylene	0	0	0	0.1	0.228	0.000091	0.0479
Biphenyl	0	0	0	1.3	0.507	1.27E-05	0.0708
Chrysene	0	0	0	0.2	0.1224	0.000127	0.0305
Dibenz(a,h)anthracene	0	0	0	0.2	0.0189	2.97E-06	0.00227
Dimethylnaphthalenes	0	1	1	30	1.75	0.000804	0.257
Fluoranthene	0	0	0	0.2	0.851	2.97E-06	0.1523
Fluorene	0	0	0	5	0.299	0.000309	0.0716
Indenol(1,2,3-cd)pyrene	0	0	0	0.2	0.161	0.000145	0.0394
Methylphenanthrenes	0	0	0	0.3	0.322	0.000244	0.075
Naphthalene	0	1	1	52	1.86	0.00226	0.385
Perylene	0	0	0	0.2	0.0675	7.09E-05	0.0152
Phenanthrene	0	0	0	0.2	0.787	0.000224	0.141
Pyrene	0	0	0	0.2	0.542	0.000226	0.117
Trimethylnaphthalenes	0	0	0	10	0.856	0.000891	0.21

(Continued)

Table 1.13 (*Continued*)
Emission Calculations and Values for Crude Oil Burns

Calculated Values

Summations of Organic Chemicals and Gases

	Burn Area Taken as (µg/m³)			Normal Threshold (µg/m³)	Equations to Calculate Equation Parameters		
VOCs	200 m²	400 m²	800 m²	Limit (µg/m³)	a	b	c
1,2,3-Trimethylbenzene	0	1	5	123	11.4	0.0106	2.53
1,2,4-Trimethylbenzene	1	6	15	123	22.4	0.0239	4.58
1,3,5-Trimethylbenzene	0	1	8	123	17.3	0.0191	4.28
1,4-Diethylbenzene	0	1	3	260	4.66	0.00529	0.947
2,2,3-Trimethylbutane	0	0	3	2,850	25	0.0256	7.49
2,2,4-Trimethylpentane	0	1	6	1,230	5.41	0.0131	1.66
2,2,5-Trimethylhexane	0	0	0	925	8.49	0.00806	2.58
2,2-Dimethylbutane	0	0	35	1,550	61	0.105	19.3
2,2-Dimethylpentane	0	0	22	1,440	52.3	0.0799	16.5
2,2-Dimethylpropane	0	0	2	1,500	25.2	0.0271	7.93
2,3,4-Trimethylpentane	0	0	8	1,230	14	0.0249	4.53
2,3-Dimethylbutane	0	0	89	1,550	168	0.308	57
2,3-Dimethylpentane	0	0	84	1,445	173	0.294	56.8
2,4-Dimethylhexane	0	0	30	1,230	72.2	0.109	22.7
2,4-Dimethylpentane	0	0	48	1,445	99	0.164	32
2,5-Dimethylhexane	0	0	22	1,230	40.5	0.0787	14.3
2-Ethyltoluene	0	1	4	123	5.98	0.00826	1.47
2-Methylbutane	0	0	1,202	1,500	2,221	4.58	821

(*Continued*)

Table 1.13 (*Continued*)
Emission Calculations and Values for Crude Oil Burns

Calculated Values

Summations of Organic Chemicals and Gases

	(µg/m³)			(µg/m³)	Equations to Calculate		
	Burn Area Taken as			Normal Threshold	Equation Parameters		
	200 m²	400 m²	800 m²	Limit (µg/m³)	a	b	c
VOCs							
2-Methylheptane	0	0	106	1,230	240	0.384	77.4
3-Methylhexane	0	0	245	1,445	526	0.896	175
3-Methylpentane	0	0	399	1,550	822	1.41	272
4-Ethyltoluene	1	2	4	2,850	4.79	0.0051	0.85
4-Methylheptane	0	1	27	1,230	30.1	0.063	9.44
Benzene	0	1	11	1.6	72	0.0242	14.1
Butane	0	0	903	1,900	1,700	3.31	604
c-1,3-Dimethylcyclohexane	0	7	91	2,000	82.4	0.21	28
c-1,4/t-1,3-Dimethylcyclohexane	0	9	34	2,000	22.4	0.0626	6.74
c-2-Butene	0	0	4	1,100	4.73	0.0108	1.6
Cyclohexane	0	0	410	1,030	726	1.43	256
Cyclopentane	0	0	148	1,720	262	0.526	93.8
Decane	0	0	29	935	97	0.0899	24.5
Dodecane	0	0	14	740	27.1	0.0368	7.43
Ethylbenzene	0	2	18	434	25	0.0391	6.69
Heptane	0	0	576	1,640	1,170	2.11	400
Indan (2,3-Dihydroindene)	0	1	2	83	2.64	0.00305	0.557
Isobutane (2-Methylpropane)	0	0	313	1,670	414	1.05	165

(Continued)

Table 1.13 (*Continued*)
Emission Calculations and Values for Crude Oil Burns

Calculated Values

Summations of Organic Chemicals and Gases

	(µg/m³)			(µg/m³)	Equations to Calculate		
	Burn Area Taken as			Normal Threshold	Equation Parameters		
	200 m²	400 m²	800 m²	Limit (µg/m³)	a	b	c
VOCs							
iso-Butylbenzene	0	0	2	260	3.48	0.00574	1.06
Isoprene (2-Methyl-1,3-Butadiene)	0	0	11	13	17.4	0.0314	5.51
iso-Propylbenzene	0	0	0	246	21.4	0.0178	6.41
m,p-Xylene	0	14	57	434	88.6	0.109	20.8
Methylcyclohexane	0	0	827	1,610	1,660	3.03	571
Methylcyclopentane	0	0	343	2,687	2,090	2.9	713
Naphthalene	0	0	4	52	5.92	0.00991	1.7
n-Butylbenzene	0	0	1	260	3.28	0.003	0.806
Nonane	0	0	92	1,050	232	0.328	70.5
n-Propylbenzene	0	1	4	246	6.85	0.0073	1.52
Octane	0	0	210	1,400	513	0.776	162
o-Xylene	0	3	10	434	26	0.0186	5.38
p-Cymene (1-Methyl-4-iso-propylbenzene)	4	5	7	140	2.52	0.0055	0.0125
Pentane	0	0	1,383	1,770	2,590	5.05	920
Propane	0	0	18	4,508	733	0.789	236
Propene	0	0	24	2,615	21.8	0.062	8.28
Total	0	0	7,332	100,000	13,400	24	4,430
Undecane	0	0	21	830	50	0.0525	12.4

TABLE 1.14

Emission Calculations and Values for Diesel Fuel Burns

Substance	Distance from Fire Taken at 300 m				Equations to Calculate		
	Values When Burn Area Taken as 200 m²				Equation Parameters		
	Concentration (mg/m³)	% of Limit	Safe Distance (m)	Normal Threshold Limit (µg/m³)	a	b	c
Total particulates	<0	<0	90	0.25	1.998	0.00817	0.642
PM-10	**0.14**	**32**	**140**	0.15	1.019	0.00532	0.329
PM-2.5	**0.07**	**91**	**320**	0.035	1.44	0.00523	0.412

Summations of Organic Chemicals

	(µg/m³)			(µg/m³)	Equations
Total VOCs	**21**	**3**	**80**	1,110	Concentration $= a + b*\text{size} - c*\ln(\text{distance})$
PAHs	**0**	**0**	**30**	10	Concentration is mg/m³ for particulates
Fixed gases	**86**	**0**	**0**	3,370	Concentration is µg/m³ for gases and organics
Carbonyls	**110**	**8**	**0**	280	Size is area of fire in m²
					Distance is distance from edge of fire in m

Calculated Values

Summations of Organic Chemicals and Gases

	(µg/m³)			(µg/m³)	Equations to Calculate		
	Burn Area Taken as			Normal Threshold	Equation Parameters		
	200 m²	400 m²	800 m²	Limit (µg/m³)	a	b	c
Fixed Gases							
Sulphur dioxide	0	0	0	20	0.5329	0.001	0.173
Carbon dioxide	86	131	220	10,000	14.9	0.224	−4.56
Carbon monoxide	0	0	0	100	0.87	−48.5	88.4

(Continued)

TABLE 1.14 (*Continued*)

Emission Calculations and Values for Diesel Fuel Burns

Calculated Values

Summations of Organic Chemicals and Gases

	(µg/m³) Burn Area Taken as			(µg/m³) Normal Threshold	Equations to Calculate Equation Parameters		
	200 m²	400 m²	800 m²	Limit (µg/m³)	a	b	c
Carbonyls							
Acetaldehyde	0	0	0	200	0.499	0.0325	18.4
Acetone	4	16	39	170	14.7	0.0573	3.84
2-butanone	102	111	127	350	115.1	0.0407	3.64
Butyraldehydes	0	4	18	350	22.5	0.0344	5.68
Formaldehyde	4	26	69	260	35.4	0.107	9.18
Proprionaldehyde	0	7	22	350	19.6	0.0371	4.85
PAHs							
1-Methylnaphthalene	0	0	2	20	1.79	0.00435	0.585
1-Methylphenanthrene	0	0	1	0.3	0.698	0.00182	0.238
2,3,5-Trimethylnaphthalene	0	1	10	10	9.51	0.0218	3.05
2,6-Dimethylnaphthalene	0	0	3	30	2.87	0.00657	0.923
2-Methylnaphthalene	0	0	2	20	2.37	0.00568	0.775
Acenaphthene	0	0	6	5	5.67	0.0132	1.88
Acenaphthylene	0	0	10	2	11.3	0.0248	3.65
Anthracene	0	0	4	0.2	4.22	0.0101	1.39

(*Continued*)

TABLE 1.14 (*Continued*)

Emission Calculations and Values for Diesel Fuel Burns

Calculated Values

Summations of Organic Chemicals and Gases

	(μg/m³) Burn Area Taken as			(μg/m³) Normal Threshold	Equations to Calculate Equation Parameters		
	200 m²	400 m²	800 m²	Limit (μg/m³)	a	b	c
PAHs							
Benz(a)anthracene	0	0	0	0.2	0.315	0.000762	0.105
Benzo(a)pyrene	0	0	0	0.2	0.379	0.00114	0.141
Benzo(b)fluoranthene	0	0	1	0.2	0.536	0.00129	0.178
Benzo(k)fluoranthene	0	0	0	0.2	0.0137	4.48E-05	0.00441
Benzo(e)pyrene	0	0	0	0.2	0.213	0.000644	0.079
Benzo(g,h,i)perylene	0	0	0	0.1	0.326	0.000808	0.109
Biphenyl	0	0	2	1.3	1.86	0.00435	0.603
Chrysene	0	0	0	0.2	0.325	0.00078	0.108
Dibenz(a,h)anthracene	0	0	0	0.2	0.378	0.00078	0.108
Dimethylnaphthalenes	0	0	2	30	1.62	0.0037	0.52
Fluoranthene	0	0	2	0.2	1.95	0.00463	0.647
Fluorene	0	0	0	5	0.101	0.000195	0.0283
Indenol(1,2,3-cd)pyrene	0	0	0	0.2	0.261	0.000635	0.0873
Methylphenanthrenes	0	0	0	0.3	0.276	0.000772	0.0921
Naphthalene	0	0	2	52	2.01	0.00541	0.674
Perylene	0	0	0	0.2	0.0486	0.000116	0.0162
Phenanthrene	0	0	2	0.2	1.62	0.00375	0.527
Pyrene	0	0	2	0.2	1.99	0.00465	0.66
Trimethylnaphthalenes	0	0	1	10	0.8	0.0024	0.269

(*Continued*)

TABLE 1.14 (*Continued*)

Emission Calculations and Values for Diesel Fuel Burns

Calculated Values

Summations of Organic Chemicals and Gases

	(µg/m³)			(µg/m³)	Equations to Calculate		
	Burn Area Taken as			Normal Threshold	Equation Parameters		
	200 m²	400 m²	800 m²	Limit (µg/m³)	a	b	c
VOCs							
1,2,3-Trimethylbenzene	0	0	2	123	2.99	0.00443	0.853
1,2,4-Trimethylbenzene	13	34	76	123	15.3	0.105	4
1,2-Diethylbenzene	0	1	2	123	0.894	0.003	0.254
1,3,5-Trimethylbenzene	0	0	1	123	5.55	0.00442	1.45
1,3-Butadiene	0	3	13	400	6.49	0.0244	2.27
1,3-Diethylbenzene	0	0	1	260	0.623	0.00104	0.129
1,4-Diethylbenzene	0	0	0	260	3.57	0.00179	0.836
1-Butene/2-Methylpropene	2	10	26	500	7.5	0.0404	2.43
1-Heptene	2	6	14	1,500	2.14	0.0202	0.717
1-Hexene/2-Methyl-1-Pentene	0	1	2	1,500	1.01	0.00241	0.228
1-Methylcyclohexene	0	1	3	1,500	1.13	0.00563	0.392
1-Methylcyclopentene	0	0	1	1,500	0.238	0.00116	0.0442
1-Nonene	0	0	4	1,500	4.09	0.0088	1.33
1-Octene	0	0	0	1,500	0.777	0.000651	0.164
1-Pentene	3	8	18	1,500	1.55	0.0248	0.635
2,2,3-Trimethylbutane	0	0	1	1,230	0.694	0.00125	0.208
2,2,4-Trimethylpentane	0	0	1	1,230	3.23	0.00263	0.801
2,2,5-Trimethylhexane	0	1	2	1,230	1.09	0.00323	0.314

(*Continued*)

TABLE 1.14 (*Continued*)
Emission Calculations and Values for Diesel Fuel Burns

Calculated Values
Summations of Organic Chemicals and Gases

	(µg/m³)			(µg/m³)	Equations to Calculate		
	Burn Area Taken as			Normal Threshold	Equation Parameters		
	200 m²	400 m²	800 m²	Limit (µg/m³)	a	b	c
VOCs							
2,2-Dimethylbutane	0	0	1	1,230	1.69	0.00274	0.475
2,2-Dimethylpropane	0	0	1	1,230	0.335	0.00145	0.0886
2,3,4-Trimethylpentane	0	0	1	1,230	1.92	0.00285	0.542
2,3-Dimethylbutane	0	3	10	1,230	3.35	0.0158	1.1
2,3-Dimethylpentane	0	0	6	1,230	7.62	0.0145	2.33
2,4-Dimethylhexane	0	0	2	1,230	2.23	0.00445	0.646
2,4-Dimethylpentane	0	0	2	1,230	3.26	0.0062	1.02
2,5-Dimethylhexane	0	0	1	1,230	1.12	0.00228	0.298
2-Ethyltoluene	0	0	1	123	3.32	0.00295	0.857
2-Methyl-1-Butene	0	0	1	1,230	0.951	0.00207	0.275
2-Methyl-2-Butene	0	0	2	1,230	1.67	0.00406	0.53
2-Methylbutane	0	0	29	1,230	43.1	0.0762	13.2
2-Methylheptane	0	2	10	1,230	7.87	0.0205	2.45
2-Methylhexane	0	4	20	1,230	13.4	0.0399	4.44
2-Methylpentane	0	7	21	1,230	15.7	0.0366	4.17
3,6-Dimethyloctane	0	3	13	1,230	−0.0342	0.0259	1.27
3-Ethyltoluene	0	0	1	434	5.74	0.004	1.44
3-Methylheptane	0	1	6	1,230	4.9	0.0124	1.51

(*Continued*)

TABLE 1.14 (Continued)

Emission Calculations and Values for Diesel Fuel Burns

Calculated Values

Summations of Organic Chemicals and Gases

	(µg/m³) Burn Area Taken as			Normal Threshold	Equations to Calculate Equation Parameters		
	200 m²	400 m²	800 m²	Limit (µg/m³)	a	b	c
VOCs							
3-Methylhexane	0	0	36	1,230	34.1	0.0889	12.2
3-Methylpentane	0	1	6	1,230	7.11	0.0139	2.09
4-Ethyltoluene	0	0	1	1,230	2.84	0.00266	0.717
4-Methylheptane	0	1	4	1,230	1.62	0.00668	0.49
Benzene	0	7	33	1.6	27.4	0.0649	8.15
Butane	0	0	11	1,900	19.6	0.0286	5.55
c-1,3-Dimethylcyclohexane	0	3	12	2,000	5.81	0.022	1.95
c-1,4/t-1,3-Dimethylcyclohexane	0	1	4	2,000	2.46	0.00776	0.837
c-2-Butene	0	1	2	1,100	0.673	0.00265	0.205
c-2-Heptene	0	0	0	1,100	2.02	0.00134	0.53
c-2-Hexene	0	0	2	1,100	1.91	0.00492	0.697
c-2-Pentene	0	1	1	1,100	0.596	0.00233	0.178
Cyclohexane	0	2	12	1,030	8.27	0.024	2.74
Cyclohexene	0	0	2	600	1.55	0.00346	0.479
Cyclopentane	0	1	3	1,720	2.6	0.00684	0.811
Cyclopentene	1	1	1	1,030	0.229	0.0016	0.0066
Decane	0	0	0	2,000	23.1	0.0124	6.05
Dodecane	0	0	7	2,000	139	0.121	40.1

(Continued)

TABLE 1.14 (*Continued*)

Emission Calculations and Values for Diesel Fuel Burns

Calculated Values

Summations of Organic Chemicals and Gases

| | (µg/m³) | | | (µg/m³) | Equations to Calculate | | |
| | Burn Area Taken as | | | Normal Threshold | Equation Parameters | | |
	200 m²	400 m²	800 m²	Limit (µg/m³)	a	b	c
VOCs							
Ethylbenzene	0	0	3	434	6.53	0.00714	1.69
Heptane	0	8	47	1,640	32.2	0.096	10.9
Hexylbenzene	0	0	4	600	4.55	0.00942	1.38
Indan (2,3-Dihydroindene)	0	0	1	83	0.761	0.00181	0.191
Indene	0	0	1	123	0.309	0.00142	0.0972
Isobutane (2-Methylpropane)	0	0	0	1,610	7.22	0.00282	1.58
m,p-Xylene	0	2	20	434	29.7	0.0458	8.13
Methylcyclohexane	0	6	39	1,610	27.9	0.0806	9.44
Methylcyclopentane	0	2	7	2,687	5.21	0.0131	1.55
Naphthalene	0	0	5	52	10.8	0.0146	3.05
n-Butylbenzene	0	0	0	260	1.63	0.00128	0.433
Nonane	0	0	10	1,050	19.6	0.0284	5.68
n-Propylbenzene	0	0	1	246	1.77	0.00178	0.435
Octane	0	6	22	1,400	13.9	0.041	4.31
o-Xylene	0	0	13	434	20.9	0.0356	6.3
p-Cymene (1-Methyl-4-iso-propylbenzene)	0	0	0	140	1.02	0.000282	0.275
Pentane	0	1	24	1,770	29.2	0.0587	9.1

(*Continued*)

TABLE 1.14 (*Continued*)

Emission Calculations and Values for Diesel Fuel Burns

Calculated Values

Summations of Organic Chemicals and Gases

| | (µg/m³) | | | (µg/m³) | Equations to Calculate | | |
| | Burn Area Taken as | | | Normal Threshold | Equation Parameters | | |
	200 m²	400 m²	800 m²	Limit (µg/m³)	a	b	c
VOCs							
Propane	0	0	0	4,508	19.5	0.002	4.5
Propene	0	9	27	2,615	10.2	0.0436	3.25
Propyne	0	0	1	1,000	0.874	0.00155	0.236
sec-Butylbenzene	0	0	1	123	0.882	0.00158	0.247
Styrene	0	5	13	123	3.96	0.021	1.37
t-1,2-Dimethylcyclohexane	0	2	6	1,600	2.86	0.0111	0.933
t-2-Butene	0	0	1	1,600	0.898	0.00256	0.281
t-2-Heptene	0	0	2	1,600	1.89	0.00392	0.553
t-2-Hexene	0	4	17	1,600	0.377	0.032	1.53
t-2-Octene	0	23	68	1,600	4.58	0.112	4.67
t-2-Pentene	0	2	5	1,600	2.24	0.00797	0.677
t-3-Heptene	0	0	0	1,600	85.4	0.0688	25.7
tert-Butylbenzene	0	0	1	123	1.37	0.0026	0.411
Toluene	0	11	39	123	34.6	0.0696	8.94

Analysis of the VOC data shows VOCs to be close to being a matter of concern; however, it should be noted that the level of VOCs is much higher (as much as three times higher as measured in some tests) when oil is evaporating in the absence of burning than when burning. Carbonyls are another emission of concern, although they are significantly below health concern levels for the scenarios in Tables 1.13 and 1.14. The level of concern is the percentage of the regulated level attained by the emission. For example, if a regulated level is 75 μg/m^3 and the calculated value is 150, then the level of concern is given as 200%. There is no health concern for fixed gases such as carbon dioxide or carbon monoxide at levels measured at burns to date.

Safe distances downwind from a crude oil burn (based on PM-2.5 concentrations) can be calculated as follows:

$$\text{Safe distance}(m) = \exp\left[\frac{12.5 + 0.0347 \times \text{fire size}(m^2)}{4.79}\right]$$

Safe distances downwind from a diesel fire can be calculated as follows:

$$\text{Safe distance}(m) = \exp\left[\frac{1.44 + 0.0052 \times \text{fire size}(m^2)}{0.412}\right]$$

Note: To convert feet to metres, multiply by 0.3048.

To convert metres to feet, multiply by 3.28084.

Note that the level of PM-2.5 measured for diesel emissions is the same as the PM-10 level or exceeds it. This indicates that most of the matter consists of PM-2.5. As PM-2.5 is the emission of highest concern, this becomes the most important factor for calculation of safe distances.

Based on these data, safe distances have been calculated for a variety of fire sizes for diesel fuel and crude oil. These calculations are given in Table 1.15 (Fingas 2017).

TABLE 1.15
Safe Distance Calculations (Based on PM-2.5 Concentrations)

Type and Area	Burn Area (in ha [acres])	Safe Distance (in km)	Safe Distance (in mi)
Crude Oil Burns			
Small area, 250 m^2	0.25 (.6)	0.09	0.06
Full boom pull, 500 m^2	0.5 (1.2)	0.5	0.3
Large boom pull, 750 m^2	0.75 (1.9)	3.2	2
Diesel Burns			
Small area, 250 m^2	0.25 (.6)	0.8	0.5
Full boom pull, 500 m^2	0.5 (1.2)	2	1.2

1.5 ASSESSMENT OF FEASIBILITY OF BURNING

1.5.1 Burn Evaluation Process

When an oil spill occurs, information must be obtained on the spill location, weather conditions and any other relevant conditions at the site. The necessary questions to be asked before deciding to use ISB are outlined in Figure 1.24.

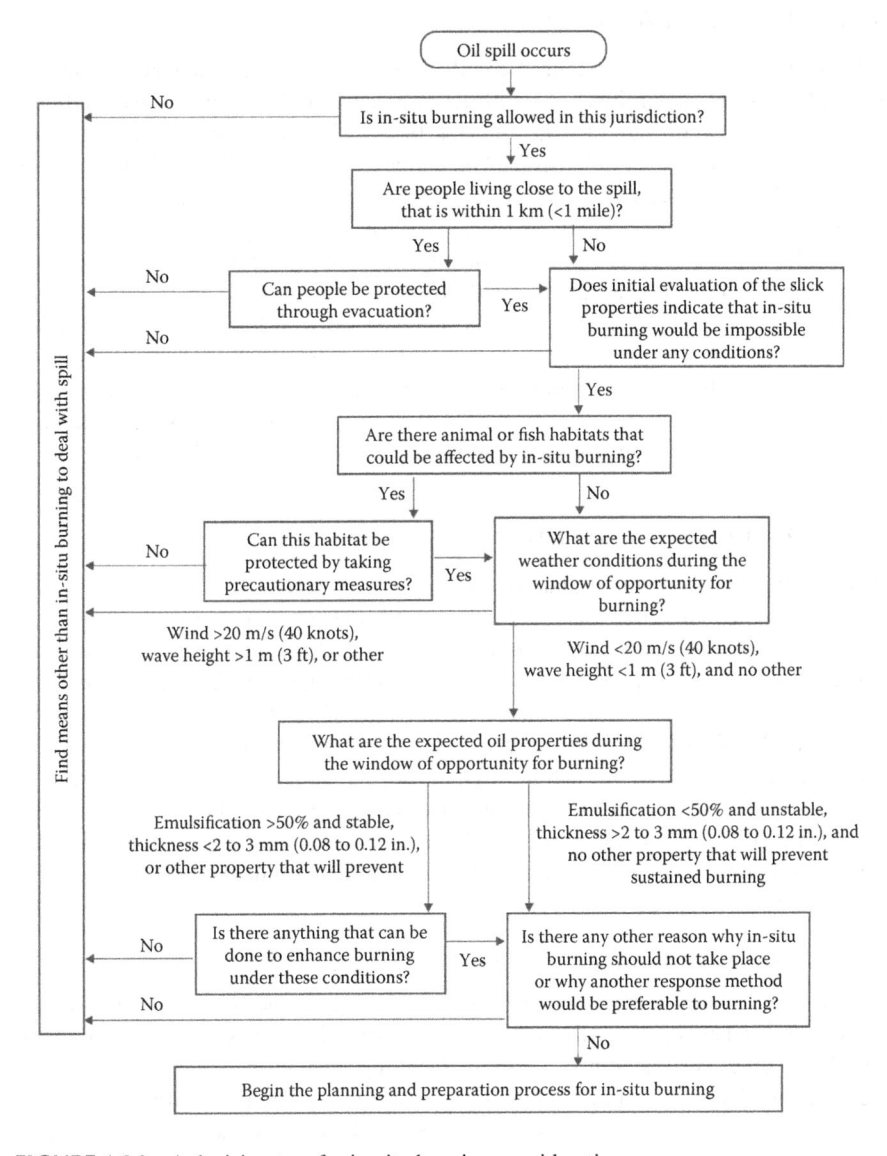

FIGURE 1.24 A decision tree for in-situ burning considerations.

1.5.2 AREAS WHERE BURNING MAY BE PROHIBITED

Burning may be prohibited within a specified distance of human habitation, for example, within 1 km and within a specified distance of the shoreline; of petroleum loading, production or exploration facilities; or of a nature preserve, bird colony or national or state/provincial parks. Burning may also be prohibited over a marine park or preservation area and over areas designated as military target areas or former areas of munitions dumping.

1.5.3 REGULATORY APPROVALS

The regulatory approvals required for ISB vary among different jurisdictions. In general, the legal constraints and liabilities associated with ISB are not well defined. The situation is aggravated by the fact that the public is reluctant to accept regulations that allow any kind of burning. The public must be provided with information about the issues associated with ISB in order to accept regulations allowing it. This information must include a comparison of the risks of burning with the risks associated with other clean-up options, and the results of simply leaving the spilled oil and not treating it at all (Fingas and Punt 2000).

In general, regulatory agencies are most concerned with how the burn will affect air quality (Snider 1994). Most jurisdictions stipulate air quality levels that cannot be exceeded no matter what is being burned. Some jurisdictions have modified the air quality limits for special cases, such as ISB of oil during an emergency. There are differences for land burns, and these are discussed in greater detail in Chapter 2.

1.5.4 ENVIRONMENTAL AND HEALTH CONCERNS

The primary environmental and health concerns related to ISB are the emissions produced by the fire. The measurement of emissions and calculations using equations developed from emission data has revealed several facts about the quantity, fate and behaviour of the basic emissions from burning. Overall, emissions are now understood to the extent that emission levels and safe distances downwind can be calculated for fires of various sizes and types. A typical crude oil burn (500 m^2) would not exceed health limits for emissions beyond about 500 m from the fire. The emissions produced by in-situ burns are discussed Section 1.4. People and the environment can be protected by ensuring that the burn is kept the minimum distances away from populated and sensitive areas. Procedures for calculating these safe distances were given earlier in this chapter.

1.5.4.1 Safety of Response Personnel

During in-situ burn operations, all response personnel must be fully trained in the operational and health and safety procedures associated with any equipment or operation being used. Personnel involved in the planning stage of the operation and for the deployment of vessels, booms and ignition devices must also be well trained. General health and safety guidelines are discussed below. These guidelines should be used to develop site-specific plans once it has been decided that ISB will take place.

1.5.4.2 Public Health

In general, and depending on weather conditions, ISB should not be carried out within 1 km of heavily populated areas. Weather conditions to be considered include the presence or absence of an inversion and the wind direction. According to monitoring of oil fires done up until 1998, ground-level emissions from crude oil fires have never exceeded 25% of established human health concern levels more than 1 km away from the fire (ASTM F1788 2017). Therefore, if no significant air turbulence or ground-level atmospheric inversions occur, burning can be conducted close to populated areas. In sparsely populated areas, it may be best to evacuate residents close to the burn site. Methods are now available for calculating emission concentrations and safe distances downwind from in-situ oil burns, and these are summarised below.

Several studies on human health effects have been conducted, typically using animal models. Jaligama et al. (2015) conducted studies on mice using particulate material collected from the Deepwater Horizon burns and found that doses caused cytotoxicity and generated reactive molecules in the mice models. The particulate matter also had a number of effects such as weight loss and promotion of allergic responses. These studies used high doses of particulate matter that would be typical of exposure directly over the fire.

1.5.4.3 Air Quality

The major barrier to the acceptance of ISB of oil spills is the lack of understanding of the resulting combustion products and the belief that it is just transferring pollution to the sky. It should be noted that emissions from oil fires are much smaller than typical emissions from other types of burning, for example, biomass burning (Kasische and Penner 2004).

Several types of emissions are formed and released when oil is burned. The atmospheric emissions of concern include the smoke plume, particulate matter precipitating from the smoke plume, combustion gases, unburned hydrocarbons, organic compounds produced during the burning process and the oil residue left at the burn site. Although consisting largely of carbon particles, soot particles contain a variety of absorbed and adsorbed chemicals. Complete analysis of the emissions from a burn involves measuring all these components. The emphasis in sampling has been on air emissions at ground level as these are the primary human health concern and the regulated value.

It should be noted that the monitoring of emissions conducted at past burns was as comprehensive as possible, and the best field samplers and instrumentation available at the time were used. Measurement techniques have progressed over the years, however, and continue to improve. In addition, the data from these burns are so extensive that not even overall summaries can be provided here. The summarised data appear in the references cited in this section and provided at the end of the chapter. The analysis carried out on the Deepwater Horizon burns in 2010 is consistent with the older test burn data.

1.5.4.4 Water Quality

Research has shown that ISB of oil does not release any more oil components or combustion by-products into the water column than are present if the oil is left unburned on the water surface (Fingas et al. 2001b). Water samples from underburning oil have been analysed and very low levels of organic compounds were detected (Daykin et al. 1995; Environment Canada 1993, 1997). Only low levels of hydrocarbons have been found, at concentrations that would not result in fish mortality, even in a confined body of water. No PAHs have been detected in water samples from underburning oil. Toxicity tests of the water column were also conducted, and no toxicity was noted. One study used a confined water system for allowing sampling after ISB. Seawater samples and oil were collected prior to and immediately after ISB, and chemical analysis was conducted. Acute toxicity tests with the marine copepod *Calanus finmarchicus* and Microtox bioassay was performed to establish LC_{50}/EC_{50} values of the water. The results were compared to regular Water Accommodated Fractions (WAF) systems, and indicated no increase in toxicity in the underlying water after (Faksness et al. 2012). Comparison studies using Ethoxyresorufin-O-deethylase (EROD) – a biomarker used in fish as an indicator of chemical exposure and burning showed the least effect of tested countermeasures (Cohen et al. 2006).

The burning process leaves a residue, however, that is primarily composed of oil with little removed other than some of the more volatile materials (Fingas et al. 1998, 2001b). The residue contains large quantities of PAHs, although usually less than the original oil, and it may also contain a slightly higher concentration of metals. The residue consists of unburned oil, oil depleted of volatiles, re-precipitated soot, and partially burned oil. It appears to be similar to weathered oil of the same type and is typically viscous and dense. Several tests have shown that burn residue is no more aquatically toxic than other weathered oils and, in fact, is much less toxic than fresh oils of the same type. There is evidence that the metals contained in the original oil (usually 10–40 ppm of vanadium, chromium and nickel) become concentrated in the burn residue (ASTM F1788 2017).

The density of this residue depends on how heavy the original oil is and the completeness of the burn, although it will never be denser than the heaviest hydrocarbons found in the original oil. A very efficient burn of a heavier crude oil produces a dense residue that may sink and pose a threat to benthic species. Sinking is rare, however, and has been recorded in only a few of about 600 burns worldwide, including those burns from the Deepwater Horizon. Aquatic toxicity tests performed on samples of residue have shown very low toxicity (Daykin et al. 1995). Residues can be collected in a backup boom using sorbents or a skimmer can be used to collect lighter residues.

Another concern is that burning will raise the water temperature below the oil, as extreme temperature changes can affect marine species (Environment Canada 1997). Measurements during burn trials, however, show no significant increase in water temperature, even during some burns in shallow, confined test tanks. Thermal transfer to the water is limited by the insulating oil layer and is actually the mechanism by which the combustion of thin slicks is extinguished.

1.5.4.5   Effects on Birds and Other Biota

Wildlife on land is generally not affected if burning is conducted more than 1 km away from shore or sensitive areas. It has also been observed that birds avoid the burning site and therefore are unlikely to be affected by the burn. Similarly, marine species should not be affected as the water column normally does not become contaminated and the water temperature does not change within a few centimetres below the slick. Benthic species may be affected by the sinking of heavy burn residue.

1.5.4.6 Infrastructure Concerns

Oil slicks should not be burned close to infrastructures such as buildings, docks, lighthouses, oil platforms and vessels that originally contained the oil.

1.5.5 Weather and Ambient Conditions

Weather conditions such as wind speed, gusts, shifts in wind direction, wave height and geometry and water currents can all jeopardise the safety and effectiveness of a burn operation. Strong winds can make it difficult to ignite the oil during ISB. Once the oil is ignited, high winds can extinguish the fire or make it difficult to control. In general, oil can be successfully ignited and burned safely at wind speeds less than 20 m/s (40 knots) (ASTM F 1990 2017; Fingas and Punt 2000). Tank tests have shown that at wind speeds greater than 15 m/s (30 knots), the flames do not propagate upwind (USCG 2003). During a test in England, however, oil burned in winds up to 25 m/s (50 knots) (Guénette and Thornborough 1997).

The effects of air and water temperatures on the ability to ignite and burn oil slicks is not well documented; however, tank tests have shown that air temperatures of −11°C to 23°C and water temperatures of −1°C to 17°C did not affect the ability of a slick to burn (Tennyson 1994) While no testing has been done on the effect of rain on burning, rain would probably lower the efficiency of the burn due to the cooling effect of the water.

High sea states can make it difficult to contain oil. Waves higher than 1 m can cause the oil to splash over the containment boom (Fingas and Punt 2000). High waves can also contribute to the emulsification of oil, which could make it more difficult to ignite.

Tests in ice-covered areas have shown that ice coverage has a minimal effect on the ability of a slick to burn (Buist et al. 1994). In fact, ice is typically used as a natural method to contain oil for burning. There is more on ice situations in Chapter 3. Further reflection of heat from ice walls and cavities can increase the propensity to burn and the burn rate (Farahani et al. 2015; Shi et al. 2015, 2016a, 2017).

Burning can only be done safely at night if oil conditions, weather conditions and sea conditions are well known. Towing booms at night would be unsafe under most conditions. Burning at night would be a relatively safe choice in the case of a thicker, uncontained spill at sea, especially if the spill is offshore and its extent is well known. Some nearshore spills and spills in marshes have been burned at night, which is a relatively safe practice because the concentrations and location of the oil are known and precautions can be taken to ensure that the fire does not spread to surrounding areas.

1.6 BURNING ON WATER

A plan is followed using pre-established scenarios, check lists and safety procedures. In most cases, containment will be required either because the slick is already too thin to ignite or will be too thin within hours.

The basic processes are shown in Figure 1.1. Personnel and equipment are transported to the site. In most cases, a fire-resistant boom is deployed downwind of the spill and a tow is begun. When enough oil collected in the boom, it is ignited using an igniter. The boom tow is resumed and continued until the fire is extinguished or the tow is stopped for operational reasons. The burning and progress of the tow are monitored by personnel on aircraft and on a larger ship from which an overview of the slick and conditions is possible. The monitoring crew can also direct the boom tow vessels to slick concentrations upwind. During the burn, monitoring normally includes estimating the area of oil burning at specific time intervals so that the total amount burned can be estimated. The amount of residue is similarly estimated. Particulate matter downwind might be monitored to record the possible exposure levels.

The burn could be stopped in an emergency by releasing one end of the boom tow or by speeding up the tow so that oil is submerged under the water. If the burning stops because there is not enough oil in the boom, the tow can be resumed going downwind and then turning around into the wind before re-igniting. After the burn operation is finished for the day or for the single burn, the burn residue should be removed from the boom. As the burn residue is very viscous, a heavy-oil skimmer may be required if there is a large amount of material. A small amount of residue can be removed by hand using shovels or sorbents.

The burning that took place at Deepwater Horizon was an example of a successful burn campaign at sea. Indeed, there were about 400 successful burns carried out during the Deepwater Horizon spill and this removed a significant part of the oil on the water. The basic technique was to collect oil in a fire-resistant boom (hereinafter called fireboom) and then ignite the oil and slowly pull the fireboom forward to push the oil to the rear or wait to see if the winds and currents were doing this. The oil was spotted using a fixed-wing aircraft. Two shrimp boats (about 100-ft-long) towed about 150 m (500 ft) of fireboom at about ½–¾ knot to avoid loss of the oil through entrainment under the boom. The tow lines were about 100 m (about 300 ft) for the safety of the tow crews. This is shown in Figure 1.25. Once sufficient oil had been collected for a burn and marine and air monitoring approved, ignition was requested. A small boat carrying two persons approached from upwind and an igniter was dropped over the edge of the boom. The igniters were made from a plastic jar (about 1 L) of diesel fuel, a marine flare and some Styrofoam floats. The flare, once activated, burned down to the bottle of gelled diesel fuel, which started burning and acted as a primer to ignite the oil.

Once lit, the heavy, weathered oil burned until most oil was removed. The burn was monitored from the air by trained observers and from larger vessels in the area. The amount burned was gauged by estimating the burning area in the boom and multiplying by the burning rate.

Many precautions were taken during the burn. Extensive training was given to the crews and several practice sessions were undertaken. Particulate emissions from the burns were monitored.

Oil can also sometimes be burned without containment and by using natural containment features such as oceanic fronts, ice or shorelines to contain oil.

FIGURE 1.25 A small burn in the Deepwater Horizon spill clean-up program. Note the boats pulling the boom in a catenary configuration. The small boat to the right carries the ignition and monitoring crew, while the supply ship at the lower end is the mother boat and monitoring vessel.

1.7 BURNING ON LAND AND WETLANDS

Burning on land or wetlands is the subject of Chapter 2 of this book. Burning oil on land is an older and more frequently used technique than ISB on water. Burning on land involves several different considerations than burning on the sea. First, the effect on soil vegetation is a prime consideration corresponding to the vegetation present. Certain types of vegetation are very sensitive to fire; others are not. Second, the location of the proposed burn should be considered in terms of the burn history or use of prescribed burning in the area in the past. Prescribed burning is extensively used around the world for a variety of purposes including control of vegetative species and removal of fire hazards. Many guidelines exist at local, state and provincial levels on the use of prescribed burning and these should be consulted for given areas. A third consideration is the exposure of the sub-surface environment to heat as indicated by temperature. This is influenced by soil moisture and depth of oil coverage. This is especially important in terms of plant propagation from roots and tubers. If these are permanently damaged by the fire, then re-vegetation of the area is very slow or non-existent. A fourth consideration for burning on land is the amount of oil penetration before and after the proposed fire. If there is little penetration before the fire, then burning is more useful.

In terms of operations, many of the same considerations in this overall burning section apply to land as might apply to burning on water. There are several important differences to consider, however. First, the ease of ignition and oil thickness may not apply if there is combustible material such as dry grass available on the land. Burning in cases where there is dried vegetative material or wood in the target area is simply a matter of igniting that material. Both the dried vegetative material and oil will burn.

The procedures for lighting a fire on land differ from that on water. First, a firebreak must be established around the entire perimeter of the proposed burn. Sometimes natural barriers, such as rivers, roads and so on, can serve as firebreaks. Once the firebreak is established, one proceeds to the furthermost upwind position and uses an igniter to start the fire. This is typically a drip torch. Once started, the fire is monitored, especially close to the firebreaks at the downwind side.

Means should be available to extinguish fires and unwanted fire propagation. This is usually in the form of a fire truck with trained firefighters. After the main fire is out, it requires monitoring for several hours until all hot spots are cool and there is no danger of flare-ups.

1.8   BURNING ON OR IN MARSHES

Burning on or in marshes is similar to the wetlands is noted above. Chapter 2 in this book includes this topic. Burning marshes often results in less ecological damage than other methods of dealing with the oil (Zengel et al. 2018).

1.9   BURNING IN OR ON ICE

Chapter 3 covers this topic. Many test burns have been conducted on or among ice floes. The ice serves as a natural barrier to the spreading of the oil. Much of the early burn work was carried out as a countermeasure for oil in ice.

1.10   EQUIPMENT – SELECTION, DEPLOYMENT AND OPERATIONS

This section outlines the types of equipment used in responding to a spill with ISB and the steps involved in deploying and operating this equipment. This equipment includes containment booms; other containment and burning equipment; igniters; aircraft and response vessels; treating agents; monitoring, sampling and analytical equipment; and residue recovery equipment. This section is intended to assist response personnel in the selection and deployment of equipment for particular response situations.

1.10.1   BURNING WITHOUT CONTAINMENT

Controlled burning of uncontained slicks is sometimes possible if the slick is thick enough and all other safety factors are considered (Ross 1986). Because it takes time to get containment booms to a site, it may be advisable to ignite and burn as much of the slick as possible as a first response if it is already fairly thick and then bring in containment booms to thicken the remaining parts of the slick for a second burn. Uncontained oil can be ignited with an igniter at the location where the oil is thickest.

When burning an uncontained slick, personnel must ensure that there is no direct link between the oil to be burned and the source of the oil, for example, the tanker or platform on the sea, to prevent the fire from spreading to the source. The safest and quickest option is to move the source away from the slick. When the spill originates from a platform or other fixed source, the portion of the slick that is to be burned should be moved away from the source and the slick around the source should be isolated using containment booms.

Several oil spills or blowouts have accidentally caught fire while uncontained and have burned well (McKenzie 1994). During the Deepwater Horizon, burns it was found that several patches of oil would burn without containment. Figure 1.26 shows such an uncontained burn during the Deepwater Horizon countermeasures. Figures 1.27 and 1.28 show accidental and uncontained burns. While it is not known what conditions are best for burning uncontained oil, emulsified oil may stop or retard the spreading of uncontained oil while it burns (McKenzie 1994). In a large burn, large volumes of air are drawn into the fire, which is referred to as a 'fire storm'. This may provide enough force to prevent the oil from spreading.

FIGURE 1.26 Two un-contained burns that were ignited at the same time as the forward (left) contained burn. Sometimes oil is thick enough to burn without containment, as shown here.

FIGURE 1.27   A ship on fire. Note that the oil is burning on the water without containment.

FIGURE 1.28   An overview of an offshore platform on fire, with oil on the water burning without containment.

In remote areas, natural barriers such as shorelines, offshore sand bars or ice can sometimes be used to contain oil in order to burn it. The shorelines must consist of cliffs, rocks, gravel, or sandy slopes to resist burning and there must be a safe distance between the burning oil and any combustible materials, such as wooden structures, forests or grass cover. On land, containment generally occurs naturally. In populated areas, the weather conditions must be such that the smoke plume will drift away from the populated centres. Zones of convergence on the sea can also be used to contain oil. Local oceanographers should be consulted to determine the location of these zones. The Coast Guard and local fishermen are also familiar with currents in an area.

1.10.2 Oil Containment and Diversion Methods

As discussed earlier, an oil slick should be at least 0.5–3 mm thick in order quantitatively remove significant amounts of oil. It is not fruitful to burn thin slicks anyway. Several methods for increasing the thickness of a slick to this level or to maintain a thickness at or above this level are discussed in this section.

1.10.2.1 Fire-Resistant Booms

The biggest concern with containment booms for ISB is the ability of the boom's components to withstand heat for long periods of time. Few fire-resistant booms are commercially available because the market is small and the cost of production is high. Fire-resistant booms cost considerably more than conventional booms. These booms are constantly being tested for fire resistance and for containment capability and designs are modified in response to test results.

The fire resistance of these booms has been extensively tested at the U.S. Coast Guard Fire and Safety Test Detachment in Mobile, Alabama. These booms have also been tested for strength, integrity and oil containment capabilities during tow tests at the OHMSETT facility in Leonardo, New Jersey.

The different types of fire-resistant boom are water-cooled booms, stainless steel booms, thermally resistant booms and ceramic booms. Fire-resistant booms, especially stainless steel booms, require special handling because of their size and weight. Thermally resistant booms are similar in appearance and handle like conventional booms, but they are built of many layers of fire-resistant materials. The various types of fire-resistant boom are shown in Figure 1.29.

Fire-resistant booms developed by Environment Canada in the late 1970s consisted of a series of ceramic, stainless steel designs or those that used air or water sprays to contain oil during burning (Buist et al. 1983; Meikle 1983). Figure 1.30 shows an early ceramic boom being tested at OHMSETT. Environment Canada also worked with conventional booms using water cooling systems and with log booms.

In the early 1980s, Dome Petroleum Ltd. further modified the stainless steel boom developed by Environment Canada. The Dome boom consisted of 1.5 m vented stainless steel flotation units with a pentagonal cross section. A stainless steel panel attached to the top of each unit creates the freeboard and a PVC-coated nylon skirt attached to the bottom of the float provides the draft. The flotation sections were attached using 0.75 m flexible panels constructed of stainless steel mesh

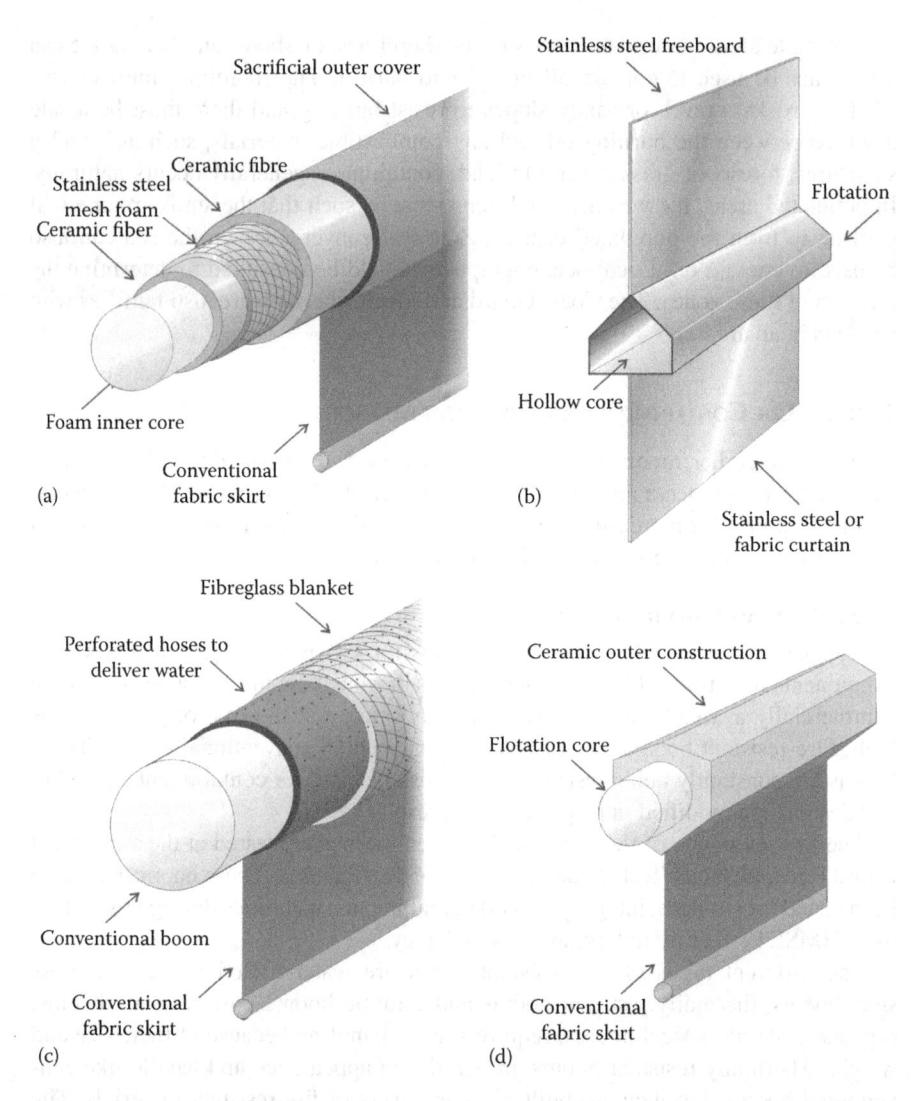

FIGURE 1.29 Types of fire-resistant booms. (a) Thermally resistant fibre-based boom; (b) stainless steel boom design; (c) water-cooled boom cover; (d) ceramic boom.

encased in a Fibrefax blanket with a PVC-coated nylon skirt. The Dome boom was designed to be used for more than one in-situ burn incident. Fire-resistant booms manufactured today are generally designed to survive several burns at one site but are then disposed of or refurbished.

A standard has been promulgated by ASTM to test the durability of fire-resistant booms for ISB (ASTM 2152 2013). The standard is a minimum 5-h test involving three 1-h burning periods with two 1-h cool-down periods between the burning periods. Booms are tested in a test tank with crude oil or diesel fuel. Oil is pumped into the centre of the boom at a predetermined rate and is burned.

FIGURE 1.30 A ceramic fire-resistant boom being tested at OHMSETT.

The oil is continuously fed into the boom for a 1-h burn and then is shut off, allowing the burn to die out. The boom then cools for 1 h and is tested for two additional 1-h burn/1-h cooling sessions. At the start of the third burn, oil is pumped into the boom to test for gross leakage. Several booms were tested in this manner. An analogous test was developed using propane and conducted at the OHMSETT test facility (Buist et al. 2001). In testing in 2000, the SWEPI fireboom, designed for an ice-infested environment, passed the basic test requirements as did 4 protective covers, including one from Oil Stop and three from Applied Fabric Technologies. Figures 1.31 through 1.38 shows booms being tested at Mobile, Alabama.

The Marine Spill Response Corporation (MSRC) conducted at-sea towing tests of four fire-resistant booms: the American Marine (3M) Fireboom, the Applied Fabrics PyroBoom, the Kepner Plastics SeaCurtain FireGard and the Oil Stop Auto Boom Fire Model (Nordvik et al. 1995). The purpose of these tests was to evaluate the relationship between boom performance and buoyancy-to-weight ratio, tow speed and sea state. The booms were towed in a U configuration at tow speeds of between 0.25 and 1.25 m/s (0.5 and 2.5 knots). The results of these tests showed that the higher the buoyancy-to-weight ratio of the boom, the faster the boom can be towed before it will submerge. In general, fire-resistant booms have a lower buoyancy-to-weight ratio than conventional booms. It was also found that three of the four booms tested exhibited mechanical failure at high tow speeds. The report further concluded that the mechanical integrity, sea-keeping performance and ease of deployment and recovery of commercially available fire-resistant boom must be improved. Some of the manufacturers did make the suggested improvements.

FIGURE 1.31 The FESTOP fireboom shown after a short demonstration fire.

FIGURE 1.32 The Hydro-fire boom shown before testing.

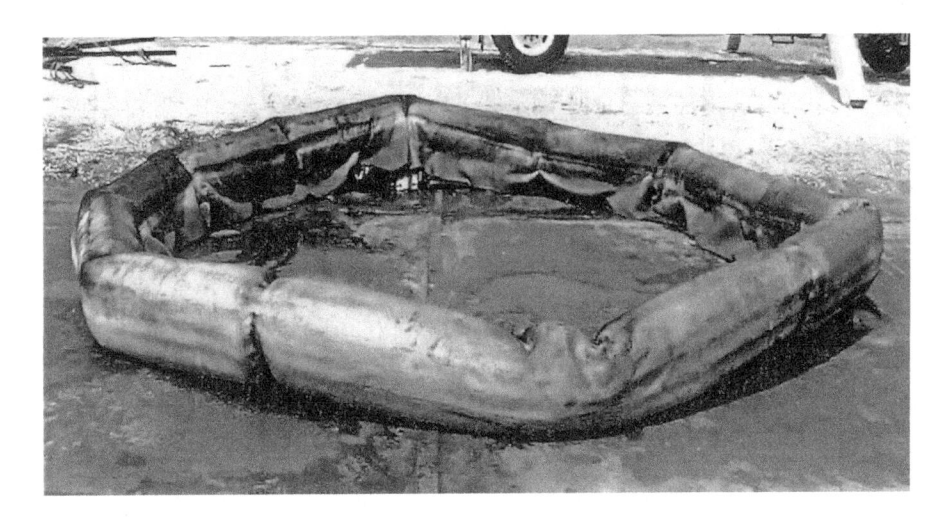

FIGURE 1.33 The Hydro-fire boom shown after the ASTM test.

FIGURE 1.34 The PyroBoom shown before testing.

FIGURE 1.35　The PyroBoom shown after the ASTM test.

FIGURE 1.36　The PocketBoom shown before testing.

FIGURE 1.37 The PocketBoom shown after the ASTM fire test.

FIGURE 1.38 A boom undergoing tests in a propane fire at OHMSETT.

The United States Coast Guard and the USMMS evaluated the containment behaviour of the fire-resistant booms currently on the market in a test tank and compared these results with previous at-sea performance results (Bitting and Coyne 1997; Cunneff et al. 2000). These studies determined the tow speeds at which the booms first began to lose oil (called 'first loss') and the speed at which a continuous, significant loss occurs (called 'gross loss'). It also determined the rate of loss of oil at specific tow speeds and the tow speed at which the boom physically failed, that is, became submersed or suffered structural damage. The results of these tests are summarised in Table 1.16.

The following are the conclusions of these tests:

- In terms of oil containment, the performance of the fire-resistant booms was similar to conventional, non-fire-resistant booms, with first losses occurring at tow speeds of 0.44–0.52 m/s (0.85–1.0 knots) in calm waters. These losses were relatively unaffected by regular waves and were reduced slightly by short-crested waves.
- The physical failure of fire-resistant booms was also similar to that of conventional booms with critical tow speeds between 1 and 1.5 m/s (2 and 3 knots), with the exception of the Spill-Tain boom, for which the critical tow speed exceeded 3 m/s (6 knots).
- The critical tow speeds determined during the at-sea tests were lower by 0.25–0.75 m/s (0.5–1.5 knots) than the critical tow speeds determined during tank tests.
- From the limited data available from the in-tank and at-sea tests, an increase in the buoyancy-to-weight ratio of the boom appears to increase the boom's ability to contain oil at higher than normal tow speeds.

The use of fire resistant booms at the Deepwater Horizon spill was well-documented. Table 1.17 shows the results of the use at this site (Mabile 2010). The findings of these at-sea burns have since become incorporated into procedures (Midlinx 2013).

The following is a brief description of the fire-resistant booms currently available – either on the market or in existing inventories. Detailed specifications for these booms can be found on the manufacturers' websites.

American fireboom (http://elastec.com/oilspill/fireboom/americanfireboom/ index.php) has flotation sections made of rigid ceramic foam surrounded by two layers of stainless steel knitted mesh, a high temperature-resistant ceramic textile fabric and a PVC outer cover that also forms the skirt. This boom is normally deployed from a container or tray.

Auto boom fire model (not in current production) is an inflatable boom with an internal water-cooling system. The flotation chamber is insulated with a ceramic blanket covered with a stainless-steel mesh. The skirt is made of a polyurethane fabric. This boom can be stored and deployed from a reel. Before the boom is placed in the water, however, the water-cooling system must be connected on a large, flat area.

TABLE 1.16
Performance of Fire Resistant Booms

Boom Type	Loss	First and Gross Loss Tow Speeds, m/s (knots)				Loss Rate Test, L/min at Tow Speed m/s (knots)		
		Wave Conditions[a]				First Loss + 0.05 m/s (0.1 knots)	First Loss + 0.15 m/s (0.3 knots)	Critical Tow Speed m/s (knots)
		C	1	2	3			
PyroBoom	First gross	0.51 (1.00) 0.62 (1.20)	0.37 (0.72) 0.48 (0.93)	0.55 (1.07) 0.67 (1.30)	0.49 (0.95) 0.57 (1.10)	246 at 0.57 (1.10)	534 at 0.67 (1.30)	1.03 (2.75)
Spill-Tain	First gross	0.44 (0.85) 0.54 (1.05)	0.21 (0.40) 0.31 (0.60)	0.44 (0.85) 0.54 (1.05)	0.45 (0.88) 0.55 (1.07)	27 at 0.49 (0.95)	178 at 0.59 (1.15)	3.08 (>6.00)
American Marine/3M	First gross	0.44 (0.85) 0.57 (1.10)	0.37 (0.72) 0.46 (0.90)	0.45 (0.87) 0.59 (1.15)	0.46 (0.90) 0.59 (1.15)	64 at 0.49 (0.95)	303 at 0.59 (1.15)	1.16 (2.25)
Dome Boom	First gross	0.49 (0.95) 0.68 (1.32)	0.38 (0.75) 0.54 (1.05)	0.49 (0.95) 0.62 (1.20)	0.52 (1.00) 0.64 (1.25)	32 at 0.54 (1.05)	151 at 0.64 (1.25)	1.03 (2.00)
Oil Stop	First gross	0.46 (0.90) 0.63 (1.22)	0.41 (0.80)	0.55 (1.07)	0.52 (1.00)	74 at 0.51 (1.00)	286 at 0.61 (1.20)	1.80 (3.50)

[a] Wave conditions.

C = Calm water, no waves generated.

1 = Wave 1, regular sinusoidal wave, H1/3 = 25 cm (9.8 in), L = 4.9 m (16 ft).

2 = Wave 2, regular sinusoidal wave, H1/3 = 33.8 cm (13.3 in), L = 12.8 m (42 ft).

3 = Wave 3, regular sinusoidal wave, H1/3 = 22.6 cm (8.9 in), L not calculated.

TABLE 1.17

Fire Resistant Booms Used at Deepwater Horizon Burns

Boom Name	Manufacturer	Size (inches)	Fire Resistance Component	Reinforce-ment	Buoyancy to Weight Ratio	Number of Systems Used	Longest Burn (hours - minutes)	Typical Reuse	Maximum Reuse	Re-Furbishment	Deployment	Handling	Containment
Hydro-Fire	Elastec/American Marine Inc.	14 top and 18 skirt	Water-cooled jacket	Regular boom	6.3:1	27	11 h and 48 min	10–14	14	Possible	Easy	Requires inflation as deployed	High level of integrity
American Marine	Elastec/American Marine Inc.	12 top and 21 skirt; 18/24	Ceramic fibre	Stainless steel mesh	3.8:1	37	11 h and 21 min	~5		Partially	OK	Rigid	Good, containment increased with use
PyroBoom		11 top and 19 skirt	Silicone-coated barrier and stainless Steel floats		3.3:1	13	3 h & 13 min			Kits		Heavy and rigid	Good except under higher winds and waves
Oil Stop Fire		11 top and 19 skirt	Ceramic insulation	Inflatable membrane		3	27 min			None		Requires inflation as deployed	Failed
Kepner Fire		11 top and 15 skirt	Ceramic insulation			2							Problems

FESTOP fireboom is a stainless-steel fireboom available in two sizes that is claimed to withstand temperatures up to 1,260°C. The company is located in France. A photograph of this boom appears in Figure 1.31 and shows this boom after a short demonstration burn.

The *Hydro-fire boom* (http://elastec.com/oilspill/fireboom/index.php) is a water-cooled, inflatable boom that is sometimes stored on and deployed from a reel. A 150-m length of boom can be stored on a reel with sections (30 m). Figures 1.32 and 1.33 show this boom before and after the ASTM test burn.

PyroBoom (http://www.appliedfabric.com/content/pages/fireproof-barriers. php) (DESMI – Applied Fabric Technologies) is a fence boom with a free-board constructed of a patented refractory material and a skirt made of a urethane-coated material. Hemispherical stainless-steel floats are attached to either side of the fence portion. This boom can be stored in a container and deployed from a large flat area or can be deployed from a reel system, which in turn is stored in a container. The boom is shown in Figures 1.34 and 1.35 before and after the burn test.

PocketBoom (http://www.appliedfabric.com/content/pages/pocketboom.php) (Applied Fabric Technologies) is a stainless steel boom that is similar to the design of the Dome Boom but in a small version. Figures 1.36 and 1.37 show this boom before and after testing.

SeaCurtain FireGard (Kepner Plastics) uses a heavy-gauge stainless steel coil covered with a high temperature refractory material to make up the flotation sections of the boom. The skirt is made of a polyurethane-coated polyester or nylon fabric. The stainless-steel coil causes the boom to self-inflate during deployment, but the boom must be manually compacted during recovery. The boom is no longer manufactured.

Spill-tain fire proof boom is a stainless-steel boom constructed in sections connected by hinges. Floats, made of stainless steel filled with closed cell glass foam, are located at the midway point of the stainless-steel panels so that the lower half of the panel forms the skirt and the upper half forms the freeboard. This boom is stored and deployed from a folded position. Larger sizes of the boom would require a boat hoist or crane for deployment. This boom is no longer actively listed as an oil spill product.

1.10.2.2 Conventional Booms

Conventional booms cannot usually be used to contain burning oil as the construction materials either burn or melt, compromising the boom's ability to contain the oil. Test on conventional booms shows that life in a burn may only be a matter of seconds. It is often much quicker to get a conventional boom to a spill site, however, as they are much less expensive and very few fire-resistant booms are stockpiled at spill response depots.

Conventional booms can be used to corral a slick and contain it until a fire-resistant boom can be obtained. These booms can also be used to contain and thicken a slick to an acceptable burning thickness and then burn it, thus sacrificing the boom. The overall burn efficiency of this method is questionable, however, as the boom will not remain intact for very long once the oil is burning. When the boom fails, the slick could spread and quickly become too thin to sustain burning.

Logs or other floating material can sometimes be used as temporary booms. In narrow rivers, dams can be constructed across the upper layer of water to contain or divert the oil for burning.

1.10.2.3 Boom Configurations and Towing

The size of boom required for an in-situ burn depends on the amount of oil to be burned. Generally, the oil in the boom should fill no more than one-third of the area of the catenary. If the boom is too long, it will be difficult to control and the stress on the boom may be too great. If the boom is too short, the catenary may not be large enough to contain the burned oil. In general, the length of boom used ranges from 150 to 300 m (Fingas and Punt 2000; Marine Research Associates 1998). Most commercial booms come in standard lengths of 15 or 30 m. The relationship between the boom length and the area of oil that can be contained is shown in Figure 1.39. The overall height of the boom should be equal to the maximum expected wave height (short period waves, not swell) from peak to trough.

An important factor when containing oil is the direction and speed at which the boom is being towed. The distance from the burn to the tow vessels should be far enough that the burn does not pose any danger to the tow vessel or personnel onboard the vessel. Temperature profile tests performed during the NOBE trials showed that the air and water temperature ahead of the burn levels off very quickly (Environment Canada 1997). Therefore, unless the tow line is very short (only a few metres), the heat from the fire is not an issue. As well, since the boom is being towed upwind, the smoke from the burn should not reach the tow vessels.

Tow lines from tow boats should generally be at least 75 m long. The boom must always be towed into the wind so that the smoke will go behind it. As tow speeds are measured relative to the current, the boom may have to be towed very slowly or even downwind to maintain a low enough speed relative to the current while towing into the wind. If the boom is towed too slowly, however, the burn will begin to move up towards the end of the boom.

In general, the boom must be towed at a speed of less than 0.4 m/s (0.7 knots) relative to the current in order to prevent the oil from splashing over the boom or becoming entrained beneath the boom. The towing speed may have to be increased periodically if the burn begins to fill more than two-thirds of the boom catenary (Fingas and Punt 2000). If contained oil does become entrained in the water column below the boom or splash over the boom, it will resurface or pool directly behind the apex of the boom. This oil could be reignited by burning oil inside the boom or by oil that splashes over the boom.

Another important factor in ensuring that the oil is properly contained for burning is the configuration of the boom. Booms can be towed in various configurations, depending on the equipment available and the weather and sea state conditions. The various conventional configurations for towing oil spill booms are shown in Figure 1.40.

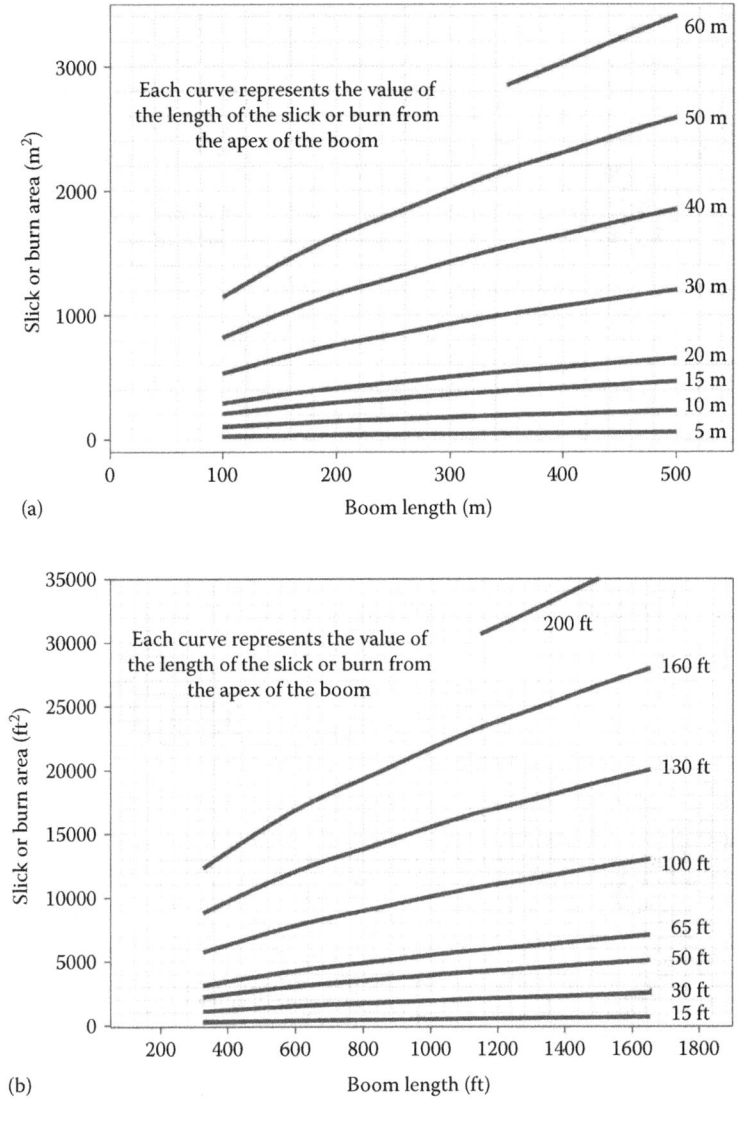

FIGURE 1.39 Nomogram to calculate burn or slick area. (a) Metres and (b) feet.

The standard configuration is a length of fire-resistant boom connected with tow lines to two vessels at either end of the boom to tow the boom in a catenary or U shape, as shown in Figure 1.40a. This is illustrated by the Deepwater Horizon burn in Figures 1.41 and 1.42. A tether line or cross bridle is often secured to each side of the boom several metres behind the towing vessels to ensure that the boom maintains the proper U shape, as shown in Figure 1.40b and d. This tether line or cross bridle is very useful in maintaining the correct opening on the boom tow as well as preventing the accidental formation of the J configuration. The tether line can also

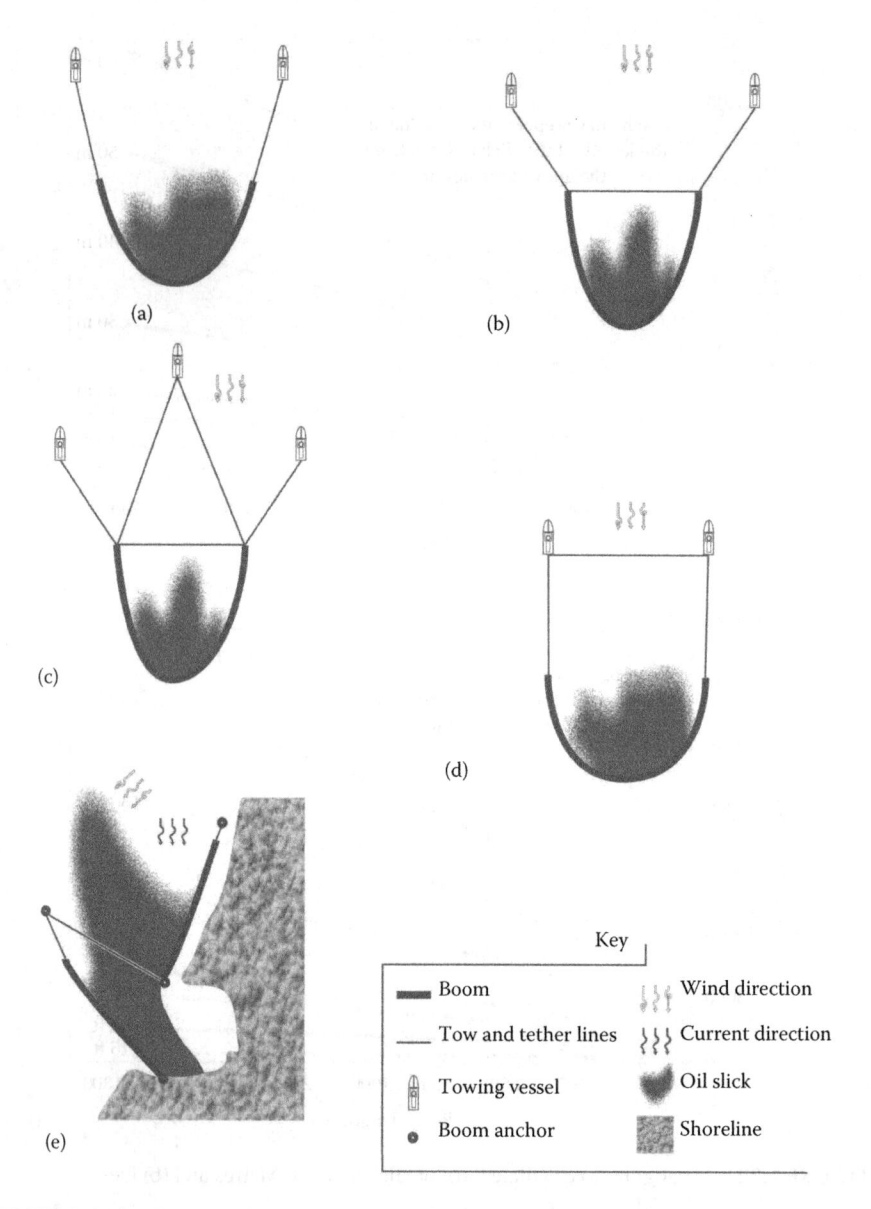

FIGURE 1.40 Boom towing configurations for in-situ burning. (a) Standard 'U configuration'; (b) 'U configuration' with tether line or bridle; (c) 3-tow vessel 'U configuration' with tether line or bridle; (d) boom with tether line or bridle at boats; (e) double diversion.

FIGURE 1.41 A photo of a boom tow during the Deepwater Horizon burns. Note that the smoke is off to the side, which could be changed by slowly changing the direction of the tow.

FIGURE 1.42 A photo of another boom tow during the Deepwater Horizon burns. The vessels in the rear of the tow are backup and monitoring boats.

be attached to the vessels, as shown in Figure 1.40b. The advantage of this method is that boat operators can detach the tether line very quickly in case of an emergency.

With the standard U configuration, it can be difficult to ensure that the two towing vessels maintain the same speed. To overcome this problem and to increase control over the boom configuration, three vessels can be used, as shown in Figure 1.40c. One vessel tows the boom by pulling from the centre using tow lines at each end of the U, while the other two vessels pull outward from the ends of the boom to maintain the U shape. This configuration was used during the NOBE tests in 1993. During these tests, 210 m of boom was towed in a modified U configuration. A 45-m tether line or cross bridle was attached across the ends of the U. One vessel towed the boom using two 120-m lines attached to the ends of the U. The U was kept open by lines towed from two other vessels in an outward direction at an approximately 45° angle. The towing speed was maintained at 0.25 m/s (0.5 knots) throughout the burn (Figure 1.43).

Bitting and co-workers tested a number of these configurations and found that many of the proposed configurations in this subsection were viable (Bitting et al. 2001).

If the oil is near shore, a boom or booms can be used to divert it to a calm area, such as a bay, where the oil can be burned. An example of this method using two booms is shown in Figure 1.40c. Diversion booms must be positioned at an angle relative to the current that is large enough to divert the oil, but not too large that the current would cause the boom to fail. The boom must be held in place either by anchors, towing vessels or lines secured to the shoreline.

In nearshore situations, anchors can be used to secure booms in a stationary position. It is important, however, that a proper anchor is used, particularly in high currents, to

FIGURE 1.43 The 3-point tow used during NOBE, viewed from the central tow vessel. One of the tow vessels opening the boom is off to the left.

ensure that the boom stays in place for the duration of the burn. Various types of anchors suitable for anchoring containment booms are available (Fingas and Punt 2000).

1.10.2.4 Novel Containment Configurations for In-Situ Burning

A number of boom configurations or containment methods have been proposed in the literature or at workshops. Most of these have not been tested or have not been tested quantitatively. Log booms, which are illustrated in Figure 1.44a, have been used several times in Northern Canada. In fact, the first documented in-situ burn was conducted successfully using a log boom on the Mackenzie River in 1958 (McLeod and McLeod 1972). Although log booms burn, if the boom maintains its buoyancy ratio, there is sufficient time to conduct a burn lasting several hours. The major problem with log booms is the leakage between sections. The gaps between sections are usually sealed with fire-resistant material such as fibreglass cloth.

Booms can also be used to divert oil slicks rather than to contain them. Diversion modes are usually used when the current is too fast for the oil to be contained in a U configuration, that is, greater than 0.4 m/s (0.75 knots). Conventional booms can be used to divert oil so that the oil is actually burned beyond the boom or contained by a natural barrier, such as the shoreline. One such method involves concentrating and 'funnelling' the oil through an opening created by two booms, as shown in Figure 1.44b, so that the burning takes place mostly behind the boom. As far as is known, this type of configuration has never been tested even in model form. A boom with solid flotation sections would have to be used because any flame impingement on an inflatable boom causes rapid failure. Despite the apparent weaknesses, the proposal has merit in that it would only be used in a situation where complete containment is not necessary and losses, even failures, would not cause major problems. The rear opening would have to be wide enough to avoid buildup of oil in front of the boom and narrow enough to ensure that the oil slick is thick enough to sustain burning even with the re-spreading that would occur behind the boom.

A modification of this configuration is the use of paravanes, rigid metal boom-towing sections that attach at the rear mouth of the conventional boom. This is illustrated in Figure 1.44c. This is also an untested concept, but it has the advantage of having relatively fire-resistant paravanes at the mouth of the boom. Thus, if fire does propagate inside the boom, there would be no catastrophic boom failure.

The use of corrugated steel sheets as a temporary fireboom has also been proposed (Marine Research Associates 1998). The corrugated sheets could be fastened to metal stakes in shallow water, as shown in Figure 1.44d, or coupled to drums for application in deeper waters, as shown in Figure 1.44e. As this has never been tested, it is not known how long the corrugated steel would withstand the heat flux of the fire, although it might withstand it for at least a few hours.

1.10.2.5 Deployment of Booms for In-Situ Burning

The deployment procedures for fire-resistant containment booms depend on the type of boom used. The water-cooled booms are either inflatable or flexible in some way and therefore they can be stored on and deployed from a reel. However, these booms sometimes require a large flat area for the proper installation of the water-cooling equipment as the boom is removed from the reel. Stainless steel booms and thermally resistant

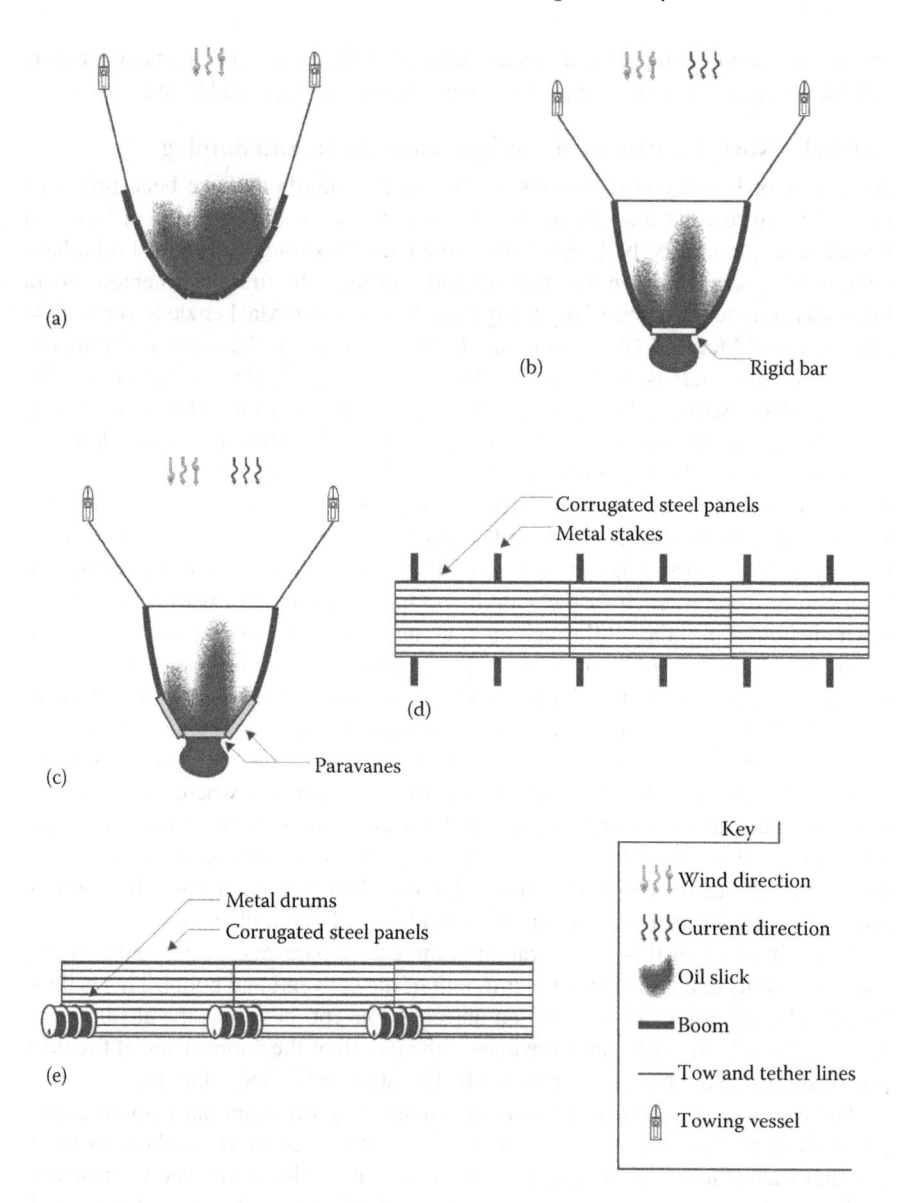

FIGURE 1.44 Untested containment/diversion configurations for in-situ burning. (a) Standard 'U configuration' using log booms; (b) open barrier with tether line or bridle; (c) open barrier with paravanes; (d) shallow-water fence-type boom; (e) temporary fence boom for calm water.

booms are rigid and therefore must be stored in sections in a container. They also require a large flat area to lay out and connect the sections. Because of their rigidity and weight, a winch or crane is normally required to assist in deploying and recovering these booms. Figure 1.45 shows the storage of fire-resistant boom in a container.

After floating in the water for some time, some containment booms become water-logged, making them much heavier than when they were deployed. The vessel used to recover the boom must therefore be stable enough to handle this weight, especially if a crane or winch is being used. Because of the added difficulty in handling some fire-resistant booms, they may be damaged during deployment and recovery. Care must be taken to ensure that the boom is moved slowly and handled carefully. For example, the cinch and choker attachment of a crane can damage a boom; it is therefore better to use a web belt to lift the boom. It is also much easier to deploy and recover the boom if a powered reel is used.

Containment booms normally come in sections that are joined by a connector. Many of the commercially available fire-resistant booms are being designed with standard connectors as prescribed by ASTM or to accommodate adapters that fit such standard connectors (Ross 2017). It has been found that these connectors must be stainless steel, as aluminium will melt. These connectors allow different types of booms to be joined easily and securely. In any case, if more than one type of boom is used for containment, the connectors on these booms should be checked first to ensure that they can be properly joined.

If a burn is to be performed nearshore, that is, within 5 km, the boom can be deployed from shore and then towed out in a straight line. For this reason, the ASTM

FIGURE 1.45 Fire-resistant boom on a reel inside a container. The boom may be deployed directly from the reel. (Courtesy of DESMI – Applied Fabric Technology Inc., Orchard Park, NY.)

standard for fire-resistant booms indicates that a fire-resistant boom section that is at least 150 m long must be able to withstand towing in a straight line at 2.5 m/s (5 knots) for a period of 2 h (Allen 1994).

If the burn is to take place too far from shore for the boom to be deployed from the shoreline, the boom must be deployed from a vessel. Again, because a fire-resistant boom is quite cumbersome, a large deck area is normally required for boom deployment.

The following is a typical procedure for deploying a boom in open water from a vessel using a standard U configuration:

- The deployment vessel situates itself far enough downwind from the oil so that there is enough time to deploy the boom before approaching the oil.
- The deployment vessel aligns itself so that its bow is facing upwind.
- Before the first part of the boom is deployed from the deck, a tow line for the towing vessel is attached to the end.
- The boom is deployed off its stern so that the wind causes the boom to trail behind the vessel.
- When the last section is deployed, the end of the boom is attached with a tow line to the deployment vessel, which now becomes one of the towing vessels.
- The tow line at the other end of the boom is then attached to a second towing vessel.
- The second towing vessel heads upwind until the proper U configuration is formed.

If a tether line or cross bridle is used across the opening of the U (see Figure 1.40b–d), this line should be attached to the end of the boom or tow line closest to the deployment vessel before the last section is deployed. Once the U is formed, a third vessel brings this line across to the other end of the boom or tow line and connects it. If a third tow vessel is used for stability, as shown in Figure 1.40d, the tow lines for this third vessel should also be attached as the boom is deployed and then attached to the third vessel, which then situates itself between and ahead of the other two tow vessels.

The method for deploying a diversion barrier in a river (Figure 1.40e) is very different from deploying a containment boom in a U configuration in the open ocean. The boom must be held in place at an angle relative to the current that is large enough to divert the oil but not too large that the current would cause the boom to fail. The boom must therefore be secured in place either with lines to the shoreline or towing vessels, or by anchoring the boom on the river bottom.

Unless the boom can be fixed to both shorelines, it is normally more secure to use anchors. In fact, the Canadian Petroleum Association has found that two anchors placed in series are usually required to prevent the boom from moving in high current situations (PROSCARAC 1992). The proper deployment of anchors in order to hold the boom can be difficult, as they must be deployed slowly and systematically in order to properly set in the river bottom. The anchors should be securely in place before the boom is deployed. The Canadian Petroleum Association has developed a detailed guideline for the deployment of anchors and a diversion boom in fast flowing rivers. This guideline is summarised in Figures 1.46 and 1.47.

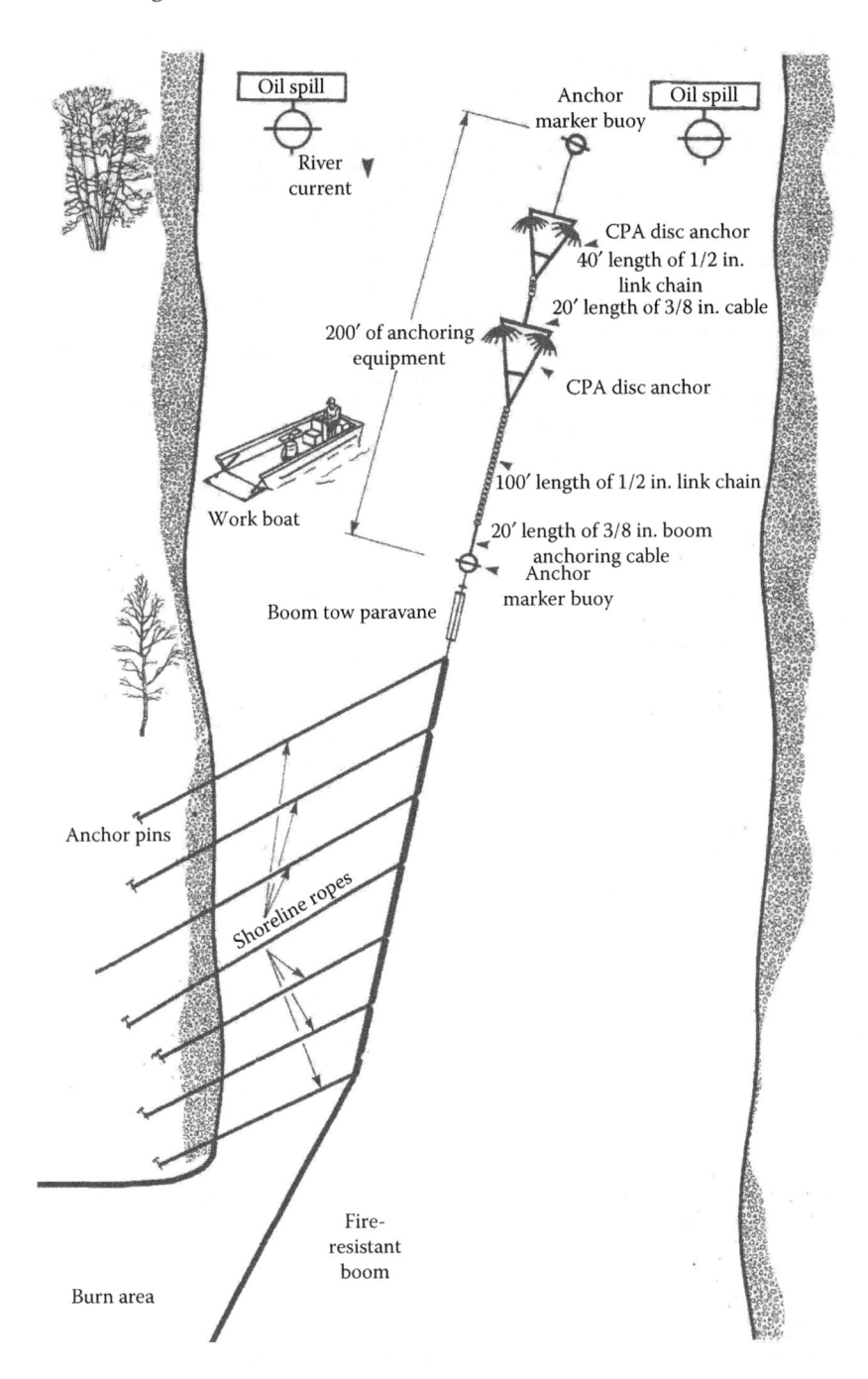

FIGURE 1.46 River boom deployment schematic. (Adapted from PROSCARAC [Prairie Regional Oil Spill Containment and Recovery Advisory Committee], *Anchor Design and Deployment Review*, Edmonton, Canada, 1992.)

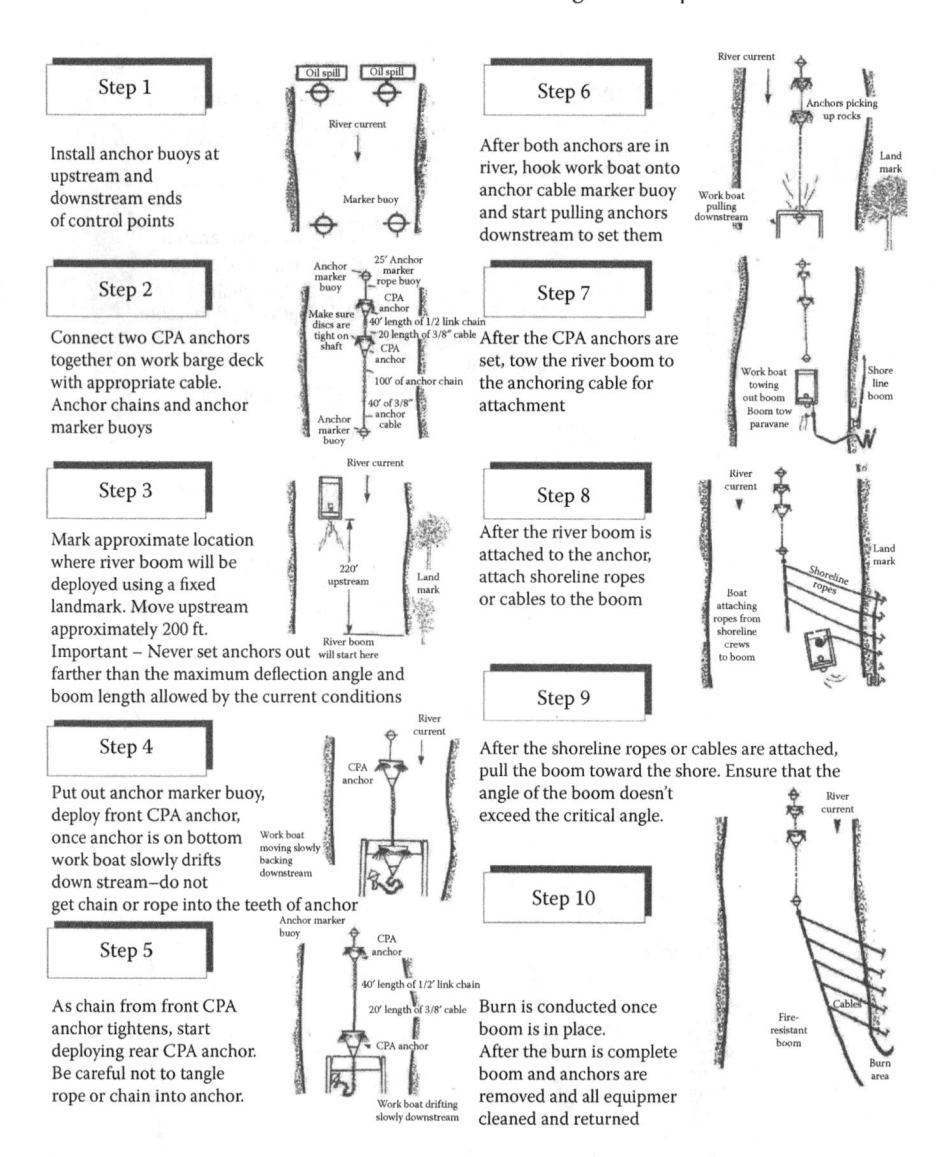

Step 1

Install anchor buoys at upstream and downstream ends of control points

Step 2

Connect two CPA anchors together on work barge deck with appropriate cable. Anchor chains and anchor marker buoys

Step 3

Mark approximate location where river boom will be deployed using a fixed landmark. Move upstream approximately 200 ft.
Important – Never set anchors out farther than the maximum deflection angle and boom length allowed by the current conditions

Step 4

Put out anchor marker buoy, deploy front CPA anchor, once anchor is on bottom work boat slowly drifts down stream–do not get chain or rope into the teeth of anchor

Step 5

As chain from front CPA anchor tightens, start deploying rear CPA anchor. Be careful not to tangle rope or chain into anchor.

Step 6

After both anchors are in river, hook work boat onto anchor cable marker buoy and start pulling anchors downstream to set them

Step 7

After the CPA anchors are set, tow the river boom to the anchoring cable for attachment

Step 8

After the river boom is attached to the anchor, attach shoreline ropes or cables to the boom

Step 9

After the shoreline ropes or cables are attached, pull the boom toward the shore. Ensure that the angle of the boom doesn't exceed the critical angle.

Step 10

Burn is conducted once boom is in place.
After the burn is complete boom and anchors are removed and all equipmer cleaned and returned

FIGURE 1.47 River boom deployment procedures. (Adapted from PROSCARAC, [Prairie Regional Oil Spill Containment and Recovery Advisory Committee], *Anchor Design and Deployment Review*, Edmonton, Canada, 1992.)

1.10.2.6 Backup Booms

A backup boom can be placed 200–300 m behind the burn to contain any oil that has been entrained or has splashed over a fire-resistant boom during the burn. A conventional boom that is not fire-resistant can be used as any burning stray oil would typically be extinguished on its own or by the fire-extinguishing vessel before it reaches this boom. Figure 1.48 shows a backup boom after two burns.

FIGURE 1.48   A view of the backup boom used during NOBE. The boom contains slight amounts of burn residue.

It has also been found that oil escaping from the fire-resistant boom usually pools directly behind the boom because of eddies formed in this area. This oil usually remains in this area for some time and therefore can become re-lighted or remain lighted. If this oil escapes from this area, it spreads and becomes too thin to sustain burning and can therefore be safely collected in the backup boom.

1.10.2.7   Alternatives to Booms

A number of ideas have been proposed to replace fire-resistant booms when burning oil on water. Marine Research Associates has proposed the use of modified barges to contain the oil for burning. Some of these are shown in Figure 1.49 (Marine Research Associates 1998). One concept involves cutting the centre tanks from a barge or extensively modifying a barge without centre tanks so that only wing tanks remain. The barge would be towed at the apex of a boom and oil contained within the centre of the barge, as illustrated in Figure 1.49a. A design for a barge with inflatable sides is illustrated in Figure 1.49b, and another design which uses forced air to enhance burning is illustrated in Figure 1.49c. These concepts and several variations of these are analysed in detail in the report, which shows that the barge concepts should provide a stable burn platform and a more extended life over a fire-resistant boom. These concepts would be very costly to implement and result in large, heavy devices.

Bubble barriers have also been relatively effective at containing oil when tested in calm waters under actual operation situations, although they have never been used in conjunction with burning. A bubble barrier consists of an underwater air delivery system which creates a curtain of rising bubbles that deflects the oil. This concept is illustrated in Figure 1.50. Work on bubble barriers has shown that the horsepower requirement is high, with a very large compressor needed for barriers longer than about 100 m (Fingas and Punt 2000). Testing has also shown that a large blower can power a bubble barrier using a firehose as outlet. The maximum length of the barrier in this case varies from 50 to 150 m (Alyeska 1998).

Environment Canada has also worked on the development of a water jet barrier which could potentially be used for ISB (Punt 1990). The design developed consists of high-pressure hoses connected to a water pump. Each arm of the barrier is formed

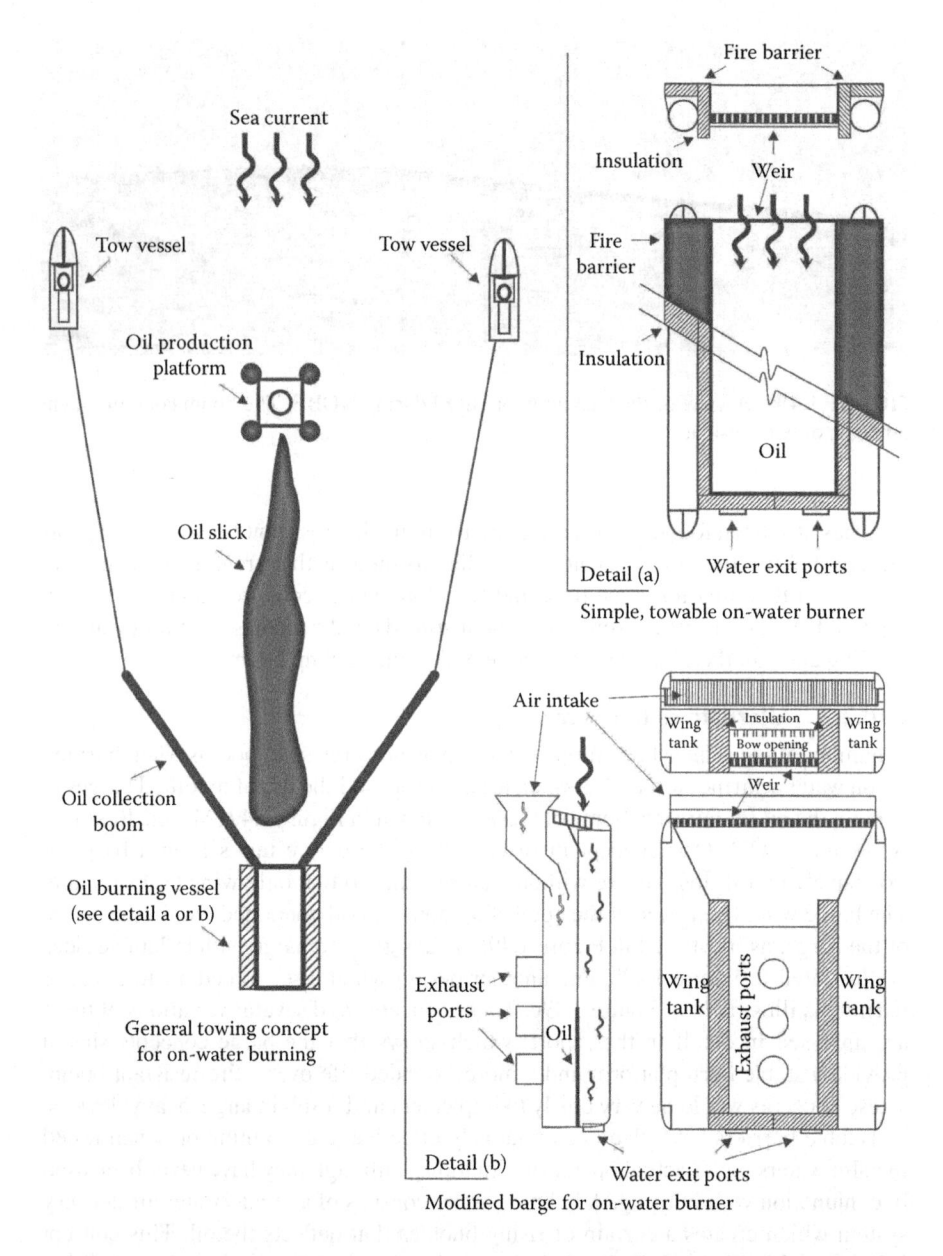

FIGURE 1.49 Novel concepts for burning oil on water. (Adapted from Marine Research Associates. 1998. *Technology Assessment and Concept Evaluation for Alternative Approaches to In-Situ Burning of Oil Spills in the Marine Environment,* for U.S. MMS, Herndon, VA.)

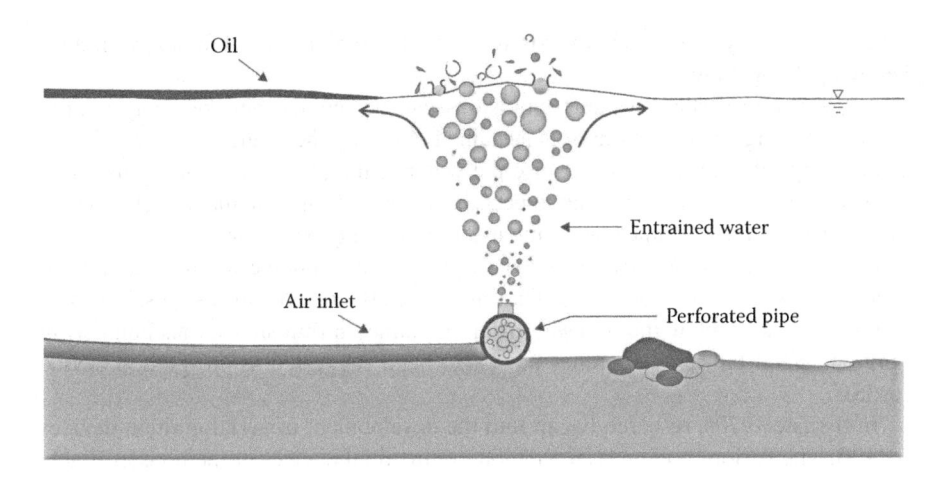

FIGURE 1.50 Concept of a bubble barrier.

by two hoses, each with four evenly spaced sets of opposing jets. The force from the water jets would hold the oil in the V formed by the barrier arms. This containment would allow oil to be safely burned. It was also felt that air entrained by the water jets would increase the efficiency and cleanliness of the burn. Unfortunately, these claims have not been fully tested due to mechanical problems and difficulties in manoeuvring the barrier using its current configuration.

Cooper et al. describe a concept of a small oleophilic skimmer which is coupled directly to a burner to dispose of recovered oil (Cooper et al. 2012). Prototype units were tested at bench and meso-scale tank tests.

1.10.3 IGNITION DEVICES

A variety of ignition devices or methods, both commercial and non-commercial, have been used to ignite oil slicks, although the methods of igniting oil on water have not been well documented (ASTM F 1990 2012; Buist et al. 2016). Many of the methods used are modifications of ignition devices used for other purposes; however, there are currently 2 commercial handheld devices on the market.

In general, an ignition device must meet two basic criteria in order to be effective. It must apply sufficient heat to produce enough oil vapours to ignite the oil and then keep it burning, and it must be safe to use. Safety issues to be considered when operating ignition devices are outlined in Section 1.15.1.3.

Research has shown that, to a certain extent, the thicker the slick, the more easily and quickly it will ignite. The main factor, the lighter (i.e. the more volatile or less weathered) the oil, the more easily it will ignite. For heavy oils, more heating time is required to produce enough ignitable vapours. It is suggested that for heavy oils, a primer, preferably diesel fuel or kerosene, is used to soak in the oil for a few minutes before applying an igniter.

As discussed earlier, unstable emulsions can be ignited, but they may require additional energy before burning is sustained. On the other hand, stable emulsions can be very difficult to ignite because the water in the oil acts as a heat sink and a high amount of energy is required to heat the water and vaporise the oil before burning can be sustained. Primers are quite useful in igniting emulsions.

Commonly available devices, such as propane and butane torches, have been used in the past to ignite oil slicks. They are more effective on thick slicks; however, as torches tend to blow the oil away from the flame on thin slicks, thus hampering ignition. Weed burners or torches have also been suggested as an ignition device for ISB.

In the late 1970s, research began into the development of aerial ignition devices for ISB. The various commercial and non-commercial devices or methods available for igniting oil slicks and the operational procedures for their use are discussed in this section.

1.10.3.1 Helicopter-Borne Ignition Devices

Sophisticated commercial devices used for igniting oil slicks are the Helitorch igniters and other helicopter-slung drip torches. These are helicopter-slung devices that dispense packets or globules of burning, gelled fuel and produce an 800°C flame that lasts up to 6 min (Allen 1986; ASTM F1990 2012; Buist et al. 2016). This type of igniter was designed for the forestry industry and is used extensively for forest fire management. Two helicopter-based systems suitable for igniting in-situ burns are the Simplex Heli-torch manufactured by Simplex Manufacturing of Portland, Oregon and drip torches such as the Universal Drip Torch available from Universal Helicopters of Deer Lake, Newfoundland, or Canadian Helicopters of Prince George, British Columbia. These helicopter-borne devices and the devices in operation are shown in Figures 1.51 through 1.54. The Simplex Helitorch was used effectively during the NOBE in-situ burn exercise off the

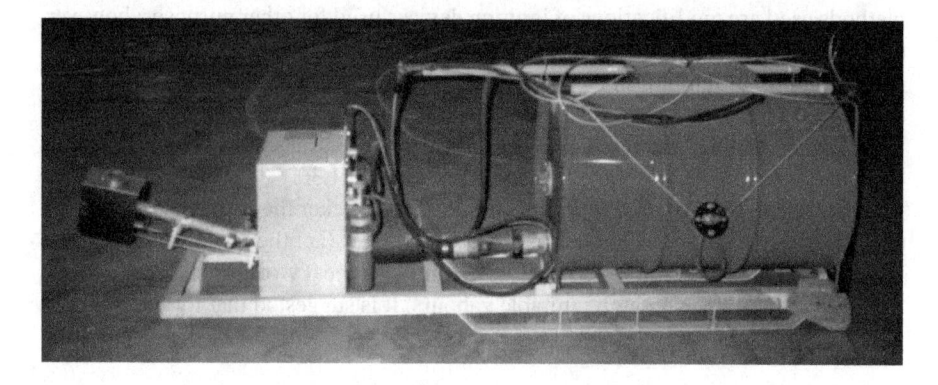

FIGURE 1.51 The Simplex Helitorch used at the Newfoundland test burn.

FIGURE 1.52 A Helitorch being used to ignite back fires to fight a forest fire. (Courtesy of George Kournounis, Ontario Ministry of Natural Resources, Peterborough, Canada.)

FIGURE 1.53 The universal drip torch.

FIGURE 1.54   A drip torch being used to light back fires in a forest firefight.

coast of Newfoundland in 1993 (Lavers 1997). Simplex information can be found
at http://www.simplexmfg.net/.

While the two units are assembled differently, they operate in a similar way. Both
have a 205-L fuel barrel connected to a fuel pumping and ignition system. On the
Simplex torch, all parts are mounted on an aluminium frame to which the slinging
cables are attached. The pumping and ignition system of the Drip Torch are attached
to the fuel transport pipe which is connected with a hose to the opening of the barrel.
The pipe with all the attachments is mounted on top of the barrel with clips, and
the whole system is slung by cables running from the pipe. The components of a
Helitorch are illustrated in Figure 1.55.

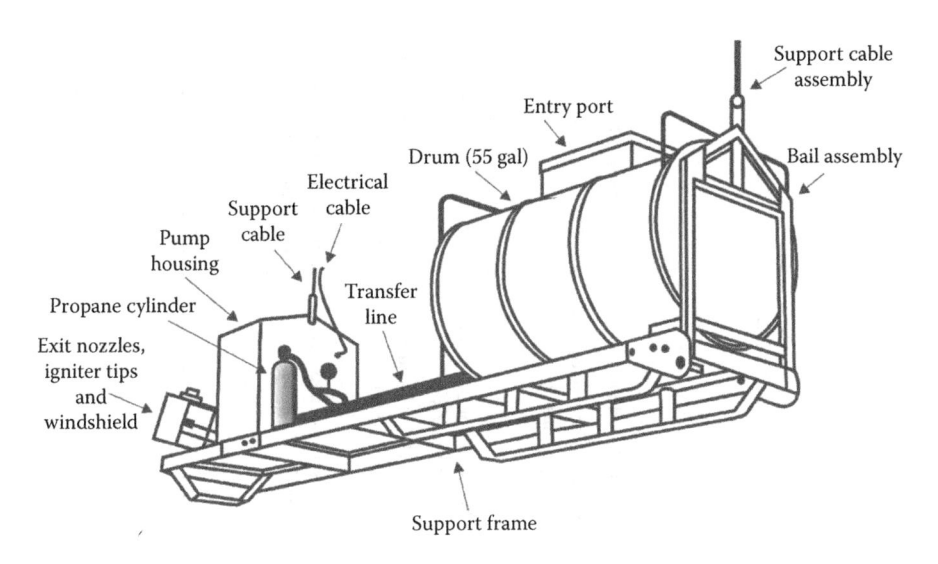

FIGURE 1.55 Helitorch components.

The fuel used in the Helitorch system is a mixture of a powdered gelling agent with either gasoline, jet fuel or a diesel/gas mixture. SureFire, an aluminium soap, is the most commonly used gelling agent. Alumagel is another type of gelling agent that was used to make Napalm for military purposes. It is currently available only through military surplus. The SureFire powder is more readily available and gels faster than Alumagel. An improved version of SureFire gel, known as SureFire II, is now available. This new product mixes easier, gels faster and at a lower temperature and remains in suspension longer than the original product. SureFire and SureFire II are available from Simplex Manufacturing in Portland, Oregon. Gelling times and other information are given in Tables 1.18 and 1.19.

When preparing to operate a Helitorch, the gelling agent and fuel must be mixed in a secure area well away from any ignition sources. Therefore, the first step is to set up a mixing area where the fuel is mixed with the gelling agent and a loading area where the barrels are loaded onto the Helitorch system. These two areas should be at least 30 m apart and 150 m away from the helipads and helicopter re-fuelling areas. They should also be well away from any ignition sources and upwind from the burn area. The general setup of these areas is shown in Figures 1.56 and 1.57. These areas must be used solely for the work associated with the Helitorch and should not be combined with other helicopter operations or other work associated with the burn. No personnel other than the Helitorch crew should be allowed in these areas unless authorised by the Helitorch supervisor.

The organizational structure for all those involved in operating the Helitorch system during an in-situ oil spill burn includes a Helitorch supervisor, a safety officer, a hook-up operator and three personnel to carry out fuel mixing. This is typical of the structure described in Helitorch operation manuals, which are written mainly for controlled burning on land, that is, forestry operations that require additional team members. For small spills, where very few drums of gelled fuel are needed, this team

TABLE 1.18

Gelling Times of Some Oils (Using SureFire)

Fuel Type	Mixing Time Required to Attain Desired Viscosity[a]	Effect of Air Temperature	Stability of Gelled Fuel	Comments
Jet fuel A	Mixes very slowly (20–120 min)	Recommended when air temperatures are high, because of stability, but longer mixing time is required.	Most stable – lasts 4–5 weeks, 2–3 days at higher than 20°C (68°F).	
Jet fuel B	Mixes quickly (8–18 min)	Increase in air temperature has little effect on mixing time.	Stable for 2–3 weeks or 2–3 days at higher than 20°C (68°F).	
Regular or unleaded gasoline	Mixes fairly quickly (10–45 min)	Recommended for air temperatures below 10°C (50°F) to ensure continuous ignition.	Stable for 2–3 weeks or 2–3 days at higher than 20°C (68°F).	
70% diesel/30% gasoline	Mixes slowly (10–110 min)	Increase in air temperature affects mixing time.	Stable for 2–3 weeks or 2–3 days at higher than 20°C (68°F).	Percentage of gasoline should be increased below 10°C (50°F).

[a] Depends on Temperature and Mixing Ratio.

TABLE 1.19
SureFire Gel Ratios

Using the following table select an appropriate mixing ratio, then locate the graph for the type of fuel to be gelled. The time for the gelled fuel to reach the acceptable viscosity can then be determined from the mix type and air temperature.

Mixture	Mixing Ratio (Weight of Sure Fire/Volume of Fuel)			
	g/L	lb/U.S. gal	lb/imp. gal.	
A	5.9	0.05	0.059	Dotted line
B	7.9	0.066	0.079	Dashed line
C	9.9	0.083	0.099	Dashed line −2
D	11.9	0.1	0.119	Solid line

For air temperatures in degree Celsius

For air temperatures in degree Fahrenheit

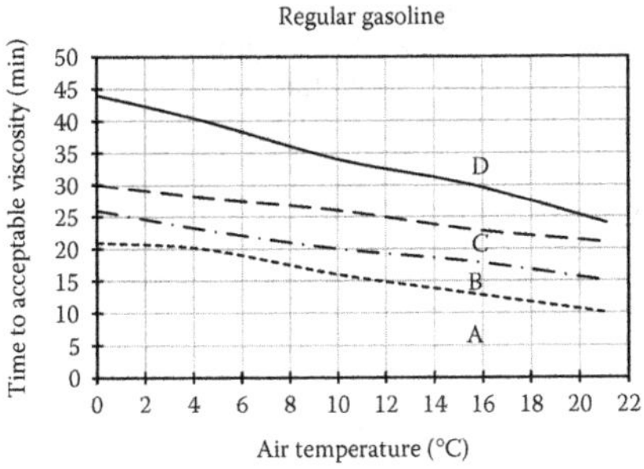

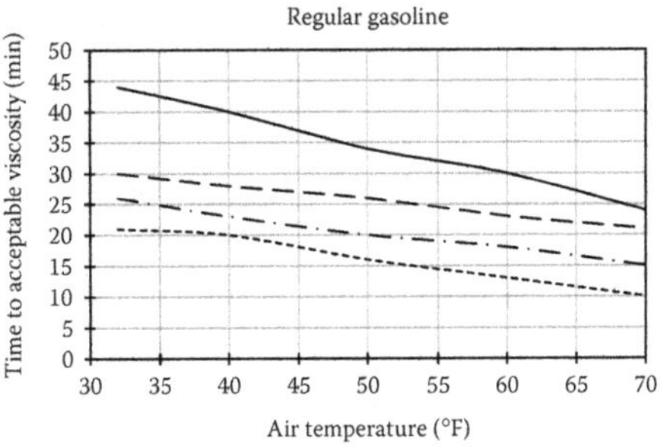

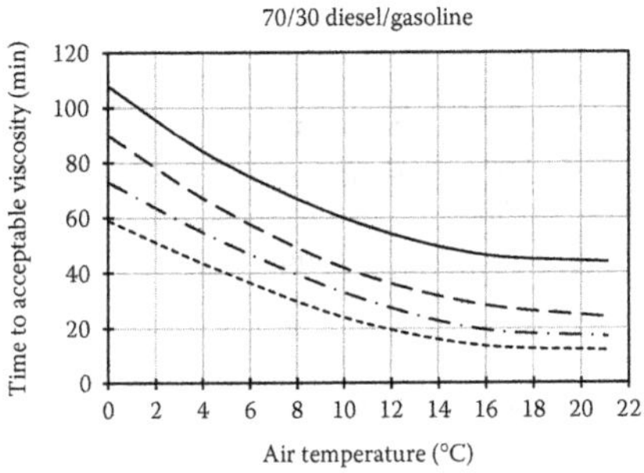

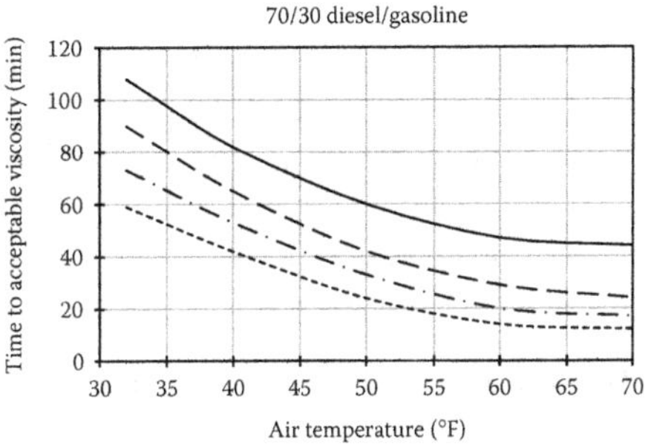

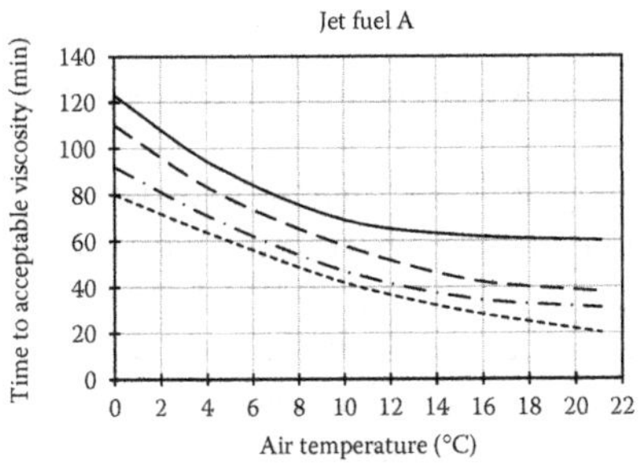

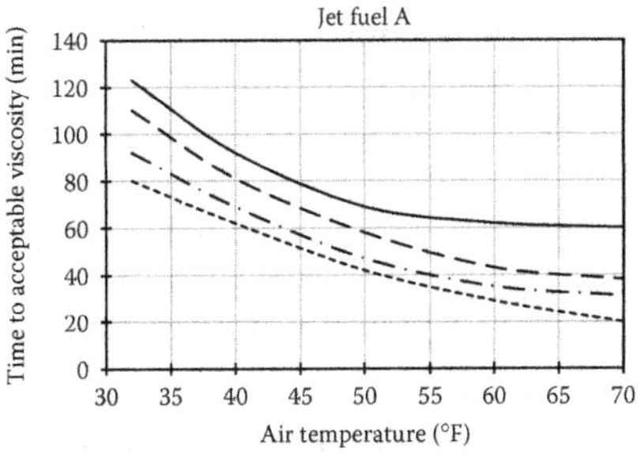

Regular gasoline

Regular gasoline

70/30 diesel/gasoline

70/30 diesel/gasoline

Jet fuel A

Jet fuel A

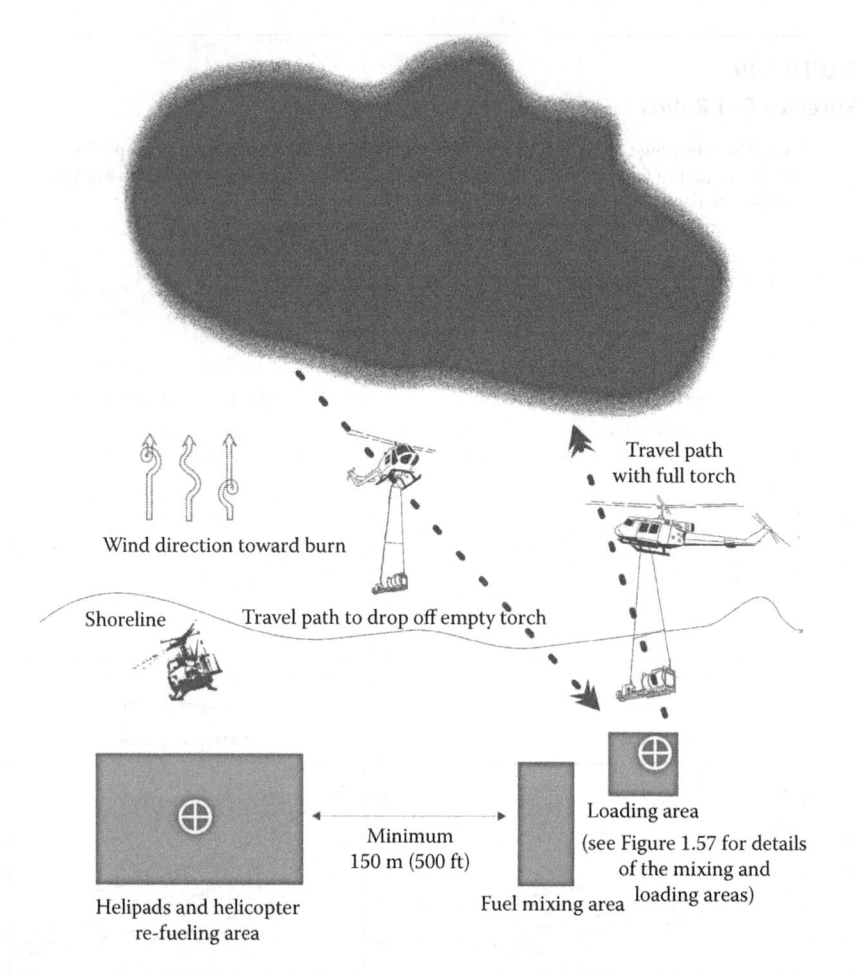

FIGURE 1.56 Location of fuel mixing and Helitorch loading areas.

could be further simplified to the following three persons: (1) the Helitorch supervisor, who would also perform the duties of the safety officer and the hook-up operator, (2) one fuel mixer and (3) the pilot.

The mixing of gelling agent and fuel, the loading of the fuel and the hook-up of the Helitorch to the helicopter should be done on land unless the burn site is too far from land for the helicopter to ferry the Helitorch, that is, more than 20 km. In this case, the fuel and agent should be mixed at a land-based site, and the barrels of gelled fuel should be stored on a ship in an area approved for fuel storage. This area must be above deck in a contained, ventilated area, well away from any ignition sources. A loading area should be set up on the ship, where the barrels of gelled fuel will be loaded onto the Helitorch system and hooked up to the helicopter. In this case, any preliminary testing and preparations for the ignition procedure should be done at a land base.

The fuel is mixed with the gelling agent directly in the specialised barrels that come with the Helitorch unit, using the raised hatch opening in the barrel.

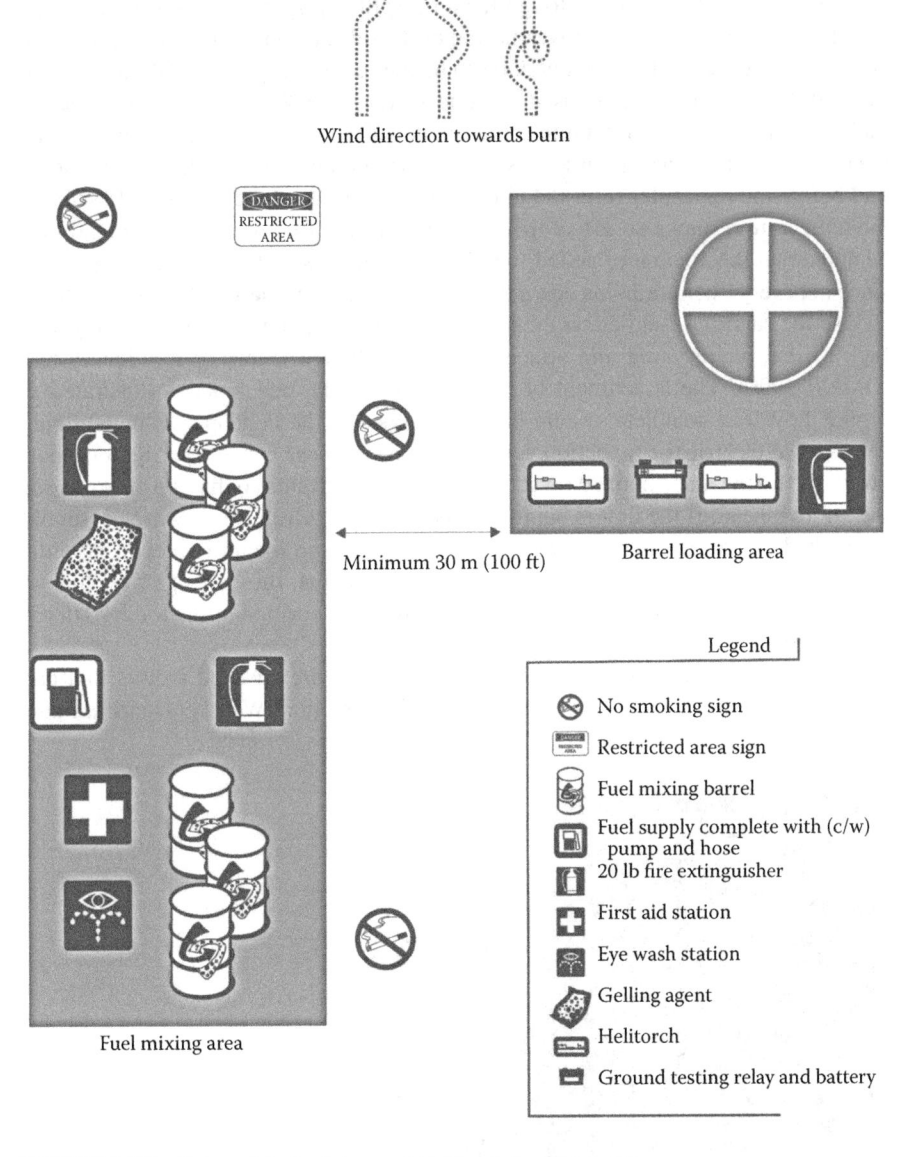

FIGURE 1.57 Setup of fuel mixing and Helitorch loading areas.

The required ratio of gelling agent to fuel depends primarily on the type of fuel and the air temperature. In general, the lower the flash point of the fuel, the less gelled agent is required. The gelling times of various types of fuel when mixed with the SureFire brand of gelling agent are shown in Table 1.17. In most cases, regular gasoline is recommended as it is often the most readily available fuel. The mixing ratios should be determined using the information provided in Table 1.18. Mixing times at various temperatures are also given in these tables.

The amount of fuel needed to ignite an oil spill is primarily related to the number of slicks and the degree of weathering of the oil. The amount of fuel should not normally be related to the amount of oil to be burned. During the NOBE burn test in 1993, 20 L of gelled fuel were used to ignite a slick of 50,000 L. One barrel of gelled fuel containing 180 L could ignite approximately 450,000 L of oil covering the same area as during this trial. Figure 1.58 shows a Helitorch being discharged of excess fuel before the helicopter returned to base. The volatility of the type of oil used and the temperature may also affect the amount of gelled fuel required. It should also be noted that the amount of gelled fuel dropped depends on the individual operator, since not every operator holds down the ignition switch for the same amount of time.

Use the carrying handles on the barrel to transport it and its contents, the gelled fuel, to the loading area and attach it to the Helitorch frame or ignition system (OMNR 1990). The attachment of the Helitorch to the helicopter is illustrated in Figure 1.59. The complete system is then attached to the helicopter using slinging cables. The electrical connection runs along one of these cables. For ignition purposes, the torch can be hooked up at right angles to the frame so that the pilot can see the ignition head. If the unit is being transported a long distance, however, it should be hooked up parallel to the frame to reduce the drag on the unit and conserve the helicopter's fuel. Before the ignition preparation begins, the helicopter should set down on a helipad on a ship near the site to change the position of the torch so that it is perpendicular to the frame.

Before the Helitorch is deployed, wind conditions are checked so the pilot can approach the burn from an upwind or crosswind direction. Water currents are also

FIGURE 1.58 A helicopter discharging excess gelled igniter before returning to base. This was after igniting the first burn at NOBE.

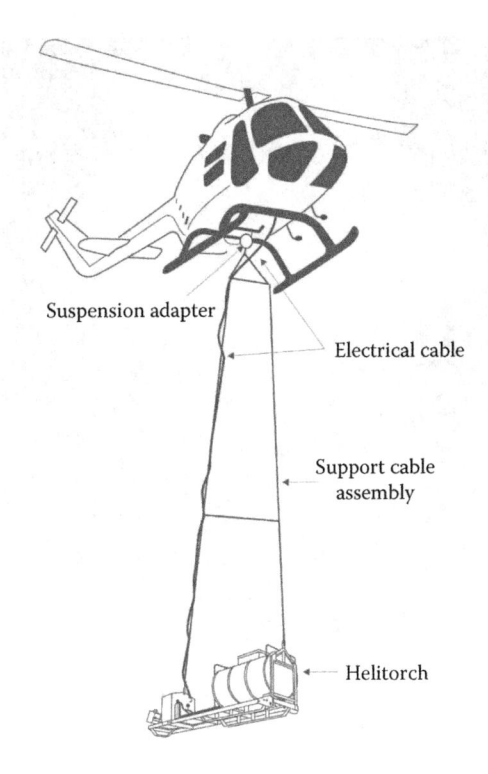

FIGURE 1.59 Mounting configuration of a Helitorch to a helicopter.

checked to ensure that the burning gel will not drift towards any vessels involved in the burn operation. A test drop can be carried out. If this indicates that the gelled fuel is igniting and falling properly, the pilot positions the helicopter over the desired location, fires the torch on a slow pass and then leaves the area. If igniting a fuel with a high flash point, the pilot may have to hover over the burn area and release multiple balls of burning fuel to concentrate the fire in one location.

1.10.3.2 Commercial Ignition Devices

At this time, there are two commercial ignition devices, the Safe Start Igniter from Elastec and the Fire Start (final name not yet decided) from Desmi (ASTM F1990 2017; Buist et al. 2016; Lane 2017). The Safe Start Igniter is illustrated in Figures 1.60 and 1.61 (Elastec 2017). This device uses a marine safety flare to ignite a jar of gelled fuel or fuel. The marine safety flare has a pull-off portion which ignites the remaining flare. The flame from the flare impinges on the bottle of fuel, which ultimately ruptures and leaks fuel and in turn ignites from the flame.

The Desmi Fire Start is an electrically initiated device that contains a military-like burning agent propellant and that can also contain a primer. The device is shown in Figure 1.62 and a helicopter-borne dispenser for this device is shown in Figure 1.63.

FIGURE 1.60 The front of the Safe Start igniter from Elastec. The device floats on water in the horizontal position. (Courtesy of Elastec, Carmi, IL.)

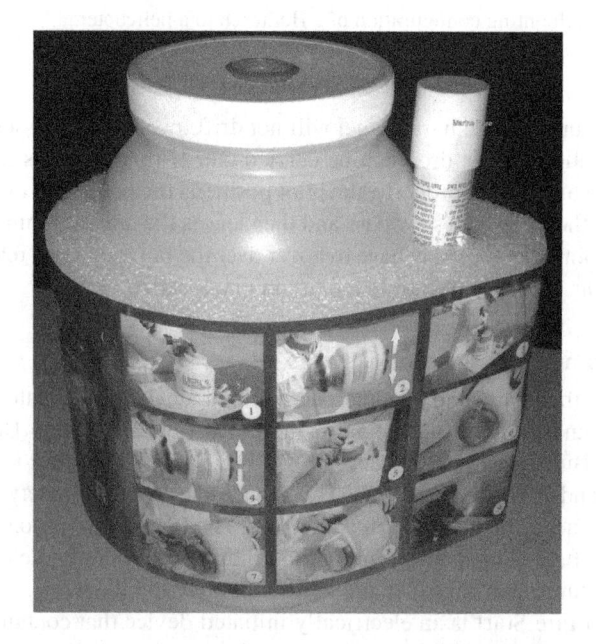

FIGURE 1.61 The rear of the Safe Start igniter from Elastec. This shows the position of the fuel jar and the marine flare. The simplified instructions are also shown. (Courtesy of Elastec, Carmi, IL.)

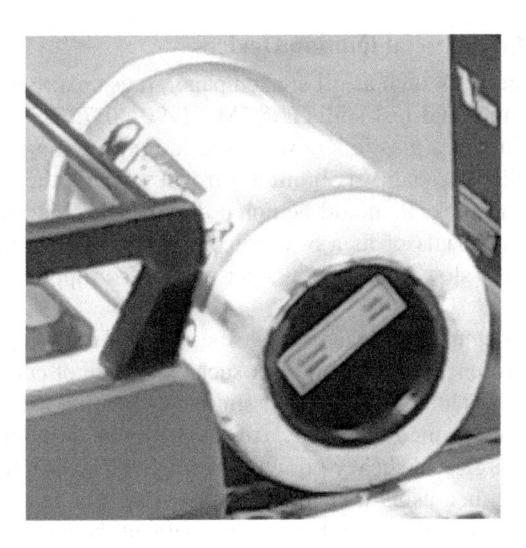

FIGURE 1.62 The Desmi Fire Start igniter. This device is initiated electronically with a specified delay.

FIGURE 1.63 The Desmi Fire Start device dispenser. The dispenser holds 15 igniters and activates the devices on exit from the dispenser.

1.10.3.3 Non-Commercial Ignition Devices

Simple ignition methods such as oil-soaked paper, rags or sorbent have been used to ignite oil at actual and test spills (ASTM F1990 2017; Buist et al. 2016). For example, gelled fuel in a plastic bag was used to ignite some of the oil from the *Exxon Valdez* spill. The bag was ignited, thrown towards the slick from a boat and floated into the slick. It should be noted that diesel oil is preferable to gasoline for soaking materials or as a base for the gelled fuels in handheld igniters because diesel burns slower, making it safer and it also supplies more pre-heat to the slick.

As noted earlier, ignition of heavier oils is best carried out using a primer such as diesel fuel and kerosene, and a small wick such as a piece of cardboard or sorbent (Fingas et al. 2003). This enables a start similar to lighting a candle. The flames then spread to the un-primed oil nearby. An illustration of such an ignition is given in Figure 1.16. In large-scale heavy oil, ignition might be accomplished by applying a bit of primer and then using a handheld igniter. Use of a gelled fuel igniter was found inadequate to directly ignite heavy fuels without the use of a primer (Fingas et al. 2003).

A variety of handheld igniters have been devised for igniting oil slicks (Buist et al. 2016). These are meant to be thrown into a slick from a vessel or helicopter. These devices often have delayed ignition switches to allow enough time to throw the igniter and, if required, allow it to float into the slick. These igniters use solid propellants, gelled fuel, gelled kerosene cubes, reactive chemical compositions or a combination of these, and burn for 30 s to 10 min at temperatures from 1,000°C to 2,500°C (Energetex Engineering 1981).

Some igniting devices use reactive metals and therefore do not have to be lit before being deployed. The Kontax igniter is an example of such a self-igniting device. It was tested and marketed in the 1970s (Buist et al. 2016). This device consisted of a metal cylinder filled with calcium carbide with a metal bar coated with sodium metal running through the middle. When the device is thrown into the spill, the sodium metal reacts with the water to produce heat and hydrogen. The calcium carbide reacts with the water to produce acetylene. The hydrogen ignites and in turn ignites the acetylene. The flame from the burning acetylene sustains long enough to heat the oil and produce vapours that are subsequently ignited. The main concern with this type of device is safety. The chemicals must be stored in a very dry place as accidental exposure to water would cause them to ignite.

In the late 1970s, during offshore oil exploration activities in the Beaufort Sea, researchers began investigating the use of aerial ignition devices for ISB of oil spills. This work led to the development of two Canadian igniters: (1) the DREV Igniter and (2) the Dome Igniter. The DREV igniter was initially designed in the early 1980s by the Canadian Defence Research Establishment in Valcartier, Quebec (DREV), in conjunction with the Environmental Protection Service of Environment Canada (Energetex Engineering 1981; Twardawa and Coutour 1983). Several configurations of the igniter were built, some intended for deployment on pools of shallow water on ice. This igniter has also been referred to as the EPS Igniter, the AMOP Igniter, the DREV/ABA Igniter and the Pyroid. It was manufactured by Astra Pyrotechnics, Ltd. (formerly ABA Chemical Ltd.) of Guelph, Ontario, but is no longer in production. The advantage of this type of igniter was that it was built by a licensed pyrotechnic company using approved components and was licensed to be transported by truck or air freight. Figure 1.64 shows this igniter being tested on water.

As shown in Figure 1.65, the DREV igniter was an air-deployable igniter with a pyrotechnic device sandwiched between two square flotation pads. Before tossing the device from the aircraft into the slick, the operator pulled the firing switch, which strikes a primer cap. The system had a 10-s delay mechanism that allowed time for the device to be thrown and to settle into the slick. After the delay, an initial fast-burning ignition composition was ignited and in turn ignited a rocket motor propellant consisting mainly of 40%–70% ammonium perchlorate, 10%–30% magnesium or aluminium metal and 14%–22% binder. This produced a ring of fire with temperatures close to 2,300°C that burned for 2 min, long enough for the surrounding oil to vaporise and ignite.

The Dome igniter was developed by Dome Petroleum Ltd. in Calgary, Alberta, in conjunction with Energetex Engineering of Waterloo, Ontario (Energetex Engineering 1982). It has also been known as the Energetex Igniter or the Tin Can Igniter and was intended to be manufactured on site. This unit was not a

FIGURE 1.64   A 'DREV' igniter being tested on open water. This igniter was a pyrotechnic device and is equipped with safety features such as a delay fuse.

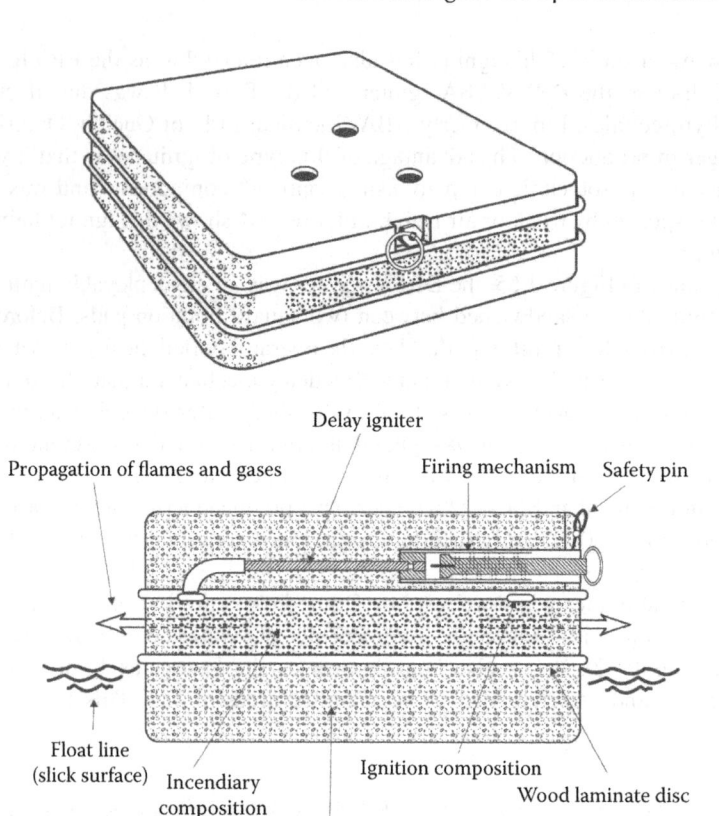

Delay igniter

Propagation of flames and gases   Firing mechanism   Safety pin

Float line
(slick surface)   Incendiary
composition

Ignition composition

Wood laminate disc

Flotation pad

FIGURE 1.65   DREV igniter.

manufactured unit and was intended to be 'homemade'. As shown in Figure 1.66, the wire-mesh fuel basket, which contained a solid propellant and gelled kerosene, was surrounded by two metal floats. An electric ignition system activated a fuse wire allowing about a 45-s delay. The fuse then ignited a thermal igniter wire, which ignited the solid propellant and finally ignited the gelled kerosene. The gelled kerosene burned at temperatures of 1,200°C–1,300°C for about 10 min, allowing the oil to vaporise and burn.

The drawback of both the DREV and the Dome igniters is that one igniter is required for each slick or part of a slick to be burned. For large oil slicks and oil in melt pools, several igniters may be required, which is costly and time-consuming.

Another technique for igniting in-situ oil fires is the use of lasers. In the 1980s, Environment Canada sponsored research by Physical Sciences Inc. of Andover, Massachusetts (Frish et al. 1986, 1989). This involved testing various laser techniques

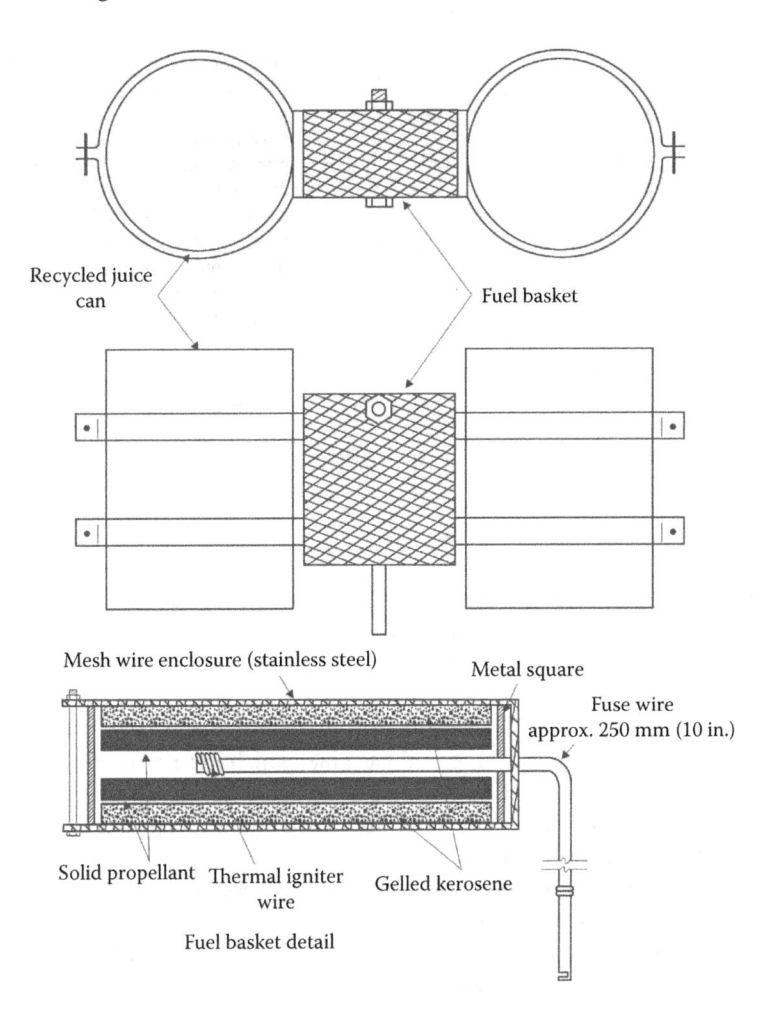

FIGURE 1.66 Dome igniter.

for igniting a variety of types of oil at different temperatures. The most successful technique in laboratory tests was to use a continuous-wave CO_2 laser to heat a localised area of the oil slick. The laser heats the oil to a temperature above its fire point. The heating time varies from a few seconds to more than 30 s depending on the type of oil, degree of weathering and the oil temperature. The oil vapours are then ignited by a spark produced just above the oil surface by a focused high-power pulse beam from a second laser. A laser-focusing telescope with focusing mirrors is used to aim this second laser. Despite the success of the early tests, this device was not fully developed due to lack of funding.

A handheld igniter, designed by Simplex and Spiltec, was used during ISB tests in 1996 off the shores of Great Britain (Guénette and Thornborough 1997). This igniter

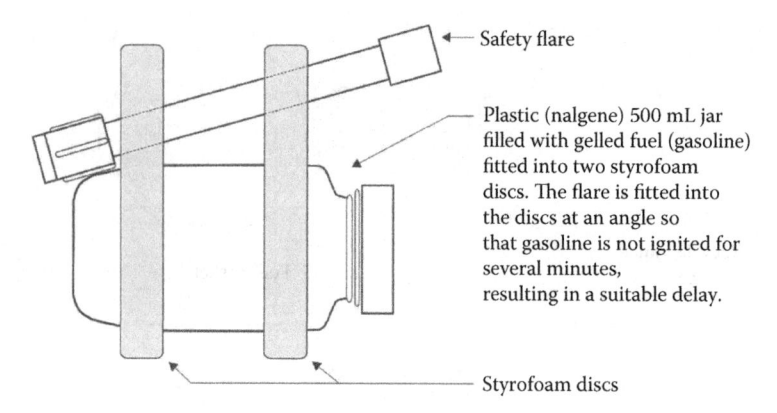

Safety flare

Plastic (nalgene) 500 mL jar
filled with gelled fuel (gasoline)
fitted into two styrofoam
discs. The flare is fitted into
the discs at an angle so
that gasoline is not ignited for
several minutes,
resulting in a suitable delay.

Styrofoam discs

FIGURE 1.67   Handheld igniter.

consists of a 1-L polyethylene 'Nalgene' bottle filled with gasoline gel. The gel was made by mixing 1 L of gasoline with 0.01 kg of SureFire fuel gelling agent, which is the agent used in the Helitorch. This bottle and a standard 15-cm marine handheld distress flare are secured side-by-side within two polystyrene foam rings. The flare is lit and thrown into the slick, where it burns for approximately 60 s before melting the plastic bottle and lighting the gelled gasoline, which in turn lights the oil. Such a device, which is relatively easy to make and to deploy, is shown in Figure 1.67. A similar igniter was used during the successful burns on the Deepwater Horizon. One such igniter is shown in Figure 1.12.

Goodman et al. (2014) proposed a new air-transportable igniter. The chemical initiator is the potassium permanganate/ethylene glycol system. The potassium permanganate is injected with ethylene glycol immediately prior to deployment using a specialised dispenser. The potassium permanganate and ethylene glycol react exothermically. The reaction ignites a capsule containing the potassium permanganate after a delay. The size of the delay depends on the particle size of the crystals and the ratio of the two components. The proposed igniter uses a capsule with the potassium permanganate. A prototype dispenser has been built that injects the proposed igniter and drops it for testing purposes. A dispenser that attaches to an aircraft door or window would allow the proposed igniter to be deployed from an aircraft (Figure 1.68).

1.10.4 TREATING AGENTS

In general, as a burn becomes hotter and thus more efficient, the emissions from the burn are reduced. Work has been done to investigate the use of chemical additives to enhance burning. A number of agents can be used; however, none of these is readily available or has proven to be effective for the task. Agents include emulsion breakers, ferrocene, combustion promoters and sorbents. Work showed that combining chemicals that suppress smoke emissions with those that break emulsions and promote combustion is ineffective (Bech et al. 1992). However, the agents worked

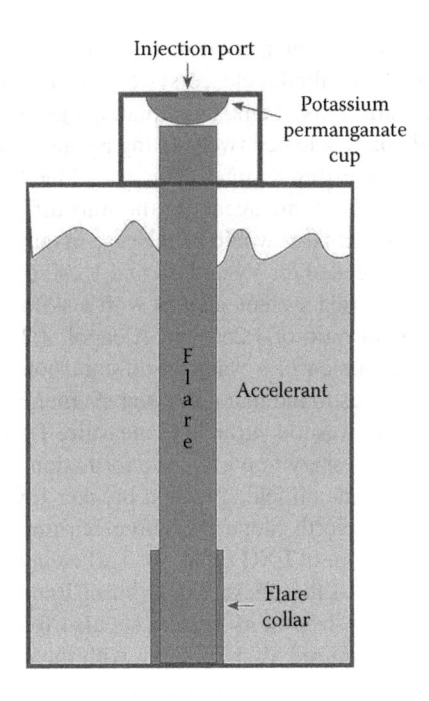

FIGURE 1.68 Proposed handheld igniter using glycol and permanganate to initiate the flare. (From Goodman, B.T. et al., Initiating in-situ burning of difficult-to-ignite oil spills via and aircraft-deployable igniter system, *IOSC*, 1821–1833, 2014.)

well separately. Chemicals could also be added to oil before transport so that it will burn more efficiently if spilled.

Emulsion breakers and inhibitors are formulated to break water-in-oil emulsions or to prevent them from forming. They have not been used extensively in field trails and rarely in actual spills. Some information is available on specific formulations of these agents, but the formulations vary extensively and many are not specifically patented. Only three products, Gamelin EB439, Vytac DM and Breaxit OEB-9, are specifically marketed for oil spills (Walker et al. 1993). Another product, Alcopol 60, has also been used extensively in field trials in the past. Many products of this type are marketed for use in breaking emulsions that occur in petroleum production, but most have never been applied to oil spills (Ross et al. 1992).

Several tests of emulsion breakers or inhibitors have been conducted. The results of some of these tests may not be useful, however, as they did not focus on the fact that there are several stability classes or water-in-oil states, that is, stable emulsions, mesostable emulsions, unstable emulsions and entrained water. Furthermore, some testing may not have used proper analytical methods to evaluate the effectiveness. The action required of the product must also be considered when developing effectiveness tests. It has been shown that some products inhibit emulsification better than they break an emulsion that is already formed (Fingas and Fieldhouse 2011). It is therefore appropriate to have two types of tests for each of these functions. In addition, some emulsion breakers are used on the open sea, which is called an open system, and others are used in conjunction with skimmers, tanks and pumps, with little water present, which is called a closed system. Thus, a total of four different tests are required to test all facets of emulsion treating agents.

Environment Canada has evaluated two treating agents in tests that are designed to measure each of the four testing regimes (Fingas and Fieldhouse 1994). Different results were obtained with the same agents in the four different tests. In breaking stable emulsions in open systems, as would be the case in the open sea, the minimum ratio of 1:300 (wt:wt) was needed for Vytac DM and 1:200 for Alcopol 60. In breaking stable emulsions in a closed system such as with a skimmer or a closed vessel, Vytac required a minimum ratio of 1:250 and Alcopol, 1:280. Much less agent is required to inhibit the formation of a water-in-oil emulsion than to break such an emulsion. Furthermore, it was found that meso-stable emulsions required much less agent, although this amount was too variable to measure. Tests were also conducted to determine the amount necessary to prevent the formation of emulsions.

Several combinations of the oilfield emulsion breaker EXO 0894 were tested to break emulsions of Alaska North Slope oil before burning (Buist et al. 1995). It was found that 500–5,000 ppm of EXO 0894 was sufficient to break emulsions that contained up to 65% water so that these would burn. Emulsions containing more water would not burn. These laboratory-scale tests also found that at least 1 hour of mixing time was often required after spraying with the emulsion breaker before the emulsion would break. When used in tests, emulsion breakers have been applied using hand sprayers. In actual situations, it has been proposed that dispersant application equipment would be used.

Buist et al. used a herding agent to thicken oil for burning in pack ice (Aggarwal et al., 2017; Buist et al. 2006, 2007, 2008, 2010a, b, 2011, 2017; Cooper et al. 2017; Fritt-Rasmussen et al. 2017; Potter et al. 2017). Tests at various scales showed that the herders were effective at reducing the slicks to burnable areas in pack ice. The United States Navy (USN) herding agent was tested in loose drift ice in Svalbard on a 0.1 and a 0.7 m^3 spill. Both were subsequently burned, the smaller one with 80% efficiency and the larger one with 90% efficiency. It should be noted that the herder worked in calm conditions but a wind of 1.5 m/s was sufficient to overcome the effect. Further tests on silicone-based (Silsurf A 108 and Silsurf A004-D) and hydrocarbon-based (USN) herders showed that these had potential. The tests showed that the herder Silsurf A 108 performed the best of all herding agents and on calm water would suffice to herd oil to sufficient thicknesses for burning. The presence of breaking or cresting waves disrupted the herder monolayer. Analysis of the most common agent shows that it is a silicon surfactant (Billings et al. 2016; Bullock et al. 2017).

Ferrocene is a chemical that can reduce or eliminate soot production from burns (Mitchell 1990, 1991, 1992, 2014). Tests have shown that ferrocene, if it can be mixed with the oil, is highly effective at percentages from 1% to 2%. The problem with ferrocene is that it is denser than oil and water so it must be pre-mixed just before burning, which is very difficult to do outside a pan test burn. Ferrocene can now be encapsulated so that it does float and even be added to the fire once in progress. The difference in smoke emission with and without addition of ferrocene can be seen in Figures 1.69 and 1.70.

FIGURE 1.69   A typical burn without ferrocene.

FIGURE 1.70   A burn with a full dose of ferrocene carried out at the same test site as shown in Figure 1.69.

In the past, several combustion promoters, usually agents that would act as both a wicking agent like a sorbent and an auxiliary fuel, have been tested and shown to be marginally useful (Thompson et al. 1979). None of these agents is currently available. Some have suggested that such agents may be useful in burning uncontained slicks, but further research is required on these agents before they can be applied to actual in-situ burn situations.

Sorbents such as peat moss have proven useful in burning by acting as wicking agents (Coupal 1972; Shi et al. 2016b). It was shown that such agents could reduce the minimum burning thickness and increase the efficiency of a burn. Sorbents may allow un-contained burning to be conducted in marginal conditions, but again more research is needed. Breitenbeck (2001) studied the use of various materials as wicking agents, including bagasse, corn cobs, kenaf, peanut hulls, rice hulls, rice straw and wheat straw. It was found that kenaf, a natural fibre similar to jute, was best for the purpose. A study of the shape factor also showed that ellipsoids were best. In burning diesel fuel, the kenaf ellipsoids would enable ignition as low as 0.25 mm and a burn down to 0.08 mm. Even at thicker starting slicks, the final thickness was only 0.05–0.2 mm. Similar tests on cold water showed similar results: The temperature had little effect. In burning weathered crude oil, using kenaf allowed the burn down to 0.07–0.2 mm. Shi et al. (2016b) studied the use of peat moss, straw, sorbent pads and jute cloth on burning of Alaska oils with various amounts of water. They found that burn efficiencies were vastly improved when using wicking agents, particularly for jute cloth with 40% water in the oil.

A variation on the theme of wicking agents is to use blankets and coils or other heat-reflecting objects submerged in the oil (Arsava et al. 2017). These submerged objects were found to increase burning rate and efficiency by reflecting heat to the fire.

1.10.5 Support Vessels/Aircraft for At-Sea Burns

Vessels and aircraft play an important role in a successful in-situ burn operation. Vessels are required to bring equipment and personnel to the burn site, to tow booms, and to carry monitoring equipment (Fingas and Punt 2000). Barges and small boats may also be required for standby fire safety operations, monitoring, recovering residue and storing equipment and residual oil. Tug boats may be required if a tanker must be moved away from the burn area.

A sufficient number of vessels must be available to transport and deploy the length of containment boom required at the burn site. The vessels must have a large enough deck to carry the boom as well as any equipment and materials required for handling the boom. They must also be able to move steadily at a slow speed (<0.5 m/s [1 knot]) and have bow-thrusters for easy manoeuvring and moving quickly in reverse if required. When containment booms are used in open water, two vessels are required to carry, deploy, recover and tow each end of the boom, depending on the configuration.

For safety reasons, any vessels used in a burn operation must be large and stable enough to carry the necessary equipment in all possible sea states including storm conditions. A vessel with an onboard crane and one or more tugger winches is recommended for handling equipment on deck and for recovering oil from the water. Separate, smaller tow vessels can be used to tow the boom.

Fixed-wing aircraft and/or helicopters may also be required to perform surveillance of the spill site, carry monitoring equipment and perform ignition and extinguishing operations. For safety reasons, twin-engine helicopters are recommended for Helitorch operations. If a single-engine helicopter must be used, it should be equipped with floats to allow emergency landing on the water. This is not a requirement for twin-engine helicopters. When using more powerful twin-engine helicopters in ignition operations, however, the oil must be ignited high enough above the slick to ensure that the down draft from the helicopter does not extinguish the burn.

For all aircraft operations, reliable air-to-ground communications are essential to coordinate operations. During Helitorch operations, this includes communications between the base ship, the helicopter and the fireboom deployment vessels. A safety standby boat having communications with the helicopter may also be desirable under certain circumstances.

Any vessel used as a floating base for helicopter operations must have a heli-deck with a nearby fuel storage area and be equipped for onboard firefighting operations. If using a Helitorch or other helicopter-deployed igniter, and the distance from shore is too far for safe helicopter transit from a land base, another vessel may be required to store the gelled fuel and for Helitorch re-fueling operations.

When burning against a shoreline without the use of deflection or containment booms, only one helicopter (preferably a twin engine) is required to carry the Helitorch and conduct ignition operations. If booms are needed, vessels or aircraft are required to transport the equipment to the site. Vessels and aircraft may not be needed to hold the boom in place, however, as this can be done with anchors.

A vessel with a low freeboard to allow for easy access to the water surface is recommended for recovering oil residue using skimmers. A sea truck or landing craft used in conventional oil spill response is ideal for access to the water surface. The amount of residue that can be recovered depends on the displacement of the boat used and the size of tank and cargo that can be safely carried on deck considering vessel stability.

1.10.6 Equipment Check List

Before starting any in-situ burn response operation, it must be ensured that all the required equipment is available. To assist in determining the type and specifications of the equipment that may be required for a burn operation, an at-sea equipment check list has been included in Table 1.20.

TABLE 1.20
Burn Equipment Check List

Vessels and Aircraft

- Tow vessels
- Command vessel
- Surveillance aircraft
- Helicopter for igniter

Safety Equipment

- Fire pump for each tow boat
- Fire hoses
- Fire nozzles
- Fire extinguishers
- First aid kits
- Fire blankets for tow boats
- Extra radios

Containment Equipment

- Full length of fire-resistant boom
- Extra lengths
- Towing paravanes
- Towing cables
- Bridles
- Attachment shackles
- Anchors – if needed
- Equipment for backup boom if needed

Ignition Equipment

- Hand-held igniters
- Helitorch and accessories

Residue Cleanup Equipment

- Sorbents
- Shovels or bailers
- Drums or recovery collections containers
- Heavy oil skimmer if – necessary
- Pumps and hoses for skimmer

General Supplies

- Burn plan
- Safety plan
- Radios
- Contact lists

Helitorch Equipment

- Helitorch unit
- Helicopter connecting harness
- Fuel gellant
- Fuel mixture
- Fire extinguishers
- Hard hat
- Gloves
- Goggles
- Protective clothing
- Safety boots
- Respirators
- Propane bottle

Monitoring Equipment

- RAM and/or DataRAM
- PAH sampling pump/filters
- Summa cannister
- Recording notebook, pens

Personal Protection Equipment

- Respirators
- Boots, gloves
- Special clothing
- Duct tape for sealing
- Goggles

Personal Cleanup Equipment

- Sorbents, rags, towels
- Citrus cleaner
- Garbage bags
- Soap, warm water
- Extra clothing

1.11 MONITORING, SAMPLING AND ANALYTICAL EQUIPMENT

Monitoring the emissions during an in-situ burn operation can provide continuous feedback about whether the burn is progressing properly and safely. A well-planned monitoring program, in which data are recorded before, during and after a burn, also helps answer any questions that come up after a burn operation is completed (Midlinx 2013). It is generally recommended that, if possible, the following sampling and monitoring be performed for any in-situ burn operation:

- Real-time monitoring of PM-2.5 particulate matter in the smoke
- Real-time monitoring of VOCs in the smoke
- Soot sampling for analysis for organic compounds and PAHs
- Residue sampling for analysis for organic compounds and PAHs

If it is determined that burning can be done safely and will likely result in the least overall environmental impact, operations should not be delayed because of monitoring and sampling activities.

1.11.1 VISUAL MONITORING

Visual monitoring is not as effective as monitoring using instruments. Obviously, gases and light concentrations of particulate matter cannot be seen. The trajectory of the smoke plume can be observed; however, and its passage over land, population centres and other points of concern can be noted, timed and recorded. This information is necessary if there is ever a question of exposure to emissions after an in-situ burn incident. The prime areas of deposition should be surveyed after a burn to check for soot deposits. If soot is found, it should be sampled for possible analysis.

1.11.2 REAL-TIME MONITORING

In general, real-time monitoring of emissions should be performed downwind of the fire and at a point closest to populated areas. Studies of the emissions from in-situ oil burns indicate that the main public health concern is particulate matter in the smoke plume as this is the first emission that normally exceeds recommended health concern levels.

For monitoring of particulate matter, it is generally accepted that the concentration of small respirable particles having a diameter of 10 μm or less (PM-10) should be less than 150 μg/m^3 for a 24-h period and PM-2.5 should be less than 35 μg/m^3. This is the standard set by several national authorities including the National Institute of Occupational Safety and Health (NIOSH).

The second emission of concern is PAHs on the particulate matter. VOCs are a tertiary concern.

The devices currently used to carry out real-time monitoring of particulates are the DustTrak, MiniRAM and DataRAM aerosol monitors, which are capable of detecting the PM-10 and PM-2.5 particulates emitted by a burn. Figure 1.71 shows a cluster of particle-measuring instruments; these are mostly DataRAMs and the old RAMs. It is important to note that the concentrations of particles downwind are very variable over time. A reading can be over the recommended maximum value

FIGURE 1.71   A cluster of particle-measuring instruments which are mostly DataRAMs and the older RAMs. These are set under a test burn.

one instant and then at baseline values the next. Furthermore, the background values must be measured and subtracted from the current value. As both the MiniRAM and DataRAM measure humidity as particulate (which it is), the instructions state that these instruments should not be used in locations where there is high humidity. This certainly applies to locations on boats and near the sea. Experimentation has shown that high humidity can lead to readings as much as five times the maximum exposure value, although the data can be corrected for this. In both cases, the real-time value on the instrument is noted only for interest. The instrument readings should be electronically recorded, and averages calculated from the recorded and corrected data. The DataRAM has an internal recorder.

There are no reliable real-time or near real-time methods for monitoring PAHs. There are many methods for sampling particulates using pumps and filter papers, however, and some portable devices are also available.

Real-time monitoring of VOCs can be done, but it is fraught with difficulties and inaccuracies. VOCs are sampled in many ways; however, the use of evacuated metal cylinders, known as Summa canisters, is easy and yields accurate results.

1.11.3   Sample Collection and Analysis

There are several methods for collecting and analysing samples to be used for evaluating the effectiveness of ISB. Not all these methods are required in an actual emergency burn situation, but depending on the circumstances, regulations and/or the specific operational plan, some or all of them may be required.

The secondary emissions of concern from an in-situ burn are the PAHs associated with the particulate matter. There are several simple methods for collecting these particles for subsequent laboratory analysis. Simple sampling pumps can also be used to confirm particulate counts as well as to trap particles. Analysis of the trapped particles is complex and must be done by a laboratory with the required equipment and experience in PAH analysis.

VOCs are a third emission of concern. These can be sampled using evacuated metal canisters known as Summa canisters, which are opened for a specified time to collect a representative sample of the gas. The compounds must be analysed by a specialised laboratory with the required equipment and experience in analysing VOCs from Summa canisters. Figure 1.72 shows Summa canisters along with other basic equipment for monitoring emissions.

Full sampling and analysis of chemical spills or situations such as burns is performed regularly by agencies such as the Emergency Response Team (ERT) of the U.S. Environmental Protection Agency (U.S. EPA) and the Emergencies Science Division (ESD) of Environment Canada. These organizations may be able to assist in monitoring burns. Figures 1.73 and 1.74 shows a cluster of instruments put out by Environment Canada and U.S. EPA to monitor a test burn. The residue should be sampled and, if possible, quantified, as shown in Figure 1.75. It may also be appropriate to sample the smoke plume for a broad range of subsequent analyses. Figure 1.76 shows a remote sampling helicopter used for this purpose at the NOBE burns.

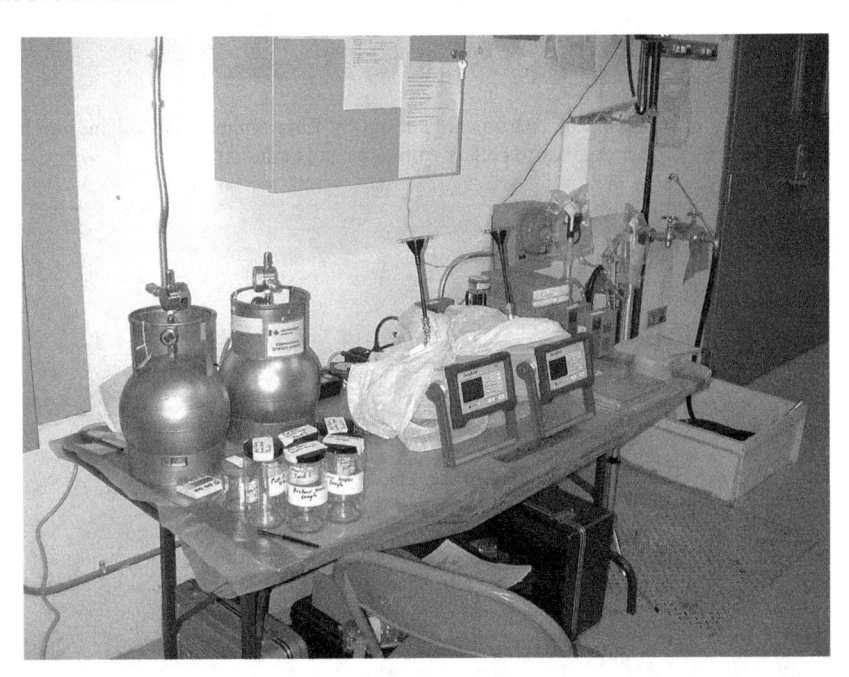

FIGURE 1.72 A collection of instruments and devices for measuring the emissions from burning. The metal canisters are Summa canisters, the instruments are DataRAMs for measuring particulate concentrations and at the far right of the table are pumps for taking particulates for subsequent PAH analysis. The jars in the front are for sampling the oil and the subsequent burn residue.

FIGURE 1.73 A collection of instruments put out by Environment Canada and the U.S. Environmental Protection Agency to collect emissions and emission data.

FIGURE 1.74 Another collection of instruments put out by Environment Canada and the U.S. Environmental Protection Agency to collect emissions and emission data.

FIGURE 1.75 A sample of residue collected after a test burn. This sample will be weighed to establish efficiency and sampled for further chemical analysis.

FIGURE 1.76 A remote-controlled sampling helicopter used to sample the smoke plume at a burn.

1.11.4 DATA ANALYSIS

Analysis should be performed on the electronically recorded real-time particulate data. First, a baseline of background values should be established. This background should then be subtracted from the entire dataset. This baseline may change throughout the burn as is evidenced by the data trend moving up or down throughout the monitoring period. If the background does change, it is more complex to subtract because it changes at each point in time. The background data can be subtracted by using a spreadsheet program that uses the slope of the line to subtract the background at each point in time. Second, the data should be averaged over the time period that the data was taken. Third, the data needs to be corrected to reflect a 24-h period, which is the time period over which the maximum exposure is usually specified. For example, if the average particulate concentration was 100 $\mu g/m^3$ over a 6-h period, the 24-h value is 25 $\mu g/m^3$, assuming there is no other source of particulates. Because of these necessary data manipulations, data from real-time monitoring of burn emissions must be regarded with caution and cannot be used to establish that a burn is either safe or unsafe before the appropriate analysis.

1.12   FINAL RECOVERY OF RESIDUE

The oil residue left after a burn is usually a heavy, tar-like material which is very viscous and adhesive, similar to a highly weathered oil. The greater the burn efficiency, the higher the density and viscosity of the residue. The burn residue from some types of oil may sink in the water column (Fritt-Rasmussen et al. 2013). This behaviour should be determined in advance for common crude and bunker oils being transported in American or Canadian waters.

The decision to recover the residue mechanically or leave it to break down biologically depends on the total volume of the residue, whether the residue is dense enough to sink, and where it is expected to go if left alone. Other considerations include the immediate availability of equipment and personnel who may be deployed in other recovery efforts. On land, residue is readily recovered using mechanical means.

At sea, residue is best recovered using a vessel with low freeboard, which provides easy access to the water surface. A sea truck, powered barge or landing craft used in conventional oil spill response is ideal for this purpose. The amount of residue that can be recovered depends on the displacement of the vessel and the size of tank and other equipment that can be safely carried on the deck. Figure 1.77 shows the collection of burn residue using sorbents. Depending on sea conditions and the dimensions and displacement of the powered barge, such a vessel could carry an estimated 1–5 tons of residue.

Recovering residue is simplified if the recovery vessel can be operated from a shore base. The vessel can be launched from shore and the recovered residue can be removed using a vacuum truck on shore. If the residue is too viscous to remove using vacuum devices, it can be removed manually. When conducting a burn on the open ocean, launching and retrieving a boat to recover residue can be difficult. Unless the burn site is within reasonable distance of shore, the residue recovery vessel must be deployed from one of the larger vessels towing the fireboom. This vessel must be equipped with a suitably sized crane to launch and retrieve the residue boat and have enough tankage or deck space to hold the recovered residue.

Transferring the recovered residue to a larger vessel could be difficult, especially if the larger ship has a high freeboard. The residue tanks should therefore be carried on the ship with the lowest transfer height. Residual oil can also be collected in a backup boom and recovered using sorbents or skimmers suitable for use with

FIGURE 1.77 The collection of residue from the last burn at NOBE. The residue is being collected using sorbents.

heavy oil. Depending on the volume, the residue can be recovered or transferred using either a vacuum suction system or a submersible pump, or it can be manually transferred with shovels and buckets.

Residual oil can also be collected in a backup boom and recovered using sorbents or skimmers suitable for use with heavy oil. Depending on the anticipated volume and properties of the residue, the collected residue could be transferred using either a vacuum suction system, a submersible pump such as the many heavy oil pumps now available, or manually using shovels and buckets.

Another option is to herd the residue into one area using pumps or water hoses deployed from a small boat. Once herded, it may be possible to re-ignite the residue or to ignite it with newly collected oil to further reduce the volume of residue to be recovered. Because of the small areas involved, handheld igniters are more suitable than helicopter-borne devices for re-igniting residue.

1.13 POSSIBLE SPILL SITUATIONS

As no two oil spill situations are the same, it is helpful to look at several possible scenarios when developing response techniques for spill situations: burning at sea, burning in a protected bay, burning on a river, burning in melt pools in the Arctic and burning in an intertidal zone (Fingas and Punt 2000).

The strategies listed in Table 1.21 can best be implemented using specific tactics. These tactics are illustrated separately in Figures 1.78 through 1.87 and summarised in Table 1.22. Each of these tactics has specific advantages and limitations (Fingas and Punt 2000).

TABLE 1.21

Specific Spill Scenarios and Burning Strategies

Scenario 1

Burning at Sea	Strategy
Location: At sea	**General**
Position: Offshore	Verify wind and current direction to ensure that burning the slick will not affect people, property, or environmentally sensitive areas.
Proximity of oil to source: A large slick of oil well away from the source without any trail leading back to the source	As a first response, as much of the slick as possible can be burned without using containment. This will require a helicopter with a heliorch. Several ignition points may be required to burn all parts of the slick that are burnable.
Condition of oil: The oil in the centre of the slick is more than 3 mm (0.12 in.) thick and is not emulsified	Depending on the size of the slick and distance from land, a ship stationed near the slick may be required to refuel the helicopter and heliorch.
Weather and sea state: Calm conditions	Once the slick will no longer burn, containment can be used to further thicken the remaining oil and attempt to burn it again.

Containment Configuration

For the second stage of burning, ideally a fire-resistant boom should be used in the U configuration towed by two vessels. If a fire-resistant boom is not available, a conventional boom can be used with the understanding that the boom would be sacrificed and that its containment ability will be severely limited as the burn proceeds.

Depending on the amount of oil to be burned, manageable sections of the slick (about one-third of the boom's U area) should be carved off from the main slick using the boom and transported away from the slick for burning.

The slick should be approached from downwind and the boom should be towed into the wind during burning.

Protection

Aircraft overflights should be carried out to ensure that burning is under control and that sensitive areas are not being affected.

A standby boat should be nearby for helicopter rescue.

Aircraft with extinguishing foam or water-bombing capability should be available.

During containment operation, towing vessels should have water spray guns ready to protect them from flames.

Accident Response

During containment operation, tow vessels disconnect boom towing lines and sail away upwind from the burning oil or they should speed up to entrain oil, thus reducing slick thickness and extinguishing the burn.

Notice of a floating hazard is filed to ships in the area.

Aircraft with extinguishing foam or water-bombing capability fly over burn.

(Continued)

TABLE 1.21 (Continued)
Specific Spill Scenarios and Burning Strategies

	Strategy
Scenario 2	
Burning at Sea	**General**
Location: At sea	As a first response, send tugs out to the site to move the tanker away from the main part of the slick. Surround the tanker with
Position: Offshore	containment boom to prevent further seepage from the area and fully separate the vessel from the main slick. Water cannons can be
Proximity of oil to source: A large	used to separate any sheen connecting the tanker to the main part of the slick.
slick of oil with a trail leading	Verify wind and current direction to ensure that burning the slick will not affect people, property, or environmentally sensitive areas.
back to the tanker from which it	As much of the slick as possible can be burned without using containment. A Helitorch should be used for ignition. Several ignition
was spilled	points may be required to burn all parts of the slick that are burnable. Depending on size of slick and distance from land, a ship
Condition of oil: The oil in the	stationed near the slick may be required to refuel the helicopter and Helitorch.
centre of the slick is more than	Once the slick will no longer burn, containment can be used to further thicken the remaining oil and attempt to burn the slick again.
3 mm (0.12 in.) thick and is not	
emulsified	**Containment Configuration**
Weather and sea state: Calm	For the second stage of burning, ideally a fire-resistant boom should be used in the U configuration towed by two vessels. If a
conditions	fire-resistant boom is not available, a conventional boom can be used, with the understanding that the boom will be sacrificed and
	that its containment ability will be severely limited as the burn proceeds.
	Depending on the amount of oil to be burned, manageable sections of the slick (about a third of the U area) should be carved off from
	the main slick using the boom and transported away from the slick for burning.
	The slick should be approached from downwind and during burning, the boom should be towed into the wind.
	Protection
	Aircraft overflights should be carried out to ensure that burning is under control and that sensitive areas are not being affected.
	Vessels with water sprayers can be situated around the tanker to prevent any flames from reaching it. A standby boat should be
	situated nearby for helicopter rescue.
	Aircraft with extinguishing foam or water-bombing capability should be available.
	During containment operation, towing vessels should be equipped with water spray guns to protect vessels from flames.

(Continued)

TABLE 1.21 (*Continued*)
Specific Spill Scenarios and Burning Strategies

Accident Response

During containment operation, tow vessels should disconnect boom towing lines and sail upwind from the burning oil or they should speed up to entrain oil, thus reducing slick thickness and extinguishing the burn.

Notice of a floating hazard is filed to ships in the area.

Aircraft with extinguishing foam and/or water-bombing capability fly over burn.

Scenario 3

Burning at Sea   **Strategy**

Location: At sea   **General**

Position: Offshore   An emulsion breaking treating agent should be applied to the parts of the slick that have stable emulsions.

Proximity of oil to source: A large   Verify wind and current direction to ensure that burning the slick will not affect people, property, or environmentally sensitive
slick of oil well away from the   areas.
source without any trail leading   Using containment boom with an overall height of at least 1 m (3.3 ft), small sections of the slick should be pulled away from the
back to the source   main slick and burned.

Condition of oil: The oil in the   Monitor wave heights and try to burn during times when waves are less than 1 m (3.3 ft) or, if possible, tow contained portion to an
slick is less than 2 mm (0.08 in.)   area where waves are less than 1 m (3.3 ft) high.
thick and some parts of the slick   Ideally, a helicopter with a Helitorch would be required to burn the contained oil.
are emulsified   Depending on the size of the slick and distance from land, a ship stationed near the slick may be required to refuel the helicopter
Weather and sea state: Winds   and Helitorch.
approximately 15 m/s (30 knots)
and waves occasionally greater   **Containment Configuration**
than 1 m (3.3 ft)
                                      Because several burns will have to take place, a fire-resistant boom in the U configuration towed by two vessels should be used.
                                      Manageable sections of the slick (about a third of the U area) should be carved off from the main slick using the boom and
                                      transported away from the slick for burning.
                                      The slick should be approached from the downwind side and boom should be towed into the wind during burning.

(*Continued*)

TABLE 1.21 (*Continued*)
Specific Spill Scenarios and Burning Strategies

Protection

Aircraft overflights should be carried out to ensure that burning is under control and that sensitive areas are not being affected.

A standby boat should be nearby for helicopter rescue.

Aircraft with extinguishing foam or water-bombing capability should be available.

Towing vessels should have water spray guns to protect vessels from flames.

Accident Response

During containment operation, tow vessels disconnect boom towing lines and sail away upwind from the burning oil or they should speed up to entrain oil, thus reducing slick thickness and extinguishing the burn.

Notice of a floating hazard is filed to ships in the area.

Aircraft with extinguishing foam and/or water-bombing capability fly over burn.

Scenario 4

Burning in Protected Bay

Location: Protected bay

Position: Nearshore, close to a small populated area

Proximity of oil to source: Well away from the source without any trail leading back to the source

Condition of oil: Slick less than 2 mm (0.08 in.) thick

Weather and sea state: Calm conditions

Strategy

General

If the shoreline around the bay is too sensitive to allow for burning, the oil should be pulled out of the bay using containment boom and burned away from the shoreline. A Helitorch can be used for igniting the burn.

If combustible materials are well away from the edge of the shoreline or the shoreline can be protected, the oil can be burned within the bay using the shoreline and/or containment booms to concentrate and contain the oil for burning. A Helitorch can be used for ignition, but if accuracy is a concern, handheld igniters should be used, thrown from a boat and allowed to float into the slick.

Verify wind and current direction to ensure that burning the slick would not affect people, property, or environmentally sensitive areas.

Containment Configuration

If oil is to be burned outside the bay, booms should be used in a U configuration to bring the oil out of the bay and away from the shoreline for burning. If possible, the burning should take place within a fire-resistant boom and the slick should be lighted with a Helitorch. Boom should be towed into the wind during burning.

If burning is to take place in the bay, boom should be used in a diversion mode to direct the oil towards a calm part of the bay to concentrate it for burning. The slick can be lighted with either a Helitorch or an igniter thrown into the slick from a vessel.

(*Continued*)

TABLE 1.21 (*Continued*)
Specific Spill Scenarios and Burning Strategies

Protection

Aircraft overflights should be carried out to ensure that burning is under control and that sensitive areas are not being affected.

A standby boat should be nearby for helicopter rescue, if a Helitorch is being used.

Aircraft with extinguishing foam or water-bombing capability should be available.

For offshore burning, towing vessels should be equipped with water spray guns to protect vessels from flames.

Within the bay, burning should take place at low tide if possible and the shoreline should be soaked with water before and during the burn. Water sprayers can be located on shore to divert flames from shoreline.

If possible, fire trucks should be placed on the shoreline in case flames reach combustible material on the shoreline.

Accident Response

For offshore burning, tow vessels disconnect boom towing lines and sail upwind from the burning oil or they should speed up to entrain oil, thus reducing slick thickness and extinguishing the burn.

Notice of a floating hazard is filed to ships in the area.

Aircraft with extinguishing foam and/or water-bombing capability fly over burn and fire trucks are available on the shoreline.

Scenario 5

Burning on River

Location: River

Position: Nearshore, away from amenities and populated areas

Proximity of oil to source: Distant – no trail back to the source

Condition of oil: Slick less than 2 mm (0.08 in.) thick

Weather and sea state: Calm conditions, current more than 0.5 m/s (knots)

Strategy

General

Before burning can take place, the oil should be diverted to a calm part of the river (slow current area, a point or bay area) where the shoreline is free of combustible materials or can be protected from the flame.

Both the shoreline and containment booms should be used to concentrate and contain the oil for burning.

A Helitorch can be used for ignition, but if accuracy is a concern, handheld igniters should be used, thrown from a boat and allowed to float into the slick.

Verify wind and current direction to ensure that burning the slick will not affect people, property, or environmentally sensitive areas.

Containment Configuration

Boom should be used in a diversion mode to direct the oil towards a calm part of the river to concentrate it for burning.

If containment boom is required during the burning phase, a fire-resistant boom should be used when possible.

(*Continued*)

TABLE 1.21 (*Continued*)
Specific Spill Scenarios and Burning Strategies

	Strategy
Protection	
	Aircraft overflights should be carried out to ensure that burning is under control and that sensitive areas are not being affected.
	Aircraft with extinguishing foam or water bombs should be available.
	The shoreline should be soaked with water before and during the burn. Water sprayers can be located on shore to divert flames from shoreline.
	If possible, fire trucks should be available on the shoreline in case flames reach combustible material on shore.
Accident Response	
	Aircraft with extinguishing foam and/or water-bombing capability should fly over burn and fire trucks should be available on shore.

Scenario 6

Burning in Melt Pool in Arctic

Location: Arctic, slicks of oil in several melt pools

Position: Nearshore, away from amenities and populated areas

Proximity of oil to source: Distant – no trail back to the source

Condition of oil: More than 3 mm (0.12 in.) thick and emulsification that has remained stable over several days

Weather and sea state: Calm conditions, wind speeds approximately 20 m/s (40 knots)

	Strategy
General	
	An emulsion breaking treating agent should be applied to the parts of the slick that have stable emulsions.
	Verify wind and current direction to ensure that burning the slick will not affect people, property, or environmentally sensitive areas.
	A Helitorch should be used to ignite the oil in each melt pool.
	Depending on size of slick and distance from land, a ship stationed near the slick may be required to refuel the helicopter and Helitorch.
Containment Configuration	
	Containment boom should not be required as the melt pools should act as natural containment.
	Water spray can be used to push oil to one side of the pool during the burn to keep the thickness at a burnable level.
Protection	
	Aircraft overflights should be carried out to ensure that burning is under control and that sensitive areas are not being affected.
	Standby boat should be nearby for helicopter rescue.
	Aircraft with extinguishing foam or water-bombing capability should be available.
Accident Response	
	Aircraft with extinguishing foam and/or water-bombing capability fly over burn.

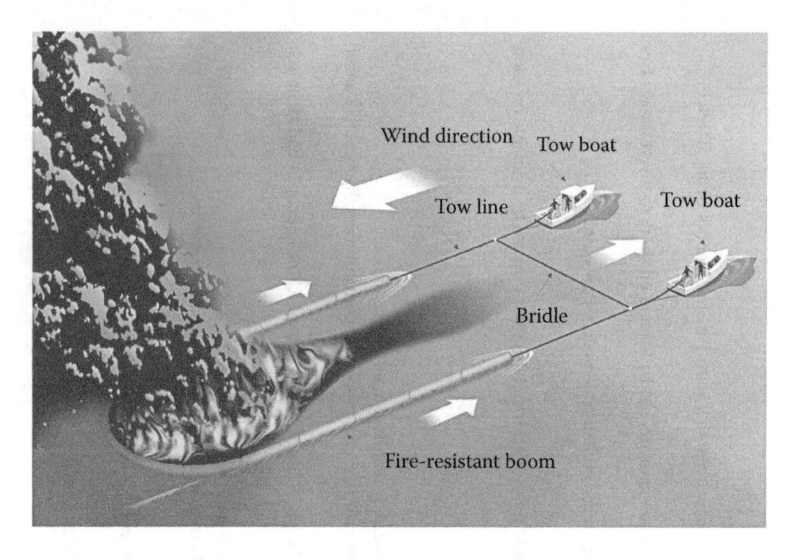

FIGURE 1.78 Use of towed boom to burn oil directly.

FIGURE 1.79 Use of towed boom to collect, tow away and then burn oil.

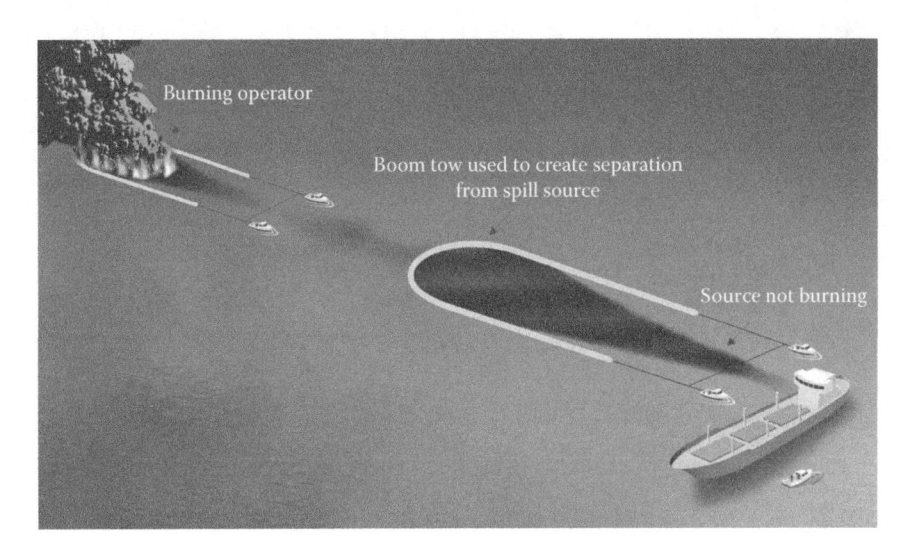

FIGURE 1.80 Use of towed boom to separate the spill source from fire.

FIGURE 1.81 Use of towed boom to protect amenities.

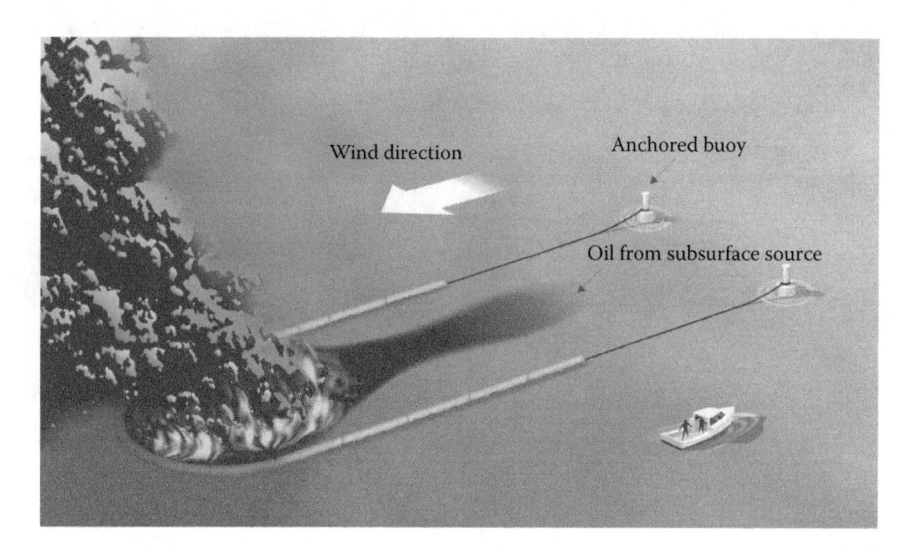

FIGURE 1.82   Use of anchored fire-resistant boom.

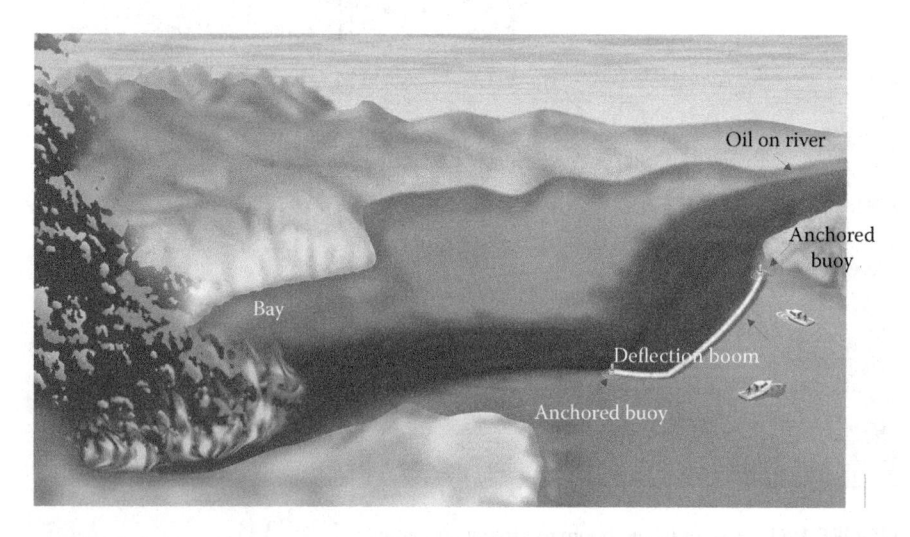

FIGURE 1.83   Fire-resistant boom used in deflection mode.

FIGURE 1.84 Burning against shorelines.

FIGURE 1.85 Burning on ice.

Shallow water

Temporary steel boom built
with galvanized steel sheets
and metal stakes

FIGURE 1.86   Temporary fire-resistant boom used in deflection mode.

Helitorch

Uncontained oil slick

Tow boat

FIGURE 1.87   Unconfined burning at sea.

TABLE 1.22

Tactics for Dealing with Oil in Various Situations

Tactic	Applications
Towed boom, burn in tow (shown in Figure 1.78)	Burning source
	Separation between source and oil
	Source separated
Towed boom, collect and burn (shown in Figure 1.79)	Source of oil is not burning
	Oil near populated or sensitive areas
Towed boom, source separated (shown in Figure 1.80)	Source of oil is not burning
Boom used to separate source (shown in Figure 1.81)	General or protect amenities
Anchored boom (shown in Figure 1.82)	Rivers, estuaries, or in shallow water
	Over subsurface sources or blowouts
Deflection boom (shown in Figure 1.83)	Oil deflected away from amenities
	Oil deflected to burn area
Burning against shoreline (shown in Figure 1.84)	Remote shoreline with no hazards
In ice (shown in Figure 1.85)	Oil is thick enough to burn
Temporary steel boom (shown in Figure 1.86)	Oil can be contained in shallows
Uncontained burning (shown in Figure 1.87)	Oil is thick enough to burn

The well-known tactic of using a towed fireboom to collect and burn oil directly in the boom is shown in Figure 1.79. As with all booms, this technique has a relative current limitation of 0.4 m/s (0.7 knots) before oil is lost under or over the boom. This can be overcome on the open ocean by towing at the relative velocity, despite the surface current. This means that if the actual current exceeds 0.4 m/s (0.7 knots), the boom tow could be slipping down current. Another limitation of this method is that the fire could propagate to the source of the oil or endanger the tow boats and their crew.

Collecting the oil separately, towing the boom away from a non-burning source and then burning the oil is shown in Figures 1.79 and 1.80. This approach prevents the fire from spreading to the oil source. Another advantage is that the oil can be collected using a conventional boom and then transferred to a fire-resistant boom for actual burning. Since a fire-resistant boom is more expensive and harder to deploy than a conventional boom, this option has some practical and economic benefits. The use of a towed boom to protect amenities from a burning source of oil is shown in Figure 1.81. Using an anchored boom to burn oil is shown in Figure 1.82. This tactic poses no risk to tow boats and their crew. The boom may not maintain correct alignment with the wind and current; however, and the relative velocity of the surface current and the boom are also considerations.

The use of an anchored deflection boom to direct oil away from amenities or toward burn areas is shown in Figure 1.83. The burning of oil against a shoreline is shown in Figure 1.84. This can only be done if there is no combustible material such as trees and buildings on the shoreline. In addition, highly adhesive oil residue may be left on the shoreline, which may be difficult to remove. Burning oil in ice is illustrated in Figure 1.85. The natural containment of ice can serve to thicken oil sufficiently for ignition and burning to take place. This technique has often been used to burn oil spills in the Arctic. These techniques are described in more detail in Chapter 3.

Oil can be contained in shallow water using a temporary steel boom, as shown in Figure 1.86. The boom is constructed of corrugated steel sheets and metal stakes. As a portion of the corrugated steel is in the water, heat is dissipated and the sheet metal should remain intact long enough for the oil to be burned. It is important to stress that this method has not been extensively tested and backups should be in place in case of failure.

Finally, burning uncontained oil is shown in Figure 1.87. While this method is simple and economical, the oil must be thick enough to support ignition and burning, which is rare for most uncontained spills of crude oil.

1.14   POST-BURN ACTIONS

1.14.1   Follow-Up Monitoring

The site must be surveyed immediately after the burn to ensure that no burning materials remain in the area. This could include thick patches of escaped oil, parts of the boom or burning organic matter. After this immediate surveillance, the residue should be recovered quickly before it sinks. Areas where residue may have sunk should be carefully documented as this could adversely affect the benthic environment. The area should be surveyed and the amount of unburned oil remaining should be estimated. This value and the amount of residue are important in estimating the overall mass balance.

Analysis of particulate matter, PAHs and VOCs at the downwind locations should be completed if these are sampled and these results included in the final burn report. In the case of VOCs, a background sample must be collected on a day when burning is not taking place and when the wind is blowing in a similar direction as on the day of the burn.

A report on the actions taken during the burn should be prepared at this time to ensure that others can learn from the burn and that a good record remains if there are any questions on efficiency or other issues.

1.14.2   Estimation of Burn Efficiency

Burn efficiency is measured as the percentage of oil removed compared to the amount of residue left after the burn. The burn efficiency, E, can be calculated by the following equation, where v_{oi} is the initial volume of oil to be burned and v_{of} is the volume of residual oil remaining after burning:

$$E = \frac{v_{oi} - v_{of}}{v_{oi}}$$

In this equation, the initial volume of oil, v_{oi}, can be estimated in a number of ways. If the spill source is known, as in the case of a vessel or coastal storage depot, the volume spilled can be estimated from the tank size and the amount of oil remaining in the tank.

In the case of an offshore rig, the pumping rate can be used to estimate the initial volume. If the source is unknown or the volume of oil released from the source cannot be estimated, the volume of the slick can be estimated either visually using objects of known dimensions, for example, response vessel or containment boom, or using timed overflights, aerial photographs or remote sensing devices. This area together with an estimate of the average thickness of the oil, performed either visually by taking samples or by remote sensing, can then be used to estimate the volume of the slick.

It should be noted that this equation does not take into account the volume of oil lost through soot produced from the burn, which is a small amount and difficult to measure. This estimate does also not include any residue that has sunk or cannot be collected.

If the residue remains afloat, it can be recovered either by skimmers or sorbents. The volume of residual oil remaining after burning, v_{of}, can be estimated by measuring the volume or weight recovered. Figure 1.88 shows some residue after a burn. If the residue cannot be recovered, the volume of the residue slick can be measured by estimating its area and thickness, in the same way described for estimating the initial volume of oil. The volume of any tar balls in the residue should also be taken into account.

Panetta et al. (2017) used an acoustic system to measure slick thickness and a camera system to measure area, yielding a real-time system to measure burn rates. The method is restricted, however, to measuring burn rates on small experimental spills because of the proximity needed for sensors and the practicality of applying these sensors.

If some or all of the residue sinks, which is rare, the amount of oil that burned ($v_{oi} - v_{of}$) can be estimated using the fact that, for most oils and conditions, an oil slick burns at a rate of 1–4 mm/min. Table 1.23 shows the range of burning rates.

FIGURE 1.88 The burn residue from one NOBE burn this is less that about 0.1% of the oil released.

TABLE 1.23

Burning Properties of Various Fuels

Fuel	Burnability	Ease of Ignition	Flame Spread	Burning Rate[a] (mm/min)	Sootiness of Flame	Efficiency Range (%)
Gasoline	Very high	Very easy	Very rapid, through vapours	4	Medium	95–99
Diesel fuel	High	Easy	Moderate	3.5	Very high	90–98
Light crude	High	Easy	Moderate to high	3.5	High	85–98
Medium crude	Moderate	Easy	Moderate	3.5	Medium	80–95
Heavy crude	Moderate	Medium	Moderate	3	Medium	75–90
Weathered crude	Moderate	Add primer	Slow	2.5–3	Low	60–90
Crude oil with ice	Low	Difficult, add primer	Slow	2	Medium	50–90
Light fuel oil	Low	Difficult, add primer	Slow	2.5	Low	50–80
Heavy fuel oil	Moderate	Add primer	Slow	2–3	Low	60–90
Dilbit	Moderate	Easy if fresh	Moderate	2–3	Medium	40–60
Weathered dilbit	Moderate	Add primer	Slow	2–3	Medium	50–70
Bitumen	Low	Add primer	Slow	2–3	Low	40–70
Emulsified oil	Low	Add primer	Slow	2–3	Low	30–70
Waste oil	Very low	Add primer	Slow	1–2	Medium	15–50

[a] Typical rates only – to get the rate in $L/m^2/h$ multiply by 60.

The amount burned can be estimated using this range, the area of the slick on fire and the total time of the burn.

Research has shown that burn efficiency is not affected by the oil properties, but depends primarily on the thickness of the slick and oil type. Regardless of the initial thickness of the oil, the final thickness will be in the order of 0.5–1 mm. As such, a much greater burn efficiency is achieved when burning a 20-mm thick slick than a 2-mm thick slick. The burn efficiency also depends on the flame-contact probability. This is a random parameter that can be controlled by proper containment, but it is also affected by wind speed and direction. The burn efficiency can be reduced if the thickness of the slick is inconsistent; that is, the flame reaches patches that are too thin to sustain burning or if the slick is not continuous.

1.14.3 CALCULATION OF BURNING AREA IN A BOOM

The value of the burn area inside a fire-resistant boom is needed to estimate the amount of oil burned (ASTM F3195 2016; Fingas 2014). Towed booms assume a catenary configuration similar to a suspended chain, the classic catenary curve. To calculate the area inside a catenary boom is not a simple matter because catenaries can assume several different forms. Catenary curves were analysed for shape and it was found that the typical oil-filled boom would be categorised as a variation on a flattened catenary curve or a semi-flattened catenary. These curves are analysed and examples calculated and compared to graphical models. On the basis of this, a typical nomogram was created to estimate the area of oil in a semi-flattened catenary-shaped oil containment boom.

A generalised nomogram can be produced. Figure 1.89 shows the recording chart. Table 1.23 shows the burning rates that can be used for each oil and Table 1.24 shows the fill volumes as recorded in Figure 1.89 to calculate the burn volumes. Figure 1.90 shows how to convert pool regression rates to volumetric rates.

1.14.4 BURN RATE

It is generally accepted that an oil slick burns at a slick thickness reduction rate of 1–4 mm/min. See Table 1.23 for rates of specific oils. This range translates to about 1,000–5,000 $L/m^2/day$. During the final stages of burning, when the slick becomes very thin, the rate decreases until the slick becomes less than about 0.5 mm thick.

Like the burn efficiency, the burn rate is somewhat independent of the physical conditions and properties of the oil. Oil emulsification can reduce the burn rate, as does ice in the burn area, because the water in the oil increases the amount of heat required for burning and thus reduces the rate at which the burn spreads.

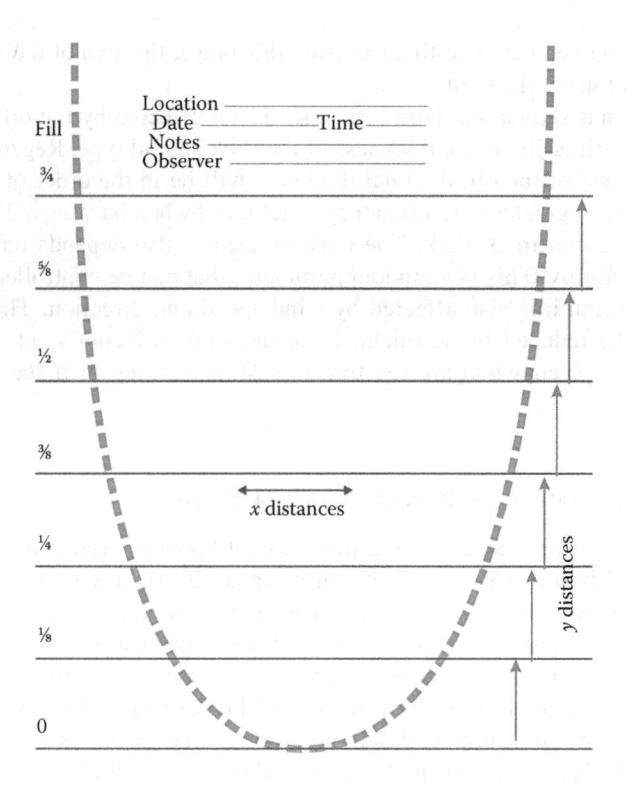

FIGURE 1.89   Chart used to record and estimate boom area burning.

TABLE 1.24
Fill to Area Conversions for a Crude Oil-Filled Catenary Boom

Metric Units

		Boom Size – 150–50 m Opening			200–66 m Opening		
	Fill	Length (m)	Width (m)	Burn Area (m²)	Length (m)	Width (m)	Burn Area (m²)
3/4	Three quarters	51	48	2,020	68	64	3,590
5/8	Five eighths	43	46	1,610	57	61	2,860
1/2	One half	34	44	1,220	45	59	2,170
3/8	Three eighths	26	41	860	35	55	1,530
1/4	One quarter	17	38	530	23	51	940
1/8	One eighth	9	32	220	12	43	390

(Continued)

TABLE 1.24 (*Continued*)
Fill to Area Conversions for a Crude Oil-Filled Catenary Boom

U.S. Customary Units

		Boom Size – 500–166 ft Opening			700–233 ft Opening		
	Fill	Length (ft)	Width (ft)	Area (ft²)	Length (ft)	Width (ft)	Area (ft²)
3/4	Three quarters	165	156	21,000	231	218	41,200
5/8	Five eighths	137.5	149	16,800	193	209	32,900
1/2	One half	110	142	12,700	154	199	24,900
3/8	Three eighths	82.5	132	9,000	116	185	17,600
1/4	One quarter	55	122	5,500	77	171	10,800
1/8	One eighth	27.5	102	2,300	39	143	4,500

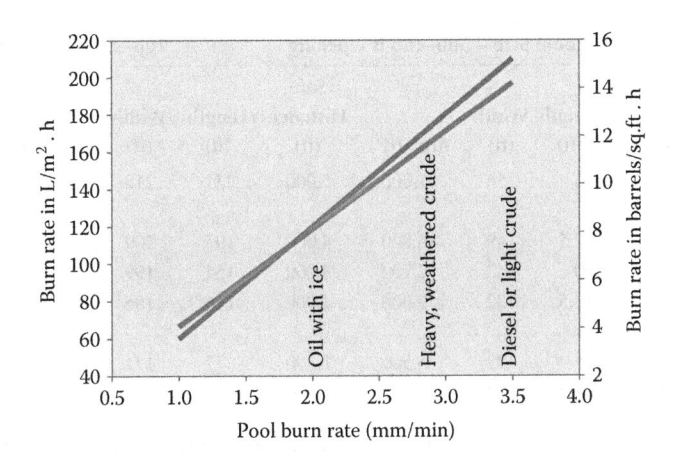

FIGURE 1.90 Nomogram to convert pool regression rates to volumetric burn rates.

1.15 HEALTH AND SAFETY PRECAUTIONS DURING BURNING

1.15.1 WORKER HEALTH AND SAFETY PRECAUTIONS

To protect the health and safety of workers involved with ISB, a thorough HASP must be established and be well understood by all personnel involved before the operation begins. As with any operation in which health and safety are issues, workers are responsible for their own safety and for the safety of their co-workers. To assist in the development of proper HASPs for ISB, much of the information required can be obtained from firefighting associations. Table 1.25 shows safe distances from various areas of boomed burns.

TABLE 1.25
Safe Distances for a Crude Oil-Filled Catenary Boom

Metric Units

Fill	Boom Size – 150–50 m Opening				200–66 m Opening			
	Length (m)	Width (m)	Burn Area (m²)	Safe Distance (m)[a]	Length (m)	Width (m)	Burn Area (m²)	Safe Distance (m)
3/4 Three quarters	51	48	2,020	1,500	68	64	3,590	2,700
5/8 Five eighths	43	46	1,610	1,200	57	61	2,860	2,100
1/2 One half	34	44	1,220	900	45	59	2,170	1,600
3/8 Three eighths	26	41	860	600	35	55	1,530	1,100
1/4 One quarter	17	38	530	400	23	51	940	700
1/8 One eighth	9	32	220	200	12	43	390	300

U.S. Customary Units

Fill	Boom Size – 500–166 ft Opening				700–233 ft Opening			
	Length (ft)	Width (ft)	Area (ft²)	Safe Distance (ft)	Length (ft)	Width (ft)	Area (ft²)	Safe Distance (ft)
3/4 Three quarters	165	156	21,000	5,000	231	218	41,200	8,900
5/8 Five eighths	137.5	149	16,800	4,000	193	209	32,900	6,900
1/2 One half	110	142	12,700	3,000	154	199	24,900	5,300
3/8 Three eighths	82.5	132	9,000	2,000	116	185	17,600	3,600
1/4 One quarter	55	122	5,500	1,300	77	171	10,800	2,300
1/8 One eighth	27.5	102	2,300	700	39	143	4,500	1,000

[a] Safe distances are typical downwind distances where fine particulates (2.5 µm) are less than the prescribed level (<35 µg/m³).

The side distances in a constant wind are estimated to be 1/10 of these values, The values for diesel or diesel-like crude oils are 8.5 times the crude oil values.

1.15.1.1　Preventing Unwanted Ignition and Secondary Fires

Once the operation begins, the burn must be closely monitored to allow response personnel to determine if the burn situation must be reassessed, the plan needs to be modified, or the burn must be controlled or terminated. If on the sea, surveillance of the burn area should be arranged to provide essential information to the tow operators such as the thickness and frequency of slicks in the path of the boom tow or containment area, the precise direction of the smoke plume, the area of oil burning and whether this is increasing or decreasing. If on land, surveillance of the area around the burn, before, during and after the burn is essential.

At sea, two surveillance tactics should be considered: (1) aerial surveillance and (2) surveillance from a larger vessel. The increased visibility from aircraft, particularly helicopters, ensures the safety of the burn operation. However, a larger vessel not only provides a good view of the tow operation from the surface but can also be equipped with extra fire monitors for firefighting capability. This vessel also provides a means of rescue if one of the tow vessels fails.

Any potential difficulties in a burn operation, such as encountering thick burnable slicks that could burn out of control, should be anticipated and avoided. The fire could propagate ahead of the tow vessels or to amenities that can be burned. Other difficulties that should be avoided are the loss of significant amounts of burning oil behind the boom. These burning patches could also cause problems downwind. This can be avoided by having an extra fire-resistant boom downwind to catch any burning patches or vessels with fire monitors to extinguish them.

Flames spread very rapidly through vapours, as fast as 100 m/s or 200 knots. If burning a highly volatile oil such as a fresh, very light crude, gasoline or mixtures of these in other oils, vapour flame spread could occur and cause serious injury. This is referred to as vapour flashback. This can only be avoided by carefully assessing the properties and characteristics of the oil to be burned. When burning these very light mixtures, it must be ensured that no people are in the area. These circumstances are rare because normally, by the time responders have reached an oil spill, the volatile fraction of the oil has been removed. In any case, all burn personnel should be familiar with the hazards and with the difference between the speed of flames spreading on a pool and through a vapour cloud.

Burning should not be attempted on a slick that could flash back to the source of the spill such as a tanker or towards populated areas. This can usually be prevented by removing or isolating the source from the part of the slick to be burned or separating manageable sections of the slick with containment booms and burning these sections within the boom well away from the main source of the slick. In tanker spills, the source can be moved away using tug boats, which can be brought to the site more quickly than containment booms. When this is not possible, containment booms can be used to isolate the main part of the slick from the source. Precautions must also be taken to prevent the fire from spreading to nearby combustible material such as grass cover, trees, docks, buildings and operational vessels.

Perhaps the best way to prevent unwanted or uncontrollable burns is to carve off a manageable section of oil from a large slick and pull it well away from the main slick or other combustible material before igniting it. This oil can be collected using conventional booms and then transferred to fire-resistant booms in an area where it is safe to burn. If oil is close to shore, deflection booms can be used to deflect oil toward a calm area such as a bay where it can be safely burned. Exclusion booms can be used to keep oil away from areas where it is not wanted.

A number of techniques can be applied to prevent secondary fires, fire spreading to unwanted areas, and flashback of the fire to workers. If a boom is used, it must be towed properly. It is important to recognise that a boom fails when towed at a speed faster than about 0.4 m/s (0.8 knots) and that the boom should always be towed into the wind. On most oil slicks, flames do not spread across an oil slick at a rate faster than about 0.2 m/s (0.4 knots). Thus, in a typical situation in which the boom is steadily towed at least at the flame-spreading speed, flames will not reach the boom tow vessels, even at low winds. Caution should be taken, however, because winds can change rapidly. Burns should not be conducted if the tow boats are actually in thick oil or could pass through it.

Operators of a boom tow should be knowledgeable about how to control the area of the burn by increasing or decreasing the tow speed. At excessive tow speeds, the oil is lost through the boom apex as a result of boom failure, entrainment under the boom or loss over the top of the boom. At a towing speed that is too slow, the oil, and therefore the fire, slowly spreads to the boom opening, towards the towing vessels. The movement of oil back and forth in the boom is also influenced by the amount of oil encountered. If more oil is encountered than can be burned in the area of the boom, measures must be taken to prevent the fire from spreading towards the tow vessels. If no safe action is possible, the fire may have to be extinguished or the boom tow dropped.

Once the oil is burning, extinguishment may not always be straightforward or easy. Several tow control methods have been suggested to extinguish the fire within a towed fire-resistant boom. The first method is to release one end of the boom tow and let the oil spread until it is too thin to burn (Fingas and Punt 2000). Second, if the tow speed is increased to greater than containment velocities (0.4 m/s or 0.8 knots), oil submerges under the boom and the fire is often extinguished. Since this method has not been tested and may be hard to carry out, it is not suggested as the primary technique. Another suggested method is to slow down the towing rate, thereby reducing the encounter rate. Several methods of extinguishing fires were tried at the Deepwater Horizon burns and the above methods and others were found to work.

It is recommended that fire-extinguishing equipment be available during the burn. One dedicated fire-extinguishing equipment vessel should be positioned beside the boom containing the burn. During burn operations at sea, those who must be near the burn, such as the tow-boat operators, can be protected by ensuring that fire monitors of sufficient capacity are available. These monitors can be left on to ensure they are ready if needed. Extra fire monitors and experienced crews should be available on the surveillance vessel to assist if a fire spreads. The fire can also be extinguished by using a firefighting foam made for liquid fuel fires. To ensure safety, at least two of these extinguishing methods should be ready at a burn site. When burning is done close to shore, fire trucks and crews can be stationed at strategic points on land to fight unwanted secondary fires.

1.15.1.2 Boom Handling

When booms are being moved and recovered, personnel should avoid cables under tension such as the boom towing lines or tugger winch cables when in use. Personnel should also avoid standing in the coil or bight of a rope or cable lying on deck, which could tighten around a leg or foot and drag a person overboard.

Crane operations onboard ship are particularly dangerous as the roll of the ship may cause the load to swing like a pendulum on the crane wire. Anything being lifted by crane should have two handling lines attached to control the load. Only the crane operator, the signal person, and the two persons holding the load control lines should be involved in the operation. All other personnel should stay well away from the load while it is being lifted. The signal person is in charge of the operation. All personnel must maintain visual contact during the work. Hand signals should be reviewed and understood before operations begin. Figures 1.91 and 1.92 show the deployment of an inflatable boom; Figure 1.93 shows the deployment of a rigid fire-resistant boom.

Communications between the vessel bridge and the deck supervisor should be clear. Hand signals should be understood by all participants. It is recommended that a trained spill response team leader should supervise the entire operation from a safety point of view to detect any unsafe situations as they arise.

Recovering the boom after the burn has been completed is difficult and extremely messy work as the boom is usually waterlogged and covered with a tar-like residue (Fingas and Punt 2000). Workers should wear protective clothing with neoprene gloves, rubber boots and eye goggles. Cuffs should be taped with duct tape. Appropriate decontamination materials are also required for cleaning personnel after the work is completed. Sorbent materials, rags, paper and fabric towels, citrus cleaners, soap and warm water, hand cream, garbage bags and containers should all

FIGURE 1.91 Deployment of a boom from the rear of a supply vessel. This is a common way to deploy boom. (Courtesy of Elastec, Carmi, IL.)

FIGURE 1.92   Deployment of a boom by a straight-line tow. The boom could have been deployed from a dock or a ship. (Courtesy of Elastec, Carmi, IL.)

FIGURE 1.93   Deployment of a boom using a crane.

be available onboard the vessel. Any cleaning materials used should be collected after the burn for proper disposal.

1.15.1.3 Ignition Operation Safety

The following are some general safety issues that relate to ignition devices:

- The operators must fully understand the operational and safety instructions for the specific device being used. This includes understanding the safe operating procedures, training requirements, disposal requirements for spent igniters, and requirements for retrieving and handling igniters that misfire.
- The device should be protected against accidental activation.
- Handheld igniters should have a delay mechanism that postpones the ignition of the device for at least 10 s from the time of activation. This delay allows time to activate and throw the device and for it to float into the slick.
- For helicopter ignition systems, specific helicopter safety precautions must be followed, as well as the specific precautions for Helitorch systems outlined in Section 1.7.3.
- Any device deployed from a helicopter should not require the use of open flames or sparks within the aircraft.

1.15.1.4 Helitorch or Drip-Torch Safety

Because the safety aspects associated with Helitorch or Drip-torch setup and deployment are multifaceted, strict coordination among the various persons involved in the operation is extremely important (Fingas and Punt 2000). Safety issues are associated with helicopter operations, shipboard operations (if the fuel is being stored onboard and/or the Helitorch is being deployed from a ship), and the storage, mixing, transporting and loading of flammable liquids.

Under no circumstances should any untrained persons be involved in the Helitorch operation. In particular, those responsible for preparing, deploying and igniting the Helitorch must be fully trained in helicopter safety and the grounding procedures when transferring fuel. They must also be aware of the volatility of the fuel mixtures used and understand that static charges can occur when fuelling and moving equipment (Fingas and Punt 2000; Lavers 1997).

The Helitorch is ignited by the helicopter pilot. The door on the pilot's side of the helicopter can be removed on some aircraft to give the pilot a clear view of the Helitorch. The Helitorch control switch (toggle switch) should be mounted directly on the cyclic stick at a point where the pilot can comfortably operate it.

The attachment of the Helitorch frame to the helicopter is crucial from a safety point of view. The device must remain stable when carried from the helicopter's cargo-hook, but it must also detach quickly if it needs to be jettisoned in the event of an emergency.

If the Helitorch is deployed from a ship where space for manoeuvring a helicopter is limited, the following precautions should be taken:

1. When the Helitorch is ready for pickup and the helipad is clear of equipment, the Helitorch supervisor radios the pilot with a request to move into position and pick up the torch.
2. When the helicopter returns for refueling, it hovers over the helipad so that the Helitorch can be disconnected. The helicopter then moves away from the ship and assumes the hover position. The helicopter is not permitted to land until the Helitorch and all other equipment and obstructions are removed from the helipad.
3. A three-person fire safety crew should be available onboard the ship at all times, as well as a dedicated 68-kg fire extinguisher. Two 9-kg dry chemical fire extinguishers suitable for extinguishing fuel fires, a first-aid burn kit and a spill clean-up kit for any fuel spills should be available at both the mixing and the loading areas. Personnel must wear fire protective clothing, goggles, a respirator and gloves when mixing and dispensing the gelled fuel and testing the system.
4. The Helitorch must be maintained in good working order at all times. The valve that prevents the fuel from exiting the torch after the pilot has released the toggle switch can become clogged by dust or grit and remain partially open. The valve should therefore be checked and cleaned if necessary before each flight. As a further precaution, it is also recommended that the valve be thoroughly cleaned after every third or fourth refueling of the Helitorch and that the O-ring in the valve be replaced as soon as it shows any sign of degradation. In general, all parts of the Helitorch equipment must be cleaned regularly and any faulty parts replaced at the first sign of wear and tear or any other problem. Spare parts for the torch must always be available at the burn site.
5. All personnel involved in operating the Helitorch must also be aware of the dangers of dealing with highly flammable gelled fuels. As such, proper grounding procedures must be used during the mixing of the fuels, when the fuel barrels are attached to the torch system, and when the torch is attached to the helicopter. It should also be noted that the helicopter picks up static as it flies through the air. The helicopter should therefore also be grounded as soon as it lands, before the torch is unhooked from the cargo hook. The Helitorch barrels must be filled using a non-sparking pump in a well-ventilated area to dissipate fumes. If mixing is done by hand, a wooden or aluminium paddle should be used to prevent sparking. The proper grounding procedures to be followed in the mixing area are shown in Figure 1.94.
6. Before the Helitorch is deployed, the water currents and wind conditions should be noted to determine the safest location for the ignition. A preflight test must also be carried out at this time to test the cargo hook, fuel pump, propane discharge, sparkers and the toggle switch connected to the pilot's cyclic stick.

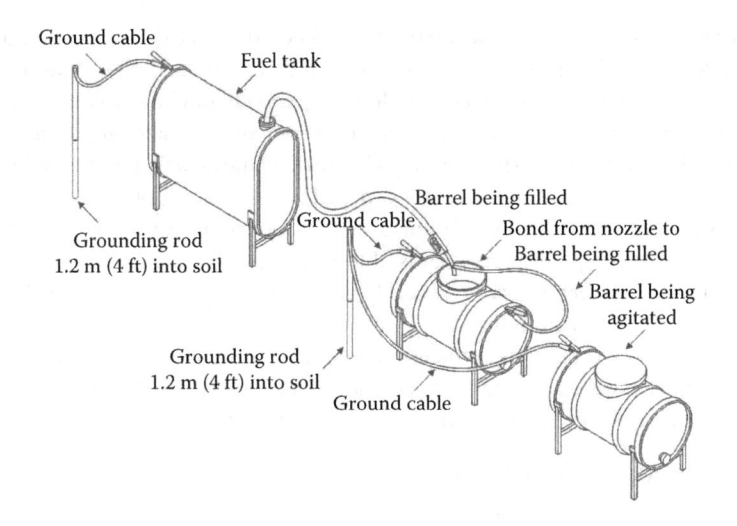

FIGURE 1.94 Grounding and bonding procedures for mixing Helitorch fuel.

7. Before igniting the slick, a predetermined location should be chosen to perform a test drop of a small amount of ignited gelled fuel. Wind and current direction should be checked again to ensure that the burning gelled fuel does not drift towards any of the operational vessels. If the test burn indicates that the gelled fuel is igniting and falling properly, the pilot positions the helicopter over the desired location, fires the torch on a slow pass and then leaves the area. If igniting a fuel with a high flash point, the pilot may have to hover over the burn area and release multiple balls of burning gelled fuel in order to concentrate the fire in one location.

8. When the ignition session is completed, the pilot disengages the Helitorch circuit breaker to isolate the toggle switch so that no burning gelled fuel is accidentally dropped. The helicopter then returns to the land- or ship-based Helitorch deployment site. When the helicopter lands, the recovery crew should stabilise and secure the Helitorch before the helicopter pilot disconnects the cargo hook. This is especially important when the gelled fuel barrel is empty, because the torch system can easily be blown off the helipad by the downdraft of the helicopter's rotors.

1.15.1.5 Exposure of Personnel to Burning Operations

Crews in vessels involved in tow operations are in danger of being exposed to fire or flames if the fire should move up the boom. This could occur if thick patches of oil are encountered and the flame spreads along this thicker patch. The flame velocity is about 0.02–0.16 m/s. The flames would not spread towards the tow vessels if the boom is moving at a speed of at least 0.4 m/s (0.8 knots) in an upwind direction. Because winds can change rapidly, this fact should not be taken as an assurance of safety. In highly variable winds, caution must be taken to ensure that thick concentrations of oil are not encountered at low boom-tow speeds.

Any crews working alongside the burn could be exposed to high concentrations of particulate matter, PAHs and/or VOCs if the wind changes and blows towards them. For this reason, operational vessels should not operate behind the tow-boat positions.

On land, fires can move very rapidly if there is combustible material such as trees and grass available. A fire break should always be made in the area in advance of ignition.

Helitorch personnel are not directly exposed to the dangers of burning operations other than being exposed to small amounts of vapours from the fuel used for gelling. If necessary, respirators can be used to minimise this exposure. The Helitorch operator in the helicopter is not physically exposed to any dangers, other than those normally associated with flying.

When booms and other equipment are handled, the appropriate personal protective equipment must be worn. This includes safety boots, hard hats, goggles, neoprene gloves, life jackets, chemical-resistant clothing and foul weather gear.

1.15.1.6 In-Situ Burn Training

All personnel involved in a burn in most jurisdictions must complete a 40-h hazmat courses. Personnel involved in a burn should be familiar with the technology and procedures in this section.

It is recommended that experienced boom operations staff attend at least a one-day course on the use of booms for ISB and that an additional day be spent on practicing towing booms and releasing oil from booms such as might be required in an emergency. Personnel who are not totally familiar with boom deployment and operations should spend at least 1 week in training and practice.

All members of the Helitorch operating team require extensive training. Only a highly experienced lead person, such as the Helitorch supervisor, should be used to provide training. Operators and ground support personnel should generally participate in at least 3 days of training, including several practice runs.

1.15.1.7 Vessel Safety

The size, structure and navigational equipment of any vessels used in an in-situ oil burn must be suited to the wind, sea state, carrying requirements and visibility conditions expected during the burn operation. For operations on the open water, vessels should have a reliable positioning system, such as GPS, a compass or gyrocompass, working radar, working depth sounder, HF radio, VHF radio and telephone.

Under the Canada Shipping Act, each vessel is legally required to have the appropriate safety equipment in accordance with the size and type of vessel and the type of operation being undertaken. This includes life boats, life rafts, life-saving rings, flares, firefighting equipment, life jackets, survival suits and navigation lights.

Any vessel chartered in Canada or the United States should possess a valid Coast Guard inspection certificate. A survey by a qualified ship surveyor or naval architect is recommended before chartering a vessel. The Ship Safety Branch of the Canadian Coast Guard should be consulted for more information about these requirements.

1.15.1.7.1   Burning Directly Inside a Ship

A special situation is needed if the ISB is to be carried out in a stranded vessel (ASTM 2533 2017). Safety is, of course, the primary criteria. The oil must be accessible to ignition and accessible to air. Explosives or industrial cutting equipment may be used to allow oil to flow from tanks to spaces where it will be burned and to increase the ventilation area. This should be conducted by salvage and explosive experts. Typically, the planned burn would take place in the ship's hold(s) and explosives would be used to open passage from lubrication and fuel tanks to the hold. Lubrication and fuel tanks generally do not have sufficient exposure to the air to allow for burning. Oxygen from the air is necessary for burning. Studies have shown that the area of ventilation is a critical regulating factor in the burning of oil directly on ships and in other confined spaces. The rate of burning is generally calculated based on the area of ventilation openings in the case of low wind situations. Studies have shown that top and side openings combined yield better ventilation than top openings alone. The presence of two openings allows for air circulation over the area of fire. Small-scale studies have shown that a minimum of 10% ventilation is needed to prevent extensive coking. The 10% refers to the area of ventilation compared to the surface area of oil available to burn. An area of more than 20% ventilation has been shown to result in little coking during test burns. External winds assist in providing additional ventilation, despite the semi-closed conditions that may exist. One study showed a threefold increase in burn rate with wind increase from 0 to 11 m/s (ASTM 2533 2017).

During the burn process, some localised oil may become superheated. When the heating is sufficient, flash evaporation of a component of this oil may occur and the surrounding boiling oil can erupt upwards towards the top ventilation port. This could result in oil being splashed onto other parts of the vessel or sea. This phenomenon has been observed in test situations with crude oil.

The safety of the proposed operation should be the primary consideration (ASTM 2533 2017). The vessel should be stable and relatively stationary during the preparation and burn phases. Figure 1.95 shows such a burn. The operation should only be contemplated if the operation will not result in flashback to other sources of fuel. The fire should be prevented from spreading to other combustible material in the area, including trees, docks and buildings.

Preparation of the vessel for burning by using explosives and subsequent burning of the oil weaken the ship's structure. Burning in ships should be considered only if there is no potential for future salvage of the vessel or if the trade-off between future salvage potential and removing the oil is favourable. The use of preparation and burning may weaken the structure sufficiently to result in breakup of the vessel. A breakup may result in the release of oil. Salvage experts and experts on ship design should be consulted where possible before proceeding with the preparation for ignition and burn. They should also be consulted after the burn regarding options to deal with the remaining vessel. The vessel may not be seaworthy, towable or even in condition to allow ship-breaking in place.

FIGURE 1.95 The deliberate burning of oil in the *New Carissa* after her grounding off the coast of Oregon.

1.15.1.8 Aircraft Safety

All flying operations must be carried out in accordance with federal flight regulations (Federal Aviation Agency [FAA] in the United States or Transport Canada). All aircraft associated with an in-situ burn should be chosen carefully to suit the required tasks. Flight plans should be well thought out to take into consideration wind, visibility, cloud types and height, the presence or forecasted presence of fog, precipitation, sea state and other relevant weather conditions.

For helicopter-borne igniter operations, the helicopter must have sufficient lift capacity to carry a pilot, co-pilot and a Helitorch full of fuel and be equipped with a cargo hook able to sling the Helitorch as well as jettison it. The pilot must test the jettison mechanism before each Helitorch operation. For safety reasons, a twin-engine helicopter is preferred, particularly for nearshore operations. These helicopters are more powerful than single-engine machines and can therefore gain altitude more quickly. If a single-engine helicopter is used, it must be equipped with floats to facilitate emergency landings. The helicopter must comply with Transport Canada or U.S. transport regulations regarding helicopter maintenance and the operation being undertaken.

Only the pilot and co-pilot or one other person if required for the ignition activation should ride in the helicopter during the Helitorch operation. All persons in the helicopter should wear a survival suit. During nearshore operations, updraft and downdraft winds against cliffs must be considered. Emergency landing locations for the helicopter should be identified in advance through site surveillance in case of mechanical difficulty.

It is recommended that when helicopter services are being arranged, the performance capability of the aircraft and its suitability for its intended use be confirmed with the helicopter pilot and/or helicopter operator.

1.15.2 PUBLIC HEALTH AND SAFETY PRECAUTIONS

The public should not be exposed to emissions exceeding the recommended human health concern levels. The most concern would be exposure to PM-2.5 particulates greater than 35 $\mu g/m^3$ over a 24-h period. This can be determined by using the formulae provided in Section 1.10.4 to calculate minimum safe burn distances and by monitoring the particulate levels using the methods outlined in this chapter.

It is important to note that atmospheric inversions can occur that increase ground-level concentrations to high levels, and that the smoke plume itself might drop to ground level at higher elevations further inland. Monitoring must be done to ensure that this situation does not occur. If there is the potential of this occurring, the burn should not be started. If a burn is already started and the plume drops to ground level, the situation should be immediately assessed to determine whether the burn should be stopped, people evacuated and/or the plume could drop again. Any people who may be affected by the burning, even if only remotely, must be briefed so that they are aware of the activity and the possible need to evacuate the area on short notice.

If burning near land, sufficient personnel must be available on land, in good communications with the burn command vessel. The land-based personnel should monitor the smoke plume and stay in contact with local weather officials to be informed of any potential changes that could cause the plume to directly affect people on the ground.

If burning against or very near the shore, additional precautions must be taken to ensure that the fire does not spread from the oil to other combustible material. The fire should be monitored from shore by personnel with the ability to put out any potential fires. Trees and other combustibles near the shore might be wetted down as an extra precaution.

1.15.3 ESTABLISHING SAFETY ZONES

An important part of the safety program for an in-situ burn operation is establishing minimal safety zones. This has been accomplished in several ways, including the use of values that are larger than the measured hazardous distances, calculated as shown above, and by the use of smoke plume modelling. An extensive section on calculating safe distances is given in Section 1.10.4.

Smoke dispersion modelling has been used frequently in the past decade to establish safe zones and obtain permits for large industrial sources. Specialised models have been developed that can also be applied to ISB. Although models are not intended to replace monitoring, they provide an important tool for assessing the impact of smoke both before and after a burn.

The smoke model A Large Outdoor Fire Plume Trajectory (ALOFT) was developed by the National Institute for Standards and Technology (McGrattan 1999). It is designed to run on a Windows-based PC and thus could be used as an immediate tool for predicting safety zones. The model has been used to prepare tables of safe distance predictions for typical fires. The model also incorporates the effects of surface roughness and the mixing layer depth, which is the depth of atmospheric mixing, or the atmospheric boundary layer. It might also be viewed as the height of the clouds.

It is important to recognise the limitations of each type of hazard zone estimation. Differing weather conditions can change the concentrations of particulate matter dramatically. In many cases, the plume drops to ground level. Weather officials should be consulted for possible wind changes, atmospheric inversions and other factors that can change the trajectory and impact of the plume.

1.15.4 MONITORING BURN EMISSIONS

A major barrier to the acceptance of ISB of oil spills is the lack of understanding of the resulting combustion products.

Several types of emissions are formed and released when oil is burned. The atmospheric emissions of concern include the smoke plume, particulate matter precipitating from the smoke plume, combustion gases, unburned hydrocarbons, organic compounds produced during the burning process, and the oil residue left at the burn site. Although consisting largely of carbon particles, soot particles contain a variety of absorbed and adsorbed chemicals. Complete analysis of the emissions from a burn involves measuring all these components. The emphasis in sampling has been on air emissions at ground level as these are the primary human health concern and the regulated value. This section will focus on these emissions.

There have been some attempts to monitor emissions remotely. Vodacek et al. (2002) were able to monitor biomass burning remotely using a visible detector monitoring the strong emission lines of potassium (K) at 766.5 and 769.9 nm. This was tested using the Airborne Visible/Infrared Imaging Spectrometer (AVIRIS) satellite spectrometer and monitoring a fire on the ground. This has not yet been extended to oil burns at sea.

Note that the monitoring of emissions conducted at past burns was as comprehensive as possible and the best field samplers and instrumentation available at the time were used. Measurement techniques have progressed over the years, however, and continue to improve. In addition, the data from these burns are so extensive that not even encapsulating summaries can be provided here. The summarised data appears in the references cited in this section (and listed at the end of this chapter) and qualitative statements about that data will be made here.

1.15.4.1 Summary of Measurements

In general, real-time monitoring of emissions should be performed downwind of the fire and at a point closest to populated areas (Fingas et al. 2001a). Studies of the emissions from in-situ oil burns indicate that the main public health concern is particulate matter in the smoke plume as this is the first emission that normally exceeds recommended health concern levels. For monitoring of particulate matter, it is generally accepted that the concentration of small respirable particles having a diameter of 2.5 μm or less (PM-2.5) should be less than 35 μg/m^3 for a 24-h period. This is the standard set out by many national agencies, including the National Institute of Occupational Safety and Health (NIOSH) and described in the U.S. Code of Federal Regulations.

The devices currently used to carry out real-time monitoring of particulates are the DustTrak, MiniRAM and DataRAM aerosol monitors, which are capable of detecting the PM-2.5 particulates emitted by a burn (Lambert et al. 1998). It is important to note that the concentrations of particles downwind are highly variable over time. A reading can be over the recommended maximum value one instant and then at baseline values the next. Furthermore, the background values must be measured and subtracted from the current value. As many instruments measure humidity as particulate (which it is), the instructions state that these instruments should not be used in locations where there is high humidity without taking corrective action. This certainly applies to locations on boats and near the sea. Experimentation has shown that high humidity can lead to readings as much as five times the maximum exposure value, although the data can be corrected for this. In both cases, the real-time value on the instrument is noted only for interest. The instrument readings should be electronically recorded and averages calculated from the recorded and corrected data.

1.15.4.2 Particle Measuring Summary

The DustTrak, MiniRAM and DataRAM are commercially available piece of equipment commonly used in the occupational health and safety industry. The portable monitors (MIE Inc, Bedford, Massachusetts) allow measurement of aerosols and particulates continuously. The advantage of time information is the potential to correlate particulates with specific burn events, such as when the burn is initiated or extinguished.

1.15.4.3 Sampling Particulates Using Filters

Particulate levels from a burn can be most accurately determined by collecting a representative sample on a quartz fibre filter using a high-volume sampling pump (Lambert et al. 1998). The accumulation of particulate on the filter can be measured by differential weighing. The concentration can be calculated by dividing the weight collected by the volume of air. An added advantage of this particulate sampling method is that, after weighing, the collected particulates can be analysed for PAH compounds by gas chromatography, following a solvent extraction procedure. Other burn products of interest, such as metals, could also be analysed.

A high-volume sampler (greater than 200 L/min capability) is necessary for collecting particulate at a burn site to collect enough sample over the relatively short duration of the burn. The flow must be measured in order to calculate the concentration. The flow decreases as the filter is loaded. For this reason, a flow rate must be recorded at both the initiation and conclusion of sampling, while the filter is in place. The flow rate is usually determined as a function of the back pressure created by the pump, although it is sometimes measured by an in-line mass flow meter.

All current high-volume samplers operate on AC power due to the current required to run the pump. The unit has either a power switch or is controlled by AC supply. There is generally a voltage regulator that can be adjusted externally. The frame for the conventional quartz fibre filter is designed to hold either a 4 in. diameter filter circle or an 8×10 in. filter sheet. The frame holds the filter in place while the pump creates a negative pressure on the bottom side of the filter, creating a flow of air through the filter that allows air to pass through but not the particulate. In most cases, the TSP fraction is being collected, for which a filter with 0.8-micron (μ) pore size is used. The collected sample can be used to determine particulate levels by differential weighing and/or can be analysed for various burn products, usually PAHs.

Various types of high-volume samplers are capable of collecting TSP samples. The conventional TSP sampler, which has an 8 × 10 in. frame and all-weather housing, is the most popular for permanent or semi-permanent sampling locations. More portable samplers have a tripod stand rather than the all-weather housing. These models are preferred when frequent relocation is required, but they are less weather-resistant. For use in isolated areas, compact samplers are available that are carried in one hand, such as the GMW Handi-Vol 2000.

Most of the samplers that use an 8 × 10 in. filter frame operate in the range of 500 L/min. When a 4 in. filter frame is used, the flow rate is usually closer to 200 L/min. In some samplers, such as those used for pesticides (PS-1), an additional sample collection medium is employed. After the particulate is collected on the particulate filter, the air passes through a flow-through sorbent medium that collects airborne vapours. These samplers can be effectively used to collect particulate samples, with or without the second sampling medium.

1.15.4.4　Methodology of VOC Sampling Using Summa Canisters

The Summa canister is one method used to collect a metered amount of whole air for laboratory analysis (Fingas et al. 2001a; Li et al. 1992). Air is collected in these evacuated, stainless steel canisters to be analysed for VOCs. In conventional high-volume sampling methods, the VOCs are lost either during sampling or in transit. By contrast, the Summa canister method ensures that most of the VOCs are captured and remain stable between the sample collection on-site and the subsequent laboratory analysis. The quantity of VOCs found in air samples collected close to oil burns varies, depending on several factors, including fuel composition and distance from the burn. A precision restrictor valve on the canister provides the controlled flow required when sampling during the course of a burn or pre-burn.

The Summa canister is a spherical, polished, stainless steel container with a single, manually controlled valve. The canister must be cleaned and evacuated by an accredited laboratory before use. A pre-cleaned and pre-calibrated flow restrictor valve is affixed in order to metre the flow into the canister. No restrictor valve is necessary to collect an instantaneous grab sample. These canisters are most commonly available in sample volumes of 6 L, although 1 and 20 L sizes, as well as less common sizes, are also available.

For accurate time-averaged sampling, flow must remain constant throughout the sample collection period. When using a flow restrictor, air should not be drawn into the canister until ambient pressure is attained, since the flow rate begins to change as the back pressure is reduced. This occurs when the pressure inside the canister is approximately half (0.53) ambient pressure. The time taken to reach this point varies with the flow rate. Since the flow rate does not fall off sharply until close to ambient pressure, sample collection can continue until the sample volume (calculated from the flow rate) reaches 75% of the canister capacity without gross variance in actual volume collected. Flow rate should be calculated based on the sampling period.

Flow restrictors are calibrated using a dry standard method capable of measuring accurately down to 1 mL/min.

Locate the canister in an area that is representative of the sampling area. To begin sample collection, fully open the valve on the canister. Note the time opened. A grab sample can be collected in less than 1 minute. For time-averaged sample collection, continue collecting the sample until either the sampling is completed or the maximum calculated volume of the canister (75% of capacity) is reached. Fully close the valve. Record the time closed and calculate the sampling time. Remove the flow restrictor valve carefully and replace the brass cap. Carefully clean (or bake out) the flow restrictor assembly before using the canister again.

Both the extraction and VOC analysis of the contents of the Summa canisters should be performed by an accredited laboratory. The canister must then be cleaned and re-evacuated before it is used to collect more samples.

The main limitation of Summa canisters is that the analysis of the canisters must be done off-site so there is no on-site indication of the quality of the sample collected. Proper pre-checks and precautions do not prevent an occurrence during the sampling period that may compromise the flow of sample into the Summa canister.

When sampling in humid ambient conditions at temperatures below 5°C, water vapour could freeze inside the orifice because of the adiabatic expansion and resultant cooling effect when air passes through the restricted opening, effectively closing the valve until ambient temperatures rise.

1.15.4.5 Combustion Gas Measurement

Combustion gases of concern include carbon dioxide, carbon monoxide, and sulphur dioxide.

Carbon dioxide: Carbon dioxide is the result of combustion and is found in increased concentrations around a burn (Avens et al. 2011; Goldthorp et al. 1999). Normal atmospheric levels are about 300 ppm and levels near a burn can be around 500 ppm, which presents no danger to humans. The three-dimensional distributions of carbon dioxide around a burn have been measured. Concentrations of carbon dioxide are highest at the 1 m level and fall to background levels at the 4-m level. Concentrations at ground level are as high as 10 times that in the plume and distribution along the ground is broader than for particulates. Carbon dioxide can be measured in many ways; real-time instruments generally measure it using an infrared technique, discrete samples can be taken and quantified by gas chromatography and infrared open-path instruments can provide real-time measurement.

Carbon monoxide: Carbon monoxide levels are usually at or below the lowest detection levels of the instruments and thus do not pose any hazard to humans. The gas has only been measured when the burn appears to be inefficient, such as when water is sprayed into the fire. Carbon monoxide appears to be distributed in the same way as carbon dioxide. Measurements of carbon monoxide can be done using similar techniques as for carbon dioxide.

Sulphur dioxide: Sulphur dioxide, per se, is usually not detected at significant levels or sometimes not even at measurable levels in the area of an in-situ oil burn. Sulphuric acid, or sulphur dioxide that has reacted with water, is detected at fires and levels, although it is not of concern. It appears to correspond to the sulphur content of the oil. Sulphur dioxide itself, although not detected, can be measured using specialised sensor type instruments or reactive tape instruments. Sulphuric acid aerosols can be measured by titrating caustic solutions through which the sample air was drawn (impinger method) or using a reactive tape instrument.

1.15.4.6 Monitoring Poly Aromatic Hydrocarbons on Particulates

PAHs are aromatic compounds found in crude oil and are often produced as a result of combustion. Many PAHs are toxic to humans and the environment, particularly the larger PAHs. Crude oil burns result in PAH downwind of the fire, but the concentration on the particulate matter is often an order of magnitude less than the concentration in the starting oil and sometimes several orders of magnitude less. Diesel contains low levels of PAHs with smaller molecular size, but results in more PAHs of larger molecular sizes after burning. Larger PAHs are either created or concentrated by the fire. Larger PAHs, some of which are not even detectable in the diesel fuel, are found both in the soot and in the residue. The concentrations of these larger PAHs are low and often just above detection limits. Overall, studies have shown that more PAHs are destroyed by the fires than are created.

The Soxhlet extraction method can be used to extract PAHs from the whole filter sheets more easily than microwave extraction (Wang et al. 1999). The extracts are dried by filtering through anhydrous sodium sulphate and concentrated to approximately 1–2 mL. The concentrated extracts are then quantitatively transferred to a pre-conditioned 1.5 g silica gel micro-column topped with 1 cm anhydrous sodium sulphate for sample clean-up. The eluent is collected in a pre-calibrated centrifuge tube and concentrated under a stream of nitrogen to appropriate volume. Finally, the concentrated eluent is spiked with an internal standard d_{14}-terphenyl and made up to the accurate pre-injection volume (0.5–1.0 mL) for GC analysis.

The analysis of target PAHs and other hydrocarbons is performed on a gas chromatograph by a qualified laboratory.

1.15.4.7 Carbonyls

Carbonyls such as aldehydes and ketones are created by oil fires, but they do exceed health concern levels only very close to fires (Lambert et al. 1998; Zervas 2005). Monitoring for carbonyls is conducted using a specialised sorption tube (DNPH) and sampling pump. Analysis is conducted in the laboratory.

The methods are detailed and require experienced laboratory personnel but are not fraught with particular difficulties. Accuracies are ensured by using standards and internal standards. The condition of the sample tubes is important, and sample tubes must be kept frozen before use.

The particular limitation that is noted is that the sensitivity of the method depends on the amount of soot collected. Small samples often have insufficient material to allow proper detection of PAHs.

REFERENCES

Aggarwal, S., Schnabel, W., Buist I., Garron, J., Bullock, R., Perkins, R., Potter, S., and Cooper, D. 2017. Aerial application of herding agents to advance in-situ burning for oil spill response in the arctic: A pilot study, *Cold Reg. Sci. Tech.*, 135, 97–104.

Allen, A.A. 1986. Alaska clean seas survey and analysis of air deployable igniters, *AMOP*, 2, 353–373.

Allen, A.A. 1994. *In-situ Burning Manual, An Economical Solution for Oil Spill Control*, Elastec Inc., Carmi, IL.

Allen, A.A., Mabile, N.J., Jaeger, D., and Costanzo, D. 2011. The use of controlled burning during the Gulf of Mexico deepwater horizon MC-252 oil spill response, *IOSC*, abs194.

Alyeska Pipeline Service Company. 1998. Supplemental information document #1, section 8—burning, in *Addendum to the Prince William Sound Contingency Plan*, Alyeska, Valdez, AK.

API. 2015a. *Field Operations Guide for In-Situ Burning of Inland Oil Spills*, American Petroleum Institute Technical Report 1251, Washington, DC.

API. 2015b. *Field Operations Guide for In-Situ Burning of On-Water Oil Spills*, American Petroleum Institute Technical Report 1252, Washington, DC.

API. 2016. *In Situ Burning: A Decision Maker's Guide*, American Petroleum Institute Technical Report 1256, Washington, DC.

Apicella, B., Barbella, R., Ciajolo, A., and Tregrossi, A. 2002. Formation of low- and high-molecular-weight hydrocarbon species in sooting ethylene flames, *Combust. Sci. Tech.*, 174, 309–324.

Arai, M., Saito, K., and Altenkirck, R.A. 1993. Flame propagation over a layer of crude oil floating on water, *Int. Chem. Eng.*, 33(1), 129–135.

Arsava, K.S., Rangwala, A.S., and Hansen, K. 2017. An offshore in-situ burn enhanced by immersed objects, *AMOP*, Environment Canada, Calgary, AB, pp. 833–843.

ASTM 2152. 2013. *Standard Guide for In-Situ Burning of Oil Spills on Water: Fire-Resistant Containment Boom*, ASTM International, Conshohocken, PA.

ASTM 2533. 2017. *Guide for in Situ Burning of Oil in Ships or Other Vessels*, ASTM International, Conshohocken, PA.

ASTM F 1990-17. 2017. *ASTM Standard Guide for In-Situ Burning of Oil Spills: Ignition Devices*, ASTM International, Conshohocken, PA.

ASTM F1788. 2017. *Standard Guide for In-Situ Burning of Oil Spills on Water: Environmental and Operational Considerations*, ASTM International, Conshohocken, PA.

ASTM F3195. 2016. *Standard Guide for Estimating the Volume of Oil Consumed in an In-Situ Burn*, ASTM International, Conshohocken, PA.

Aurell, J., and Gullett, B.K. 2010. Aerostat sampling of PCDD/PCDF emissions from the Gulf oil spill in-situ burns, *Environ. Sci. Technol.*, 44(24), 9431–9347.

Aurell, J., and Gullett, B.K. 2013. Emission factors from aerial and ground measurements of field and laboratory forest burns in the Southeastern U.S.: PM2.5, black and brown carbon, VOC, and PCDD/PCDF, *Environ. Sci. Technol.*, 47(15), 8443–8452.

Avens, H.J., Unice, K.M., Sahmel, J., Gross, S.A., Keenan, J.J., and Paustenbach, D.J. 2011. Analysis and modeling of airborne BTEX concentrations from the Deepwater Horizon oil spill, *Environ. Sci. Technol.*, 45(17), 7372–7379.

Balthasar, M., and Frenklach, M. 2005. Detailed kinetic modeling of soot aggregate formation in laminar premixed flames, *Combust. Flame*, 140, 130–145.

Balthasar, M., Mauss, F., and Wang, H. 2002. A computational study of the thermal ionization of soot particles and its effects on their growth in laminar premixed flames, *Combust. Flame*, 129(1), 204–216.

Barakat, M., Souil, J.-M., Breillat, C., Vantelon, J.-P., Knorre, V., and Rongère, F.-X. 1998. Smoke data determination for various types of fuel, *Fire Safety J.*, 30(3), 293–306.

Bech, C., Sveum, P., and Buist, I. 1992. In-situ burning of emulsions: The effects of varying water content and degree of evaporation, *15th Arctic and Marine Oil Spill Program Technical Conference, Edmonton,* Canada, pp. 547–558.

Bellino, P.W., Rangwala, A.S., and Flynn, M.R. 2013. A study of in-situ burning of crude oil in an ice channel, *Proc. Combust. Inst.*, 34(2), 2539–2546.

Billings, S., Bullock, R., Aggarwal, S., and Perkins, R. 2016. Analysis of a chemical herding agent in water, *Proceedings of the Thirty-Ninth AMOP Technical Seminar*, Environment and Climate Change Canada, Ottawa, Canada, pp. 728–734.

Bitting, K., Gynther, J., Drieu, M., Tideman, A., and Martin, R. 2001. In-situ burning operational procedures development exercises, *Proceedings of the Twenty-Fourth AMOP Technical Seminar*, Environment and Climate Change Canada, Ottawa, Canada, pp. 695–706.

Bitting, K.B., and Coyne, P.M. 1997. Oil containment tests of fire booms, *Proceedings of the Twentieth AMOP Technical Seminar*, Environment and Climate Change Canada, Ottawa, Canada, pp. 735–754.

Booher, L.E., and Janke, B. 1997. Air emissions from petroleum hydrocarbon fires during controlled burning, *Am. Indus. Hyg. Assoc. J.*, 58(5), 359–365.

Brandvik, P.J., Daling, P.S., Faksness, L.-G., Fritt-Rasmussen, J., Daae, R.L., and Leirvik, F. 2010a. *Experimental Oil Release in Broken Ice – A Large-Scale Field Verification of Results from Laboratory Studies of Oil Weathering and Ignitability of Weathered Oils*, Joint Industry Project, JIP-26, SINTEF, Trondheim, NO.

Brandvik, P.J., Fritt-Rasmussen, J., Danilof, R., and Leirvik F. 2010b. Using a small scale laboratory burning cell to measure ignitability for in-situ burning of oil spills as a function of weathering, *AMOP*, 2, 755–786.

Brandvik, P.J., Fritt-Rasmussen, J., Daniloff, R., Leirvik, F., and Resby, J.L. 2010c. *Establishing, Testing and Verification of a Laboratory Burning Cell to Measure Ignitability for In-situ Burning of Oil Spills*, Joint Industry Project, JIP-20, SINTEF, Trondheim, NO.

Breitenbeck, G.A. 2001. *Devices to Support In-Situ Burning of Oil on Water*, LSU Report 01-003, Baton Rouge, LA.

Buist, I. 2010. *Field Testing of the USN Oil Herding Agent on Heidrun Crude in Loose Drift Ice*, JIP-6, SINTEF, Trondheim, NO.

Buist, I., McCourt, J., Morrison, J., Schmidt, B., DeVitis, D., Nolan, K. et al. 2001. Fire boom testing at OHMSETT in 2000, *Twenty-Fourth Arctic and Marine Oilspill Program Technical Seminar Proceedings*, Environment Canada, Ottawa, Canada, pp. 707–728.

Buist, I., Potter, S., Belore, R., Guarino, A., Meyer, P., and Mullin, J. 2010a. Employing chemical herders to improve marine oil spill response operations, *AMOP*, 2, 1109–1133.

Buist, I., Potter, S., Nedwed, T., and Mullin, J. 2007. Field research on using oil herding surfactants to thicken oil slicks in pack ice for in-situ burning, *AMOP*, Environment Canada, Edmonton, AB, pp. 403–418.

Buist, I., Potter, S., Nedwed, T., and Mullin, J. 2008. Herding agents thicken oil spills in drift Ice to facilitate in-situ burning: A new trick for an old dog, *International Oil Spill Conference Proceedings*, American Petroleum Institute, Savannah, GA, pp. 673–682.

Buist, I., Potter, S., Nedwed, T., and Mullin, J. 2011. Herding surfactants to contract and thicken oil spills in pack ice for in situ burning, *Cold Reg. Sci. Technol.*, 67, 3–23.

Buist, I., Potter, S., and Sørstrøm, S.E. 2010b. Barents sea field test of herder to thicken oil for in situ burning, *AMOP*, 2, 725–742.

Buist, I., Potter, S., Zabilansky, S.L., Meyer, P., and Mullin, J. 2006. Mid-scale test tank research on using oil herding surfactants to thicken oil slicks in pack ice: An update, *Proceedings of the Arctic and Marine Oilspill Program Technical Seminar No. 29*, Vol. 2, Environment Canada, Ottawa, Canada, pp. 691–709.

Buist, I.A., Cooper, D., Trudel, K., Zabilansky, L., and Fritt-Rasmussen, J. 2017. Ongoing research on herding agents for in situ burning in arctic waters: Laboratory and test tank studies on windows-of-opportunity, *IOSC*, American Petroleum Institute, Long Beach, CA, pp. 2915–2934.

Buist, I.A., Glover, N., McKenzie, B., and Ranger, R. 1995. In-situ burning of Alaska north slope emulsions, *Proceedings of the 1995 International Oil Spill Conference*, American Petroleum Institute, Washington, DC, pp. 139–148.

Buist, I.A., Pistruzak, W.M., Potter, S.G., Vanderkooy, N., and McAllister, I.R. 1983. The development and testing of a fireproof boom, *Proceedings of the Sixth Arctic and Marine Oilspill Program (AMOP) Technical Seminar*, Environment Canada, Ottawa, Canada, pp. 70–78.

Buist, I.A., Potter, S.G., and Lane, P. 2016. *Historical Review and State of the Art for Oil Slick Ignition for ISB*, Joint Industry Project Report, Washington, DC.

Buist, I.A., Potter, S.G., Trudel, S.K., Walker, A.H., Scholz, D.K., Brandvik, P.J., Fritt Rasmussen, J., Allen, A.A., and Smith, P. 2013. *In-Situ Burning in Ice-Affected Waters: A Technology Summary and Lessons from Key Experiments*, Joint Industry Project Report, London, UK.

Buist, I.A., Ross, S.L., Trudel, B.K., Taylor, E., Campbell, T.G., Westphal, P.A., Meyers, M.R., Ronza, G.S., Allen, A.A., and Nordvik, A.B. 1994. *The Science, Technology and Effects of Controlled Burning of Oil Spills at Sea*, MSRC Technical Report Number 94–013, Washington, DC.

Bullock, R.J., Aggarwal, S., Perkins, R.A., and Schnabel, W. 2017. Scale-up considerations for surface collecting agent assisted in-situ burn crude oil spill response experiments in the arctic: Laboratory to field-scale Investigations, *J. Environ. Manag.*, 190, 266–273.

Bullock, R.J., Perkins, R.A., Aggarwal, S., Schnabel, W. and Sartz, P. 2016. Arctic in-situ burn experiments: Laboratory, meso-, and field-scale observations and scale-up considerations, *Proceedings of the Thirty-Ninth AMOP Technical Seminar, Environment and Climate Change Canada*, Ottawa, Canada, pp. 784–794.

Bullock, R.J., Perkins, R.A., Aggarwal, S., Schnabel, W., and Sartz, P. 2017. Environmental partitioning of herding agents used during an in-situ burning field study in Alaska, *IOSC*, American Petroleum Institute, Long Beach, CA, pp. 2935–2954.

Chatris, J.M., Quintella, J., Folch, J., Planas, E., Arnalddos, J., and Casal, J. 2001. Experimental study of burning rate in hydrocarbon pool fires, *Combust. Flame*, 126, 1373–1383.

Cohen, A., Gagnon, M.M., and Nugegoda, D. 2006. Oil spill remediation techniques can have different impacts on mixed function oxygenase enzyme activities in fish, *Bull. Environ. Contam. Toxicol.*, 76(5), 855–862.

Cooper, D., Buist, I., Belore, R., Nedwed, T., and Tidwell, A. 2012. One-step offshore collection and removal: Combining an oleophilic skimmer and floating burner, *Proceedings of the Thirty-fifth Arctic and Marine Oilspill Program (AMOP) Technical Seminar on Environmental Contamination and Response*, Environment Canada, Ottawa, Canada, pp. 537–550.

Cooper, D., Buist, I., Potter, S., Daling, P., Singsaas, I., and Lewis, A. 2017. Experiments at sea with herders and in situ burning (HISB), *IOSC*, American Petroleum Institute, Long Beach, CA, pp. 2184–2203.

Coupal, B. 1972. *Use of Peat Moss in Controlled Combustion Techniques*, Environment Canada, EPS Report 4-EE-72-1, Environment Canada, Ottawa, Canada.

Crawley, F.K. 1982. The effects of the ignition of a major fuel spillage, *Institution of Chemical Engineers Symposium Series No. 7*, Institute of Chemical Engineers, London, UK, pp. 125–145.

Cunneff, S., Devitis, D., and Nash, J. 2000. Test and evaluation of six fire resistant booms at OHMSETT, *Spill Sci. Technol. Bull.*, 6(5), 353–355.

Dagaut, P., and Cathonnet, M. 2006. The ignition, oxidation and combustion of kerosene: A review of experimental and kinetic modeling, *Prog. Energ. Combust.*, 32(1), 48–92.

Dagaut, P., El Bakali, A., and Ristori, A. 2006. The combustion of kerosene: Experimental results and kinetic modelling using 1- to 3-component surrogate model fuels, *Fuel*, 85(7–8), 944–956.

Dahm, C.N., Candelaria-Ley, R.I., Reale, C.S., Reale, J.K., and Van Horn, D.J. 2015. Extreme water quality degradation following a catastrophic forest fire, *Freshwater Biol.*, 60(12), 2584–2599.

Dave, D., and Ghaly, A.E. 2011. Remediation technologies for marine oil spills: A critical review and comparative analysis, *Am. J. Environ. Sci.*, 7(5), 424–440.

Daykin, M.M., Kennedy, P.A., and Tang, A. 1995. *Aquatic Toxicity from In-situ Oil Burning – Newfoundland Offshore Burn Experiment (NOBE)*, Environment Report, Environment Canada, Ottawa, Canada.

Eddings, E.G., Yan, S., Ciro, W., and Sarofim, A.F. 2005. Formation of a surrogate for the simulation of jet fuel pool fires, *Combust. Sci. Tech.*, 177(4), 715–739.

Elam, S.K., Tokura, I., Saito, K., and Altenkirck, R.A. 1989. Thermal conductivity of crude oils, *Exp. Therm. Fluid Sci.*, 2(1), 1–6.

Elastec. 2017. *Elastec Safe Start Igniter for Controlled Burns*, Elastec, Carmi, IL.

Energetex Engineering. 1981. *Arctic Field Trials of the DREV/AMOP Incendiary Devices*, Environment Canada, Environmental Protection Service, Manuscript Report EE-17, Ottawa, Canada.

Energetex Engineering. 1982. *Environmental Testing of Dome Air-Deployable Ignitor*, Dome Petroleum, Calgary, AB.

Environment Canada. 1993. *NOBE: Newfoundland Offshore Burn Experiment*, Environment Canada, Ottawa, Canada.

Environment Canada. 1997. *Compilation of Physical and Emissions Data*, Newfoundland Offshore Burn Experiment (NOBE) Report, Environment Canada, Ottawa, Canada.

Evans, D.D., Mulholland, G.W., Lawson, J.R., Tennyson, E.J., Tebeau, P.A., Fingas, M.F., and Gould, J.R. 1991. Burning of oil spills, *IOSC*, American Petroleum Institute, San Diego, CA, pp. 677–680.

Faksness, L.-G., Hansen, B.H., Altin, D., and Brandvik, P.J. 2012. Chemical composition and acute toxicity in the water after in-situ burning: A laboratory experiment, *Mar. Pollut. Bull.*, 64(1), 49–55.

Farahani, H.F., Shi, X., Simeoni, A., and Rangwala, A.S. 2015. A study on burning of crude oil in ice cavities, *Proc. Combust. Inst.*, 35(3), 2699–2708.

Fay, J.A. 2006. Model of large pool fires, *J. Haz. Mat.*, 136(2), 219–232.

Fingas, M. 2011. In-situ burning, Chapter 23, in *Oil Spill Science and Technology*, M. Fingas (Ed.), Gulf Publishing Company, New York, pp. 737–903.

Fingas, M. 2016a. *In-situ Burning of Spilled Oil: Good Practice Guidelines for Incident Management and Emergency Response Personnel*, IPIECA, International Association of Oil & Gas Producers, London, UK, 52 p.

Fingas, M. 2016b. The fate of polycyclic aromatic hydrocarbons resulting from in-situ oil burns, *Proceedings of the Thirty-Ninth AMOP Technical Seminar*, Environment and Climate Change Canada, Ottawa, Canada, pp. 795–819.

Fingas, M. 2017. In-situ burning: An update, Chapter 10, in *Oil Spill Science and Technology*, 2nd ed., M. Fingas (Ed.), Gulf Publishing Company, Cambridge, MA, pp. 483–576.

Fingas, M., and Fieldhouse, B. 2009. Studies on crude oil and petroleum product emulsions: Water resolution and rheology, *Colloids Surf. A: Physicochem. Eng. Asp.*, 333, 67–81.

Fingas, M., and Fieldhouse, B. 2011. Studies on water-in-oil products from crude oils and petroleum products, *Mar. Poll. Bull.*, 64, 272–283.

Fingas, M.F. 2002. In-situ burning of orimulsion: Small scale burns, *Proceedings of the Twenty-Fifth AMOP Technical Seminar, Environment Canada*, Ottawa, Canada, pp. 809–817.

Fingas, M.F. 2010. Soot production from in-situ oil fires: Review of the literature and calculation of values from experimental spills, *Proceedings of the Thirty-Third AMOP Technical. Seminar on Environmental Contamination and Response, Environment Canada,* Ottawa, Canada, pp. 1017–1054.

Fingas, M.F. 2014. Calculating the amount of oil burned in a boom: Catenary calculations, *Proceedings of the Thirty-Seventh AMOP Technical Seminar on Environmental Contamination and Response,* Environment Canada, Ottawa, Canada, pp. 737–753.

Fingas, M.F., Ackerman, F., Lambert, P., Li, K., Wang, Z., Bissonnette, M.C. et al. 1995a. The Newfoundland Offshore Burn Experiment: Further results of emissions measurement, *Proceedings of the Eighteenth Arctic and Marine Oil Spill Program Technical Seminar,* Environment Canada, Ottawa, Canada, pp. 915–995.

Fingas, M.F., Ackerman, F., Lambert, P., Li, K., Wang, Z., Nelson, R. et al. 1996a. Emissions from mesoscale in-situ oil (diesel) fires: The mobile 1994 experiments, *Proceedings of the 19th Arctic and Marine Oil Spill Program (AMOP) Technical Seminar,* Environment Canada, Ottawa, Canada, pp. 907–978.

Fingas, M.F., Ackerman, F., Li, K., Lambert, P., Wang, Z., Bissonnette, M.C. et al. 1994a. The newfoundland offshore burn experiment - NOBE - Preliminary results of emissions measurement, *Proceedings of the Seventeenth Arctic and Marine Oil Spill Program (AMOP) Technical. Seminar,* Vancouver, Canada, pp. 1099–1164.

Fingas, M.F., and Fieldhouse, B. 1994. Studies of water-in-oil emulsions and techniques to measure emulsion treating agents, *Proceedings of the Seventeenth Arctic and. Marine Oil Spill Program Technical Seminar, Environment Canada,* Ottawa, Canada, pp. 213–244.

Fingas, M.F., Fieldhouse, B., Brown, C.E., and Gamble, L. 2004. In-situ burning of heavy oils and orimulsion: Mid-scale burns, *Proceedings of the Twenty-Seventh AMOP Technical Seminar, Environment Canada,* Ottawa, Canada, pp. 207–233.

Fingas, M.F., Goldthorp, M., Lambert, P., Wang, Z., Li, K., Ackerman, F. et al. 2001a. *Monitoring Emissions from the In-Situ Burning of Oil Spills on Water,* Environment Canada Manuscript Report Number EE-167, Ottawa, Canada.

Fingas, M.F., Halley, G., Ackerman, F., Nelson, R., Bissonnette, M.C., Laroche, N. et al. 1995b. The Newfoundland offshore burn experiment, *IOSC,* pp. 123–132.

Fingas, M.F., Halley, G., Ackerman, F., Vanderkooy, N., Nelson, R., Bissonnette, M.C. et al. 1994b. The Newfoundland Offshore Burn Experiment (NOBE): Experimental design and overview, *Proceedings of the 17th Arctic and Marine Oilspill Program Technical Seminar, Environment Canada,* Ottawa, Canada, pp. 1053–1061.

Fingas, M.F., Lambert, P., Ackerman, F., Fieldhouse, B., Nelson, R., Goldthorp, M. et al. 1998. Particulate and carbon dioxide emissions from diesel fires: The Mobile 1997 experiments, *Proceedings of the Twenty-First Arctic and Marine Oilspill Program (AMOP) Technical Seminar,* Environment Canada, Ottawa, Canada, pp. 569–597.

Fingas, M.F., Lambert, P., Goldthorp, M., and Gamble, L. 2003. In-situ burning of orimulsion: mid-scale burns, *Proceedings of the Twenty-Sixth Arctic and Marine Oil Spill Program Technical Seminar,* Environment Canada, Ottawa, Canada, pp. 649–660.

Fingas, M.F., Lambert, P., Wang, Z., Li, K., Ackerman, F., Goldthorp, M. et al. 2001b. Studies of emissions from oil fires, *Proceedings of the Arctic and Marine Oil Spill Program Technical Seminar No. 24a,* Environment Canada, Ottawa, Canada, pp. 767–823.

Fingas, M.F., Li, K., Ackerman, F., Wang, Z., Lambert, P., Gamble, L. et al. 1996b. Soot production from in-situ oil fires: Review of the literature, measurement and estimation techniques and calculation of values from experimental spills, *Proceedings of the Arctic and Marine Oilspill Program Technical Seminar No. 19b,* Environment Canada, Ottawa, Canada, pp. 999–1032.

Fingas, M.F., and MacKay, R. 2003. In-situ burning of orimulsion: Small-scale burns, *IOSC,* Environment Canada, Victoria, BC, pp. 103–107.

Fingas, M.F., and Punt, M. 2000. *In-Situ Burning: A Cleanup Technique for Oil Spills on Water*, Environment Canada Special Publication, Ottawa, Canada.

Fingas, M.F., Wang, Z., Fieldhouse, B., Brown, C.E., Yang, C., Landriault, M., and Cooper, D. 2005. In-situ Burning of Heavy Oils and Orimulsion: Analysis of Soot and Residue, *Proceedings of the Arctic and Marine Oil Spill Program Technical Seminar No. 28a*, Environment Canada, Ottawa, Canada, pp. 333–348.

Frank, B., Schlögl, R., and Su, D.S. 2013. Diesel soot toxification, *Environ. Sci. Technol.*, 47(7), 3026–3027.

Frish, M., Gauthier, V., Frank, J., and Nebolsine, P. 1989. *Laser Ignition of Oil Spills: Telescope Assembly and Testing*, Environment Canada Manuscript Report EE-113, Ottawa, Canada.

Frish, M., Nebolsine, P., DeFaccio, M., Scholaert, H., Kung, W., and Wong, J. 1986. Laser ignition of arctic oil spills: Engineering design, *Proceedings of the Ninth Arctic Marine Oilspill Program Technical Seminar*, Environment Canada, Ottawa, Canada, pp. 203–221.

Fritt-Rasmussen, J., Ascanius, B.E., Brandvik, P.J., Villumsen, A., and Stenby, E.H. 2013. Composition of in-situ burn residue as a function of weathering conditions, *Mar. Pollut. Bull.*, 67, 75–86.

Fritt-Rasmussen, J., Gustavson, K., Wegeberg, S., Møller, E.F., Dyrmose, R., Nørregaard, R.D. et al. 2017. Ongoing research on herding agents for in situ burning in arctic waters: Studies on fate and effects, *IOSC*, American Petroleum Institute, Long Beach, CA, pp. 2976–2995.

Fritt-Rasmussen, J., Linnebjerg, J.F., Sørensen, M.X., Brogaard, N.L., Rigét, F.F., Kristensen, P., Jomaas, G., Boertmann, D.M., Wegeberg, S., and Gustavson, K. 2016. Effects of oil and oil burn residues on seabird feathers, *Mar. Poll. Bull.*, 109(1), 446–452.

Fu, Y., Gao, Z., Ji, J., Li, K., and Zhang, Y. 2017. Experimental study of flame spread over diesel and diesel-wetted sand beds, *Fuel*, 204, 54–62.

Glassman, I., and Hansel, J.G. 1968. Some thoughts and experiments on liquid fuel spreading, steady burning and ignitability in quiescent atmospheres, *Fire Res. Abst. Rev.*, 10, 217–226.

Goldthorp, M., Lambert, P., Fingas, M.F., Ackerman, F., Schuetz, S., Turpin, R., and Campagna, P. 1999. Duplicating conditions for field testing of carbon dioxide: A modeller's dream becomes a technician's nightmare, *AMOP*, Environment Canada, Calgary, AB, pp. 13–22.

Goodman, B.T., Davidson, R.A., Sievert, E.S., Wood, L., and Homer, V.H. 2014. Initiating in-situ burning of difficult-to-ignite oil spills via and aircraft-deployable igniter system, *IOSC*, American Petroleum Institute, Savannah, GA, pp. 1821–1833.

Gruenberger, T.M., Moghiman, M., Bowen, P.J., and Syred, N. 2002. Dynamics of soot formation by turbulent combustion and thermal decomposition of natural gas, *Combust. Sci. Tech.*, 67, 67–86.

Guénette, C.C., and Thornborough, J. 1997. An assessment of two off-shore igniter concepts, *Proceedings of the Arctic and Marine Oil Spill Program Technical Seminar No. 20b*, Environment Canada, Ottawa, Canada, pp. 795–808.

Gullett, B.K., Aurell, J., Holder, A., Mitchell, W., Greenwell, D., Hays, M. et al. 2017. Characterization of emissions and residues from simulations of the Deepwater Horizon surface oil burns, *Mar. Poll. Bull.*, 117, 392–405.

Ho, C.-C., Chan, C.-C., Chio, C.-P., Lai, Y.-C., Chang-Chien, G.-P., Chow, J.C., Watson, J.G., Chen, L.-W.A., Chen, P.-C., and Wu, C.-F. 2016. Source apportionment of mass concentration and inhalation risk with long-term ambient PCDD/Fs measurements in an urban area, *J. Haz. Mat.*, 317, 180–187.

Jaligama, S., Chen, Z., Saravia, J., Yadav, N., Lomnicki, S.M., Dugas, T.R., and Cormier, S.A. 2015. Exposure to Deepwater Horizon crude oil burnoff particulate matter induces pulmonary inflammation and alters adaptive immune response, *Environ. Sci. Technol.*, 49(14), 8769–8776.

Jézéquel, R., Simon, R., and Pirot, V. 2014. Development of a burning bench dedicated to in-situ burning study: Assessment of oil nature and weathering effect, *Proceedings of the Thirty-Seven AMOP Technical Seminar on Environmental Contamination and Response*, Environment Canada, Ottawa, Canada, pp. 556–567.

Johnson, M.R., Devillers, R.W., and Thomson, K.A. 2011. Quantitative field measurement of soot emission from a large gas flare using sky-LOSA, *Environ. Sci. Technol.*, 45(1), 345–350.

Kadota, T., and Yamasaki, H. 2002. Recent advances in the combustion of water fuel emulsion, *Prog. Energ. Combust.*, 28(5), 385–404.

Kasische, E.S., and Penner, J.E. 2004. Improving global estimates of atmospheric emissions from biomass burning, *J. Geophys. Res.*, 109, D14SO1.

Lambert, P., Ackerman, F., Fingas, M., Goldthorp, M., Fieldhouse, B., Nelson, M. et al. 1998. Instrumentation and techniques for monitoring the air emissions during in-situ oil/fuel burning operations, *AMOP*, Environment Canada, Edmonton, AB, pp. 529–567.

Lane, P. 2017. Development of an applicator and a helicopter-borne ignition device launching system for using oil herders in remote and Arctic locations, *IOSC*, American Petroleum Institute, Long Beach, CA, pp. 2017052.

Lavers, L. 1997. Newfoundland Department of Forestry, personal communication.

Law, C.K. 2005. Comprehensive description of chemistry in combustion modeling, *Combust. Sci. Tech.*, 177, 845–870.

Lemieux, P.M., Lutes, C.C., and Santoianni, D.A. 2004. Emissions of organic air toxics from open burning: A comprehensive review, *Prog. Energ. Comb.*, 30(1), 1–32.

Li, K., Caron, T., Landriault, M., Paré, J.R.J., and Fingas, M. 1992. Measurement of volatiles, semi-volatiles and heavy metals in an oil burn test, *Proceedings of the Fifteenth Arctic and Marine Oil Spill Technical Seminar*, Environment Canada, Ottawa, Canada, pp. 561–579.

Li, M., Lu, S., Chen, R., Guo, J., and Wang, C. 2016. Experimental investigation on flame spread over diesel fuel near sea level and at high altitude, *Fuel*, 184, 665–671.

Li, M., Lu, S., Guo, J., Chen, R., and Tsui, K.-L. 2015. Initial fuel temperature effects on flame spread over aviation kerosene in low- and high-altitude environments, *Fire Technol.*, 51(3), 707–721.

Mabile, N. 2010. *Fire Boom Performance Evaluation: Controlled Burning During the Deepwater Horizon Spill Operational Period*, Internal Report, BP, Houston, TX.

Mabile, N. 2012a. Controlled in-situ burning: Transition from alternative technology to conventional spill response option, *AMOP*, Environment Canada, Vancouver, BC, pp. 584–605.

Mabile, N.J. 2012b. Considerations for the application of controlled in-situ burning, *2012 Society of Petroleum Engineers—SPE/APPEA International Conference on Health, Safety and Environment in Oil and Gas Exploration and Production*, Perth, Australia, Vol. 3, pp. 2556–2575.

Maki, T., and Miura, K. 1997. A simulation model for the pyrolysis of orimulsion, *Energy Fuels*, 11(4), 819–823.

Marine Research Associates. 1998. *Technology Assessment and Concept Evaluation for Alternative Approaches to In-Situ Burning of Oil Spills in the Marine Environment*, for U.S. MMS, Herndon, VA.

Maybourn, R. 1971. The work of the IP working group on the burning of oil, *J. Inst. Petrol.*, 57, 12–20.

McCourt, J., Buist, I., and Buffington, S. 2000. Results of laboratory tests on the potential for using in-situ burning on 17 crude oils, *Proceedings of the 23rd Arctic Marine Oil Spill Program (AMOP) Technical Seminar*, Environment Canada, Ottawa, Canada, Vol. 2, pp. 917–922.

McCourt, J., Buist, I., and Buffington, S. 2005. Results of laboratory tests on the potential for using in- situ burning on 17 crude oils, *IOSC*, American Petroleum Institute, Miami Beach, FL, pp. 7229–7232.

McCourt, J., Buist, I., Mullin, J., Pratte, B., and Jamieson, W. 1998. Continued development of a test for fire booms in waves and flames, in *Proceedings of the Twenty-First Arctic and. Marine Oilspill Program (AMOP) Technical Seminar*, Environment Canada, Ottawa, Canada, pp. 505–528.

McCourt, J., Buist, I., Schmidt, W., Devitis, D., Urban, B., and Mullin, J. 1999. OHMSETTs propane-fueled test system for fire-resistant booms, *Proceedings of the Twenty-Second AMOP Technical Seminar*, Environment Canada, Ottawa, Canada, pp. 439–445.

McEnally, C.S., and Pfefferle, L.D. 2009. Sooting tendencies of nonvolatile aromatic hydrocarbons, *Proc. Combust. Inst.*, 32(1), 673–679.

McEnally, C.S., Pfefferle, L.D., Atakan, B., and Kohse-Höinghaus, K. 2006. Studies of aromatic hydrocarbon formation mechanisms in flames: Progress towards closing the fuel gap, *Prog. Energ. Combust.*, 32(3), 247–294.

McGrattan, K.B. 1999. Smoke plume trajectory modeling, *Proceedings of In-situ Burning of Oil Spills*, U.S. Minerals Management Service, Herndon, VA.

McKenzie, B. 1994. Report of the operational implications working panel, in *In Situ Burning Oil Spill Workshop Proceedings*, N.H. Jason (Ed.), NIST, Gaithersburg, MD.

McLeod, W.R., and McLeod, D.L. 1972. Measures to combat offshore arctic oil spills, *Preprints of the 1972 Offshore Technology Conference*, Houston, TX, pp. 141–150.

Meikle, K.M. 1983. An effective low-cost fireproof boom, IOSC, pp. 39–42.

Middlebrook, A.M., Murphy, D.M., Ahmadov, R., Atlas, E.L., Bahreini, R., Blake, D.R. et al. 2012. Air quality implications of the Deepwater Horizon oil spill, *P. Natl. Acad. Sci. USA*, 109(50), 20280–20285.

Midlinx. 2013. *Controlled In-Situ Burning Operations Monitoring Handbook*, Midlinx Consulting Inc. for British Petroleum, Houston, TX.

Mitchell, J.B.A. 1990. Smoke reduction from burning crude oil using ferrocene and its derivatives, *Spill Tech. News*, pp. 11–13.

Mitchell, J.B.A. 1991. Smoke reduction from burning crude oil using ferrocene and its derivatives, *Combust. Flame*, 86, 179–184.

Mitchell, J.B.A. 1992. Smoke reduction from pool fires using ferrocene and derivatives, *Proceedings of the 15th Arctic and Marine Oil Spill Program (AMOP) Technical Seminar, Environment Canada*, Ottawa, Canada, pp. 681–687.

Mitchell, J.B.A. 2014. Soot emission reductions during in situ burning, *Proceedings of the Thirty-Seventh AMOP Technical Seminar on Environmental Contamination and Response*, Environment Canada, Ottawa, Canada, pp. 568–574.

Moore, R.H., Raatikainen, T., Langridge, J.M., Bahreini, R., Brock, C.A., Holloway, J.S. et al. 2012. CCN spectra, hygroscopicity, and droplet activation kinetics of secondary organic aerosol resulting from the 2010 Deepwater Horizon oil spill, *Environ. Sci. Technol.*, 46(6), 3093–3100.

Morandini, F., Simeoni, A., Santoni, P.A., and Balbi, J.H. 2005. A model for the spread of fire across a fuel bed incorporating the effects of wind and slope, *Combust. Sci. Tech.*, 177(7), 1381–1418.

Muñoz, M., Arnaldos, J., Casal, J., and Planas, E. 2004. Analysis of the geometric and radiative characteristics of hydrocarbon pool fires, *Combust. Flame*, 139(3), 263–277.

Muñoz, M., Planas, E., Ferrero, F., and Casal, J. 2007. Predicting the emissive power of hydrocarbon pool fires, *J. Haz. Mat.*, 144(3), 725–729.

Murphy, J.J., and Shaddix, C.R. 2006. Soot property measurements in a two-meter diameter JP-8 pool fire, *Combust. Sci. Tech.*, 178(5), 865–894.

Myers-Pigg, A.N., Griffin, R.J., Louchouarn, P., Norwood, M.J., Sterne, A., and Cevik, B.K. 2016. Signatures of biomass burning aerosols in the plume of a saltmarsh wildfire in South Texas, *Environ. Sci. Technol.*, 50(17), 9308–9314.

Nakakuki, A. 2002. Heat transfer in pool fires at a certain small lip height, *Combust. Flame*, 131, 259–272.

Nordvik, A.B., Simmons, J.L., and Hudon, T.J. 1995. At-sea testing of fire resistant oil containment boom designs, *Proceedings of the Second International Oil Spill Research and Development Forum*, IMO, London, UK, pp. 479–488.

Ontario Ministry of Natural Resources (OMNR). 1990. *Specialized Fire Equipment Manual, Ignition Devices—Helitorch*, Ontario Ministry of Natural Resources, Toronto, Canada.

Panetta, P.D., Bryne, R., and Du, H. 2017. The direct quantitative measurement of in-situ burn (ISB) rate and efficiency, *IOSC*, American Petroleum Institute, Long Beach, CA, pp. 1006–1019.

Pereira Netto, A.D., Cunha, I.F., and Krauss, T.M. 2004. Persistence of polycyclic aromatic hydrocarbons in the soil of a burned area for agricultural purposes in brazil, *Bull. Environ. Contam. Toxicol.*, 73(6), 1072–1077.

Perring, A.E., Schwarz, J.P., Spackman, J.R., Bahreini, R., De Gouw, J.A., Gao, R.S., and Holloway, J.S. 2011. Characteristics of black carbon aerosol from a surface oil burn during the Deepwater Horizon oil spill, *Geophys. Res. Lett.*, 38(17), L17809.

Potter, S. 2010. *Tests of Fire-Resistant Booms in Low Concentrations of Drift Ice: Field Experiments May 2009*, Joint Industry Project, JIP-27, SINTEF, Trondheim, NO.

Potter, S., Buist, I., Cooper, S., Aggarwal, S., Schnabel, W., Garron, J., Bullock, R., Perkins, R., and Lane, P. 2017. Aerial application of herding agents can enhance in-situ burning in partial ice cover, *IOSC*, American Petroleum Institute, Long Beach, CA, pp. 2955–2975.

Prendergast, D.P., and Gschwend, P.M. 2014. Assessing the performance and cost of oil spill remediation technologies, *J. Cleaner Product*, 78, 233–242.

PROSCARAC (Prairie Regional Oil Spill Containment and Recovery Advisory Committee). 1992. *Anchor Design and Deployment Review*, Edmonton, Canada.

Punt, M. 1990. The performance of a water jet barrier in a river, *Spill Tech. News*, pp. 1–6.

Ross, S.L. 2017. *World Catalog of Oil Spill Response Products*, 11th ed., SL Ross. http://www.oilspillequipment.com. Accessed September 12, 2017.

Ross, S.L. Canevari, G.P., and Consultchem. 1992. *State-of-the-Art Review: Emulsion Breaking Chemicals*, Canadian Petroleum Association Report, Ottawa, Canada.

Ross, S.L. Environmental Research Ltd and Energetex Engineering. 1986. *In-Situ Burning of Uncontained Oil Slicks*, EE-60, Environment Canada, Ottawa, Canada.

Russel, A.G., and Brunerkreef, B. 2009. A focus on particulate matter and health, *Environ. Sci. Technol.*, 43(13), 4620–4625.

Sartz, P., and Aggarwal, S. 2016. Gaseous emissions from herding agent-mediated in-situ burning for Arctic oil spills, *AMOP*, Environment Canada, Halifax, NS, pp. 735–750.

Sartz, P., and Aggarwal, S. 2017. Ambient air quality in the vicinity of a herder mediated in-situ burn field test in Alaska, *IOSC*, American Petroleum Institute, Long Beach, CA, pp. 2017149.

Sazhin, S.S. 2017. Modelling of fuel droplet heating and evaporation: Recent results and unsolved problems, *Fuel*, 196, 69–101.

Schaum, J., Cohen, M., Perry, S., Artz, R., Draxler, R., Frithsen, J.B., Heis, D., Lorber, M., and Phillips, L. 2010. Screening level assessment of risks due to dioxin emissions from burning oil from the BP Deepwater Horizon Gulf of Mexico spill, *Environ. Sci. Technol.*, 44(24), 9383–9389.

Schnitzler, E.G., Dutt, A., Charbonneau, A.M., Olfert, J.S., and Jäger, W. 2014. Soot aggregate restructuring due to coatings of secondary organic aerosol derived from aromatic precursors, *Environ. Sci. Technol.*, 48(24), 14309–14316.

Shi, X., Bellino, P.W., and Rangwala, A.S. 2015. Flame heat feedback from crude oil fires in ice cavities, *Proceedings of the Thirty-Eighth AMOP Technical Seminar*, Environment Canada, Ottawa, Canada, pp. 767–680.

Shi, X., Bellino, P.W., Simeoni, A., and Rangwala, A.S. 2016a. Experimental study of burning behavior of large-scale crude oil fires in ice cavities, *Fire Saf. J.*, 79, 91–99.

Shi, X., Ghion, N.S., Fu, Y., Sundberg, K.T., Ramos, J.P., Stephansky, S., Ross, K., Kang, F., Raghavan, V., and Rangwala, A.S. 2016b. Influence of wicking agent on in-situ burning of water-in-oil products from Alaska North slope crude, *Proceedings of the Thirty-Ninth. AMOP Technical Seminar*, Environment and Climate Change Canada, Ottawa, Canada, pp. 751–759.

Shi, X., Ranellone, R.T., Sezer, H., Lamie, N., Zabilansky, L., Stone, K., and Rangwala, A.S. 2017. Influence of ullage to cavity size ratio on in-situ burning of oil spills in ice-infested water, *Cold Reg. Sci. Technol.*, 140, 5–13.

Smooke, M.D., Long, M.B., Connelly, B.C., Colket, M.B., and Hall, R.J. 2005. Soot formation in laminar diffusion flames, *Combust. Flame*, 143(4), 613–628.

Snider, J. 1994. Research needs associated with in-situ burning: Report of the environmental and human health panel, in *In-Situ Burning Oil Spill Workshop Proceedings*, N.H. Jason (Ed.), NIST, New Orleans, LA, pp. 3–12.

Swift, W.H., Touhill, C.J., and Peterson, P.L. 1968. Oil spillage control, *Chem. Eng. Prog. Symp. Ser.*, 65, 265–272.

Tennyson, E.J. 1994. In situ burning overview, in *In Situ Burning Oil Spill Workshop Proceedings*, N.H. Jason (Ed.), NIST, Boston, MA, pp. 21–29.

Thompson, C.H., Dawson, G.W., and Goodier, J.L. 1979. *Combustion: An Oil Mitigation Tool*, Pacific Northwest Laboratory (operated by Battelle) for U.S. Department of Energy, Division of Environmental Control Technology, Washington, DC.

Topal, M.H., Wang, J., Levendis, Y.A., Carlson, J.B., and Jordan, J. 2004. PAH and other emissions from burning of JP-8 and diesel fuels in diffusion flames, *Fuel*, 83(17–18), 2357–2368.

Twardawa, P., and Couture, G. 1983. *Incendiary Devices for the in Situ Combustion of Crude Oil Slicks*, National Defence Research Establishment, Valcartier, Canada.

Twardus, E.M. 1980. *A Study to Evaluate the Combustibility and Other Physical and Chemical Properties of Aged Oils and Emulsions*, Environment Manuscript Report EE-5, Ottawa, Canada.

Twardus, E.M., and Bruzustowski, T.A. 1981. The burning of crude oil spilled on water, *Arch. Combust.*, 1, 49–58.

USCG. 2003. *In-situ Burn Operations Manual: Oil Spill Response Offshore*, United States Coast Guard Report CG-D-06-03, Washington, D.C.

van Gelderen, L., Alva, U.R., Mindykowski, P., and Jomaas, G. 2017a. Thermal properties and burning efficiencies of crude oils and refined fuel oil, *IOSC*, Environment Canada, Calgary, AB, pp. 985–1005.

van Gelderen, L., Brogaard, N.L., Sørensen, M.X., Fritt-Rasmussen, J., Rangwala, A.S., and Jonaas, G. 2015a. Importance of the slick thickness for effective in-situ burning of crude oil, *Fire Saf. J.*, 78(2336), 1–9.

van Gelderen, L., and Jomaas, G. 2017. The parameters controlling the burning efficiency of in-situ burning of crude oil on water, *Proceedings of the 40th AMOP Technical Seminar on Environmental Contamination and Response*, Calgary, Canada, pp. 817–832.

van Gelderen, L., Jomaas, G., Fritt-Rasmussen, J., and Rangwala, A.S. 2017b. In-situ burning of crude oil on water: A study on the fire dynamics and fire chemistry in an arctic context, Technical University of Denmark, Department of Civil Engineering. (DTU Civil Engineering Report; No. R-270).

van Gelderen, L., Malmquist, L.M.V., and Jomaas, G. 2015b. Crude oil burning mechanisms: A conceptual model review, *Proceedings of the Thirty-Eighth Arctic and Marine Oil Spill Program (AMOP) Technical Seminar*, Environment Canada, Vancouver, Canada, pp. 385–400.

van Gelderen, L., Malmquist, L.M.V., and Jomaas, G. 2017c. Vaporization order and burning efficiency of crude oils during in-situ burning on water, *Fuel*, 191, 528–537.

Vodacek, A., Kremens, R.L., Fordham, A.J., Vangorden, S.C., Luisi, D., and Schott, J.R. 2002. Remote optical detection of biomass burning using a potassium emission signature, *Int. J. Remote Sens.*, 23, 2721–2726.

Walker, A.H., Michel, J., Canevari, G., Kucklick, J., Scholz, D., Benson, C.A., Overton, E., and Shane, B. 1993. *Chemical Oil Spill Treating Agents: Herding Agents, Emulsion Treating Agents, Solidifiers, Elasticity Modifiers, Shoreline Cleaning Agents, Shoreline Pre-treatment Agents and Oxidation Agents*, MSRC Technical Report Series Report, Herndon, VA, 93–015.

Walton, W.D., Twilley, W.H., Bryner, N.P., DeLauter, L., Hiltabrand, R.H., and Mullin, J.V. 1999. Second phase evaluation of a protocol for testing fire-resistant oil-spill containment boom, *Proceedings of the Arctic and Marine Oilspill Program Technical Seminar No. 22b*, Environment Canada, Ottawa, Canada, pp. 447–466.

Walton, W.D., Twilley, W.H., Hiltabrand, R.H., and Mullin, J.V. 1998. Evaluating a protocol for testing fire-resistant oil-spill containment boom, *Proceedings of the Twenty-First Arctic and Marine Oil Spill Program Technical Seminar*, Environment Canada, Ottawa, Canada, pp. 651–671.

Wang, Z., Fingas, M.F., Landraiult, M., Sigouin, L., and Lambert, P. 1999. Distribution of PAHs in burn residue and soot samples and differentiation of pyrogenic and petrogenic PAHs from PAHs: The 1994 and 1997 Mobile Burn Study, in *Diesel Fuels*, C. Song, C. Hsu, and I. Mochida (Eds.), pp. 237–253.

Wilson, J.M., Baeza-Romero, M.T., Jones, J.M., Pourkashanian, M., Williams, A., Lea-Langton, A.R., Ross, A.B., and Bartle, K.D. 2013. Soot formation from the combustion of biomass pyrolysis products and a hydrocarbon fuel, n-decane: An aerosol time of flight mass spectrometer (ATOFMS) study, *Energy Fuels*, 27(3), 1668–1678.

Xiao, J., Austin, E., and Roberts, W.L. 2005. Relative polycyclic aromatic hydrocarbon concentrations in unsteady counterflow diffusion flames, *Combust. Sci. Tech.*, 177(4), 691–713.

Xu, G., Ikegami, M., Honma, S., Ikeda, K., Ma, X., and Nagaishi, H. 2003a. Burning droplets of heavy oil residual blended with diesel light oil: Distinction of burning phases, *Combut. Sci. Tech.*, 175, 1–26.

Xu, G., Ikegami, M., Honma, S., Sasaki, M., Ikeda, K., Nagaishi, H., and Takeshita, Y. 2003b. Combustion characteristics of droplets composed of light cycle oil and diesel light oil in a hot-air chamber, *Fuel*, 82, 319–330.

Zengel, S., Weaver, J., Wilder, S.L., Dauzat, J., Sanfilippo, C., Miles, M.S. et al. 2018. Vegetation recovery in an oil-impacted and burned phragmites australis tidal freshwater marsh, *Sci. Total Environ.*, 612, 231–237.

Zervas, E. 2005. Formation of oxygenated compounds from isooctane/toluene flames, *Energy Fuels*, 19(5), 1865–1872.

Zheng, L., Ma, X., Wang, Z., and Wang, J. 2014. An optical study on liquid-phase penetration, flame lift-off location and soot volume fraction distribution of gasoline-diesel blends in a constant volume vessel, *Fuel*, 139, 365–373.

2 Burning on Land and Wetlands

Jacqueline Michel

CONTENTS

2.1   INTRODUCTION

Burning oil on land and wetlands is an older and more frequently used technique than in-situ burning on water. Land is defined as habitats with soils that are dry most of the year and support terrestrial vegetation, such as forests, grasslands and farmed fields, and unvegetated habitats such as ditches and developed areas. Wetlands are defined as habitats where water covers the soil, or is present at or near the surface of the soil, either year-round or seasonally such that the habitat supports vegetation that is adapted to growing in saturated soil conditions. Wetlands are transitional areas between aquatic and terrestrial habitats. Wetlands include swamps (dominated by trees and shrubs), marshes (dominated by grassy or herbaceous vegetation), tundra and bogs (where the soils are largely composed of decayed vegetable matter). Burning is defined as the intentional, in-situ ignition of spilled oil on land or wetlands as part of an oil spill response plan.

In this chapter, information is summarised based on the past use of burning spilled oil on land and wetlands to provide the reader with a sense of typical conditions under which it has been used historically. Guidelines are given for when burning should be considered. There is a section on planning and implementing a burn during a spill emergency. Case studies were selected based on a range of habitats and conditions, as well as availability of time-series data and photographs. The case studies and review of the literature are used to summarise the effects of burning on land and wetlands. This section also includes lessons learned from prescribed burning, extracted from the extensive body of work on fire ecology. The summary brings together the lessons learned from the literature and case studies.

2.2   PAST USE OF BURNING OIL ON LAND AND WETLANDS

Based on a review of the literature, information was found for 43 spills where burning was used as an oil spill response tactic. This information is summarised in Tables 2.1 (32 spills to wetlands) and 2.2 (11 spills to land).

These spills range widely in volume of oil burned, as shown in Figure 2.1. For many spills, other removal methods were used before and after the burn, although most burns were very efficient in terms of oil removal in the burn footprint. Figure 2.2 shows that most spills that were burned were light to medium crude oils. Three of the seven refined product spills that were burned occurred during periods of ice and snow. Figure 2.3 shows the types of environments that were oiled and burned, with salt and brackish marshes dominating, followed by freshwater marshes. Marsh environments are very sensitive to disturbances during manual and mechanical removal activities; thus, burning is more often considered as the best approach. Six of the seven burns on cultivated fields were conducted during very small spills in 1994

TABLE 2.1

Spills Where In-Situ Burning Was Conducted in Wetlands

Spill Name/Location/ Citation	Burn Date	Oil Type/Volume Spilled/Burned	Habitat/Species	Burn Area	Results by Years Post-Burn	Years to Recovery
Old Peace River Fen Pipeline, Alberta, Canada, Blenkinsopp et al. (1996)	1970; burn in mid-1971	Nipisi crude/9,540 m³/ unknown amount burned	Wet meadow/thin peat and poor fen	1.6 ha	In 1995, vegetation was present and to the edges of heavily oiled patches.	Little recovery as of 2017
Rainbow Fen Pipeline, Alberta, Canada, Blenkinsopp et al. (1996)	1970; burn in mid-1971	Nipisi crude/3,180 m³/ unknown amount burned	Poor fen	9.7 ha	In 1995, vegetation was present and to the edges of heavily oiled patches.	Little recovery as of 2017
Nipisi Bog Pipeline, Alberta, Canada, Blenkinsopp et al. (1996)	September 1972; burn in 1976	Nipisi crude/9,540 m³/ unknown amount burned	Rain-fed bog	10 ha	Burn residue was waxy, of variable thickness; subsurface soils were highly contaminated with fresh oil; some areas showed minimal vegetative recovery.	Little recovery as of 2017
Intracoastal City Well Blowout or McCormick Well Blowout/ Intracoastal City, LA, Castle (2012)	November 1975	S. Louisiana waxy crude (pour point of 80°F)/117,5000 m³ spilled/estimated 4,770 m³ burned Minor waxy residue was observed locally	Brackish marsh/*Spartina* spp.	~70 ha, including area oiled by rainout of the blowout plume, heavily coating the plant canopy	Wetlands had been burned annually by trappers, and were due for burning at the time of the blowout. Observations of a test burn showed new growth after 1 week. Survey in April 1976 showed significant re-growth in burn areas except where berms and other earthworks were constructed.	1–2 year

(Continued)

TABLE 2.1 (*Continued*)

Spills Where In-Situ Burning Was Conducted in Wetlands

Spill Name/Location/ Citation	Burn Date	Oil Type/Volume Spilled/Burned	Habitat/Species	Burn Area	Results by Years Post-Burn	Years to Recovery
Harbor Island, TX, Holt et al. (1978)	October 1976	Crude oil/60 m³/only a small amount was burned	Salt marsh/ *S. alterniflora*, black mangrove	0.1 ha heavily oiled, burned by err	0.5 year: *S. alterniflora* biomass = 60% of unoiled/unburned controls; Lowest recovery was in area of standing water; 100% mortality of mangroves in burn area.	N/A but likely <2 years
ESSO Bayway, Port Neches, TX, McCauley and Harrel (1981)	Jan 1979	Light Arabian crude/1,040 m³/ small marsh islands burned in clean-up experiment	Brackish marsh/ *S. patens*	Small marsh island, with 3 plots of 3 m²; flooded	0.6 year: Biomass in oiled/burned was 3% of unoiled/unburned controls; Burned/unoiled biomass was 1.5% of unoiled/unburned controls; Poor recovery due to persistent high water levels (3–55 cm) and low salinity (~0 ppt) post-treatments.	N/A but likely <5 years
Trans-Alaska Pipeline, Fairbanks, AK, Buhite (1979)	February 1978	Prudhoe Bay crude/2,540 m³ spilled/80 m³ burned	Ponded tundra with water depth from a few cm to 1 m	0.8 ha burned on Day 63	0.5 year: Entire area was fertilised, with 50% plant regrowth during the first growing season.	N/A but likely <5 years

(*Continued*)

TABLE 2.1 (*Continued*)
Spills Where In-Situ Burning Was Conducted in Wetlands

Spill Name/Location/ Citation	Burn Date	Oil Type/Volume Spilled/Burned	Habitat/Species	Burn Area	Results by Years Post-Burn	Years to Recovery
Black Lake, West Hackberry, LA, Overton et al. (1981)	September 1978	Light Arabian crude/11,450 m³ spilled/most burned	Lacustrine and fringing marsh	N/A	Sediment samples collected at 1, 16, 29, and 53 weeks post-spill showed only background contamination. Foliage samples collected 1 and 16 weeks post-spill showed elevated PAHs from soot deposition several km from the site; At 29 weeks, foliage samples showed no contamination.	N/A
Texaco Lafitte oil field Site 2, LA, Mendelssohn et al. (1995)	May 1983	S. Louisiana crude/45 m³/some cleaned before burn	Brackish marsh/ *S. patens, D. spicata, S. alterniflora*	N/A	11 years: No significant differences in soil TPH, live biomass, total biomass; burned area higher species richness than unoiled control (7.6 vs. 4.8), but not significant.	N/A

(*Continued*)

TABLE 2.1 (*Continued*)

Spills Where In-Situ Burning Was Conducted in Wetlands

Spill Name/Location/ Citation	Burn Date	Oil Type/Volume Spilled/Burned	Habitat/Species	Burn Area	Results by Years Post-Burn	Years to Recovery
Texaco Lafitte oil field Site 3, LA, Mendelssohn et al. (1995)	September 1986	S. Louisiana crude/0.6 m³/ unknown amount burned	Coastal brackish marsh/ *S. alterniflora, D. spicata*	N/A	8 years: Soil TPH was 162 mg/g at the burn site versus 2 mg/g at the control site (may have been a more recent spill); No significant differences in live and total plant biomass and live-to-dead biomass; species richness in oiled/ burned plots was 2.8 versus 6.6 in control plots; Overall recovery was ranked good.	<8 years
Friendship II Pipeline, Kekcse, Hungary, Nagy (1991)	January 1988	Crude/422/4.8 m³	Peat and bog wetland (mostly sedges and reeds)	5.4 ha	1.5 years: Sedge and reed vegetation recovered to near the original plant density.	1.5 years
Imperial Oil, British Columbia, Canada, Moir and Erskin (1994)	June 1990	Canadian crude oil/134 m³ spilled/ majority burned	Freshwater wetland bog	2 ha burned on Day 2; bog was flooded	Day 5: New vegetation appeared; site was seeded and fertilised. 0.75 year: Vegetation was recovering and no oil was apparent on the site or stream.	N/A

(*Continued*)

TABLE 2.1 (*Continued*)
Spills Where In-Situ Burning Was Conducted in Wetlands

Spill Name/Location/Citation	Burn Date	Oil Type/Volume Spilled/Burned	Habitat/Species	Burn Area	Results by Years Post-Burn	Years to Recovery
Pass a Loutre, Mississippi Delta, LA, Mendelssohn et al. (1995)	August 1990	S. Louisiana crude/<100 m³/most burned	Freshwater marsh/*Phragmites australis*	5.25 ha burned shortly after the spill	4 years: Soil TPH was not different for oiled/burned versus 2 control sites; live and total plant biomass and live:dead ratio were higher at the oiled/burned sites; overall recovery was ranked excellent.	<4 years
Chiltipin Creek, TX, Gonzalez and Lugo (1995); Tunnell et al. (1995); Hyde et al. (1999)	January 1992	S. Texas light crude/470 m/183 m³ 80%–85% burned Asphaltic, taffy-like residue covered the marsh surface and was manually removed	High marsh/*D. spicata, Batis maritima. Borrichia frutescens*	6.5 ha burned on Day 4, 10 ha oiled; variable water levels	1.6 years: High % cover but mostly by *D. spicate*. 2.6 years: Increase in species diversity, bare area declining; little change in TPH, but more weathered. 3.6 years: No change; apparent 'steady state'. 7 years: Increase in bare area, species diversity but affected by drought and damage from feral hogs and seismic survey.	Predicted 14–15 years based on trajectory for climax species

(Continued)

TABLE 2.1 (*Continued*)
Spills Where In-Situ Burning Was Conducted in Wetlands

Spill Name/Location/ Citation	Burn Date	Oil Type/Volume Spilled/Burned	Habitat/Species	Burn Area	Results by Years Post-Burn	Years to Recovery
Texaco Lafitte oil field Site 1, LA, Mendelssohn et al. (1995)	June 1992	S. Louisiana crude/0.2 m³/ unknown amount burned	Brackish marsh/ *S. patens, D. spicata, J. roemerianus*	N/A	2.4 years: No significant differences in soil TPH, live and total plant biomass, or species richness for oiled/ burned and control plots, but there was a trend towards lower biomass in the oiled/ burned plots; burned plots had higher live-to-dead plant biomass; overall recovery was ranked as moderate to good.	~2.5 years
Meire Grove, MN, Amoco Pipeline, Zischke (1993); Mendelssohn et al. (1995)	September 1992	Fuel oil and gasoline/400 m³/ unknown amount burned	Freshwater wetland pond/ *Typha* spp.	0.8 ha burned on Day 2 of discovery, but leaked for 10 days	*Shortly after the burn:* no. of invertebrate taxa/m² was 18 times higher at control versus oiled/burned pond. 1 year: Considerable recovery in invertebrates. 2 years: Residual signs of trampling; live plant biomass was 35 × higher and total plant biomass was 50 × higher	>2 yrs but likely <10 yr

(Continued)

TABLE 2.1 (Continued)
Spills Where In-Situ Burning Was Conducted in Wetlands

Spill Name/Location/Citation	Burn Date	Oil Type/Volume Spilled/Burned	Habitat/Species	Burn Area	Results by Years Post-Burn	Years to Recovery
Naval Air Station, Brunswick, ME, Eufemia (1994); Metzger (1995)	March 1993	JP-5 aviation fuel/ 240/80 m^3 No burn residue	Freshwater pond *T. latifolia*, *Sparganium americanum*	~1 ha burned on Day 8; water level raised prior to burn	in control pond versus oiled/burned pond; no differences in soil TPH; overall recovery was ranked poor. 0.4 year: Studies of vegetation, fish, birds, mammals, benthic community, water quality, sediment quality oiled/burned versus reference sites the following summer; no differences in plant cover or soil TPH; normal species abundance and distribution. Increase of *S. americanum* (burreed) over cattails, which was beneficial.	<0.5 year
Kolva River Basin Pipeline Spill Site 5, Komi, Russia, Hartley (1996)	1995	Crude oil/unknown volume because of multiple leaks from 1986–1994	Muskeg swamp with no outlet	6 ha burned	Burned violently for 20 h, creating so much heat that the oil was driven deep into the peat mat; burn residue on the surface was extremely viscous and oily, making further clean-up almost impossible.	N/A but likely decades

(Continued)

TABLE 2.1 (*Continued*)

Spills Where In-Situ Burning Was Conducted in Wetlands

Spill Name/Location/ Citation	Burn Date	Oil Type/Volume Spilled/Burned	Habitat/Species	Burn Area	Results by Years Post-Burn	Years to Recovery
Rockefeller State Refuge, LA, Hess et al. (1997); Pahl et al. (1997, 2003)	March 1995	Condensate/6.4 m³/ unknown amount burned No burn residue	Brackish marsh/ *S. patens, D. spicata, S. alterniflora. Scripus robustus*	40 ha burned on Day 5; some water on marsh surface; Studied oiled, oiled/ burned, and control transects	0.6 year: Burned transects: total cover 50% of other treatments; *S. patens* 14% of other treatments; *S. robustus* much higher (*D. spicata* slowed by post-burn flooding), thus stem density 30% of other treatments; soil TPH decreased to background. 2.6 years: Stem density, live biomass, total percent cover and species composition of oiled/burned and oiled similar to control.	3 years
Refugio, TX, Clark and Martin (1999)	May 1997	Refugio Light and Giddings Stream crudes/80–160 m³/ 90% burned with minor burn residue	Freshwater wetland/*Borrichia frutescens, S. spartinae*	2.4 ha burned on Day 3	Observed new crayfish burrows shortly after the burn. Wetland was used for cattle grazing.	N/A

(*Continued*)

TABLE 2.1 *(Continued)*

Spills Where In-Situ Burning Was Conducted in Wetlands

Spill Name/Location/ Citation	Burn Date	Oil Type/Volume Spilled/Burned	Habitat/Species	Burn Area	Results by Years Post-Burn	Years to Recovery
Vermillion 16 Freshwater City, LA, Henry (1997)	July 1997	Condensate, API 50/ unknown amount spilled/most burned	Brackish marsh/*Scirpus* spp, *S. patens, D. spicata*	3–4 ha burned on Day 13 after report; had been leaking 4 mo	During the burn, there was 5–10 cm of standing water in the thick vegetation. 0.5 year: Very little vegetation re-growth—the site looked like an open pond. Plant death attributed to the 4 mo of exposure to the light crude.	N/A
Chevron Pipeline MP 68, Corrine, UT, Williams et al. (2003); Michel et al. (2002)	January 2000	Diesel/1 6 m³/75%–80% burned No burn residue	Freshwater wetlands, alkali flats, snow and ice cover	5.2 ha burned on 10 March, 1.3 ha on 27 April	Burn area = 1.3 × intended area. Vegetation died in heavily oiled areas, burning not effective in removing oil penetrated into sediments or reduce toxic effects prior to burn; 4.1 ha fertilised and tilled in 2000/2001 to get PAH levels below criteria of 20 mg/kg.	N/A. but likely <5 years

(Continued)

TABLE 2.1 *(Continued)*

Spills Where In-Situ Burning Was Conducted in Wetlands

Spill Name/Location/ Citation	Burn Date	Oil Type/Volume Spilled/Burned	Habitat/Species	Burn Area	Results by Years Post-Burn	Years to Recovery
Louisiana Point, LA, Michel et al. (2002)	February 2000	Condensate/unknown amount spilled or burned; No residue	High salt marsh/ *D. spicata, Borrichia frutescens, Batis maritime, S. patens*	5.3 ha oiled, 55 ha burned on Day 3 0.5–1 cm water over marsh during the burn	0.6 year: In burned areas, total cover 64% and stem density 22% of control, *B. frutescens* and *D. spicata* much reduced. Stem density lower for all species. 1.6 years: Total cover 76% and stem density 80% of control, with stem density of *B. frutescens* at 10%, *D. spicata* at 32%, and *Batis* at 120%.	>1.6 years, but likely <5 years
Ruffy Brook, MN, Michel et al. (2002)	July 2000	Medium crude oil/>8 m³/80% burned; tar-like residue ~1 cm thick, manually removed	Ponded freshwater wetland	1.2 ha burned on Day 1; 0.3–1 m of water in pond	1 year: All herbaceous vegetation recovered; willows died (they are known to be sensitive to fire); no evidence residues sank.	<1 year

(Continued)

TABLE 2.1 (*Continued*)

Spills Where In-Situ Burning Was Conducted in Wetlands

Spill Name/Location/ Citation	Burn Date	Oil Type/Volume Spilled/Burned	Habitat/Species	Burn Area	Results by Years Post-Burn	Years to Recovery
Mosquito Bay, LA, Michel et al. (2002)	April 2001	Condensate/>160 m³/ unknown amount burned No residue	Brackish marsh/ *S. patens, D. spicata, S. cynosuroides*	15 ha oiled, 40 ha burned on Days 7 and 8; 1–10 cm water layer on marsh	After the burn, oil in burrows still present. 0.5 year: Burned/lightly and unoiled vegetation recovered with abundant fiddler crabs present, burned/heavily oiled areas along creek banks died, so did not reduce toxicity from contact with condensate prior to burn.	<0.5 year for lightly oiled areas; 1 year for heavily oiled areas
Enbridge Pipeline, Cohasset, MN, Leppälä (2004)	July 2002	Canadian crude/950 m³/475 m³ Significant residue that was thicker	Freshwater forested/ scrub-shrub with peat base	4.5 ha affected, 2.4 ha burned on Day 1, lasted 24 h	Vegetation recovery was estimated to take many years because the deep excavation post-burn, as well as the burning of trees.	Many years, likely >10 years
Chevron Texaco #2 Tank Battery, Sabine NWR, LA, Entrix (2003)	August 2002	S. Louisiana crude/24–48 m³/ unknown amount burned Pockets of oil and oil residues with nets and sorbent materials	Brackish marsh/ *S. patens, Typha latifolia*	1.4 ha burned on Day 4	0.7 year: 80%–90% cover in burn area, slight hydrocarbon odor in sediments; mean 2,150 ppm TPH 1.2 years: 95%–100% vegetation cover; mean 8 ppm TPH; no visible sheens	1 year

(Continued)

TABLE 2.1 (*Continued*)

Spills Where In-Situ Burning Was Conducted in Wetlands

Spill Name/Location/ Citation	Burn Date	Oil Type/Volume Spilled/Burned	Habitat/Species	Burn Area	Results by Years Post-Burn	Years to Recovery
Chevron Pipeline MP 68, Corrine, UT, Earthfax Engineering Inc. (2003)	November 2002	Gasoline/72/18 m³	Freshwater wetlands, alkali flats	8.4 ha affected, 5.5 ha burned on Day 5	50% evaporated, 25%–30% burned, rest in soils that degraded naturally over weeks.	N/A
Chevron Empire, LA, Myers 2006; Merten et al. (2008); Baustian et al. (2010)	October 2005	S. Louisiana crude/16–32 m³/ Some burn residue that was sticky and liquid (unburned) oil in burrows; removed with sorbents and natural flushing	Brackish marsh/*S. patens Schoenplectus americana* (chairmaker's bulrush), *D. spicata*	15.5 ha oiled; 7.9 ha burned on Days 44–45 after the initial release during Katrina; 0–10 cm water over the marsh	30 days: New vegetation 30–60 cm high. 0.75 year: Plant biomass and species composition in oiled/burned returned to control levels; however, species richness remained somewhat lower in the oiled and burned areas compared to the reference areas. 1–1.5 years: No differences between oiled/burned and control sites for sediment accretion, cellulose decomposition, and the rate of recovery from experimental disturbances (lethal and non-lethal removal of vegetation).	1 year

(*Continued*)

TABLE 2.1 (Continued)
Spills Where In-Situ Burning Was Conducted in Wetlands

Spill Name/Location/Citation	Burn Date	Oil Type/Volume Spilled/Burned	Habitat/Species	Burn Area	Results by Years Post-Burn	Years to Recovery
Bass Burn, Empire, LA. LOSCO	2005	S. Louisiana crude/unknown amount spilled or burned	Forested upland	Unknown but many ha	Area along the Mississippi River levee oiled during Hurricanes Katrina and Rita.	N/A
Bayou Sorrel, Atchafalaya Basin, LA, NOAA (2013)	January 2013	S. Louisiana crude/24 m³/4.8 m³	Forested wetland flooded by 0.3–1 m	Three patches of floating oil burned over 25 min	Burn residue floated then sank and was not recovered. Soils were saturated and cleaned by manual methods.	N/A
TPIC Delta NWR, LA, Zengel et al. (2017)	May 2014	S. Louisiana crude/16 m³/unknown amount burned	Tidal fresh marsh, *Phragmites*	6 ha oiled; 2 ha burned	2 years: Vegetation cover and snails recovered; species composition still different at oiled-burned sites compared to unoiled-unburned and oiled-unburned sites.	5–6

Source: Michel, J. and Rutherford, N, *Oil Spills in Marshes: Planning & Response Considerations*, Emergency Response Division, NOAA, Seattle, WA and American Petroleum Institute, Washington, DC, 2013.

TABLE 2.2

Spills Where In-Situ Burning Was Conducted on Land

Spill Name/Location/ Citation	Burn Date	Oil Type/Volume Spilled/Burned	Habitat/Species	Burn Area	Results by Years Post-Burn	Years to Recovery
California pipeline break	N/A	California crude, API 13/795/0.8 m³	Dry creek bed	A pool of 0.8 m³ of oil was burned	A heavy residue remained after the burn and required extensive clean-up operations.	N/A
Marathon Pipeline, Gillespie Facility, IL, May and Wolfe (1997)	December 1995	Illinois crude/0.8/0.2 m³	Cultivated field	70 m²	After the burn, the area was fertilised, limed and tilled. Normal farming activities resumed the following spring.	0.5
Marathon Pipeline, Noble Gathering Line, IL, May and Wolfe (1997)	January 1995	Illinois crude/0.8/0.5 m³	Cultivated field and drainage ditch	210 m²	After the burn, the area was tilled. Normal farming activities resumed the following spring.	0.5
Marathon Pipeline, Noble S. Gathering Line, IL, May and Wolfe (1997)	January 1995	Illinois crude/0.5/0.3 m³	Cultivated field and drainage ditch	2,140 m²	Normal farming activities resumed the following spring.	0.5
Marathon Pipeline, Patterson, IL, May and Wolfe (1997)	April 1995	Illinois crude/0.8/0.5 m³	Slough in an agricultural field	306 m²	Patches of burn residue were removed with sorbents. The slough was then fertilised, lined and tilled. Normal farming activities resumed 16 months after the spill.	1
Marathon Pipeline, Roy Gill Site, IL, May and Wolfe (1997)	March 1994	Illinois crude/0.1/0.1 m³	Cultivated field, corn stalks and stubble	27 m²	The burned area was tilled. Normal farming activities resumed that season.	0.3

(Continued)

TABLE 2.2 (*Continued*)

Spills Where In-Situ Burning Was Conducted on Land

Spill Name/Location/ Citation	Burn Date	Oil Type/Volume Spilled/Burned	Habitat/Species	Burn Area	Results by Years Post-Burn	Years to Recovery
Marathon Pipeline, Sanders Lease, IL, May and Wolfe (1997)	April 1995	Illinois crude/1.6/0.1 m³	Cultivated corn field and adjacent road	223 m²	The burned area was fertilised and tilled. Normal farming activities resumed the following spring.	1
Siroco Pipeline, Dewberry, TX, Labay (1997)	January 1997	Texas Sweet crude/8/8 m³	Forested upland and intermittent creek	300 m of creek bed	90% oil removed.	N/A
Warick Lake, Ontario, Canada, Burns (1998)	January 1983	Diesel/60/50 m³	Streambed and frozen lake	N/A	Oiled snow was burned on site in January and during spring thaw; then creek bed was burned using wood and sawdust.	N/A
Williams Pipeline, OK, Williams Pipeline Co. (undated)	August 1995	Jet fuel/12/3 m³	Open field, ditch, and small stream	1.6 km of ditch and 0.2 ha of field	Flushing and leaf blowers were used to herd the product into collection areas for burning. Field was tilled and seeded.	N/A
Williams Pipeline, MO, Williams Pipeline Co. (undated)	Auggust 1995	Unleaded gasoline/14 m³/ unknown amount burned	Open field	1,836 m² subsurface	Product in trenches was burned to prevent rapid migration towards a creek.	N/A

Source: Dahlin, J.A. et al., *Compilation and Review of Data on the Environmental Effects of In-Situ Burning of Inland and Upland Oil Spills*, Report No. 4684, American Petroleum Institute, Washington, DC, 1999, 121 p + appendices.

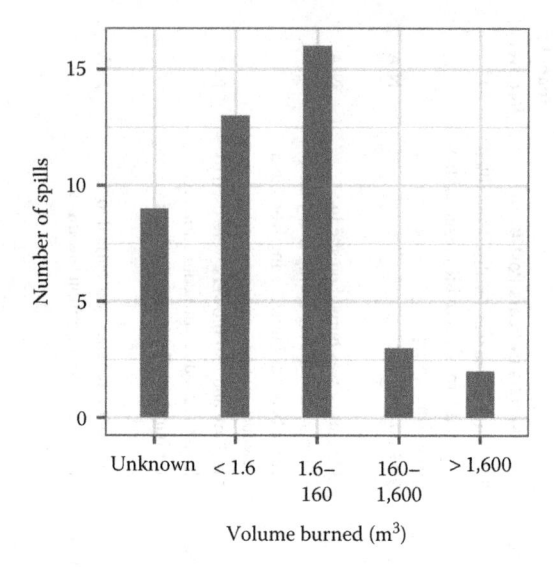

FIGURE 2.1   Range of volumes of oil burned for the 43 spills where burning was conducted on land and wetlands, as listed in Tables 2.1 and 2.2.

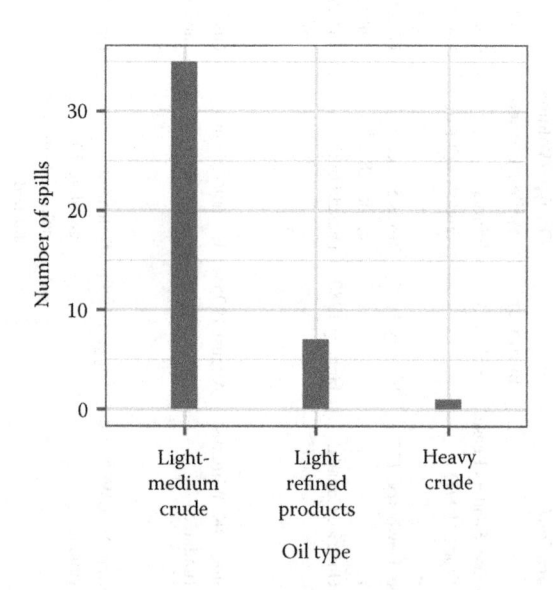

FIGURE 2.2   Types of oil burned for the 43 spills where burning was conducted on land and wetlands, as listed in Tables 2.1 and 2.2.

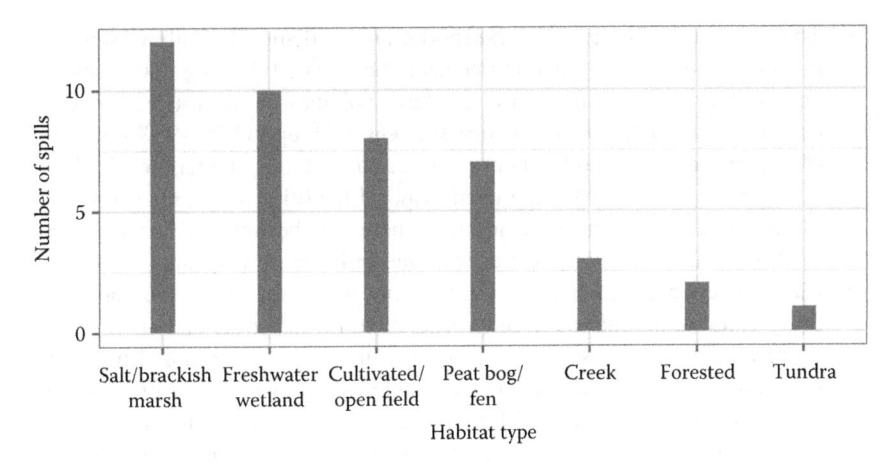

FIGURE 2.3 Types of environments for the 43 spills where burning was conducted on land and wetlands, as listed in Tables 2.1 and 2.2.

and 1995 by one oil company in Illinois. In these cases, the oiled/burned areas were tilled, sometimes fertilised and limed, and normal crops were raised the following spring, if not earlier (Dahlin et al. 1999).

2.3 WHEN TO CONSIDER USE OF BURNING OF SPILLED OIL ON LAND AND WETLANDS

The following conditions often lead to a decision to consider burning of an oil spill on land and wetlands:

- Access to the spill site is limited, making it difficult to get clean-up crews and equipment on scene. The number of people and types of equipment needed for burning oiled sites are logistically simpler than for mechanical or manual recovery, and waste handling, storage and disposal are minimal. The surrounding terrain (steep gorges and sensitive wetlands) may restrict access to the spill site. The ground may be too soft to support even foot traffic without causing significant damage (wet tundra, marshes and peat bogs). If there are no roads to a remote site, it might be possible to fly in personnel but minimal equipment.
- It is necessary to quickly remove spilled oil to prevent its spread over large areas and to reduce its impact to sensitive sites. Controlled burning removes spilled oil in hours, whereas manual or mechanical countermeasures may take days to months. Timing becomes critical under certain conditions. For example, forecasted rain may flush oil from the spill site into more sensitive areas, temporary containment structures may be predicted to fail or oil held in a small area by snow and ice may spread widely during a predicted thaw. These kinds of spill conditions trigger the need to rapidly remove as much oil as possible.

- There are limited options for transportation and disposal of oily wastes, so the amount of wastes generated needs to be reduced. During both manual and mechanical clean-up operations, large volumes of oily wastes are generated. In contrast, burning is very efficient, with up to 90%–98% removal efficacy for contained oil in hours (see the case studies in Section 2.5). The remoteness of a site from approved disposal facilities is an important factor in the decision to burn. In addition, there may be weight restrictions on roads for transport of oily wastes from the spill site for disposal.
- Clean-up countermeasures currently being used are ineffective and will likely cause additional damage to vegetation, the substrate and other natural resources. Each response countermeasure has an operational limit on how much oil it can remove. For example, skimmers are not efficient in recovering thin slicks, and flushing is not effective in removing oil adhered to vegetation. Clean-up activities such as the construction of access routes, vehicle traffic, foot traffic or the diversion of water flow can cause substantial environmental damage. Thus, it may be advisable to use less intrusive response countermeasures such as burning.

Guidelines for considering burning of oil on land and wetlands include the following:

Likely oil removal rates and amount of burn residue: The potential oil removal by burning spills on land and wetlands is related to the following conditions:
- There should be free oil on the water surface in wetlands.
- On open water, fresh crude should be at least 1 millimetre (mm), weathered oil should be 2–5 mm and heavy refined products should be 10 mm. Thicker oil burns more efficiently.
- Oil that has penetrated into the substrate or in burrows or root cavities will not likely be removed.
- Emulsified oil is more difficult to ignite and sustain a fire; laboratory studies (Buist 1998) and field experience have shown that unstable emulsions containing a significant amount of water are difficult to burn without the use of promoters or accelerators. Stable emulsions are even more difficult, though unlikely for spills on land or wetlands.
- Many oils leave residues, which may have to be removed to protect plants and animals and speed the overall rate of recovery.
- Burn residues can be heavier than water and sink; studies have shown that heavy crude oils and heavy refined products are likely to produce sinking burn residues (SL Ross 2002).

Habitat type: Habitats that are good candidates for in-situ burning include those with no vegetation (fallow agricultural fields, ditches, bare ground) or with herbaceous (grassy) vegetation such marshes, bogs, range land, grass land and tundra. These habitats generally recover within 1–5 years if the burn is conducted under appropriate conditions, particularly when there is a water layer overlying the soil. In contrast, burning is seldom considered for habitats dominated by woody vegetation, such as upland forests, taiga (moist subarctic forest dominated by conifers), swamps and mangroves.

In habitats with woody vegetation, the added fuel of the oil can increase the intensity of the fire and scorch or ignite trees and possibly kill them. In four known cases where a burn was conducted in or immediately adjacent to wooded areas, tree mortality was reported. Depending on the age of the forest, recovery can take decades. The fire ecology literature has compiled an extensive database on the fire tolerance of different plant species that can be used to predict how burning could affect oiled vegetation and vegetation adjacent to the proposed burn area that could be affected by heat stress.

Time of year: Whereas it is not possible to pick the time of year for a spill to occur, the time of year of a burn is an important factor in terms of the speed at which the vegetation will likely recover. Mendelssohn et al. (1995) reviewed 34 studies of prescribed burning and found that vegetation recovery is slowest for burns that occurred in summer and fastest for burns that occurred in winter and late spring. During the dormant season, carbohydrate reserves are stored in roots and rhizomes; thus, they are available to support regrowth immediately after the burn or during the next growing season. Burning during the growing season, after the vegetation has spent its carbohydrate resources producing vegetation, may not allow resources for regrowth right after the burn or storage of carbohydrate reserves for the next growing season.

Soil moisture and water levels during the burn: Soil heating can cause damage to the soil structure as well as destroy vegetation by permanently damaging roots and tubers. Research has shown that 60°C is a temperature above which damage to vegetation occurs (Reardon and O'Donnell 2013). Oil burning on the soil increases the heat flux to the soil. How much oil is present and the degree of soil moisture are important factors. Figure 2.4 shows a predictor of how deep a 60°C temperature profile would penetrate into the soil for different degrees of soil moisture. This figure shows that moisture content of the soil is an important factor and should be considered in making decisions. Many low shrubs have roots at depths of 2–5 cm. Grasses have roots that are quite shallow, at 0.5–2 cm. There are many differences among vegetation types. Figure 2.4 can be used to approximately gauge the effects of fire on the vegetation and soil.

The amount of standing water is also a factor in reducing thermal stress to the roots of vegetation in the burn area. Burn-tank experiments using thermocouples in potted plants set at different water depths have shown that (Lin et al. 2002, 2005):

- With a 10-cm water layer above the soil surface, average peak soil temperatures were below 35°C at all depths, and the vegetation survival and growth were similar to controls.
- With a 2-cm water layer above the soil surface, average peak soil temperatures were below 48°C at all depths and the vegetation survival and growth were similar to controls.
- When the water layer was 2 cm below the soil surface, average peak soil temperatures at the surface were >100°C at the soil surface and >40°C at 5 cm below the surface. Plants with shallow rhizomes had reduced survival.

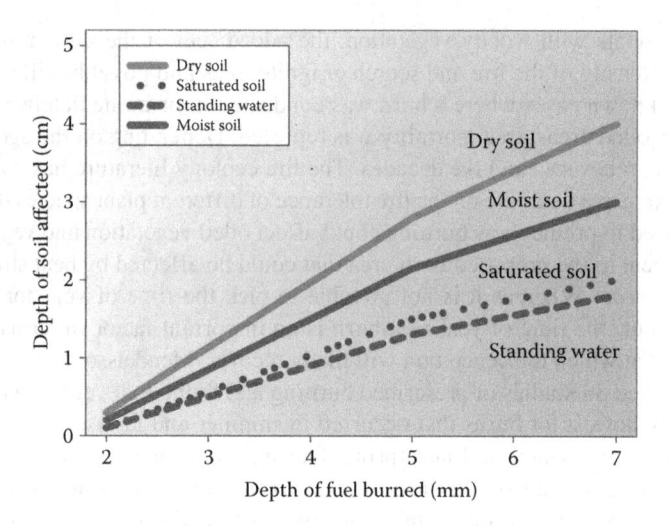

FIGURE 2.4   A simplified nomogram to predict the critical temperature of burning oil pools with different soil moisture contents. (Courtesy of Merv Fingas.)

- When the water layer was 10 cm below the soil surface, average peak soil temperatures at the surface were >300°C at the surface and >60°C at 2 cm below the surface. Plants with shallow rhizomes had almost no survival.

Soil type: Where the soils are highly organic, such as in bogs, there is a risk that the soil itself could ignite, which can: (1) lower the elevation and result in flooding of the area and preventing vegetation regrowth, (2) kill roots and the seed bank and (3) result in oil penetration even deeper into the soil.

Flooding post-burn: Certain wetland species are more likely to recover slowly if they are submerged under water for a sustained period (weeks to months) after a burn. In fact, most prescribed burns in wetlands are conducted in the fall, when water levels are low, so the vegetation has a chance to recover before spring flooding.

History of prescribed burning: Prescribed burning is extensively used around the world for a variety of purposes, including control of vegetative species and reduction of fire hazards by removal of flammable fuel such as dead or diseased vegetation. If the habitat has been recently or regularly burned, the amount of natural fuel is likely to be low. In contrast, if the habitat has not been managed under a prescribed burn plan, the amount of natural fuel can be high, increasing the risk of a hotter fire that might be more difficult to control. Also, many guidelines exist at state and provincial levels on the use of prescribed burning, and these should be consulted for guidelines on how to conduct a successful, safe burn.

Amount of oil penetration in the substrate: If there is little penetration before the fire, then burning is more effective. As shown in the case studies, burning does not remove oil that has penetrated into the soils and thus there may be a need for additional oil recovery after the burn.

2.4 PLANNING AND IMPLEMENTING BURNS OF SPILLED OIL ON LAND AND WETLANDS

Once the decision is made to use burning as an appropriate spill response measure at a site, a detailed burn plan should be prepared. Table 2.3 shows the outline of a typical burn plan. Immediately before the burn, the details are updated to provide the latest information and make a go/no go decision. API (2015) provides a field operations guide of burning of inland oil spills.

TABLE 2.3
Content of a Typical Burn Plan

Burn Site Description

Spill description and status
Maps and photographs of each burn unit
Public, wildlife and structures of concern in and adjacent to burn unit
Permits

Organizational Structure of the Burn Group

Site Safety Plan

Personal protection equipment and procedures
Medical plan
Hazards for field crews
Staging areas for crews and equipment
Burn crew evacuation route and safe refuge areas (map)
Smoke plume assessment and public evacuation zone (map)
Notification of public and aviation agencies
Evacuation zone (map)
Confirm clearance of area immediately prior to the burn

Air Quality Monitoring Plan

Parameters to be monitored
Monitoring equipment and locations (map)
Levels of concern and actions to be taken in the event of an exceedance

Fire Management Plan

Weather and environmental conditions for go/no go for the burn
Ignition method
Fire monitoring methods
Fire containment and suppression methods
Fire break type, dimensions, locations
Plan/equipment/crews to contain escaped fire
Post-burn site inspection to extinguish any hotspots

Post-Burn Monitoring Plan

Assessment and removal of any remaining oil and burn residue
Emergency stabilisation/re-planting plan (as needed)
Preparation date and approval signatures

In terms of operations, many of the same considerations discussed in Chapter 5 for burning oil on water are applicable to burning on land. However, there are several important differences to consider. The ease of ignition and oil thickness guidelines may not apply if there are combustible materials such as dried grass available in the burn area. Burning where there is dry vegetative material or wood in the target area can be simply a matter of igniting that material. Both the dried vegetative material and oil will burn. Conversely, lightly oiled, live vegetation may not burn. Refer to the case study in Section 2.5 on the brackish marsh near Empire, Louisiana, where the heavily oiled live vegetation burned intensely and completely; however, the burn could not be sustained once it reached the moderately oiled marshes. As discussed below, it is necessary to establish a fire break prior to ignition. Another difference is that it will be necessary to conduct a survey of the planned burn area to confirm that the area is clear of people and wildlife.

Selected parts of the burn plan outlined in Table 2.3 are discussed below.

2.4.1 SITE SAFETY PLAN

The most important consideration about any burn is whether it can be conducted completely safely. Models are available to assist with planning safely (FAMWEB 2015).

There is no need to take risks to the burn crew, nearby people and structures, or the environment in order to carry out a burn.

2.4.2 PERSONAL PROTECTIVE EQUIPMENT

Every person at the burn site should wear flame-resistant clothing, leather boots with metal toes and shanks, safety helmets and leather gloves. All personnel should also wear goggles or a full-facepiece respirator with a P100 filter. The latter may be carried along if the person is already wearing goggles.

2.4.3 FIRE BREAKS

There should be a fire break around any land or wetland burn to protect the surrounding area from spreading fire. The size of a fire break depends on the wind speed and direction and the vegetation height. A calculator for determining the downwind fire break is shown in Figure 2.5. The minimum width is generally 6 m. Because the downwind width of the fire break is larger than the minimum upwind distance, the fire area and its fire breaks will form a quadrilateral shape. A typical fire break sketch is shown in Figure 2.6. There are several types of fire breaks, some of which are shown in Figures 2.7 through 2.9:

- Utilising areas bare of vegetation, such as flats, roads, existing fire breaks, snow and ice and so on
- Mowing of grassy vegetation to reduce the amount of flammable vegetation
- Tilling or disking so there is no flammable vegetation present, just bare soil
- Digging a trench
- Building a dike
- Conducting a small, controlled burn to remove flammable materials surrounding the primary burn area

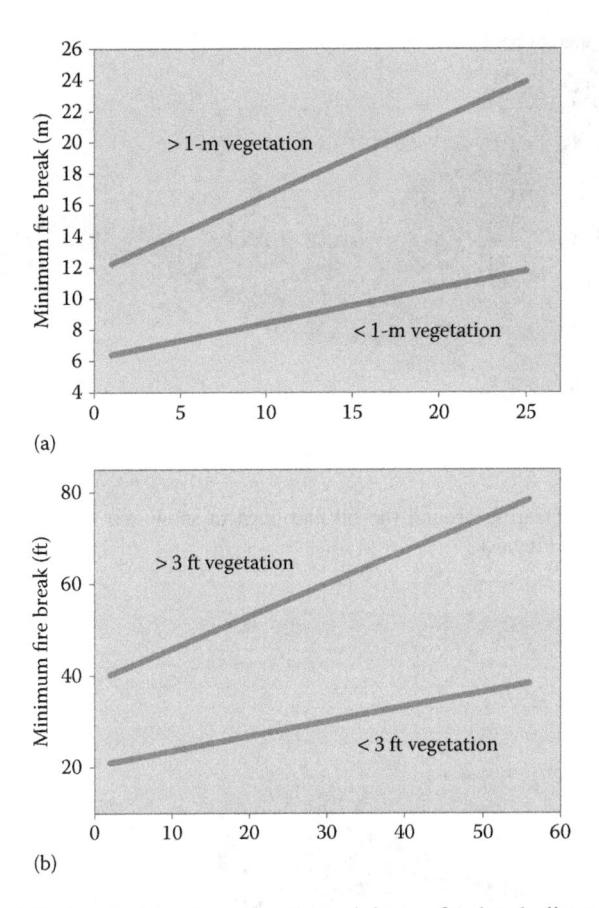

(a)

(b)

FIGURE 2.5 (a,b) A calculator for estimating minimum fire break distances. (Courtesy of Merv Fingas, derived from U.S. Department of Agriculture [USDA], *Firebreak Design Procedures. Natural Resource Conservation Service,* NE-T.G. Notice 619 Section IV, Washington, DC, 13 pp, 2010, and Smith, R., *Firebreak Location, Construction and Maintenance Guidelines Fire and Emergency Services Authority of Western Australia*, 480 Hay Street, Perth, Western Australia, 25 p, 2011.)

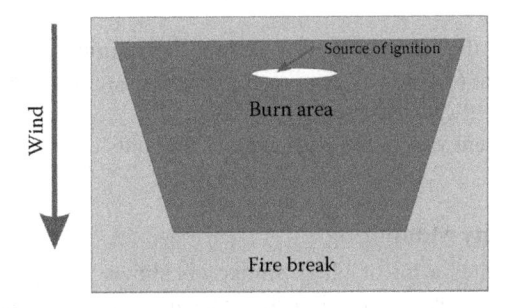

FIGURE 2.6 A schematic of a burn area and the surrounding fire break. The dimensions are exaggerated to emphasise the precautions. (Courtesy of Merv Fingas.)

FIGURE 2.7 A clearing around the oil and burn in snow and ice creates a fire break. (Courtesy of Merv Fingas.)

FIGURE 2.8 Ploughing created this fire break on farmland. (Courtesy of Merv Fingas.)

- Laying down the vegetation in a wetland by running over the vegetation with an airboat (though, as discussed in the case studies in Section 2.5, this method is not always successful)
- Wetting adjacent vegetation with copious amounts of water

2.4.3.1 Air Quality Monitoring

Burning of oil creates a lot of black smoke, as shown in Figure 2.10. Air quality monitoring is not always required in order to conduct burning of spilled oil, depending on the location, scale, conditions and proximity of potential receptors at risk. In the United States, it may be necessary to obtain a permit from the state

FIGURE 2.9   The burn of oil at this site was confined by building dikes around the area. (Courtesy of Merv Fingas.)

FIGURE 2.10   Two smoke plumes rise from a burn of JP-5 aviation fuel into a freshwater pond. (Courtesy of Steve Lehmann, NOAA, Lowell, MA.)

air quality control board or a similar agency. When monitoring air quality, either as a requirement in the permit to conduct the burn or to document air quality during the burn in the case of future concerns, it is important that the monitoring plan is well designed and implemented by trained teams. Chapter 4 includes detailed discussion of the expected air emissions from burning; Chapter 5 includes air quality concerns, which are not expected to be very different for land and wetland spills compared to on-water spills. The main air quality issues for burning of oil on land and wetlands are focused on making sure that the public is not exposed to unsafe levels of airborne contaminants. If there are such risks, air quality models and other tools can be used to predict areas that may exceed air quality criteria and establish protection evacuation zones and monitoring locations. The addition of a smoke-reducing additive, such as ferrocene, may be one option to control smoke emissions.

2.4.3.2 Fire Management Plan

Weather conditions to consider prior to the burn include the following:

- Winds <19 km/h (12 mph) for ignition of oil on land and wetlands
- Winds <40 km/h (25 mph) to sustain the burn
- A forecast shift in wind direction or speed in the next 12 h, which could change the location and width of fire breaks
- An inversion now or the forecast of an inversion in next 12 h, which could trap polluted air near the ground and increase the risk of air quality concerns
- A forecast of a weather front moving in within the next 12 h

The plan should specify the method(s) of ignition, the line of ignition and the rate at which this will be ignited. Figure 2.11 shows the use of a drip torch to ignite a prescribed burn. The drip torch is the most frequently used method of igniting prescribed burns on land. Once the fire break is established, one proceeds to the furthermost upwind position and uses an igniter to start the fire. This is typically a drip torch, Helitorch (ignited gelled fuel dispenser suspended under a helicopter), flare or diesel-soaked rags or sorbents. Once started, the fire is monitored, especially close to the fire breaks at the downwind side.

FIGURE 2.11 A drip torch being used to light a prescribed fire. (Courtesy of Cave National Park.)

Means should be available to extinguish fires and unwanted fire propagation. Fire suppression is usually provided in the form of firefighting water and trained firefighters. During many of the past burns, the local fire authorities provided this service. After the main fire is out, the site requires monitoring for several hours or days until all hot spots are extinguished and there is no danger of flare-ups.

2.4.3.3　Post-Burn Monitoring Plan

As soon as it is safe to enter the burned area, it is necessary to conduct an assessment of any burn residues, including the type, amount and need for removal. The type of burn residue can range from liquid unburned oil to sticky viscous semi-liquid, to hard dry crusts similar to peanut brittle. Buist and Trudel (1995) conducted laboratory tests where seven crude oils and one diesel product were burned at 70%–99% effectiveness. They reported that burn residues from the heavier oils formed brittle, non-sticky residues, whereas the lighter crude oils produced residues that were described as like cold roofing tar. The diesel produced a liquid residue.

On water, such as in ponds and streams, the hot burn residue can initially float; depending on the oil type, however, these residues can sink after cooling. Therefore, floating burn residues should be rapidly removed from the waterbody, which can be accomplished with hand tools, nets or sorbents. When there is little to no standing water in the burn area, semi-liquid and solid burn residues often are manually removed using hand tools, taking care to minimise disturbance of the soils. It may be necessary to access the burn area using walking boards. Table 2.4 summarises the likely behaviour of burn residues on land from different types of oil.

As discussed in the case studies in Section 2.5, any liquid oil that penetrated into the substrate (e.g. in root cavities, animal burrows and coarse-grained sediments) will not burn and should be recovered. If the site is under tidal influence, sorbents may be deployed over several tidal cycles until oil is no longer being mobilised from the site. If the site is dry, it may be appropriate to remove as much of the surface residue as possible and then till and fertilise the site to speed degradation of the residual oil. See the case studies in Section 2.5 to see how this approach was used during a diesel spill in Utah and for many of the land spills described in Table 2.2.

There may be a need for an emergency stabilisation/replanting plan to take immediate restorative actions in the burn area, considering drainage patterns, soil conditions and vegetative issues. This plan considers only the short-term to stabilise the land situation until further restoration can be carried out or the area recovers naturally. Such actions could include erosion measures such as mulching, biodegradable textiles, seeding with annual grasses and sediment fencing. Any erosion control measures need to be tailored to the specific site conditions, especially slope and soil type. Local soil conservation, forestry or range management officials can provide guidance on appropriate and effective erosion control methods.

TABLE 2.4

Predicted Behaviour of Burn Residues for Different Oil Types for Burns on Land

Oil Type	Behaviour on Burn Residue on Land
Gasoline Products	• Burns; does not leave a significant amount of residue.
Diesel-Like Products and Light Crude Oils (diesel, No. 2 fuel oil, crude concentrate, West Texas crude oil)	• Burn residue is mostly unburned oil that penetrates into the ground, root cavities and burrows with small amount of soot particles that often are enriched in heavier PAHs. • Remains liquid; able to be recovered with sorbents and flushing.
Medium Crude Oil and Intermediate Products (South Louisiana crude oil, IFO 180, lube oils)	• Burn residue may form pockets of liquid oil, solid or semi-solid surface crusts or sheets, or present as heavy, sticky coating on sediments. • Liquid oil can be flushed. Semi-solid and solid residues are typically recovered manually. • Remaining residues are tilled and fertilised in appropriate habitats.
Heavy Crude Oils and Residual Products (Venezuela crude, San Joaquin crude, No. 6 fuel oil)	• Difficult to burn; often have to add a lighter oil to initiate the burn. • Leaves heavy, sticky residue that is a mix of unburned oil and semi-solid burn residue and requires extensive clean-up. • Remaining residues are tilled and fertilised in appropriate habitats.

Source: Scholz, D.K. et al., *Risk Communication for In-situ Burning: The Fate of Burned Oil*, American Petroleum Institute Publication No. 4735, Washington, DC, 2004.

Depending on the site, vegetation adjacent to the burn may be scorched by flames or stressed from the heat generated during the burn. If this occurred, these areas should be included in the monitoring and potential revegetation plan.

2.4.3.4 Long-Term Monitoring and Rehabilitation Plan

For some burns, there may be a requirement for a long-term monitoring plan (up to 3 or even 5 years) to document rates of residual oil weathering and vegetative recovery. Such plans can include sampling of unoiled, unburned 'control' areas to provide the basis for comparing oil and vegetation metrics over time. Aerial photographs are also valuable for monitoring vegetation recovery for large burn areas to supplement data collected in small sampling plots.

When the monitoring results indicate that recovery is proceeding at a slow rate (see Figure 2.12 for a small area of a burn with residue that was suppressing vegetation growth), a rehabilitation plan can be developed to speed recovery. This plan could consist of efforts to restore vegetation by plantings, reducing oil content of the soil by tilling and addition of soil conditioners or modifying drainage patterns to control for erosion, flooding or other problems.

FIGURE 2.12 A burn of oil on land left this residue that, many years after the burn, still suppresses vegetation growth. (Courtesy of Merv Fingas.)

2.5 CASE STUDIES OF THE USE OF BURNING ON LAND AND WETLANDS

Several cases of burning in marshes are discussed in the following sections. No two oil spills are alike, and each case of burning is very different. Each case included here is briefly described and lessons learned, if available, are summarised.

2.5.1 INLAND BURNS

2.5.1.1 Crude Oil Spill, Ruffy Brook, Minnesota

On 22 July 2000, a pipeline near Clearbrook, Minnesota, failed and released over 8 m^3 of medium Bow River crude oil (API gravity of 21) into a wetland fed by Ruffy Brook. The spill affected approximately 1.2 hectares (ha) of freshwater wetland that was covered by water 30–100 cm above the soil surface. Mechanical recovery was deemed difficult to deploy and potentially damaging to the submersed and floating aquatic vegetation, so *in-situ* burning was conducted the same day of the spill. The burn lasted for 3 hours, and remaining pockets of oil were ignited over a period of 3 days. No secondary burning occurred during this operation. It is estimated that 80% of the oil was consumed during the burn. A significant amount of burn residue (in some places 1 cm thick) remained after the fire went out. The residue was picked up by hand over a period of 3 days. There is no evidence that any residue sank. The site was visited a year later, and the vegetation was found to have recovered well, with the exception of willows, a fire-sensitive species. The quick use of *in-situ* burning (ISB) prevented spreading of the oil and thereby minimising damage to the wetland. See Figures 2.13 through 2.15. Lessons learned from this spill and burn include (Michel et al. 2002):

- The burn was conducted quickly, within hours after the spill occurred; thus, most of the oil was thick and contained in a small area, increasing the efficacy of the burn.

FIGURE 2.13 An oiled area (outlined in white) from a pipeline into a freshwater slough. Arrow points to willows that were killed by the burn. (From Michel, J. et al., *Recovery of Four Oiled Wetlands Subjected to In Situ Burning*, American Petroleum Institute Publication 4724, Washington, DC, 71 p, 2002.)

FIGURE 2.14 The burn residue after the burn of the oil shown above. The residue was removed. (From Michel, J. et al., *Recovery of Four Oiled Wetlands Subjected to In Situ Burning*, American Petroleum Institute Publication 4724, Washington, DC, 71 p, 2002.)

(a)

(b)

FIGURE 2.15 (a) The Ruffy Brook burn site prior to the burn on 22 July 2000 and (b) the same site 1 year after the burn. Note that the grassy vegetation fully recovered. (From Michel, J. et al., *Recovery of Four Oiled Wetlands Subjected to In Situ Burning*, American Petroleum Institute Publication 4724, Washington, DC, 71 p, 2002.)

- Water depths in the ponded wetland were ample enough to prevent damage to the substrate and plant roots, a particularly important consideration in areas with soils of such high organic content.
- The burn residue was removed quickly, preventing its eventual sinking and contamination of bottom sediments.
- It appeared that most of the herbaceous vegetation quickly recovered after the burn. In contrast, the vegetation with low fire tolerance (willows) showed little recovery. In many wetland areas, willows are considered to be invasive; thus, the dieback of the willows may be of lesser concern.

2.5.1.2 Diesel Spill in Wetlands and Salt Flats, Northern Utah

On 21 January 2000, 16 m^3 of diesel were released from a pipeline north of Great Salt Lake in Utah, spreading over 15.5 ha of salt flat and wetlands. Manual removal efforts were of limited effectiveness, and the diesel was trapped in and under ice during freeze and thaw periods. This oil remained a risk to migratory waterfowl that were expected to arrive at the affected wetland during spring thaw; thus, burning was proposed. It took 4 weeks for the site remediation plan and fire management plan to be approved. The first burn used a Helitorch on 10 March 2000 (6 weeks post-spill) for ignition of the most affected 5 ha. The following month (late April), 1.3 ha of lightly oiled vegetation were burned using drip torches and propane torch for ignition. It was estimated that 75%–80% of the diesel in the burn footprint was removed. However, oil persisted in the soils above clean-up target levels, so 2.8 ha with the highest oiling were fertilised and tilled in the fall of 2000 and 1.4 ha again in late summer 2001. Lessons learned from this spill and burn include the following:

- Snow and ice can both help and hinder the use of *in-situ* burning on land. Snow and ice can slow the spread of oil, increasing the oil thickness and the overall efficacy of the burn. However, when the ice does not melt during the burn, it can slow the heat transfer process and prevent the oil from vaporising and burning. After the first burn, it was clear that there were areas that had been covered by ice and did not efficiently burn; thus, a second burn after the thaw was required. For spills in areas of snow and ice, sites should be surveyed during thaw periods or after the final thaw to determine the need for additional burning or other response actions.
- Snow and ice also slow natural oil weathering processes, particularly evaporation of light refined products, lengthening the window of opportunity for use of *in-situ* burning as a response option.
- Burning is not effective in removing oil that has penetrated into the soils. It is not clear whether burning had a wicking effect that drew oil to the surface at this site.
- The burn would have been more effective overall if it had been conducted soon after the release was discovered. Rapid burning would have reduced the potential for oil penetration into soils and the toxic effects to vegetation. Once it is clear that mechanical and manual removal efforts are not feasible, burning should be evaluated as a response option (summarised from Michel et al. 2002; Williams et al. 2003).

2.5.2 TIDAL MARSH BURNS

2.5.2.1 Crude Oil in Salt Marsh, Chiltipin Creek, Texas

On 7 January 1992, 460 m^3 of South Texas light crude oil were released from an underground pipeline near Chiltipin Creek, Copano Bay, Texas, affecting 10 ha of a high marsh dominated by salt grass, salt wort and shore grass. Responders used vacuum trucks to recover oil for 4 days. The oil in the marsh continued to spread below the dense vegetation. The forecast called for more rain, and it was feared that the oil would continue to spread and reach the Aransas River only about 450 m from the leading edge of the oil. Thus, a decision was made to burn the oil (which was 1–3 mm thick) even though there were only pockets of standing water (Figure 2.16). The oil was ignited with mineral spirits and burned for 21 h, removing 80%–85% of the remaining oil. An asphaltic, sticky burn residue had to be removed manually

(a)

(b)

FIGURE 2.16 Chiltipin Creek, Texas burn. (a) During the January 1992 burn and (b) post-burn in January 1993, 1 year later. The vegetation is dominated by salt grass, which is different than the species composition of unoiled-unburned areas. (Courtesy of Wes Tunnell, Texas A&M University, Corpus Christi, TX.)

using sorbents for the next 15 days. Studies were conducted to monitor recovery of the vegetation and biota and to check sediment oiling levels over a period of 5 years. By 2 years-post-burn, marsh vegetation was recovering, but with a pioneering species rather than the normal mix of species indicative of a healthy marsh (Figure 2.16). Even after 5 years, there were still more patches of bare soil compared to control areas. Without a water layer to protect the soils and roots, oil that penetrated into the soil and the subsequent burn killed most of the vegetation, and recovery was slow (summarised from Tunnell et al. 1995, Hyde et al. 1999).

2.5.2.2 Condensate Crude Oil in Brackish Marsh, Rockefeller Refuge, Louisiana

On 13 March 1995, 6.4 m³ of condensate crude oil spilled from a pipeline in the Rockefeller Refuge, Louisiana, affecting 20 ha of brackish marsh. Mechanical clean-up equipment had to be brought across wide extents of marsh, but was ineffective at collecting the oil and damaged the marsh. Prescribed burning was commonly used in the refuge to remove litter, reduce unwanted lightning fires and promote vegetative vigour (one-third of the refuge is burned annually). There were 5–10 cm of water over the marsh soil. *In-situ* burning of the oiled marsh was approved and conducted 4 days after the spill, removing the oil from 8 ha of the affected marsh. Studies conducted 3 years later concluded that the areas oiled and burned recovered better than the areas oiled but not burned. Three years after the burn, the burned areas attained the same plant density as the reference area. Lessons learned from this spill and burn include the following:

- Burning should have been considered early in the response; extensive damage occurred during the 4 days of mechanical removal efforts that were also ineffective.
- Containment boom on the marsh surface was ineffective at controlling the spread of the oil (summarised from Hess et al. 1997, Pahl et al. 1997, 2003).

2.5.2.3   Condensate Crude Oil in Brackish Marsh, Louisiana Point, Louisiana

On 23 February 2000, a release of an unknown amount of condensate crude oil was discovered from a buried pipeline about 500 m from the Sabine River channel at Louisiana Point, Louisiana, oiling an estimated 5.3 ha of brackish marsh. A fire break was created by laying down the vegetation on the downwind side and a burn was conducted 3 days after the discovery. Shortly after the burn started, however, the winds shifted from out of the south to out of the north; thus, the burn extended beyond the oiled area, with nearly 55 ha burned. Thus, the burned area was about 10 times the oiled area. Figure 2.17 shows vertical aerial photography of the area 3 months and 19 months after the burn, indicating good vegetation recovery except at the site of the pipeline break. The state of Louisiana requires aerial imagery and ground monitoring studies as part of its burn approval process.

Ground studies showed that, though there was some oil in the sediments near the release site, by 7 months later, all results were below the detection limit of 10 ppm. Vegetation recovery near the release site was slow; the very light crude oil was in contact with the marsh soils for at least 3–5 days before the burn. Lessons learned from this spill and burn include the following:

- The use of burning reduced the toxic effects of the condensate on the wetland vegetation. Although there were patches of dead vegetation in the most heavily oiled and burned areas close to the release site, areas with less oil had significant vegetative recovery by the end of the first growing season.
- *Batis* had the best recovery, whereas *Borrichia* and *Distichlis* had recovered more slowly after two growing seasons.
- The burned area was about ten times the oiled area. Fire breaks were placed to the north, which was the downwind direction at the start of the burn. However, the wind shifted during the burn, and the fire spread to the south. The burn was contained in some areas by these fire breaks that were now on the upwind sides, but the fire burned uncontrolled in downwind areas, only stopping at the water's edge or dense green vegetation (summarised from Michel et al. 2002).

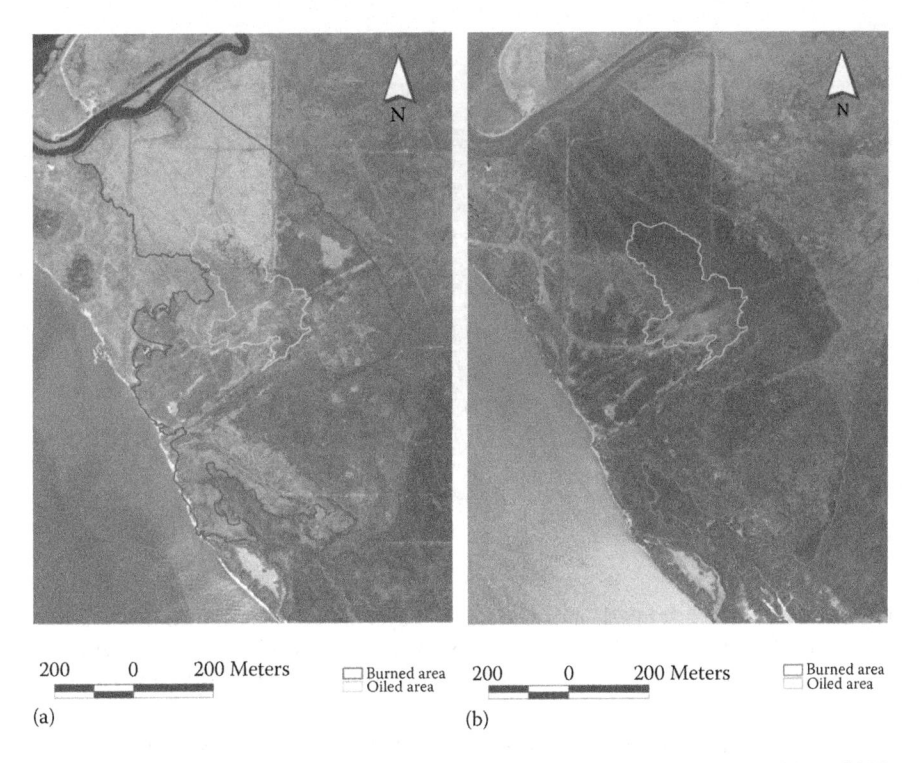

200　　0　　200 Meters　　☐ Burned area / Oiled area　　200　　0　　200 Meters　　☐ Burned area / Oiled area

(a)　　(b)

FIGURE 2.17 Vertical aerial photographs of the Louisiana Point burn site. (a) 5 May 2000, 3 months post-burn. The bright, inner line delineates the 5.3 ha oiled area; the dark, outside line delineates the 55 ha burned area. (b) 25 September 2001, 19-months post-burn. The disturbed area around the 'crater' at the leak site is visible. There is no visible difference in the unoiled/burned vegetation, compared to the adjacent unburned areas. (From Michel, J. et al., *Recovery of Four Oiled Wetlands Subjected to In Situ Burning*, American Petroleum Institute Publication 4724, Washington, DC, 71 p, 2002.)

2.5.2.4　Condensate Crude Oil in Brackish Marsh, Mosquito Bay, Louisiana

On 5 April 2001, 160 m^3 of condensate crude oil spilled in a brackish tidal marsh in Mosquito Bay, Louisiana, contaminating 15 ha, including 5 ha that were heavily oiled. After it was determined that manual removal was causing significant damage, the oiled vegetation was burned on 12 and 13 April (7 and 8 days after the spill occurred). The burn removed 90%–95% of the oil, with no burn residue, though oil that had penetrated into the burrows was not burned and had to be removed using sorbents at high tide. However, >40 ha were burned, nearly three times the oiled area; Figure 2.18 shows the extent of the first burn, which extended downwind to the edge of the water.

Lessons learned from this burn include the following:

- It is feasible and effective to burn condensate spills in marshes up to 1 week after the initial release, particularly where vegetation has trapped the oil and slowed the rate of evaporation. It should not be assumed that a condensate spill in a wetland will rapidly and completely evaporate.

FIGURE 2.18 The first marsh burn at Mosquito Bay, Louisiana, in April 2001. The arrows point to the fire break around the oiled area. Note that the vegetation burned all the way to the water's edge. (Courtesy of Steve Thum and Charlie Henry, NOAA, New Orleans, LA.)

- The burn area can greatly exceed the extent of the oil, even where the vegetation is green and in the peak of the growing season. This may be particularly true for species that have thin blades and relatively low moisture content. The marsh vegetation at the site had not been burned in several years; thus, there was abundant natural fuel to sustain a fire outside the oiled areas.
- Emergency fire breaks, such as flattening and wetting the vegetation with an airboat, may not be effective under even moderate wind conditions.
- Responders need to consider the potential consequences and back up control strategies where there is a risk of the burn spreading outside the oiled area.
- Light oils such as condensate burn completely with no residue; however, oil that has penetrated into burrows and root cavities in the soils will not be completely removed. Where the oil has penetrated the substrate, it may be necessary to deploy sorbents and containment booms to recover oil that is re-mobilised by high water or flooding by rainfall.
- Light oils such as condensate are acutely toxic to marsh vegetation where it penetrates the soils. Burning will not prevent toxic effects to vegetation that occurred prior to the burn. Therefore, it is important that burning be conducted early in an event to minimise the extent of injury to the habitat (summarised from Michel et al. 2002).

2.5.2.5 Crude Oil in a Brackish Marsh, Empire, Louisiana

On 29 August 2005 Hurricane Katrina made landfall near Buras, Louisiana, and caused multiple oil storage tanks to rupture, including one near Empire, where 600 m^3 of Louisiana Sweet crude oil were released. Most of the oil was contained in the retention pond at the facility. During Hurricane Rita (24 September 2005), 16–32 m^3 of oil were released into the adjacent brackish marsh. Of the 15.5 ha of the marsh that were oiled, 2 ha were heavily oiled and 6 ha moderately oiled, with 1–2 mm thick oil floating on 10–20 cm of water over the marsh surface. On 12 and 13 October, two burns were initiated and covered 7.9 ha of the marsh; only the areas with heavy to moderate oiling burned, removing 80%–90% of the oil in the burn footprint. Figure 2.19 shows aerial photographs of the burn area after the first burn and 5 months later. One of the important issues of concern was the potential for impaired visibility for vessels in the Mississippi River; thus, coordination with shipping interests was required. In field studies conducted 9 months and 1 year after

(a)

(b)

FIGURE 2.19 Empire, Louisiana, burn. (a) The dark areas in the marsh adjacent to the rectangular pond (indicated by an arrow) are the areas of heavy to moderate oiling and (b) the same area 5 months after the burn, showing good vegetation recovery. (Courtesy of Gary Shigenaka, NOAA, Seattle, WA.)

(a)   (b)

FIGURE 2.20   Photo-plots 6 months after the Empire, Louisiana, burn. (a) Oiled-burned site. (b) Unoiled-unburned reference site. (Courtesy of Joseph Baustian, The Nature Conservancy, Arlington, VA.)

the burn, oiled-burned sites were compared to unoiled-unburned (reference) sites. Total aboveground biomass, live biomass and dead biomass in the oil-burned areas were not significantly different than those in the reference areas after 1 year. Stem heights also showed recovery within 1 year and the number of stems of the dominant plant, *Scirpus*, in the oiled-burned areas was equal to, or greater than, that in the reference areas. Figure 2.20 shows photo-plots of an oiled-burned site compared to unoiled-unburned site 6 months after the burn. Complete recovery of the aboveground and belowground vegetation occurred within one growing season after the burn (Baustian et al. 2010). Lessons learned from this spill and burn include the following (Myers 2006, Merten et al. 2008):

- Crude oil spills in vegetation can be burned effectively over 6 weeks after the release if the oil is thick. Areas of lighter oiling in live vegetation could not sustain a fire.
- Aerial operations played a key role during both burns, helping to revise fire-control operations and re-position ground crews in real time.

2.5.2.6   Crude Oil in Tidal Fresh Marsh, Mississippi Delta, Louisiana

In late May 2014, a pipeline rupture released 16 m^3 of South Louisiana Crude oil, affecting 6 ha of tidal freshwater marsh consisting of *Phragmites australis* (common reed, Roseau cane). Due to the remote location, the degree of oiling and the difficulty of oil removal in the dense vegetation, a marsh *in-situ* burn was conducted for 2 ha of the more heavily oiled areas in early June 2014 while the marsh was flooded (water depth approximately 50 cm), removing 80%–90% of the oil (Figure 2.21).

Two years of post-burn studies, comparing oiled-burned, oiled-unburned, and reference (not oiled or burned) areas showed that oil concentrations in marsh soils were initially elevated in the oiled-burned sites; however, after 3 months, oil concentrations were similar to reference conditions and below background levels. Vegetation cover went from nearly zero after the burn to exceeding the reference and oiled-unburned sites after 2 years. However, the oiled-burned sites had very different

(a)

(b)

(c)

FIGURE 2.21 Burn of crude oil in a *Phragmites* marsh in the Mississippi River Delta, Louisiana, in June 2014. Aerial photographs on (a) the day before, (b) the day after, and (c) in March 2016. (Courtesy of Adam Davis, NOAA, Lowell, MA.)

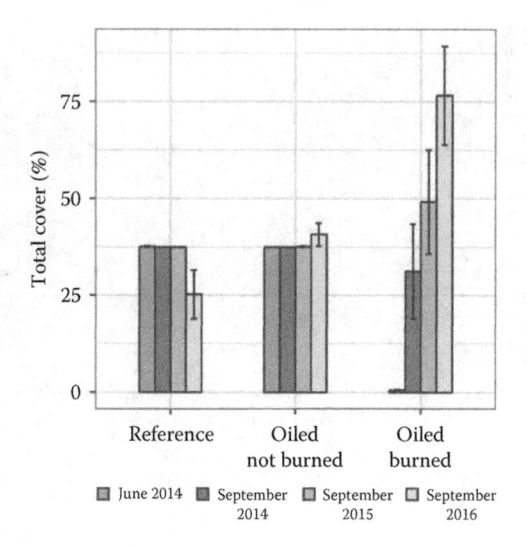

FIGURE 2.22 Total vegetation cover, 2014–2016, for the Mississippi Delta burn of tidal freshwater *Phragmites* marsh. Data are means ± 1 SE. N = 5 for all oiling/treatment classes. The oiled-burned sites exceeded the other treatment classes after 1 year. (From Zengel, S. et al., In situ burning and ecological recovery in an oil-impacted *phragmites australis* tidal freshwater marsh at Delta National Wildlife Refuge, Louisiana. *Proceedings: International Oil Spill Conference.* Long Beach, CA, pp. 2348–2368, 2017.)

vegetation species composition compared to the other sites, distributed among five freshwater species (Figure 2.22), which were considered as an increase in habitat quality. It was estimated that full recovery to dominance by *Phragmites* would take 5–6 years. Marsh snails recovered within 2 years. It was concluded that *in-situ* burning appears to be a viable response option to consider during future spills with similar marsh habitat and oiling conditions (Zengel et al. 2017).

2.6 EFFECTS OF BURNING SPILLED OIL ON LAND AND WETLANDS

Marsh burns have been conducted around the world, including recent well-documented burns in Louisiana and Texas. These burns were largely successful and provided important information on protecting the marsh plants and the best time of year to burn. Key factors that control the effects of burning oil on land and wetlands are summarised below.

2.6.1 SOIL HEATING

Soil heating can cause damage to the soil structure as well as destroying vegetation by permanently damaging roots and tubers. Research has shown that 60°C is a temperature above which damage to vegetation occurs (Reardon and O'Donnell 2013). Oil burning on the soil increases the heat flux to the soil. The amount of oil is present and soil moisture are important factors. Figure 2.4 shows a predictor of how far a 60°C temperature profile would penetrate into the soil for different degrees of soil moisture. This figure shows that moisture content of the soil is an important factor and should be considered in decision-making. There are general differences among vegetation types based on the depths of root systems. Many low shrubs have roots at depths of 2–5 cm. Grass has roots that are quite shallow, at 0.5–2 cm.

The amount of standing water is also a factor in reducing thermal stress to the roots of vegetation in the burn area. See the discussion of the burn-tank experiments using thermocouples in potted plants by Lin et al. (2002, 2005) in Section 2.3. It is preferred to have standing water or at least moist soils in the burn area to reduce the impact on vegetation and soils.

2.6.2 WATER LEVELS DURING AND AFTER THE BURN

Flooding can be a useful technique for flushing oil out of a marsh to thicken the oil for burning while protecting the roots of marsh plants. Flooding can sometimes be accomplished by building a berm across the drainage channels or by pumping water into the marsh. Care must be taken to use flood water of similar salinity to that normally in the marsh and to restore the natural drainage in the marsh after the burn. Studies have indicated that completely submerging plants for weeks after a burn can lead to poor recovery (Mendelssohn et al. 1995). Often marshes cannot be flooded, however, and thus burning could be conducted when the marsh soils are wet.

2.6.3 WILDLIFE CONSIDERATIONS

There are little data on the direct and indirect impacts of burning on land and wetlands on wildlife such as birds, terrestrial mammals, crustaceans, reptiles and amphibians. The literature has only limited mention of impacts to wildlife as a result of a burn. It is likely that mobile wildlife in the oiled area will already have been affected by the oil or moved out of the area. However, animals that live in the substrate are not able to move out of the oiled area, but they also can be protected by staying in their burrows. Examples include the report that many crayfish survived the oil and burn of a brackish wetland used for cattle grazing in Texas (Clark and Martin 1999) and the survival of fiddler crabs after the Mosquito Bay, Louisiana, burn.

The addition of oil as fuel can make a fire burn more quickly, meaning that animals may not be able to escape. During the Mosquito Bay, Louisiana, burn, where the burn spread beyond the oiled area with winds up to 40 km/h, there were reports of marsh birds being engulfed in the flames as they tried to flee the burning vegetation. Nesting birds may not abandon their nests until the fire approaches if the nests have not been directly oiled. Molting and juvenile flightless birds could be consumed in the fire. Therefore, if a burn is being considered during nesting season, wildlife agency staff members should be contacted to determine the importance of the proposed burn area as nesting habitat for birds and other wildlife. Of particular concern in the United States are species listed under the Endangered Species Act, such as gopher tortoise, whooping crane, Indiana and grey bat, listed frogs and salamanders, and many species of freshwater mussels. Wildlife agency staff members can also determine if there are likely any listed plant species in the proposed burn area. If animals and plants listed under the Endangered Species Act could be present, there will be a need to consult with the appropriate U.S. federal agency to confirm their presence in the burn area, in the area affected by the plume, or in the area affected by response actions related to the burn. The agency will advise the response on the likely impacts from the burn and provide best management practices to prevent or reduce impacts to wildlife. Such actions could include hazing of birds and mammals prior to the burn, physical removal of listed species, or delaying the burn until the wildlife migrate out of the area (this option would only be appropriate when the animals are very close to leaving the area).

2.6.4   BURN RESIDUES

Much has already been discussed in this chapter about the need to remove burn residues quickly and completely from the burn area on land and wetlands. Table 2.4 lists the likely behaviour of burn residues for different types of oil, with medium to heavy crudes and heavy residual products having the greatest potential to generate thick, sticky and persistent residues that often will sink in water once they cool. For efficient burns, the residue may represent 1%–10% of the volume of the oil burned; however, it can retard vegetation recovery and pose contact hazards for wildlife, including herbivores attracted by new vegetation in adjacent areas.

2.6.5   LESSONS LEARNED FROM PRESCRIBED BURNS

Prescribed burns, sometimes called controlled burns, are carried out frequently in many countries to maintain forests, grasslands, rangelands and wetlands. The purposes of prescribed burns vary from vegetation control to invasive species control, promoting of more desirable vegetation, habitat maintenance and wild fire prevention and control. There is an extensive body of literature of fire ecology (https://www.feis-crs.org/feis/) and expertise on the conduct of prescribed burning that should be used to support decision making for burning of oil on land and wetlands. Dahlin et al. (1999) provided the following major points based on lessons learned from prescribed burning, which were derived from Wright and Bailey (1982) and Whelan (1995):

- Soil temperature (in the root zone) influences plant survival more than surface temperature or aboveground temperature.
- Temperature and duration of the burn influence plant impact and survival more than maximum temperature.
- The organic content of the soil is important: inorganic soils are good insulators (5 cm of soil generally protects plant parts from high temperatures); organic soils, especially dry organic soils, can ignite and burn, causing severe impacts to the vegetation and the site.
- Higher fuel load and more flammable fuels cause hotter, more intense and potentially more damaging fires (fuel load refers to live and dead plant material).

- During the dormant season, plants have stored food in their roots and it is available for growth immediately after the fire or during the next growing season; thus, plant recovery is better during dormant season burns.
- During the growing season, the food has already been spent so little is left to support regrowth after a fire; thus, plant recovery can be poor during growing season burns.
- Fire can increase water yield from burned areas, increasing runoff and negatively affecting water quality and aquatic organisms in adjacent waterbodies by affecting turbidity, sedimentation, water levels, water temperature, dissolved oxygen and nutrients.
- Fire type is important. Hot, fast-moving fires, which may be more likely with the added fuel of the spilled oil, may be less damaging to the vegetation than slow-moving, low intensity fires.

Other factors derived from the experiences of prescribed burns include the potential need to control herbivores who might be attracted by the new green growth shortly after a burn; stabilise soils to prevent erosion and avoid long-term increases in water levels, which can kill plants.

2.7 SUMMARY

ISB on land can be conducted on a wide range of habitats, from ditches to wetlands. Lessons learned from ISB of spilled oil on land and wetlands are summarised below (Michel et al. 2002).

- Burning is most effective at reducing damage to vegetation and spreading of the oil to damage additional areas when it is used quickly. Oil is toxic to plants. The vegetation recovery was best when the oil was burned before it had a chance to penetrate into soils and affect plant roots.
- The window of opportunity for *in-situ* burning to be an effective means of oil removal can be days to months, depending on the spill conditions. Dense vegetation can slow evaporation, extending the window of opportunity for use of *in-situ* burning. For spills with snow and ice cover, burning may still be effective months later. Furthermore, under snow and ice conditions, it may be necessary to consider additional burns during thaw periods and after the final thaw.
- Light oils (gasoline, aviation fuel, diesel, No. 2 fuel oil, condensate and light crude oils) generally burns completely, with no residue. Medium and heavy crude oils and heavy refined products are likely to form burn residues. If these residues are thick, they have to be removed to speed habitat recovery and reduce contact risks to wildlife.
- It is best to have ample water over the soils in vegetated areas to prevent killing the roots and tubers by the high heat of the fire and reduce the risk of oil penetration into the soils. At a minimum, the soils in vegetated areas should be water saturated. More water is preferred, but not essential.

- Burning does not reduce the toxic effects of the oil that occurred prior to the burn. It can be very effective, however, at reducing the extent and degree of additional impact by quickly and efficiently removing the remaining oil.
- Responders considering the use of *in-situ* burning should be very aware of the possibility that the fire will spread to unoiled areas. Healthy, green, unoiled vegetation is not always an effective fire break, particularly down-wind and for vegetation that has low moisture content. Fires can quickly jump the kinds of fire breaks placed during spill emergencies. Responders also often cannot wait for ideal weather conditions to initiate the burn. Thus, responders should consider the consequences of burning adjacent areas in their burn plans. Where the spread of the fire is determined to be unaccept-able, then additional efforts are needed to control the spread of the fire.

REFERENCES

American Petroleum Institute (API). 2015. Field operations guide for in-situ burning of inland oil spills. API Technical Report 1251. Washington, DC: American Petroleum Institute. 74 p.

American Petroleum Institute (API). 2017. In-situ burning: A decision maker's guide. API Technical Report 1256. Washington, DC: American Petroleum Institute. 67 p.

Baustian, J., I.A. Mendelssohn, Q. Lin and J. Rapp. 2010. *In-situ* burning restores the eco-logical function and structure of an oil-impacted coastal marsh. *Environmental Management*. 46(5): 781–789.

Blenkinsopp, S., G. Sergy, P. Lambert, Z. Wang, S. Zoltai and M. Siltanen. 1996. Long-term recovery of peat bogs oiled by pipeline spills in Northern Alberta. *Proceedings: Arctic and Marine Oil Spill Program Technical Seminar*. Calgary, AB, pp. 1335–1354.

Buhite, T.R. 1979. Cleanup of a cold weather terrestrial pipeline spill. *Proceedings: International Oil Spill Conference*. Los Angeles, CA, pp. 367–369.

Buist, I.A. 1998. Window of opportunity for in-situ burning. *Workshop Proceedings In-situ Burning of Oil Spills*. New Orleans, LA: National Institute of Standards and Technology, Special Publication 935, pp. 21–30.

Buist, I., and K. Trudel. 1995. Laboratory studies of the properties of in-situ burn residues. *MSRC Technical Report Series Technical Report Series 95–010*. Washington, DC: Marine Spill Response Corporation. 110 p.

Burns, R.C. 1988. Cleanup and containment of a diesel fuel spill to a sensitive water body at a remote site under extreme winter conditions. *Proceedings of the 11th Arctic and Marine Oil Spill Program Technical Seminar*, British Columbia: Vancouver, June 7–9, 1988. Ottawa, Ontario: Environment Canada, pp. 209–220.

Castle, Robert E. RCE, Inc., Personal communication, 10 January 2013.

Clark, T., and R.D. Martin. 1999. In-situ burning: After action review (successful burn 48 hours after discharge). *Proceedings: International Oil Spill Conference*. Seattle, WA, pp. 1273–1274.

Dahlin, J.A., S. Zengel, C. Headley and J. Michel. 1999. Compilation and review of data on the environmental effects of In-situ burning of inland and upland oil spills. Report No. 4684, Washington, DC: American Petroleum Institute. 121 p + appendices.

Earthfax Engineering Inc. 2003. Final remediation report for the Chevron Pipeline Company Salt Lake to Spokane Products Systems Pipeline Milepost 69 Unleaded Gasoline Release near Tremonton. Salt Lake City, UT: Chevron Pipeline Company. 439 p.

Entrix. 2003. October 2003 Post-burn monitoring report. New Orleans, LA: Chevron Company. 37 p.

Eufemia, S.J. 1994. Brunswick Naval Air Station JP-5 aviation fuel discharge in-situ burn of fuel remaining in fresh water marsh 6–8 April 1993. *Proceedings: In Situ Burning Oil Spill Workshop*. Gaithersburg, MD: National Institute of Standards and Technology Special Publication 867. pp. 87–90.

Gonzalez, M.F., and G.A. Lugo. 1995. Texas marsh burn: Removing oil from a salt marsh using in situ burning. *Proceedings: In Situ Burning Oil Spill Workshop*. Gaithersburg, MD: National Institute of Standards and Technology Special Publication. pp. 77–85.

Hartley, A.E. 1996. Overview of the Kolva River Basin 1995 oil recovery and mitigation project. *Proceedings: Arctic and Marine Oilspill Program Technical Seminar*. Calgary, AB. pp. 1301–1307.

Henry, C.B. 1997. Vermillion oil spill: In situ burn and monitoring study. Chemistry Report IES/RCAT97.30. Baton Rouge, LA: Institute for Environmental Studies, Louisiana State University. 3 p.

Hess, T.J., L.I. Byron, H.W. Finley and C.B. Henry. 1997. The rockefeller refuge oil spill: A team approach to incident response. *Proceedings: International Oil Spill Conference*. Ft. Lauderdale, FL. pp. 817–821.

Holt, S., S. Rabalais, N. Rabalais, S. Cornelius and S.J. Holland. 1978. Effects of an oil spill on salt marshes at Harbor Island, Texas. I. Biology. *Proceedings: Conference on Assessment of Ecological Impacts of Oil Spills*. Keystone, CO, pp. 344–252.

Hyde, L.J., K. Withers and J.W. Tunnell. 1999. Coastal high marsh oil spill cleanup by burning: 5-year evaluation. *Proceedings: International Oil Spill Conference*. Seattle, WA, pp. 1257–1260.

Labay, A. 1997. Pollution complaint detailed report, Event ID 19973A332v1. Austin, TX: Texas Parks and Wildlife Department, Resources Protection Division, 2 p.

Leppälä, S. 2004. A crude oil in situ burn in a peat bog. *Proceedings: Freshwater Spills Symposium*. New Orleans, LA.

Lin, Q., I.A. Mendelssohn, N.P. Bryner and W.D. Walton. 2005. In situ burning of oil in coastal marshes. 1. Vegetation recovery and soil temperature as a function of water depth, oil type, and marsh type. *Environmental Science & Technology*. 39(6): 1848–1854.

Lin, Q., I.A. Mendelssohn, K. Carney, N.P. Bryner and W.D. Walton. 2002. Salt marsh recovery and oil spill remediation after in situ burning: Effects of water depth and burn duration. *Environmental Science & Technology*. 36(4): 576–581.

McCauley, C.A., and R.C. Harrel. 1981. Effects of oil spill cleanup techniques on a salt marsh. *Proceedings: International Oil Spill Conference*. Atlanta, GA, pp. 401–407.

Mendelssohn, I.A., M.W. Hester and J.W. Pahl. 1995. *Environmental Effects and Effectiveness of In Situ Burning in Wetlands: Considerations for Oil Spill Cleanup*. Baton Rouge, LA: Louisiana Oil Spill Coordinator's Office/Office of the Governor, Louisiana Applied Oil Spill Research and Development Program. 57 p.

Merten, A.A., C.B. Henry and J. Michel. 2008. Decision-making process to use in situ burning to restore an oiled intermediate marsh following Hurricanes Katrina and Rita. *Proceedings: Oil Spill Conference*. Savannah, GA, pp. 545–550.

Metzger, R.A. 1995. *1994 Ecological Assessment. Naval Air Station Brunswick, Maine*. Pittsburgh, PA: Halliburton NUS Corp. 28 p. + appendices.

Michel, J., Z. Nixon and H. Hinkeldey. 2002. *Recovery of Four Oiled Wetlands Subjected to In Situ Burning*. Washington, DC: American Petroleum Institute Publication 4724. 71 p.

Michel, J. and N. Rutherford, N. 2013. *Oil Spills in Marshes: Planning & Response Considerations*. Seattle, WA: Emergency Response Division, NOAA and Washington, DC: American Petroleum Institute, 129 p.

Moir, M.E., and B. Erskin. 1994. In-situ burning of oil spills on land: A case study. *Proceedings: Arctic and Marine Oilspill Program Technical Seminar*. Vancouver, BC, pp. 651–655.

Myers, J. 2006. In situ burn and initial recovery of a South Louisiana intermediate marsh. *Proceedings: Arctic and Marine Oilspill Program Technical Seminar*. Vancouver, BC, pp. 977–987.

Nagy, P. 1991. Environmental pollution caused by crude oil pipelines. *Proceedings: National Environmental Protection Conference*. Balatonaliga, HU.

National Fire and Aviation Management (FAMWEB). 2015. Modeling site. Available at http://www.fs.fed.us/fire/planning/nist. Accessed on 11 March 2017.

National Oceanic and Atmospheric Administration (NOAA). For NOAA (2013) that was the date of the report on the spill. Accessed on 11 March 2017. Bayou Sorrel Oil Spill. Incident News. Available at https://incidentnews.noaa.gov/incident/7697.

Overton, E.B., J.A. Mcfall, S.W. Mascarella, C.F. Steele, S.A. Antoine, I.R. Politzer and J.L. Laseter. 1981. Identification of petroleum residue sources after a fire and oil spill. *Proceedings: International Oil Spill Conference*. Atlanta, GA, pp. 541–546.

Pahl, J.W., I.A. Mendelssohn, C.B. Henry and T.J. Hess. 2003. Recovery trajectories after in situ burning of an oiled wetland in Coastal Louisiana, USA. *Environmental Management*. 31(2): 236–251.

Pahl, J.W., I.A. Mendelssohn and T.J. Hess. 1997. The application of in situ burning to a Louisiana coastal marsh following a hydrocarbon product spill: Preliminary assessment of site recovery. *Proceedings: International Oil Spill Conference*. Ft. Lauderdale, FL, pp. 823–828.

Reardon, J. and E.J. O'Donnell. 2013. Prediction of soil heating resulting from oil spill in-situ burning. *Proceedings: Arctic and Marine Oilspill Program Technical Seminar*, Halifax, NS, pp. 888–898.

Scholz, D.K., S.R. Warren, A.H. Walker and J. Michel. 2004. *Risk Communication for In-situ Burning: The Fate of Burned Oil*. Washington, DC: American Petroleum Institute Publication No. 4735.

S.L. Ross Environmental Research Ltd. (SL Ross). 2002. *Identification of Oils that Produce Non-Buoyant In-situ Burning Residues and Methods for Their Recovery*. Washington, DC: American Petroleum Institute Publication No. DR145.

Smith, R. 2011. *Firebreak Location, Construction and Maintenance Guidelines Fire and Emergency Services Authority of Western Australia*, 480 Hay Street, Perth, Western Australia. 25 p.

Tunnell, J.W., B. Hardegree and D.W. Hicks. 1995. Environmental impact and recovery of a high marsh pipeline oil spill and burn site, Upper Copano Bay, Texas. *Oil Spill Conference*. Long Beach, CA, pp. 133–138.

U.S. Department of Agriculture (USDA). 2010. *Firebreak Design Procedures. Natural Resource Conservation Service*. Washington, DC: NE-T.G. Notice 619 Section IV, 13 p.

Whelan, R.J. 1995. *The Ecology of Fire*. Cambridge, UK: Cambridge University Press. 346 p.

Williams, G.W., A.A. Allen, R. Gondek and J. Michel. 2003. Use of in situ burning at a diesel spill in wetlands and salt flats, Northern Utah, USA: Remediation operations and 1.5 years of post-burn monitoring. *Proceedings: International Oil Spill Conference*. Vancouver, BC, pp. 109–114.

Wright, H.A., and A.W. Bailey. 1982. *Fire Ecology, United States and Southern Canada*. New York: John Wiley and Sons. 501 p.

Zengel, S., J. Weaver, Z. Nixon, S. et al. 2017. In situ burning and ecological recovery in an oil-impacted *phragmites australis* tidal freshwater marsh at Delta National Wildlife Refuge, Louisiana. *Proceedings: International Oil Spill Conference*. Long Beach, CA. pp. 2348–2368.

Zischke, J.A. 1993. *Benthic Invertebrate Survey: Meire Grove Pipeline Project*. Report for Delta Environmental Consultants. St. Paul, MN 7 p.

3 In-Situ Burning on Ice, Snow or in the Arctic

Merv Fingas

CONTENTS

3.1 INTRODUCTION

In-situ burning is the oldest technique applied to oil spills in the Arctic. This is because logistics to carry out other types of countermeasures is often severely hampered in such remote regions. Similarly, in areas where ice and snow are prevalent, spilled oil may be contained and thus be more amenable to burning.

The first reference in the literature to the burning of oil on water was the use of a log boom to burn oil on the northern Mackenzie River in 1958 (ASTM 2013; McLeod and McLeod 1972; Swift et al. 1968). There was extensive research on in-situ burning of oil spills in the late 1970s, much of it directed at ice or snow situations, particularly in the Arctic.

Over the years, research into in-situ burning has included laboratory-, tank- and full-scale test burns. In the late 1970s, several burn tests and studies were carried out in Canada by a consortium of government and industry agencies. Oil resurfaced on first-year ice was burned in the Beaufort Sea Burn in 1975. Some tests in the early 1980s were performed by ABSORB (now Alaska Clean Seas) and the United States Minerals Management Service (USMMS) to evaluate the burning of oil in ice-covered areas. This research covered environmental and oil conditions such as sea state, wind velocities, air and water temperatures, ice coverage, oil type, slick thickness and degree of oil weathering and emulsification (Purves 1977). Table 3.1 lists some of the key tests and burns related to the Arctic and ice or snow.

Much of the technology of in-situ burning is similar in ice and snow situations as it is in the south or non-freezing conditions; however, some considerations may be different.

Ignition: Ignition is similar in hot or cold regions as ignition temperatures are typically 500°C or higher. Thus, a few degrees of ambient conditions do not make a difference.

Oil containment: Under many conditions in snow or ice, the oil is already contained. This is a distinct advantage to burning in snow or ice conditions over open water.

Logistics: Shortage of logistics is a particular reason why burning is attractive in remote and Arctic regions. Burning requires less logistics in such areas.

Emissions: The emissions from burning are similar in cold or hot regions.

Monitoring the fire and emissions: Monitoring is generally more difficult in remote and Arctic regions; however, monitoring to prevent fire spread is generally not an issue.

Protection of human health: Because of remoteness, exclusion zones are not generally an issue in Arctic or remote regions.

Overall: The situations summarised above favour the use of in-situ burning as a countermeasure for oil spills in Arctic or remote regions where there are snow and/or ice.

TABLE 3.1

History of Burning in Ice and Snow

Year	Country Location	Description	Events	Lessons	References
1952	United States, Vermont	Portland pipeline break	Spill into Black River, some burned	Ignition requires high heat and primer	Northland Journal (2016)
1958	Canada	Mackenzie River, NWT	First recorded use of in-situ burning, on river using log booms	In-situ burning possible with use of containment	McLeod and McLeod (1972)
1967	Britain	Torrey Canyon	Cargo tanks difficult to ignite with military devices	There may be limitations to burning	Swift et al. (1968)
1969	Netherlands	Series of experiments	Igniter KONTAX tested, many slicks burned	Burning at sea is possible	Batelle (1979)
1969	Finland	Sinking released diesel; fuel	Diesel burned on shore and bays	Paraffin oil used as primer	Buist et al. (2016)
1969	Finland	*Tanker Raphael*	Crude oil released by grounding	large amounts of oil burned using petrol, peat moss	Buist et al. (2016)
1970	Canada	*Arrow Tanker* sank	Limited success burning in confined pools	Confinement may be necessary for burning	Arrow Spill (1970)
1970	Sweden	Spill	Oil burned among ice and in pools	Can burn oil contained by ice	Batelle (1979)
1970	Canada	Deception Bay	Oil burned among ice and in pools	Can burn in ice and in pools	Ramseier et al. (1973)
1970	Canada	*Arrow Tanker* sank	Bunker C released	Hard to ignite oil, little burned	Arrow Spill (1970)
1970	Sweden	Vessel; collision	Bunker C trapped in sea ice	Cabosil used, burned some	Buist et al. (2016)
1972	Sweden	Ice-choked river	Diesel burned with Sanerinsull	Large amounts of oil burned	Buist et al. (2016)
1973	Canada	Rimouski – experiment	Several burns of various oils on mud flats	Demonstrated high removal rates possible, >75%	Environment Canada (1976)
1975	Canada	Balaena Bay – experiment	Multiple slicks from under ice oil ignited	Demonstrated ease of burning oil on ice	Environment Canada (1977)

(*Continued*)

TABLE 3.1 (*Continued*)
History of Burning in Ice and Snow

Year	Country Location	Description	Events	Lessons	References
1976	United States	*Argo Merchant*	Tried to ignite thin slicks at sea	Not able to burn thin slicks on open water	Batelle (1979)
1976	Canada	*Imperial St. Clair*	Diesel and gasoline incorporated into ice	Able to burn much of it, oily rags as igniters	Buist et al. (2016)
1976	Canada	Yellowknife – experiment	Test of parameters controlling burning not oil type alone	Multiple parameters controlling burning, not oil type alone	Fingas (1999)
1977	United States, Buzzards Bay	*Barge Bouchard*	*Barge Bouchard* leaked heating fuel about 1/20 burned	Aerial ignition from helicopter worked well	Fingas (1999)
1978–1982	Canada	Series of experiments	Studied many parameters of burning	Found limitations to burning was thickness	Fingas (1999)
1979	Mid-Atlantic	*Atlantic Empress* Aegean captain	Uncontained oil burned at sea after accident	Uncontained slicks will burn at sea directly after spill	Fingas (1999)
1979	Canada	*Imperial St. Clair*	Burned oil in ice conditions	Can readily burn fuels in ice	Fingas (1999)
1980	Canada	McKinley Bay – experiment	Several tests involving igniters, different thicknesses	Test of igniters, measured burn rates	Fingas (1999)
1981	Canada	McKinley Bay – experiment	Tried to ignite emulsions	Noted difficulty in burning emulsions	Fingas (1999)
1983	Canada	*Edgar Jourdain*	Vessel containing fuels and nearby fuel ignited	Practical effectiveness of burning amongst ice	Fingas (1999)
1983	United States	Beaufort Sea – experiment	Oil burned in ice	Ability to burn in ice	Fingas (1999)
1984	Canada	Series of experiments	Tested the burning of uncontained slicks	Uncontained burning only possible in few conditions	Fingas (1999)

(Continued)

TABLE 3.1 (*Continued*)
History of Burning in Ice and Snow

Year	Country Location	Description	Events	Lessons	References
May 1984	United States	Beaufort Sea – experiment	Burning with various ice coverages tested	Burning with various ice coverages possible	Fingas (1999)
June 1984	United States	OHMSETT – experiments	Oil burned among ice	Ice concentration not important, Emulsions don't burn	Fingas (1999)
1985	Canada	Offshore Atlantic – experiment	Oil in ice burned after physical experiment	Ease of burning amongst ice	Fingas (1999)
1985	Canada	Esso–Calgary – experiments	Several slicks in ice leads burned	Ease of burning in leads	Fingas (1999)
1986	Canada	Ottawa – experiments/ analysis	Analysed residue and soot from several burns	Analysis shows PAHs about same in oil and residue	Fingas (1999)
1986	United States	Seattle and Deadhorse – experiment	Test of the Helitorch and other igniters	First demonstrations of Helitorch as practical	Fingas (1999)
1986–1991	United States	NIST – experiments	Many lab-scale experiments	Science of burning, rates, soot, heat transfer	Fingas (1999)
1986–1991	Canada	Ottawa – analysis on above	Analysed residue and soot from several burns	Found PAHs and others; not major problem	Fingas (1999)
1988	Norway	U-Boom burn	Spitzbergen burn	Found U-boom good for burn	Buist et al. (2016)
1989	United States	*Exxon Valdez*	A test burn performed using a fire-proof boom	One burn demonstrated practicality and ease	Fingas (1999)
1991	United States	First set of Mobile experiments	Several test burns in newly constructed pan	Several physical findings and first emission results	Fingas (1999)
1992	United States	Second set of Mobile burns	Several test burns in pan	Several physical findings and emission results	Fingas (1999)

(Continued)

TABLE 3.1 (Continued)
History of Burning in Ice and Snow

Year	Country Location	Description	Events	Lessons	References
1992	Canada	Several test burns in Calgary	Emissions measured and ferrocene tested	Showed smokeless burn possible	Fingas (1999)
1993	Canada	Newfoundland offshore burn	Successful burn on full scale offshore	Hundreds of measurements, practicality demonstrated	Fingas (1999)
1994	United States	Third set of Mobile burns	Large-scale diesel burns to test sampler	Many measurements taken	Fingas (1999)
1994	United States	North Slope burns	Large-scale burn to measure smoke	Trajectory and deposition determined	Fingas (1999)
1994	Norway	Series of Spitzbergen burns	Large-scale burns of crude and emulsions	Large area of ignition results in burn of emulsions	Fingas (1999)
1994	Norway	Series of Spitzbergen burns	Trial of uncontained burn	Uncontained burn largely burned	Fingas (1999)
1996	Britain	Burn test	First containment burn test in Britain	Demonstrated practicality of technique	Fingas (1999)
1996	United States	Test burns in Alaska	Igniters and boom tested	Some measurements taken	Fingas (1999)
1997	United States	Fourth set of mobile burns	Small-scale diesel burns to test booms	Emissions measured and booms tested	Fingas (1999)
1997	United States	North Slope tank tests	Conducted several tests on waves/burning	Waves do not strongly constrain burning	Fingas (1999)
1998	United States	Fifth set of Mobile burns	Small-scale diesel burns to test booms	Emissions measured and booms tested	Fingas (1999)
2001	United States	Boom tests in OHMSETT	Small-scale propane tests of test booms	Tested some new fire-resistant booms	Fingas (1999)

(Continued)

TABLE 3.1 (*Continued*)
History of Burning in Ice and Snow

Year	Country Location	Description	Events	Lessons	References
2002	United States	Small-scale tests in Alaska	Tested burning in frazil and brash ice	Frazil and brash ice reduce burning rate	Fingas (1999)
2002, 2003	Canada	Small-scale heavy oil burns	Burned heavy oil and Orimulsion in test pans	Burning rate of heavy oil, ignition methods, emissions	Fingas (1999)
2006–2008	Norway	Spitzbergen burns	Test burns on oil	Oil released in ice sheet in spring	Buist et al. (2016)
2008	Norway	Use of herders for ISB	Two burns of crude oil using chemical herding agents to concentrate and contain the burn	Demonstrated effectiveness of herders in ice-affected waters	Buist et al. (2016)
2009	Norway	Use of firebooms in ice	Used fire-resistant boom to contain burning oil in 1/10th and 5/10ths concentrations	Demonstrated effectiveness of firebooms in open and very open drift ice	Buist et al. (2016)

3.2   ICE CONDITIONS UNDER CONSIDERATION

An important consideration for countermeasures is the ice conditions in which the oiling has occurred. There exists standard ice terminology which should be used for this purpose (Fingas 2015; WMO 2017). The relevant definitions and descriptions of how this affects in-situ burning will be discussed in the following sections. The ice terms will be presented alphabetically.

3.2.1   BRASH ICE

Brash ice is accumulations of floating ice made up of fragments not more than 2 m across, the wreckage of other forms of ice. This is illustrated in Figure 3.1. This is a frequent ice condition occurring during ice breakup in spring time, as well as other times when ice is under stress from winds and currents. Oil in brash ice would be considered burnable; however, some of the oil would adhere to the ice crystals and thus be unavailable for burning (Figure 3.2). Under calm wind conditions, the oil would be largely contained in such conditions.

3.2.2   BROKEN ICE

Broken ice is an unacceptable term that has made its way into oil spill literature. 'Broken ice' apparently refers to any ice that is not continuous. 'Broken ice' should not be used as it appears in no ice dictionary and does not refer to any specific condition. Use the appropriate terms in this chapter instead.

FIGURE 3.1   Brash ice consists of fragments of larger forms of ice. The small fragments in this photo are brash ice. (From http://www.photolib.noaa.gov/htmls/.)

FIGURE 3.2 Burning in brash ice. The foreground consists of pack ice. (From Buist, I.A., et al., In-situ burning in ice-affected waters: A technology summary and lessons from key experiments, Joint Industry Project Report. Washington, DC, 2013.)

FIGURE 3.3 Close ice. (From http://www.photolib.noaa.gov/htmls/.)

3.2.3 CLOSE ICE

Floating ice, in which the concentration is 7/10–8/10, composed of floes mostly in contact (Figure 3.3). Oil could be readily burned in close ice. The leads between close ice are ideal for containing oil.

3.2.4 COMPACT ICE

Compact ice is floating ice in which the concentration is 10/10 and no water is visible. Burning in compact ice would be similar to burning on ice.

3.2.5   Compacted Ice Edge

This is a clear-cut ice edge compacted by wind or current, usually on the windward side of an area of drift ice. Figure 3.2 shows an image that would be similar to a compacted ice edge and burning along that edge. A group of scientists carried out an experiment of oil under and in ice near Svalbard, Norway. The oil was allowed to surface, where it was ignited with gelled hexane (Dickins et al. 2008). The oil was Statfjord crude, 3,400 L and, once weathered, 27% was 2,480 L. The thickness was calculated to be 35 mm and covered an area of 69 m². The burn endured for 11 minutes and the 1 mm of residue yielded 106 L of 0.95 g/mL density (Fritt-Rasmussen et al. 2015). This burn reduced the volume by 96% and the burn rate was 3.1 mm/min.

Two small oil releases involving 7 and 2 m³ of oil were released into 70%–90% of ice in Svalbard, Norway (Brandvik et al. 2010). These were allowed to weather for 5 and 1 days, respectively. Both were found to be ignitable using laboratory apparatus; however, only the smaller slick was burned at sea. The larger slick was found to be too mixed with ice to be recoverable by any method.

3.2.6   Drift Ice

Drift ice is used in a wide sense to include any area of sea ice other than fast ice no matter what form it takes or how it is distributed. When concentrations are high, that is, 7/10 or more, drift ice may be replaced by the term *pack ice*. Drift ice is shown in Figures 3.4 and 3.5. Drift ice is a common condition during spring break-up and also sometimes in partial fall freeze-up conditions when there is wave action. Burning is favourable in drift ice conditions.

3.2.7   Fast Ice

Fast ice is sea ice which forms and remains fast along the coast, where it is attached to the shore, to an ice wall, to an ice front or between shoals or grounded icebergs.

FIGURE 3.4   A wide field of drift ice. (From http://en.wikipedia.org/wiki/Drift_ice.)

FIGURE 3.5 A narrow field of drift ice. This is similar to open water when considering burning; however, in some locations there are ice floes and some brash ice.

Vertical fluctuations may be observed during changes of sea level. Fast ice may be formed in situ from sea water or by freezing of floating ice of any age to the shore, and it may extend a few metres or several hundred kilometres from the coast (Figure 3.6). The fast edge is the junction or demarcation at any given time between fast ice and open water.

Fast ice may be more than one year old and may then be prefixed with the appropriate age category (old, second-year, or multi-year). If it is thicker than about 2 m above sea level, it is called an ice shelf. Burning along the edge of fast ice has taken place in the past; however, burning on fast ice is not practical and furthermore the oil is unlikely to enter a fast ice area.

FIGURE 3.6 Fast ice that has developed along a rugged shoreline. (From seaice. Alaska.edu.)

3.2.8 FIRST-YEAR ICE

First-year ice is sea ice of not more than one winter's growth, developing from young ice, with a thickness of 30 cm–2 m. First year ice may be subdivided into thin first-year ice/white ice, medium first-year ice and thick first-year ice. Oil rising under first-year ice, such as from a sub-sea blowout, will migrate through first-year ice in spring time when the brine drainage channels form. This is illustrated in Figures 3.7 and 3.8 which show an experiment in which oil was released under first-year ice and then burned on the surface after the oil rose through the ice in spring.

FIGURE 3.7   Oil was released under first-year ice in this Beaufort Sea experiment. What is shown here is the oil surfacing through the ice in spring time.

FIGURE 3.8   The oil shown in Figure 3.7 being burned. Combustion removed most of the resurfaced oil.

3.2.9 Floe

A floe is any relatively flat piece of sea ice 20 m or more across. Floes are sub-divided according to horizontal extent.

3.2.10 Frazil Ice

Frazil ice is fine spicules or plates of ice, suspended in water. Frazil ice is typically the first stage of freeze-up and can make the surface of water appear dark or grey. Frazil ice is illustrated in Figures 3.9 and 3.10. Frazil ice is encountered

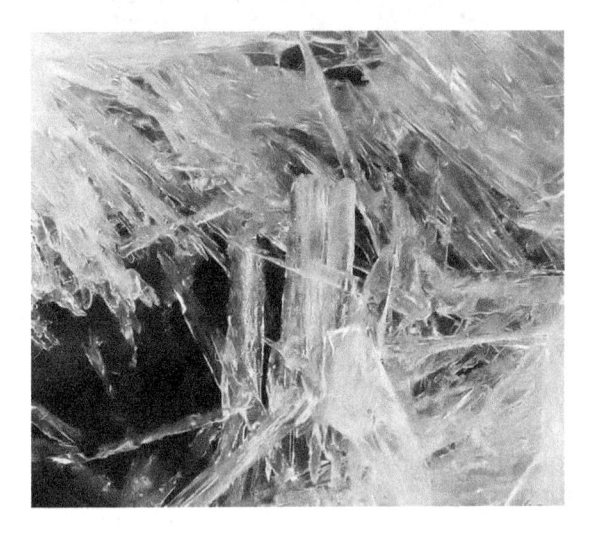

FIGURE 3.9   A close-up of frazil ice showing a cluster of individual crystals. (From https://summitcountyvoice.com/2010/05/10/the-daily-photo-frazil-ice-on-dillon-reservoir/.)

FIGURE 3.10   A view of frazil ice clusters viewed from about 3 m. The frazil ice accumulates, then grease ice and then pancake ice may result.

FIGURE 3.11 Grease ice shown at a distance of about 3 m. (From http://ccom.unh.edu/notes-arctic-swerus-c3-expedition.)

frequently in spills that occur in early freeze-up periods. Frazil ice serves to contain the oil for burning.

3.2.11 GREASE ICE

Grease ice is a later stage of freezing than frazil ice when the crystals have coagulated to form a soupy layer on the surface. Grease ice reflects little light, giving the sea a matte appearance, as shown in Figure 3.11. Grease ice serves to contain oil and allow burning.

3.2.12 LAKE ICE

Lake ice has many of the same forms as sea ice, as described in this chapter. Many burns have taken place in lake ice.

3.2.13 LEAD

A lead is any fracture or passageway through sea ice which is navigable by surface vessels. Burning in leads has occurred in the past. Leads are conducive to burning in that the lead is sheltered somewhat and the oil is contained (Brown and Goodman 1986). Figure 3.12 shows a lead; Figure 3.13 shows burning oil in a lead.

FIGURE 3.12   A lead open in pack ice at sea.

FIGURE 3.13   Burning oil in a lead. (Courtesy of Sintef.)

3.2.14 MULTI-YEAR ICE

Multi-year ice is old ice up to 3 m or thicker, that has survived at least two summers' melt. Hummocks are present that are even smoother than in second-year ice, and the ice is almost salt-free. The colour of the ice, where bare, is usually blue. Melt pattern consists of large interconnecting irregular puddles and a well-developed drainage system. Oil trapped under multi-year ice may resurface through fractures, which occur infrequently. Multi-year ice does not lend itself to in-situ burning of oil.

3.2.15　New Ice

New ice is a general term for recently formed ice which includes frazil ice, grease ice, slush and shuga. These types of ice are composed of ice crystals which are only weakly frozen together (if at all) and have a definite form only while they are afloat. As noted in the sections under each term, new ice is generally favourable for burning as oil is contained by the forms of new ice.

3.2.16　On-Ice Burning

Sometimes oil accumulates on the surface of ice. In these cases, the oil can be burned. Figure 3.14 shows the burning of oil on ice. This oil was contained by ice rubble, making the burning efficient.

3.2.17　Open Ice

Open ice is floating ice in which the ice concentration is 4/10–6/10, with many leads and polynyas, and the floes are generally not in contact with one another. Open ice is very favourable for in-situ oil burning.

3.2.18　Open Water

Open water is a large area of freely navigable water in which sea ice is present in concentrations less than 1/10. No ice of land origin is present. This is often a target area for in-situ burning.

FIGURE 3.14　Burning oil on ice. The oil is contained by ice rubble, as shown in the foreground.

3.2.19 Pack Ice

Pack ice is a term used in a wide sense to include any area of sea ice other than fast ice no matter what form it takes or how it is disposed. Pack ice is often used when concentrations are high, that is, 7/10 or more. Burning in pack ice has taken place in the past and is one of the many situations with the possibility for burning with ice present.

3.2.20 Pancake Ice

Pancake ice consists of predominantly circular pieces of ice from 30 cm to 3 m in diameter, and up to about 10 cm in thickness, with raised rims due to the pieces striking against one another. It may be formed on a slight swell from grease ice or slush or as a result of the breaking of ice rind or, under severe conditions of swell or waves, of grey ice. It also sometimes forms at some depth at an interface between water bodies of different physical characteristics, in which it floats to the surface; its appearance may rapidly cover wide areas of water. Pancake ice is illustrated in Figures 3.14 and 3.15. Pancake ice is one of the conditions allows in-situ burning as it would contain floating oil (Figure 3.16).

3.2.21 Burning in River Ice

Burning on rivers and river ice is quite an old practice and in fact the first recorded use of burning in 1952 was a spill on a river (Northland Journal 2016), as shown in Figure 3.17.

In summary, there are several situations at sea or on lakes or rivers in which in-situ burning of oil can be carried out.

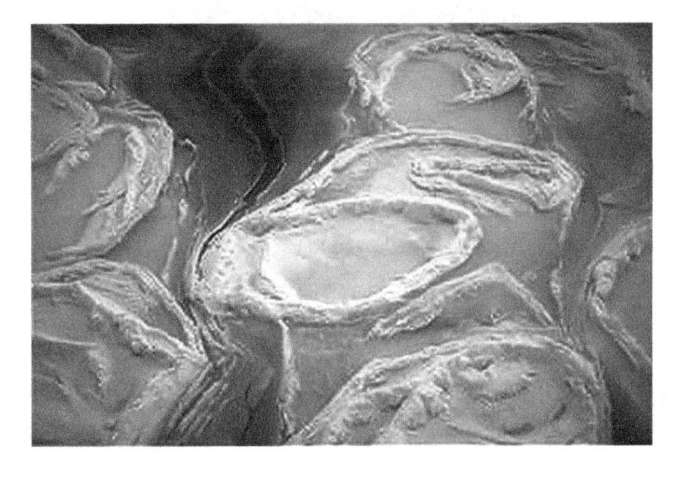

FIGURE 3.15 A close-up of pancake ice. (From http://www.photolib.noaa.gov/htmls/.)

FIGURE 3.16 Pancake ice viewed from far away. (From http://www.photolib.noaa.gov/htmls/.)

FIGURE 3.17 A burn conducted on the Black River in Vermont after a pipeline spill in 1952. (From http://www.northlandjournal.com/wp-content/uploads/2016/12/os.jpg.)

FIGURE 3.18 An oil burn on-land with snow present. This burn was of oil spilled as a result of a pipeline spill.

3.2.22 Burning on Land with Snow or Ice Present

This is a very common situation and one in which in-situ burning has been used for a very long time. The presence of snow does not prevent in-situ burning; however, if the oil is mixed with the snow extensively, ignition may be difficult. Some burns have also been carried out on tundra (Majors and McAdams 2008). A typical burn with snow present is shown in Figure 3.18.

3.3 CONDITIONS OF BURNING

3.3.1 Ignitability

Several workers studied the ignitability of oil under Arctic conditions (Brandvik et al. 2009; Fritt-Rasmussen et al. 2012, 2013; Purves 1977; Ranellone et al. 2017). These tests have shown that ignitability is relatively the same at various temperatures and that, in the Arctic, oil weathers more slowly, increasing the time span that oil can be easily ignited. Specific ignition tests showed that more time may be needed to achieve ignition at Arctic temperatures, especially if there is water and/or ice contained in the oil.

3.3.2 Burn Rates in Ice and Snow

Tests in ice-covered areas have shown that ice coverage sometimes has a minimal effect on the rate of slick burning (Fingas 2017). In the presence of ice, such as in brash ice, the burn rate may be slowed by the cooling of ice directly in the burn field (Buist et al. 2013). In situations where ice is not in the burn, burn rates can be positively affected. Reflection of heat from ice walls and cavities can increase the propensity to burn and the burn rate (Farahani et al. 2015; Fu et al. 2017; Shi et al. 2015, 2016, 2017). However, ice cavities can lower the overall efficiency as oil is trapped in ice, thus reducing the amount exposed to the fire.

3.3.3 Containment

3.3.3.1 Physical Containment

Physical containment is similar in ice and open water except that ice can impose strong forces on the boom. This is shown by the test illustrated in Figure 3.19. Tests on the use of a fireboom were carried out off Svalbard, Norway, in 2008 and 2009 (Brandvik et al. 2010; Potter 2010). Two fire-resistant booms were tested in drift ice using about 4 m^2 of oil. One commercial fire-resistant boom, the Elastec boom, was tested in about 3/10 ice. The burn produced an efficiency of about 98%. The Pyroboom was tested in less ice density and produced an efficiency of 89%. Both booms were able to handle the ice presented to them and were readily deployed.

3.3.3.2 Use of Herders

An alternative to using physical booms is to use chemical herding agents. Several studies involved the use of herding agents to attempt to thicken oil for burning in pack ice (Buist and Morrison 2005; Buist et al. 2006, 2007, 2008, 2010a, 2010b, 2011; Buist and Meyer 2012; Bullock et al. 2016, 2017). In one study, the shoreline cleaning agent EC9580 reduced the area of fluid oils somewhat, but not sufficiently enough to burn. On thicker oils (perhaps 1 mm) the

FIGURE 3.19 A test of the use of a containment boom for in-situ burning in Norway. (Courtesy of Sintef.)

shoreline cleaner agent was able to increase the thickness to about 2–4 mm. The United States Navy herder was found to be better. This was tested at several scales. Outdoor tests showed that the herders were effective at reducing the slicks to burnable areas in pack ice. It should be noted that the herder only worked in calm conditions and that a wind of 1.5 m/s was sufficient to overcome the effect. Further tests on silicone-based (Silsurf A 108 and Silsurf A004-D) and hydrocarbon-based (United States Navy [USN]) herders showed that these had greater potential than other herders. The tests showed that the herder Silsurf A 108 performed the best of all herding agents and on calm water would suffice to herd oil to sufficient thicknesses for burning. Further tests were performed using this agent (Buist et al. 2010a, 2010b, 2011). The presence of breaking or cresting waves disrupted the herder monolayer.

Buist (2010) used the USN herding agent in loose drift ice in Svalbard on a 0.1 and a 0.7 m^3 spill. Both were subsequently burned, the smaller one with 80% efficiency and the larger one with 90% efficiency. Figure 3.20 shows a burn which was contained an using oil spill herder.

Aerial techniques to apply the herder and also to ignite the slicks have been developed (Aggarwal et al. 2017). This enables operations in more remote areas. This system was successfully tested on a larger pond in Alaska using two herding agents (Siltech OP-40 and Thickslick 6535) (Aggarwal et al. 2017). A device used to both spray herder and drop igniters from a helicopter is depicted in Figure 3.21.

Laboratory and meso-scale tests showed that thicknesses up to 3–8 mm could be achieved using herders (van Gelderen et al. 2017). This is sufficient to burn and yield high efficiencies.

FIGURE 3.20   A burn on oil contained using oil spill herder. (Courtesy of Sintef.)

FIGURE 3.21   A helicopter-suspended unit intended to dispense both herder and igniters.

3.4 SUMMARY AND CONCLUSIONS

Ice and snow during cold periods in temperate climates, alpine areas or in the Arctic can serve as a natural barrier to the spreading of spilled oil. Ice and snow also function as barriers to oil penetration into soils and to soil heating from burns. Many burns have been conducted with oil in snow, on ice or among ice floes. Much of the early in-situ burning research was carried out in Canada to develop a countermeasure for spilled oil in sea ice. Burns can be conducted when the oil is

1. Contained in close pack-ice conditions (pack ice of 7/10 coverage or greater)
2. Contained in drift-ice conditions and is of a sufficient thickness to sustain a burn (drift ice of 2/10–6/10 coverage)
3. Contained in a fire-resistant boom (generally in open water up to 1/10 ice coverage)
4. Trapped along an ice floe or herded by wind and has sufficient thickness to support a burn
5. Contained in melt pools on top of ice sheets
6. Contained in open fractures or leads in ice

The containment in ice conditions often offers the responder advantages in ice and snow conditions that are not available without such ice and snow conditions.

REFERENCES

Aggarwal, S., Schnabel, W., Buist, I., Garron, J., Bullock, R., Perkins, R., Potter, S., and Cooper, D. 2017. Aerial application of herding agents to advance in-situ burning for oil spill response in the Arctic: A pilot study, *Cold Reg. Sci. Technol.* 135, 97–104.

Arrow Spill. 1970. *Operation Oil (Clean-Up of The Arrow Oil Spill in Chedabucto Bay)— Vol. 1–4*, Ottawa, ON: Transport Canada.

ASTM 2230. 2013. *Standard Guide In-situ Burning of Oil Spills on Water: Ice Conditions*, Conshohocken, PA: ASTM.

Batelle. 1979. Combustion: An oil spill mitigation tool. Washington, DC: Battelle Memorial Institute for US Coast Guard and US Department of Energy.

Brandvik, P.J., Daling, P.S., Faksness, L.-G., Fritt-Rasmussen J., Daae, R.L., and Leirvik F. 2010. Experimental oil release in broken ice: A large-scale field verification of results from laboratory studies of oil weathering and ignitability of weathered oils. Joint Industry Project, JIP-26, SINTEF, Trondheim, NO.

Brandvik, P.J., and Faksness, L.-G. 2009. Weathering processes in Arctic oil spills: Meso-sale experiments with different ice conditions, *Cold Reg. Sci. Technol.* 55(1), 60–169.

Brown, H.M., and Goodman R.H. 1986. In situ burning of oil in ice leads, *Proceedings of the Ninth AMOP Technical Seminar*. Environment Canada, Ottawa, ON, pp. 245–254.

Buist, I. 2010. Field testing of the USN oil herding agent on heidrun crude in loose drift ice, JIP-6, SINTEF, Trondheim, NO.

Buist, I., and Meyer, P. 2012. Research on using oil herding agents for rapid response in-situ burning of oil slicks on open water, *AMOP* 35:480–505.

Buist, I., and Morrison, J. 2005. Research on using oil herding surfactants to thicken oil slicks in pack ice for in-situ burning, *AMOP* 28:349–375.

Buist, I., Potter, S., Belore, R., Guarino, A., Meyer, P., and Mullin, J. 2010a. Employing chemical herders to improve marine oil spill response operations, *AMOP* 2, 1109–1133.

Buist, I., Potter, S., Nedwed, T., and Mullin, J. 2007. Field research on using oil herding surfactants to thicken oil slicks in pack ice for in-situ burning. *Proceedings of the 30th Arctic and Marine Oilspill Program (AMOP) Technical Seminar.* Vol. 1, Environment Canada, Ottawa, ON, pp. 403–418.

Buist, I., Potter, S., Nedwed, T., and Mullin, J. 2008. Herding agents thicken oil spills in drift ice to facilitate in-situ burning: A new trick for an old dog. *IOSC* 2008, 673–682.

Buist, I., Potter, S., Nedwed, T., and Mullin, J. 2011. Herding surfactants to contract and thicken oil spills in pack ice for in situ burning, *Cold Reg. Sci. Techn.* 67, 3–23.

Buist, I., Potter, S., and Sørstrøm, S.E. 2010b. Barents sea field test of herder to thicken oil for in situ burning in drift Ice, *AMOP* 2, 725–742.

Buist, I., Potter, S., Zabilansky, S.L., Meyer, P., and Mullin, J. 2006. Mid-scale test tank research on using oil herding surfactants to thicken oil slicks in pack ice: An update, *Proceedings of the Arctic and Marine Oilspill Program Technical Seminar No. 29*, Vol. 2, Environment Canada, Ottawa, ON, pp. 691–709.

Buist, I.A., Potter, S.G., and Lane, P. 2016. Historical review and state of the art for oil slick ignition for ISB. Joint Industry Project Report. Washington, DC.

Buist, I.A., Potter, S.G., Trudel, S.K., Walker, A.H., Scholz, D.K., Brandvik, P.J., Fritt Rasmussen, J., Allen A.A., and Smith, P. 2013. In-situ burning in ice-affected waters: A technology summary and lessons from key experiments, Joint Industry Project Report. Washington, DC.

Bullock, R.J., Aggarwal, S., Perkins, R.A., and Schnabel, W. 2017. Scale-up considerations for surface collecting agent assisted in-situ burn crude oil spill response experiments in the Arctic: Laboratory to field-scale investigations, *J. Environ. Manag.* 190, 266–273.

Bullock, R.J., Perkins, R.A., Aggarwal, S., Schnabel, W., and Sartz, P. 2016. Arctic in-situ burn experiments: Laboratory, meso-, and field-scale observations and scale-up considerations. *Proceedings of the Thirty-ninth AMOP Technical Seminar*, Environment and Climate Change Canada, Ottawa, ON, pp. 784–794.

Dickins, D.F., Brandvik, P.J., Bradford, J., Faksness, L.-G., Liberty, L., and Daniloff, R. 2008. Svalbard 2006 experimental oil spill under ice: Remote sensing, *Oil Weathering under Arctic Conditions and Assessment of Oil Removal by in-situ Burning. International Oil Spill Conference*, Savannah, GA, p. 681.

Environment Canada. 1976. Controlled combustion tests carried out near Rimouski. EPS 4-EC-76-2. Ottawa, ON.

Environment Canada. 1977. Probable behaviour and fate of a winter oil spill in the Beaufort Sea. EPS 4-EC-77-5. Ottawa, ON.

Farahani, H.F., Shi, X., Simeoni, A., and Rangwala, A.S. 2015. A study on burning of crude oil in ice cavities, *Proceed. Combust. Inst.* 35(3), 2699–2708.

Fingas, M. 2015. Appendix B: Ice terminology, Appendix B, in *Handbook of Oil Spill Science and Technology*, M. Fingas, (Ed.), Hoboken, NJ: John Wiley & Sons, pp. 685–680.

Fingas, M. 2017. In-situ burning: An update, Chapter 10, in *Oil Spill Science and Technology*, 2nd ed., M. Fingas, (Ed.), Cambridge, MA: Gulf Publishing Company, pp. 483–576.

Fingas, M.F. 1999. In situ burning of oil spills: A historical perspective, in *Proceedings of the Oil In-Situ Burn Workshop*, U.S. Minerals Management Service, New Orleans, LA, pp. 55–65.

Fritt-Rasmussen, J., Ascanius, B.E., Brandvik, P.J., Villumsen, A., and Stenby, E.H. 2013. Composition of in-situ burn residue as a function of weathering conditions, *Mar. Pollut. Bull.* 67, 75.

Fritt-Rasmussen, J., Brandvik, P.J., Villumsen, A., and Stenby, E.H. 2012. Comparing ignitability for in-situ burning of oil spills for an asphaltenic, a waxy and a light crude oil as a function of weathering conditions under arctic conditions, *Cold Reg. Sci. Technol.* 72, 1–8.

Fritt-Rasmussen, J., Wegeberg, S., and Gustavson, K. 2015. Review on burn residues from in situ burning of oil spills in relation to arctic waters, *Water, Air, Soil Poll.* 226(10), 329–340.

Fu, Y., Farahani, H.F., Jomaas, G., and Rangwala, A.S. 2017. Parametric study on cavity formation during in-situ burning of oils in ice, *IOSC*, 2017, 2017293.

Majors, L., and McAdams, F. 2008. Responding to spills in an Arctic oil field—Lessons learned, *Int. Oil Spill Conf.* 2008, 689–698.

McLeod, W.R., and McLeod, D.L. 1972. Measures to combat offshore Arctic oil spills. *Preprints of the 1972 Offshore Technology Conference*, Houston, TX, pp. 141–154.

Northland Journal. December 6, 2016. Remembering the Northeast kingdom oil spill of 1952, *Vermont's Northland Journal*.

Potter, S. 2010. Tests of fire-resistant booms in low concentrations of drift ice—Field experiments May 2008, Joint Industry Project, JIP-5, SINTEF, Trondheim, NO.

Purves, W.F. 1977. Techniques for igniting and burning oil on Arctic ice, 201C-1, Arctec Canada, Kanata, ON.

Ramseier, R.O., Gantcheff, G.S., and Colby, L. 1973. Oil spill at deception bay, Hudson Strait. Ottawa, ON: Inland Waters Directorate, 60 p.

Ranellone, R.T., Tukaew, P., Shi, X., and Rangwala, A.S. 2017. Ignitability of crude oil and its oil-in-water products at arctic temperature. *Mar. Poll. Bull.* 115, 261–265.

Shi, X., Bellino, P.W., and Rangwala, A.S. 2015. Flame heat feedback from crude oil fires in ice cavities, *Proceedings of the Thirty-Eighth AMOP Technical Seminar*, Environment Canada, Ottawa, ON, pp. 767–680.

Shi, X., Bellino, P.W., Simeoni, A., and Rangwala, A.S. 2016. Experimental study of burning behavior of large-scale crude oil fires in ice cavities, *Fire Saf. J.* 79, 91–99.

Shi, X., Ranellone, R.T., Sezer, H., Lamie, N., Zabilansky, L., Stone, K., and Rangwala, A.S. 2017. Influence of ullage to cavity size ratio on in-situ burning of oil spills in ice-infested water, *Cold Reg. Sci. Tech.* 140, 5–13.

Swift, W.H., Touhill, C.J., and Peterson, P.L. 1968. Oil spillage control, *Chem. Eng. Prog. Symp. Series* 65, 265.

van Gelderen, L., Fritt-Rasmussen, J., and Jomaas, G. 2017. Effectiveness of a chemical herder in association with in-situ burning of oil spills in ice-infested water, *Mar. Poll. Bull.* 115, 345–351.

WMO, World meteorological ice terminology. 2017. http://www.aari.ru/gdsidb/docs/wmo/nomenclature/WMO_Nomenclature_draft_version1-0.pdf

4 In-Situ Burning Operations during the Deepwater Horizon Spill

Neré Mabile

CONTENTS

4.1 INTRODUCTION

This book would not be complete without a chapter covering the event of the Deepwater Horizon (DWH) oil spill response in 2010 and the associated unprecedented success of the controlled in-situ burning (ISB) operations.

The DWH rig was situated in the Gulf of Mexico's Mississippi Canyon valley in the continental shelf off the coast of south Louisiana. The offshore oil well over which the rig was positioned was located on the seabed 1,522 m (4,993 ft) below the surface. On the night of 20 April , 2010, a surge of natural gas surfaced on the rig and ignited. The floating rig subsequently sank, which resulted in a continuous uncontrolled discharge of oil from the subsea well. The DWH accident occurred approximately 42 nautical miles off Southeast Pass, Louisiana. The product released was a light, sweet Louisiana crude oil. Figure 4.1 is a photo depicting the onset of the event.

As one can imagine, every possible response option was considered to capture and contain the continuously surfacing crude oil. Many different oil spill response tactics were utilised. One of the more significant response options that had unprecedented success was 'controlled in-situ burning' (CISB). This section focuses on the more than 400 CISBs that were accomplished with a relatively minimal amount of

FIGURE 4.1 Photo of the Deepwater Horizon well blowout and explosion.

equipment and personnel (Allen et al. 2011, Mabile 2010, 2012a, 2012b). A detailed account of the offshore and aerial operational tactics is explained in this chapter, along with a summary of lessons learned. An explanation of the detailed approach to the tactical burning operations and equipment that were necessary to achieve success in the offshore environment for this case is included.

During the DWH spill, many response methods and tactical protocols were utilised to optimise spilled oil removal at sea. CISB was used to remove between an estimated 220,000–310,000 barrels of oil from the Gulf of Mexico. Figure 4.2 shows the geographic location where more than 400 controlled ISBs were initiated. As shown on the map, a majority of the controlled burns were initiated northeast and northwest of the source. The latitudes and longitudes were used in reporting the approximate plume locations to the EPA. The documentation approaches and burned volume calculations methodologies that were utilised during this spill event subsequently became the latest best practices for burn volume estimation for offshore burning.

Aerial operations proved to be one of the key factors for the high rate of success during the DWH ISB operations. The tactics and strategy utilised for aerial surveillance and guidance is covered in this chapter. Daily aerial spotting to locate the concentrated oil masses each day and the monitoring of fire area proved to be key for increasing the oil encounter rate, burn volume estimation and operational success. The burned oil volume was estimated using several strategies combined with established crude oil burning rates. The burn volume protocol that was developed proved to be vital in helping to improve processes and communicate the efficacy of oil removed from the Gulf of Mexico marine environment.

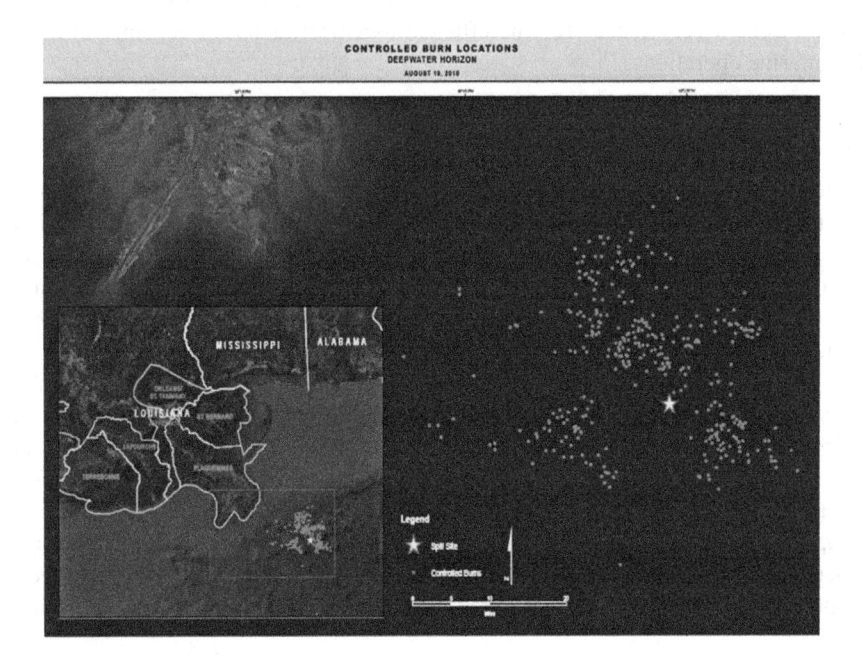

FIGURE 4.2   Locations of the Deepwater Horizon in-situ burns.

4.2 INITIAL RESPONSE ACTIONS

The first CISB task force for the DWH response was conducting a burn within 48 hours of notification that the ISB response option was approved. This quick approval was due to the fact that many of the oil spill incident command leaders engaged in the spill response had previous training in the CISB methods and realised early on its potential. The geographic location of the spill allowed for rapid logistical activity in moving the CISB resources to local coastal docks for deployment.

The decision to burn was quickly made, because the spilled oil was many miles offshore away from any public impact, it was a safe and rapid method of removing large volumes of oil and the tactical option needed minimal resources. Weighing in on the environmental cost versus benefits of CISB, the DWH incident provided favourable conditions since the spill was approximately 40 nautical miles offshore away from populated areas, wave and wind conditions were calm enough the majority of the time and the personnel and burning equipment were readily available. Another important factor driving the decision to burn included the need to supplement skimming operations to capture more oil. Burn approval was given relatively quickly by the United States Coast Guard (USCG) after a CISB safety and operations plan was successfully submitted to the incident command team.

The first test burn was conducted successfully and fireboom and other support equipment were immediately deployed to docks on the southern coast of Louisiana. Burn teams were organised and oriented to the specific tactical operations. Local fishermen were recruited and their fishing vessels contracted for the tasks of pulling fireboom to collect the floating oil for burning. This local fisherman concept proved to be a very successful and efficient logistical tactic for accomplishing the ISB marine operations.

4.3 OPERATIONS ACTIVITY

Simultaneous operations introduced many challenges, especially issues with competition for the same resources. CISB was being utilised along with on-water mechanical recovery and dispersant surface applications. Subsea oil recovery efforts were being done as well as subsurface dispersant injection applications. Managing simultaneous operations was an enormous task. Every evening operations leaders met to review current status and to plan for the next 24-hour operational period.

The ISB burn teams carried out operations simultaneously with both the mechanical skimming teams and the dispersant group. The boundaries varied during the spill response. For example, burn circles were used to place a boundary around the burn operations, as shown in Figure 4.3. At this particular time during the event, a 3- to 15-nautical-mile boundary was set for CISB. The circular boundary allowed the teams to optimise their burning operations in the vicinity of the source and outside the area where the mechanical skimming vessels were working. All teams were required to respect these boundaries, which was key in maintaining safe simultaneous operations.

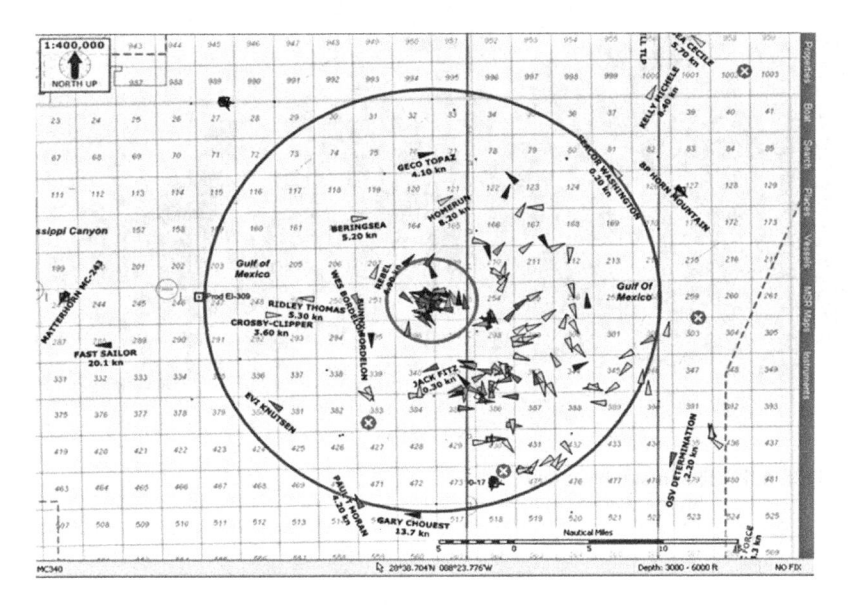

FIGURE 4.3   The burn circle 3–15 nautical miles from the spill source.

4.3.1 OPERATIONAL TACTICS

Each response option was considered for priority operations for the next day, depending on the weather, wind and sea conditions forecast. If sea conditions were rough (>1m wave height with >15 knot winds), then dispersant application was given priority due to its effectiveness by the improved mixing wave action. If seas were calm, then priority was given to mechanical recovery and CISB operations. The three surface response options were managed to cover the available surface oil at sea. With varying subsea currents affecting the subsea spill source, the surface oil had to be aerially located each day to then direct the response tools for collection. The CISB operations made optimum use of aerial surveillance to maximise burn effectiveness.

Early into the DWH spill response activity, the CISB team recruited local fishermen as part of the operations tactical team. The fishermen were naturals at quickly understanding the fireboom towing tactics and supplied their fishing vessels on contract. The local fishermen were adept at working safely out at sea for long periods of time and were already very knowledgeable about the Gulf of Mexico waters. Orientation and training sessions were conducted prior to team deployments. Figure 4.4 portrays the core DWH operations team during a training session held in Venice, Louisiana, in the early stages of the response.

At the height of the response, three CISB task forces were deployed, each supporting four to five burn teams. An example task force is depicted in Figure 4.5. Each team comprised two shrimp boats for pulling boom; one or two larger vessels

FIGURE 4.4   The in-situ burn team.

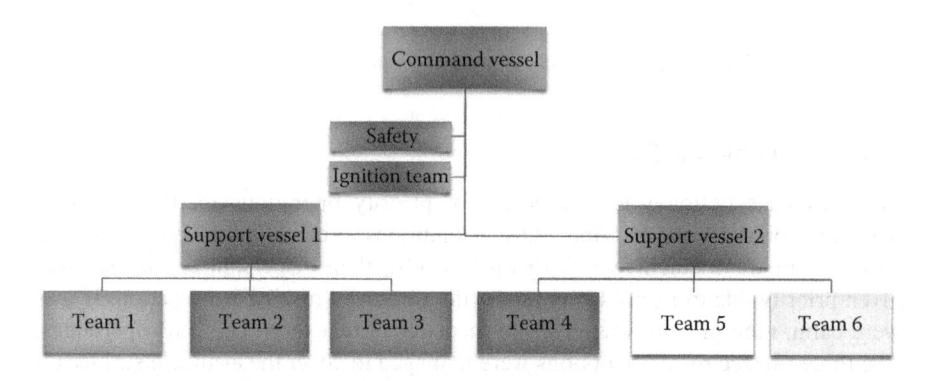

FIGURE 4.5   A typical ISB team structure for one task force.

for command, safety/fire control and boom supply/repair; and multiple smaller boats for ignition and repairs, as shown in Figure 4.6. Supporting these teams were about 10 members of the spotting team, which managed the operation of King Air aircraft to search for and initially assess oil slicks. A key to the success of ISB effectiveness was the utilisation of fixed-wing aircraft to not only spot where the larger oil concentrations were but also to monitor the oil flow and guide the CISB tactical teams to keep them in the concentrated surface oil. Since the burning operation was 35–40 nautical miles offshore, fixed-wing aircraft proved to be the best choice over helicopter transportation, especially because of the increased on-site observation and monitoring time over the spill source. Considering the difference in speed and refuelling requirements at those distances, fixed-wing aircraft can provide 7–8 times more observation time on-site. The King Air aircraft was better suited for multiple sorties per day, quiet atmosphere and comfort and ease of pilot-to-observer communications within the aircraft when directing burn-monitoring activities.

Procedure
• Fireboom (500 ft)
• 2 two boats
• Igniter boat

FIGURE 4.6   Typical ISB team configuration and operations tactics: collection and ignition.

ISB teams were also supported by a number of specialists onshore and in the field, such as those involving wildlife, geographical information, burn volume calculation, data processing, meteorology and so on. The average burn volume per controlled ISB was approximately 750 bbl and the average burn duration was 58 min. On one of the calm-water days (18 June), the CISB teams burned an estimated 50,000–70,000 bbl. of oil. Table 4.1 and Figure 4.7 provide summaries of DWH-response fireboom performance totals and overall burn volumes accomplished per day.

TABLE 4.1
Summary of the Deepwater Horizon Burns

Amount burned	35,000–50,000 m³ (220,000–310,000 barrels)
Number of fires	411 (396 effective ones)
Time of fires (range)	10 minutes to 12 hours
Dates	April 28 to August 19, 2010 (83 days)
Average burned/fire	110 m³ (700 barrels)
Average burn time	~2 hours
Most oil burned in one day	~9600 m³ (~60,000 barrels) (June 18)
Burn teams	8–12
People per burn team	7 or 8
Total people involved	Less than 100
Spotting aircraft	2 King Airs
Spotters	10
Fireboom used	7000 m (23,000 feet)
Large vessels	~10 supply boats and large shrimp boats
Small vessels	~20 rigid hull inflatable or aluminium skiffs
Igniters	1,700 handheld with gelled diesel and marine flares

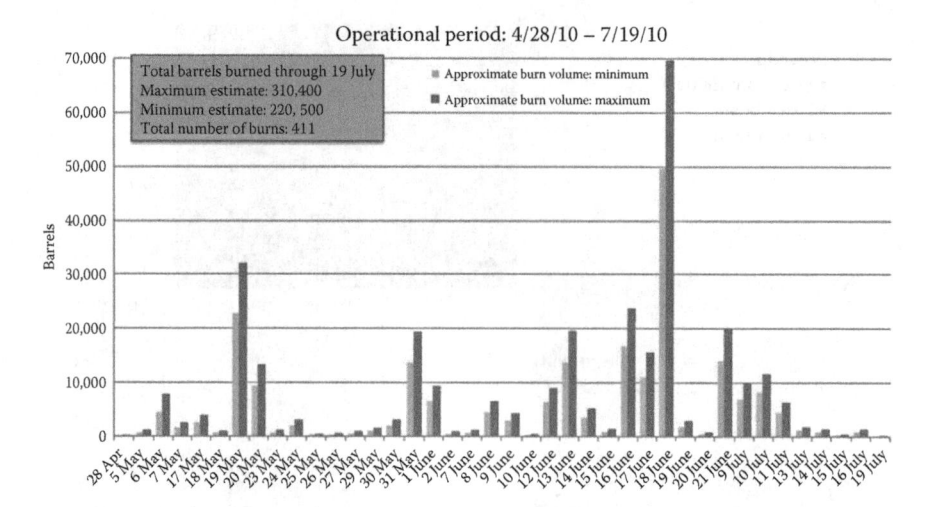

FIGURE 4.7   Burn volume estimates.

4.3.2 OPERATIONAL PERIODS

With the exception of bad weather days, the ISB task forces and all support vessels were available on location by daybreak each day for a 12-hour operational period. They experienced minor impact resulting from seismic surveys being conducted in and around the spill source area. Throughout each burn day, ISB burn teams were guided to the heaviest concentrations of oil by the spotter aircrafts. Using two King Air fixed-wing spotter planes, two separate crews and two sorties, the CISB technical advisors/spotters and documenters were able to stay on location for typically 7½ hours for continuous aerial observations. Figure 4.8 depicts the graphics of the aerial tactics utilised. The first flight departed in the morning at 0730 hours. Before that flight returned to the Houma Airport, the second spotter plane departed and flew to the burning locations to take over spotting and documenting tasks. When the first plane returned to the spill location, the second plane left to refuel. Several attempts were made to fly the spotter aircraft earlier and later to get more spotting time coverage, but the angle of the sunlight was too low to perform effective spotting and oil thickness characterisation. The CISB burn teams were directed to dark oil by spotters, as shown in Figure 4.9. Once on station, Spotters circled the area to observe oil concentrations and vector fire teams to the more concentrated oil while avoiding sheens of insignificant thickness. A log of events (times of arrival/departure for the spotter aircraft, times of ignition, durations of burn and so on) and photos were recorded and documented for each of the burn days. Toward the end of the response and after capping the subsea wellhead on 15 July 2010, the oil was more weathered and ignition became more challenging but was still accomplished for an additional 4 days.

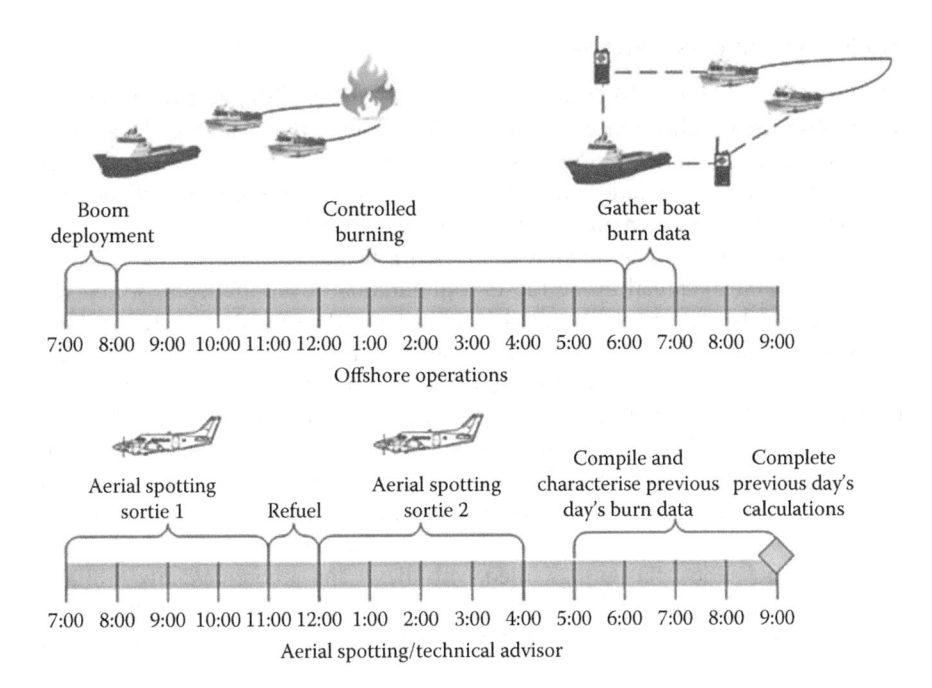

FIGURE 4.8 Aerial spotting and monitoring tactics.

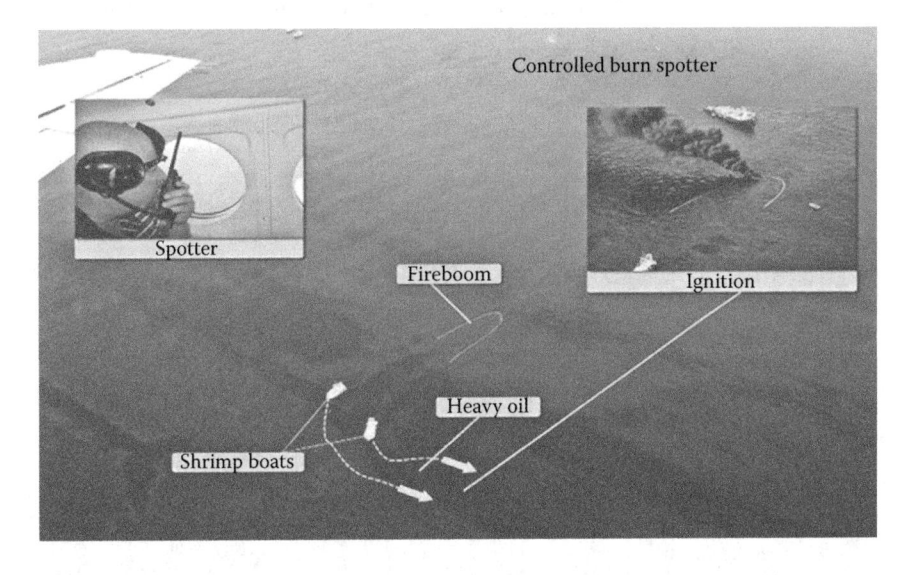

FIGURE 4.9 Spotter providing guidance and direction from an aerial view point.

4.3.3 SPOTTING AND VECTORING

The graphic in Figure 4.10 exemplifies how each of the CISB teams operated to accomplish the burning. Once the spotters in the surveillance aircraft located the more concentrated oil slick, the teams were deployed to the area by radio communications from the air. Operations teams were able to mobilise on-site quickly because they remained at sea overnight to begin operations early the next morning if conditions were favourable. The ignition team used a handheld igniter to light the oil, allowing the oil to spread inside the U-configuration of the fireboom, which was towed by two fishing vessels. Figure 4.11 shows how the teams conducted multiple burns, often simultaneously, by utilising pairs of fishing vessels to tow 500-foot sections of fireboom using approximately 300 ft of tow lines to the contain the oil for burning. The vessels towed the fireboom at ½ to ¾ knot with a swath width of approximately 150 ft to optimise oil collection. Once the oil was ignited and burning, the fishing vessels would almost come to a stop, just bumping the throttle to allow maximum burning inside the fireboom, but still advancing and collecting while burning. As shown in Figure 4.12, the handheld igniter was field fabricated using two half-gallon plastic jugs filled with diesel and a commercially available 'sure-fire' powder (brand Fire-Trol), which created a gel mixture. The gelled diesel jugs were duct-taped together with flotation material, along with a marine flare for ignition.

To facilitate identification and communications with the spotters in the aircraft above, the ISB burn team vessels were colour-coded using tarps suspended over the back deck of the boats. The automatic identification system (AIS) was continually utilised to allow for quick identification of the offshore burn vessels from the air and

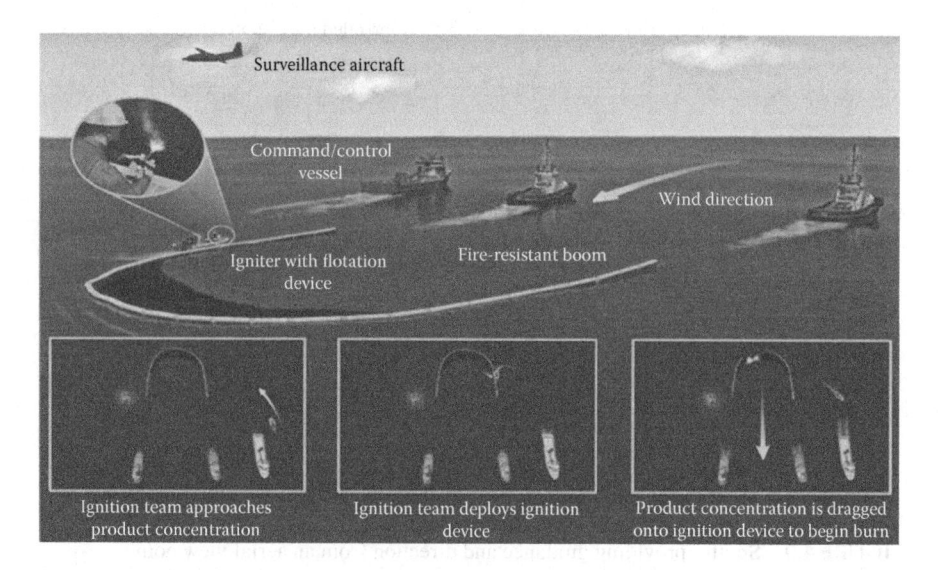

FIGURE 4.10   Controlled burn operational tactics.

Tow vessels starting a gradual
left turn intercept heavy oil
ahead

Nominal burn area
Typically approximately 650 m² (7000 ft²)
To approximately 2500 m² (27,000 ft²)
When full to leading ends of boom

Swath width
45 m (150 ft)

Tow line 90 m
(300 ft)

Fireboom 1500 m
(approximately 500 ft)

FIGURE 4.11 Boom tow procedure to collect more oil.

FIGURE 4.12 A field-fabricated burn ignitor.

confirm their positions. Implementing this method of aerial guidance for the burn teams was one of the major keys to success of the CISB operations. As shown in Figure 4.9, spotters were required to effectively communicate to both the King Air pilots and the lead vessel captain on the burn team. This important communication optimised the burn teams' time in containing as much oil as possible and became a best practice. Since the oil spread in windrows due to wind and wave action, it was difficult for the vessel crews to tell which way to navigate to stay in the higher concentration oil. Aerial observations from 1,000 to 1,500 ft made it possible to easily provide 'step' navigation communications to the burn team vessel captains in order to increase the fireboom oil encounter rate.

4.3.4 LARGE VOLUME AND DURATION BURNS

Another interesting phenomenon during the burning operations was controlled burning outside the boundaries of the fireboom collection area. This was a newly discovered method that was very successful during the DWH response. Of course, this effort was approved by the USCG and monitored closely. Use of this method allowed the task force teams to extend the burn duration and volume. Figure 4.13 depicts how this thermal phenomenon took place. If enough oil collected downstream, at the apex, outside the fireboom, sometimes ignition would occur, due to the intense heat from the already ignited oil. This allowed for burning a significant amount of additional oil, yet still in a monitored and controlled manner. The rising hot air around the intense heat of the burns created a draft, well known by firefighters, towards the fire, therefore drawing additional surface oil towards the burn and automatically feeding more oil to the existing fire. There were many simultaneous burns, and Figures 4.14 and 4.15 show examples of these burns being safely coordinated.

The DWH recovery provided an opportunity to expand and improve methods involving burn volume estimation tactics. Documenting and quantifying field data accurately and then communicating the information quickly are essential during an oil spill response operation. Timely field data were needed for decision-making, for

FIGURE 4.13   Fire draft: thermal phenomenon.

A record number of 16 burns was accomplished on June 18, 2010 50,000 to 70,000 barrels.

FIGURE 4.14   Safe and simultaneous burning operations.

FIGURE 4.15   Simultaneous burns.

increasing the responders' understanding of the situation and for providing stakeholders with information that helped them recognise logical environmental trade-offs.

4.4   BURN VOLUME ESTIMATION

During the DWH operations, responders had a unique opportunity to refine a number of processes and protocols for collecting, recording and communicating field data. One exceptionally successful endeavour to refine protocols involved field-data collection for CISB operations. CISB technical advisors applied existing methods, or protocols, for estimating burn area and calculating burn volume for dynamic controlled burns. These protocols were refined based on the scale of the response, available resources, the operational environment and simultaneous operations. One result from these efforts was the Burn Volume Estimation Protocol, which proved to be a valuable addition to the responders' toolbox. It helped to improve the understanding of the environmental impact of burned oil by quantifying and providing a performance measure that could be communicated to regulatory agencies and the public, thereby increasing stakeholder confidence. The volume estimations also provided valuable evidence regarding the effectiveness of the CISB oil removal method, which aided operations leaders in prioritising resources during simultaneous operations.

The primary challenges of estimating crude oil burn volume are as follows:

- The fire area is constantly changing; burns tend to have a changing, irregular and unpredictable shape.
- Continuous aerial monitoring is often unachievable depending upon the number and duration of burns.
- On-water responders are unable to see through dark smoke, making it difficult to estimate the perimeter (and, most important, the area) of the fire.

4.5 AIRCRAFT AND CREW

As mentioned earlier, a key to CISB effectiveness during the DWH response was the utilisation of fixed-wing aircrafts, King Air 'spotter' planes, which provided the platform for four-member crews to conduct aerial surveillance and collect field data. From these spotter planes, at an altitude of 1,000–1,500 ft, CISB technical advisors were able to spot the location of the larger oil concentrations and guide the CISB burn teams. Typically, the four-member, airborne 'spotting' teams were deployed in two shifts to search for and initially assess oil slicks to identify the darker, thicker oil and then to direct the burn teams to its location. Once a CISB operation began, these same spotting teams provided aerial support in monitoring the burn area, guiding the burn teams to keep them in the thicker oil windrows, and recording the field data needed to support documentation requirements. Each aerial surveillance team comprised two pilots, a spotter and a documenter.

4.6 SPOTTER AND DOCUMENTER

The spotter and documenter positions were filled by CISB technical specialists who had the technical expertise to observe and document multiple burn sizes and durations in sufficient detail to allow subsequent calculations of the estimated daily burn volumes. The spotter observed areas of operation from the aircraft to fulfil multiple responsibilities, including: the location of dark (thicker) oil slicks, and direction of crews to those concentrations. The spotter also had to establish direct radio communications with crews on the fireboom tow-vessels, providing methodical 'turn-by-turn' directions. At the same time, the documenter recorded all flight-related information, such as observation altitude, sea-state level, shift number and wind direction. Once the permission to burn was granted by the burn coordinator on the command vessel, the documenter recorded additional parameters, such as the location of the burn, the burn team number, burn starting time, observed burn area at different times and burn extinguishing time. While recording data to estimate the burn volume, the documenter and spotter focused primarily on observing the changing area of each burn over time.

After completing the documentation, the spotter and/or documenter provided the information on burn duration, locations, and area to a report specialist at the command centre onshore. This report specialist used the data to calculate the burn volume estimations and produce reports to be used in CISB operational shift-change meetings, communication to the public information officer and reports needed by unified command. Figure 4.16 depicts subject matter experts documenting, monitoring and calculating while airborne.

During the DWH operations, documenters were expected to accomplish the following tasks:

- Use established protocols to record observations.
- Record data clearly and concisely.
- Focus on a number of tasks simultaneously.
- Report data at the end of each operational period.
- Clearly describe the operational conditions, such as weather, altitude, sea state and so on.
- Exhibit technical competency in using scalar references to estimate burn area.
- Clearly communicate with pilots concerning directions and locations.
- Exhibit the physical tolerance to work in an airplane space for long hours (typically 2–3 hours per sortie, excluding air travel time to the site).

FIGURE 4.16   Documentation, monitoring and spotting.

4.7 SORTIE PREPARATIONS

Prior to each flight, the pilot in command conducted a pre-flight safety procedure and briefed all passengers on what to do in case of an emergency. The spotter plane team was likely to spend a total of 8 hours each day conducting surveillance missions (sorties). Therefore, each member had to bring supplies, such as those listed for inclusion in the flight kit described below:

- Flight log
- Clipboard
- Handheld very high frequency (VHF) radio
- A global positioning system (GPS) device
- Laptop computer with AIS capability
- Burn boundaries map approved by the incident commander
- Data collection sheets
- Pen and pencil
- Camera with a secure digital (SD) card that is accurately synchronised with date and time
- Notepad
- Sticky notes (Post-it™)
- Binoculars
- Water/snacks
- Polarised sunglasses

4.8 FIELD DATA COLLECTION

The burn volume data were collected from both aerial and on-water vessel observations. On-water data were collected from an observation vessel and sometimes from the command vessel. Both methods complemented each other, especially when the spotter plane returned to shore to refuel; however, aerial documentation proved to be more effective because spotters were able to circle the burn area and maintain the best view of each burn. Aerial observations allowed for a better estimation of burn area, even when multiple burns were being conducted miles apart. Figure 4.17 depicts how aerial and vessel operations were implemented.

Three sets of data were collected by aerial observers: flight log, burn area log and burn duration documentation. The flight log, shown in Figure 4.18, included information such as the following:

- The date of the flight
- Names of people on the plane (including the pilots)
- Time of departure (including destination details: from-to)
- Time of arrival at the burn site
- General observations of the field activity upon arrival
- Time of leaving the burn site
- Overall observations of the field (weather, total number of observed burns and so on)
- Time of return to the airport

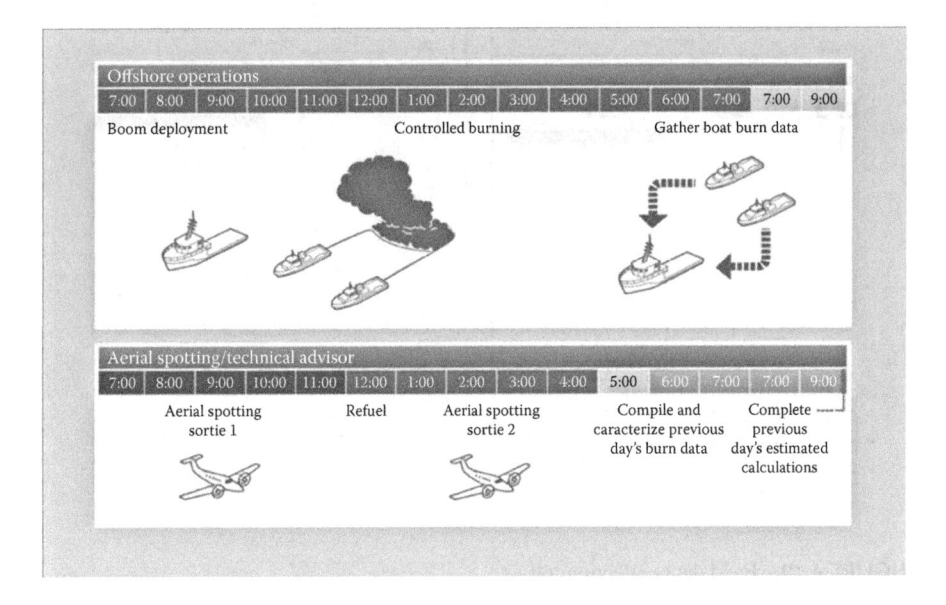

FIGURE 4.17   Aerial and vessel operations timeline.

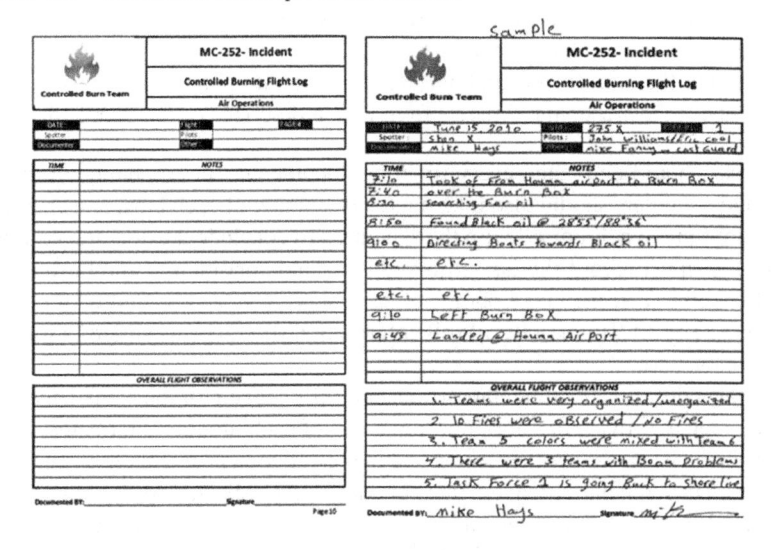

FIGURE 4.18   Sample flight log.

The burn area logs shown in Figure 4.19 were designed for recording data such as the following:

- The time of first observation
- Condition of first observation (e.g. did you see the fire or the ignition?)
- Team number (from colours of boats or from the AIS)
- Burn number (usually given by the burn coordinator following ignition and confirmation of the sustained combustion of crude oil)

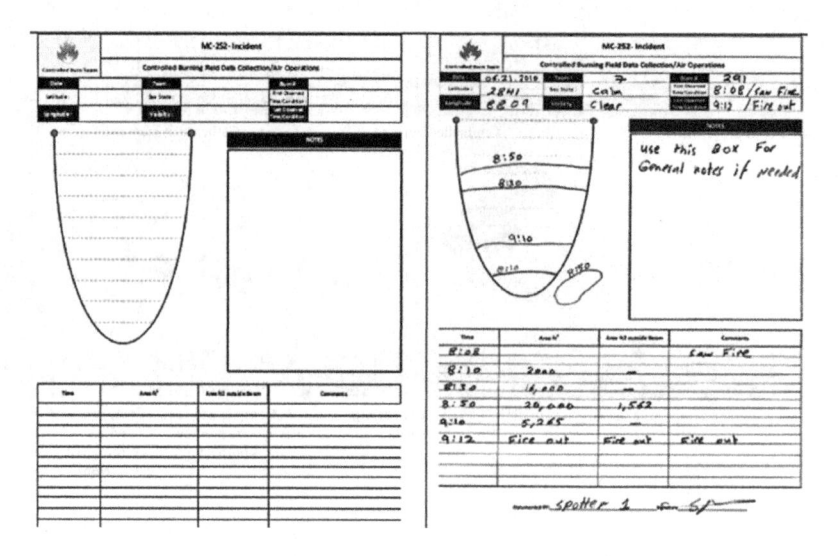

FIGURE 4.19 Field data collection log.

- Latitude/longitude of the burn (from pilots, AIS or the burn team)
- Sea state and weather/visibility
- Area of the burn (inside and outside the boom, if applicable)
- Shape of the fireboom if different from what is described on the 'standard' sheet

Burn areas should be logged every time the documenter sees a change in the burn shape. Documenters should also take photos frequently of each burn and ensure that photos are time-synchronised with the sketches. To conclude the documentation, the documenter must record the last observed time and condition of the burn (fire out or still burning) and then take photos of each page of data collected while en route back to the airport. If possible, all this information should be e-mailed to the report specialist onshore.

The burn duration documentation was carefully logged on the same sheet as the burn area log to provide the start and stop times of each uniquely characterised burn area. These start and stop times were used in the component calculations for the total estimates. Once the burn area and duration data were collected from both aerial and on-water observers/documenters, it was sent to the command centre for analysis. The flowchart in Figure 4.20 exemplifies the workflow of CISB data from both aerial and offshore operations during the DWH response.

4.9 BURN AREA ESTIMATION

The first step in estimating a burn area was to establish a scale for estimating the size of objects as observed from aircraft. During the DWH response, spotters/documenters used a number of strategies recommended in the ISB literature, including those mentioned in Environment Canada's ISB guide: 'If the … volume of oil released from the source cannot be estimated, the volume of the slick can be estimated either

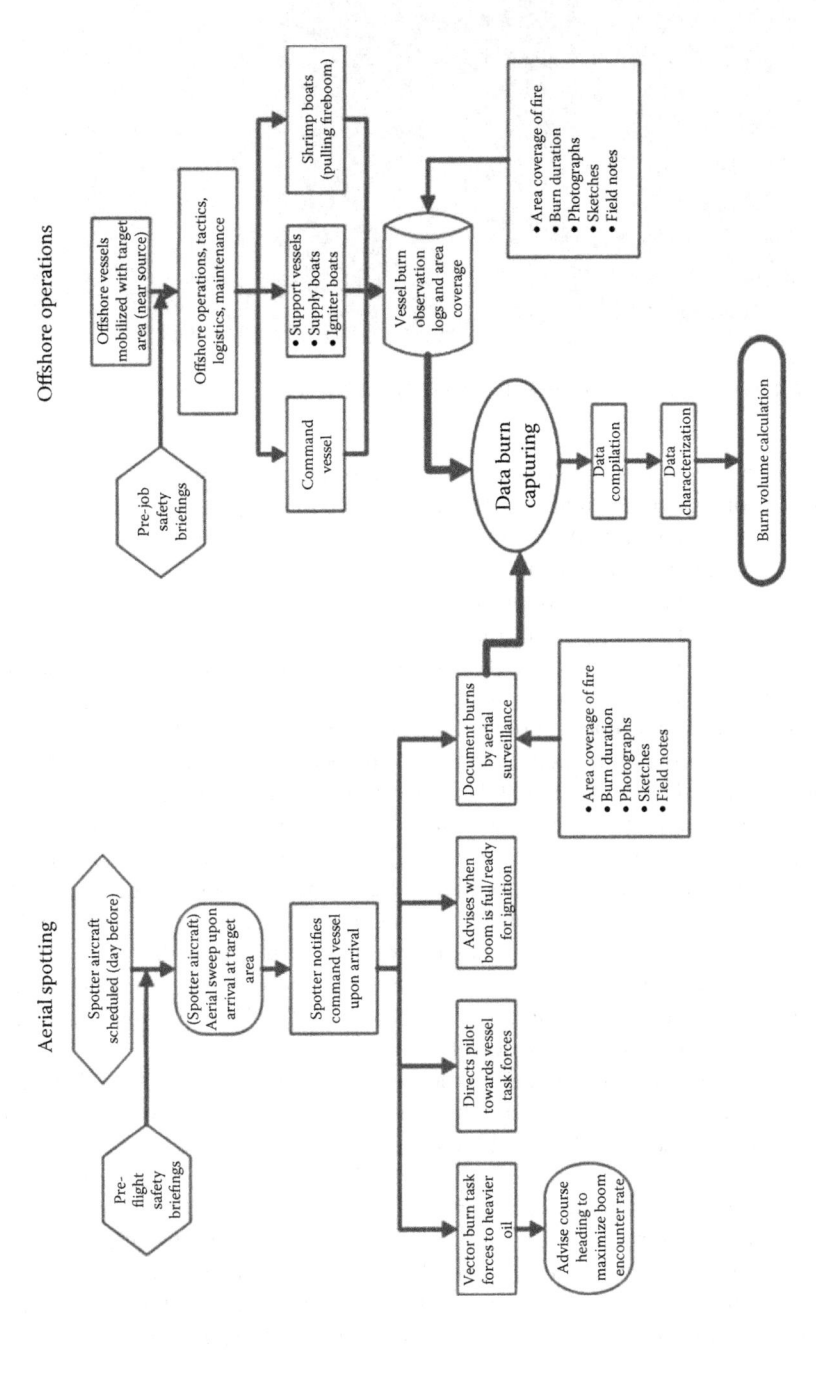

FIGURE 4.20 Data collection and burn volume calculation process.

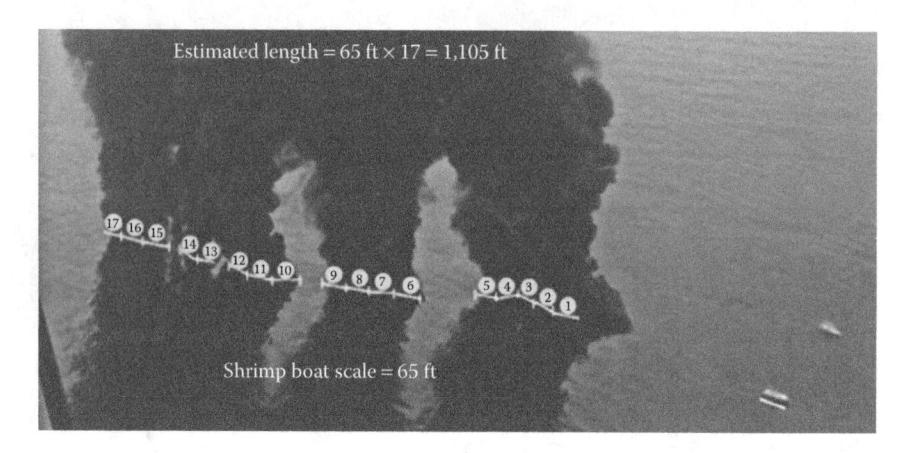

FIGURE 4.21   Area estimation based on size of known objects.

visually using objects of known dimensions, for example, a response vessel of containment boom, or using timed overflights, aerial photographs, or remote-sensing devices' (Fingas and Punt 2000). The most effective technique used during the spill was visual-reference-scaled sizing (Figure 4.21), or visual estimation of CISB size through comparison with known objects. Visual estimates can be highly accurate if the observer is trained and uses proven strategies. To estimate the burn dimensions visually from a distance, the documenter in this example compares the size of the fire with the size of shrimping vessels in the vicinity. In the example below, the vessels are 65 ft, so the estimated length of the burn is 1,105 ft, or the length of 17 vessels.

To ensure consistency of documentation, the fireboom shape had a relatively fixed configuration, as shown on the field data collection log, and documenters were able to estimate the burning area by retrofitting each burn in the defined boom shape. Sometimes the actual boom shape varied slightly and the documenter had to compensate for this in assessing fire area. Figure 4.22 shows the boom shape on the left side and depicts how time marks and area relationships were made. By knowing the length of the fireboom (500 ft), and the distance apart of the shrimping vessels towing the fireboom (150 ft), a documenter can estimate the size/area of the burn over time. The documenter was able to depict the burn size at different times; Figure 4.22 also shows how the data recording took place.

As an example, the initial duration time data point in Figure 4.22 (left side table) shows that the documenter first saw the fire at 8:08, and then 2 minutes later (at 8:10), a line was drawn on the boom along with a time stamp above it. At 8:30, the size of the fire has increased in the boom. At 9:10, the fire size decreased. Note that there was a burn outside the boom depicted on the left-hand side (at 8:50). Once the component field data is collected, it was then submitted to a technical advisor to estimate the total sum fire duration per the corresponding component areas in ft^2. A rough calculation was done early in the response to determine the total area within the U-configuration of the fireboom shape. Figure 4.23 reflects those values.

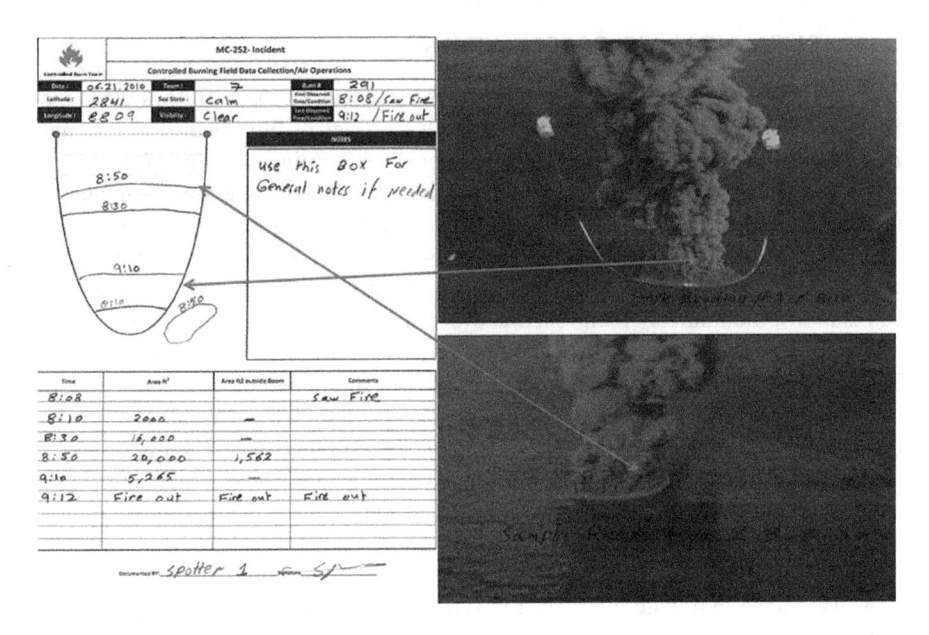

FIGURE 4.22 Field data collection log and corresponding aerial photographs from the DWH response.

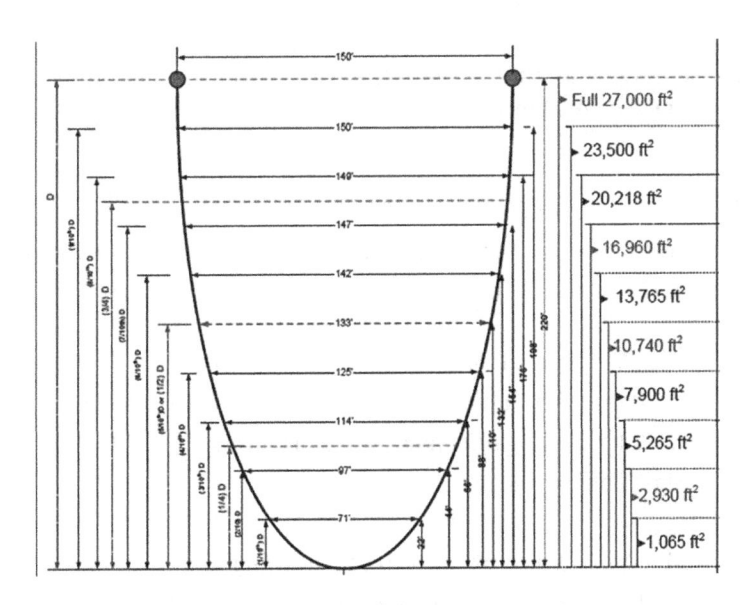

FIGURE 4.23 Fractional area coverage for a 500 ft section of fireboom. (From Allen, A.A., et al., The use of controlled burning during the Gulf of Mexico Deepwater Horizon MC-252 oil spill response, *International Oil Spill Conference Proceedings*, American Petroleum Institute, Washington, DC, abs194, 2011).

4.10   BURN VOLUME CALCULATIONS

Once the field data were submitted to the technical advisor, the burn logs were analysed and burn areas were estimated based on fractions of full boom. For instance,
if the field log showed one-third of the boom was burning, and the photographs
confirmed that fraction, then the total burning area at that time was estimated to
be $27,000 \times 1/3 = 900$ ft^2. This process was done for all field logs and observations.
The last step in estimating the area was to sum all the estimated area values and
come up with the total estimated area for that particular burn. For the burn volume
calculations, the numbers mentioned above were used in calculating the maximum
and minimum burn volume based on known rates for different oils and thicknesses.
To estimate the burn volume, the technical advisor multiplied the total burn area by
the burn rate. For the DWH response, burn rates from 0.05 to 0.07 gal/ft^2/min were
applied. The ASTM's Standard Guide for In-Situ Burning of Oil Spills on Water
(ASTM 1788 2017) establishes that 'oil burns at the rate of about 3 mm/min, which
means that the surface of the oil slick regresses downward at the rate of 3 mm/min.
This translates to a burn rate of about 5000 L/m^2/day (or 100 gal/ft^2/day)', independent of the oil's physical condition or type (ASTM 1788 2017). Since all observations were done in minute intervals, the 100 gal/ft^2/day was converted to gal/ft^2/min
as follows:

$$BR = \frac{100\,\text{gal}/\text{ft}^2}{\text{Day}} \times \frac{1\,\text{Day}}{24\,\text{h}} \times \frac{1\,\text{h}}{60\,\text{min}}, \text{ Burn Rate} = \frac{0.07\,\text{gal}/\text{ft}^2}{\text{min}}$$

The example below illustrates the complete burn calculation process for an actual
test burn during the DWH response:
   'Test Burn'

   Total burn time (t) = 28 min (16:40 – 17:08)
   Estimated area after 15 min from ignition time = $50' \times 75' = 3{,}750$ ft^2
   Estimated area after 28 min from ignition time = $25' \times 25' = 625$ ft^2

   Max. Vol. Burned (based on 0.07 gal/ft^2/min)
   3,750 ft^2 × 0.07 gal/ft^2/min × 15 min = 3,938 gal = 94 bbl.
   6,252 ft^2 × 0.07 gal/ft^2/min × 13 min = 569 gal = 14 bbl.
   Max. Total = 108 bbl.

   Min. Vol. Burned (based on 0.05 gal/ft^2/min)
   3,750 ft^2 × 0.05 gal/ft^2/min × 15 min = 2,813 gal = 67 bbl.
   6,252 ft^2 × 0.05 gal/ft^2/min × 13 min = 406 gal = 10 bbl.
   Min. Total

See Chapter 1 for a standardised approach to area calculation (ASTM F3195 2016).

4.11 SAFETY AND ENVIRONMENTAL CONCERNS

There were several areas of environmental and safety concerns during the DWH CISB response. Sea-life monitoring was one of the tactics deployed as part of the burn teams. There were as many as five sea-life observers on the vessels at any given time. As shown in Figure 4.24, these observers were part of the offshore burn teams and monitored for any turtle activities. As the size of the operation grew, the sea-life monitoring plan was augmented with additional trained and qualified turtle observers and observer trainees.

Attention to safety was always paramount during the CISB operations. The ISB fire teams demonstrated that, if the wind blew smoke plumes toward them, they could simply reposition their vessels to avoid any smoke impact. There were also no air-monitoring readings outside the established safe parameters. The only exceptions were occasional temporary excursions when exhaust vapours from the fireboom pump engine and outboard motor exhausts caused temporary high readings. These pumps were positioned on the back decks of the shrimp boats in a highly ventilated area during operations.

In the course of the approximately 400 burns, only two were intentionally extinguished. The first occurred with the longest burn of 11 h and 48 min. Although still continuing to catch oil and feed into the fireboom, crews began to show signs of fatigue and were directed to intentionally extinguish the fire by increasing towing speed and pulling the boom over the remaining fire. In a second case, to demonstrate its safety utility, teams used a deck-mounted fire water cannon on the command vessel and directed it to the fire. The water cannon easily distinguished the flames.

FIGURE 4.24 Sea life observers are part of the CISB teams.

In addition to controlling the burn, the ISB on-site leader had to ensure that crews had been sufficiently trained for their assignment, that vessels were maintained properly, that personnel were trained in vessel-to-vessel transfer and that vessel movements were coordinated. During the DWH response, all responder-vessel crews reported to the command vessel leaders and were required to participate in daily safety meetings to help ensure that they maintained a state of safety awareness. Vessel-to-vessel transfers were identified as the highest risk for personnel during the offshore operations activity. During daily work shifts, personnel had to routinely shuttle from the command vessel onto smaller vessels and back again after the work activity was completed. This safety hazard was highlighted during every safety meeting held before the start of work shifts. Operations activities were also organised to minimise the need for transfers by assigning the right number of resources and levelling in the right places. Personnel were continually reminded of all identified hazards and were reminded to always utilise the 'buddy system' to look after each other during daily activities, especially transferring to other vessels at sea. Figure 4.25 demonstrates one of the tools utilised to manage risks and communicate those risks to on-site personnel.

Air quality and potential human exposure in and around the offshore fire areas was a concern during the DWH response. During the CISB operations, fireboom towing vessel captains easily manoeuvred their crafts to avoid smoke plumes and maintained a distance of at least three 'fire diameters'. The crews of these vessels had received training specific to controlled ISB operations, including safety protocols and the use of personal protective equipment (PPE). In addition, industrial hygiene specialists were utilised for personnel monitoring around controlled ISB operations. The vapour concentrations and potential for fire spreading were assessed immediately before each burn was ignited. Also during the DWH response, National Oceanic and Atmospheric Administration (NOAA) measured particulates and found that the air near the smoke plume was well within an acceptable range and that the plumes dispersed quickly. According to NOAA, the highest level of submicron particulate mass that was measured downwind of the spill was 20 μgm^3 on June 8, which is well below the National Ambient Air Quality Standard (35 μgm^3, $PM_{2.5}$) for particulate matter less than 2.5 μm in diameter. Essentially, the overall air quality near the DWH site in the Gulf of Mexico was similar to that of a major city (NOAA 2010). If vessels avoided positioning downwind of a controlled ISB, then the concern about exposure was minimised. The air quality was also monitored by government and private groups along the shoreline to detect and record any potential impacts to populated areas. For instance, the Florida Department of Environmental Protection and the U.S. Environmental Protection Agency (EPA) monitored four sites along the Florida coast using evacuated Summa canisters, shown in Figure 4.26, to sample for the benzene, toluene, ethylbenzene and xylenes (BTEX) volatile organic compounds (VOCs). In 2003, the USCG asserted, 'In general... the smoke plume is not a safety threat to the public nor to the environment because it has very low toxicity and readily dissipates' (USCG 2003).

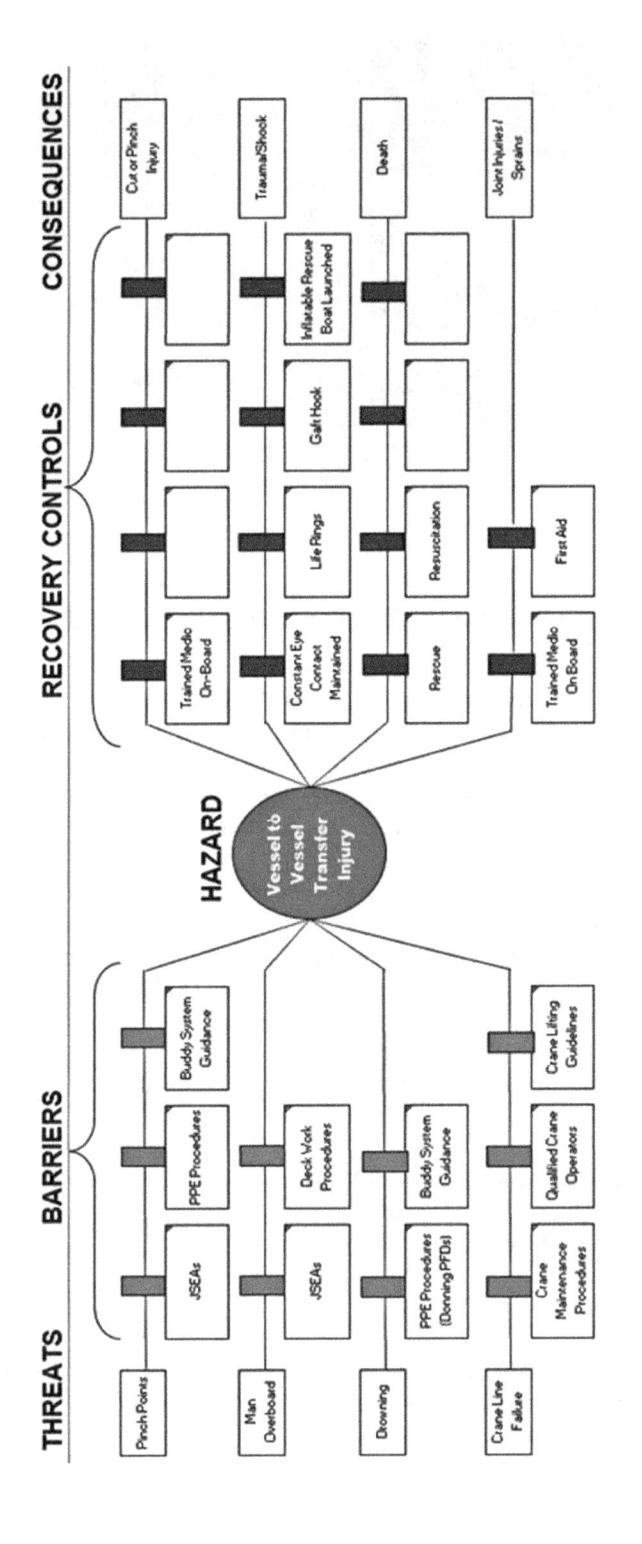

By: Nere' Mabile 05/15/10

FIGURE 4.25 Bow tie analysis (risks).

FIGURE 4.26 Summa canisters: sampling for BTEX and volatile organic compounds.

4.12 FIREBOOM PERFORMANCE AND LESSONS LEARNED

This chapter addresses the practical aspects and lessons learned from working with a fireboom on offshore waters during the DWH spill response. This assessment of boom performance was taken from burn data logs initially collected and prepared by DWH responders, first-hand observations from BP personnel, on-site fireboom manufacturer representatives, Obrien's Response Group team members, USCG supervisors, shrimp boat vessel captains and aerial surveillance spotter and guidance personnel.

This section also addresses the different types of fireboom systems used and their durability, characteristics and overall performance. Throughout the spill response, fireboom was widely requested from manufacturers and oil spill response organizations' (OSROs') stock to meet the tactical needs of the DWH burning operations. It is important to note that this chapter accounts for the performance of boom systems available at the time of the incident.

The different types of fireboom systems utilised during the DWH response included:

- American marine, Elastec/American Marine, Inc. (formerly known as 3M)
- Oil Stop: AMPOL, Oil Stop Division
- PyroBoom: Applied Fabrics Technologies, Inc.
- Hydro-Fire Boom: Elastec/American Marine, Inc.
- Kepner fireboom: Kepner Plastics

In the beginning, burns would typically last in the range of one hour. However, as more burns occurred, the technique was refined. On 16 June, a burn of 11 h and

48 min in duration became the longest continuous burn time recorded. Collectively, the burns made it possible to efficiently remove significant amounts of oil from the marine environment (an estimated 220,500–310,400 barrels). What is significant during this operation was the sheer number of controlled burns conducted, which provided a unique opportunity to repeatedly test and evaluate fireboom equipment. These fires were of a much greater intensity and size than can be generated in any test facility.

The performance evaluation of firebooms deployed during this DWH spill was based on several American Petroleum Institute (API) and ASTM guidelines. It should be noted that all of the fireboom systems used during this spill were used repeatedly until significant repair or replacement was required. Some firebooms performed better than others and some were more 'user friendly'. Some were easier to deploy, recover and repair, while others were difficult to handle and showed significant damage after only a short exposure to intense fires.

Some fireboom systems did better at oil retention and wave performance than others. Fireboom performance was not only affected by fire intensity but also by fatigue stress on boom components and connectors while deployed in varying sea states. For convenience, time saving and to minimise damage, booms remained at tow behind vessels throughout the night until operations began the next day. A gentle, straight-line tow throughout the night was generally less stressful on the firebooms. Those booms that became brittle during their burns usually suffered additional damage whether towed through the night or recovered on deck.

The parameters used for evaluating fireboom performance are listed below and take into account the observations made by personnel on site during the controlled burning activity. Some fireboom systems have already been modified in recent years based on the lessons learned during the DWH event.

4.12.1 PARAMETERS

1. Burn duration and number of systems used
2. Visual observations of oil retention/wave performance
3. Repair and durability
4. Handling and operational observations
5. Logistics, shipping (air lift capability)
6. Burn volumes accomplished per system type

4.12.2 TYPES OF FIREBOOM

The two basic types of fireboom systems utilised during the DWH response involved both non-water-cooled and water-cooled systems. Non-water-cooled booms have a permanent, solid flotation in the form of metallic or ceramic floats covered or attached to a fire-resistant fabric. Water-cooled booms incorporate inflatable buoyancy chambers allowing them to be stored and recovered onto powered reels.

These booms have pumping systems to distribute sea water to an outer fabric, saturating and cooling the boom during a burn.

4.12.3 HYDRO-FIRE BOOM (WATER-COOLED)

Hydro-Fire Boom systems feature a sectional inflatable boom covered in a fire blanket that is continually soaked with sea water during burning and is mounted on a powered reel for both deployment and recovery. This system is readily transportable by C-130 aircraft. (Several systems, for example, were shipped from Brazil to the Gulf of Mexico in one aircraft during the response.) As seen in Figure 4.27, the boom features a stainless steel top tension cable and a series of individually inflated segments that are insulated by the water-cooled blanket. Five 100-ft boom sections make up a single fireboom system. As shown in Figure 4.28, the Hydro-Fire Boom is deployed apex first so two sides of the boom are inflated at the same time. Pumps on each of the boom-towing vessels provide cooling seawater to the boom's outer fabric. With the series of inflatable segment design, should a failure occur in any one of the segments, the boom does not lose its entire flotation integrity.

The Hydro-Fire Boom system specifications are comprised of 5 sections of 100 ft (30 m) fireboom each with 14 in. flotation and 18 in. skirt; 1 boom reel with brake and air inflation system; 2 high-flow water pumps with flow meters, filters, and pressure gauges. The weight of the boom is 8 lbs./ft (12 kg/m) and tow lines are 300 ft. The water feed lines (tethered with the tow lines) used during the DWH response were 3 in., but with improved pump capacity since then, they are now 4 in. in diameter.

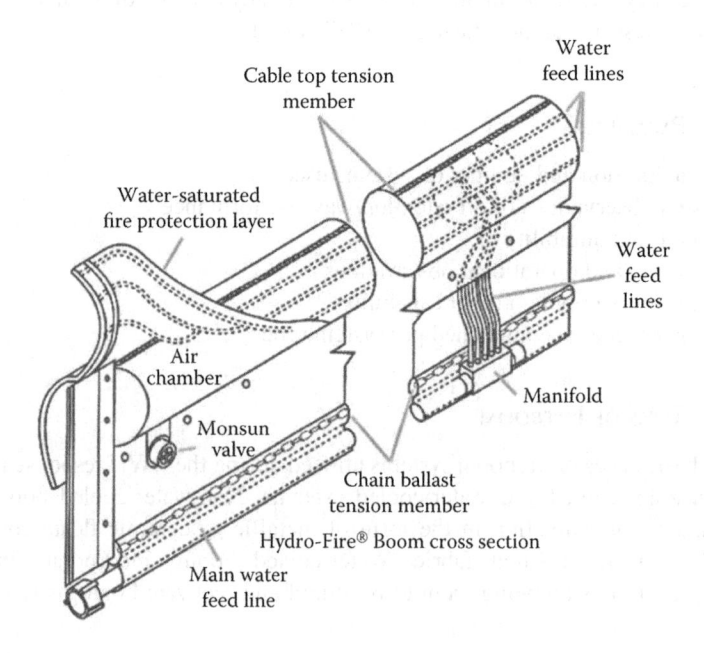

FIGURE 4.27   Cross section drawing of the Hydro-Fire Boom.

FIGURE 4.28   Hydro-Fire Boom being deployed from reel.

4.12.4   AMERICAN MARINE: NON-WATER-COOLED

The system is shown in Figure 4.29; it is a non-water-cooled, ceramic fireboom with a high-temperature solid flotation core. The high temperature core is surrounded by stainless steel mesh and ceramic fabric components to withstand 2000°F. A sacrificial outer cover provides protection and ease of handling during storage and deployment.

FIGURE 4.29   Non-water-cooled boom system.

The American Marine boom was originally developed to support offshore oil exploration activities in Alaska during 1990. During the DWH recovery, many systems were shipped to the Gulf from Alaska. Two different sizes of this boom were utilised, one with 12″ flotation and another with 18 in. flotation.

4.12.4.1 Specifications for the American Marine Fireboom

Overall size 30 in., flotation 12 in., skirt 21 in.
Overall size 32 in., flotation 18 in., skirt 24 in.

4.12.5 PyroBoom System: Non-Water-Cooled

The typical PyroBoom 'burn kit' consists of 500 ft (150 m) of PyroBoom, a fence-type boom consisting of high-temperature fabric and stainless steel flotation chambers bolted to its sides. During the first burning operations only 200 ft were available from the manufacturer. During the DWH response, BP purchased boom from Africa and placed additional orders with the manufacturer.

The PyroBoom construction is portrayed in Figure 4.30. The fireboom features a silicone-coated refractory barrier fabric and stainless steel float shells filled with glass foam. Boom components are assembled using ASTM connectors and off-the-shelf fasteners. The original booms provided during this response arrived with aluminium connectors; however, post-spill orders are being made with stainless steel connectors.

4.12.5.1 Specifications for the PyroBoom

Freeboard 11 in., draft 19 in., 8.9 lbs./ft (13.3 kg/M).

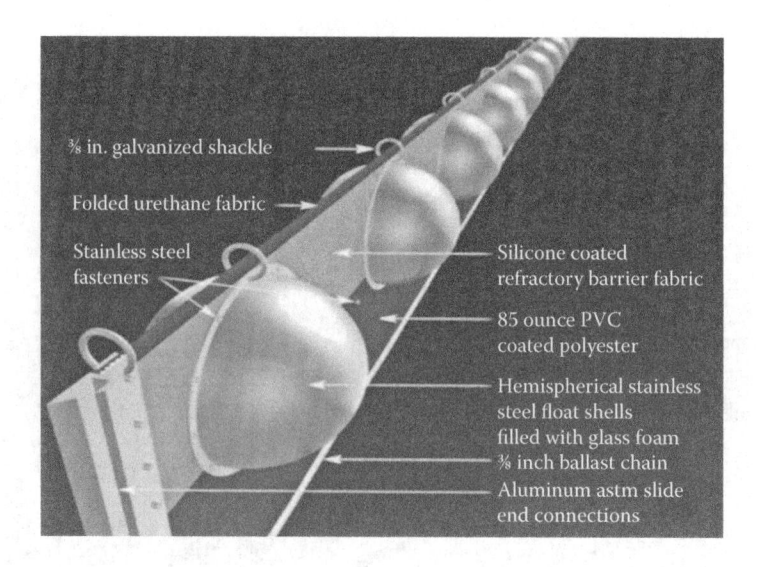

FIGURE 4.30 Schematic of the PyroBoom.

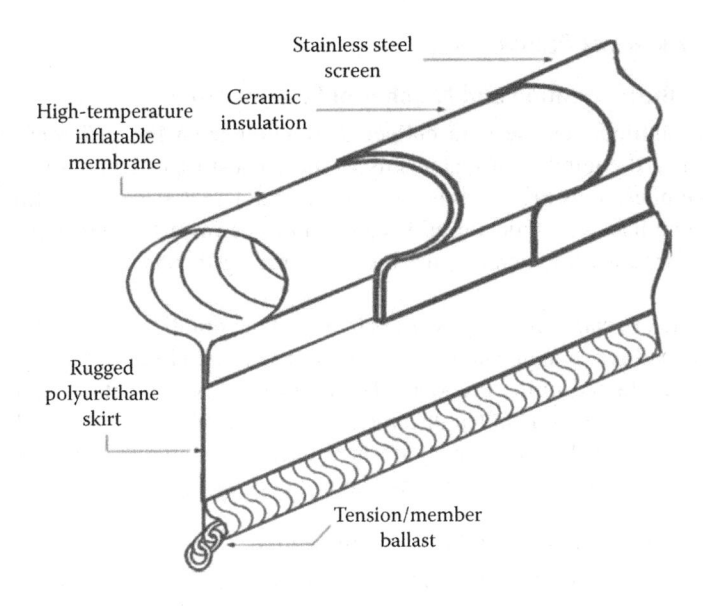

FIGURE 4.31 Schematic of the Oil Stop system.

4.12.6 Oil Stop Fireboom System (Non-Water-Cooled)

The Oil Stop fireboom has a multi-layer construction consisting of a high-temperature inflatable membrane, covered by a ceramic insulation, encapsulated with a stainless steel screen. A sketch of the fireboom design is shown in Figure 4.31.

4.12.6.1 Specifications for the Oil Stop Fireboom

Single-point inflation Harbor Model, approximately 8 lb./ft.
Single-point inflation Harbor Model, 12 in. × 18 in.
Reel with 500 ft Harbor Fire Boom and 600 ft single-point inflation guide
Boom: approximately 10′ × 7′ × 7.5′–8,500 lbs.

4.12.7 Kepner Fireboom System

The Kepner fireboom design is a non-water-cooled type with the size specification listed below. The small number of Kepner fireboom systems used had been manufactured approximately 20 years ago.

4.12.7.1 Specifications for the Kepner Boom

Model #BTTB1115 FG FireGuard Fire Containment Boom
Float diameter: 11 in.
Skirt length: 15 in.
Two, 250 ft section lengths per system

4.12.8 Fireboom Performance

4.12.8.1 Burn Duration and Number of Systems Used

Table 4.2 summarises the data collected by comparing the different fireboom types against the number of systems used, the longest reported continuous burns and the number of average barrels burned per fireboom system. Some fires within the fireboom had very short durations and had to be restarted. The data captures only the durations for 'continuous' burns occurring during the controlled burn operations.

Many factors came into play when achieving long duration burns. The sea state and winds had a major impact on the length of burns and the capability of keeping the oil contained in the fireboom. The oil properties encountered were also of concern as the water content (emulsion) varied considerably from day to day. Some days the oil was thick and relatively fresh, at times concentrated along natural convergence zones and other times thin and spread out over larger areas.

4.12.8.2 Oil Retention and Wave Performance

The Hydro-Fire Boom systems maintained a high level of containment integrity for extended periods of time and were able to repeatedly collect large amounts of oil. These systems endured some of the longest and largest burns experienced.

In general, booms with ceramic floatation systems became less capable of retaining oil with each burn. But an interesting observation was that the American Marine/3M boom developed a buildup of oil residue that would impregnate the fire-resistant fabric. This would enhance the oil-holding proprieties of the boom and increase the number of times it could be used. This probably extended its containment capability and allowed for more burning time. The more modern versions of the American Marine systems proved to be reliable as well and allowed multiple burns. In general, non-water-cooled or dry fabric booms suffered more than water-cooled booms when exposed to wave action following a burn. The more flexible American Marine boom fared better than others in this respect.

TABLE 4.2
Fireboom Performance Summary

Factors	Hydro-Fire Boom	American Marine/3M	PyroBoom	Oil Stop	Kepner
Number of systems used	27	37	13	3	2
Longest continuous burn	11 h 48 min	11 h 21 min	3 h 13 min	27 min	43 min
Average maximum/ minimum number of barrels burned per system	5,173/3,775	3,916/2,800	1,750/1,238	28/11	296/211

An attempt to use the three reel-mounted Oil Stop fireboom systems was made during May 2010 but was unsuccessful. (It should be noted that the small amount of Oil Stop fireboom used on the response, obtained from local OSROs, was manufactured 12–14 years prior.) The first Oil Stop boom system deployed sank within a short time. The next boom system deployed accomplished a 27 min burn, but after a couple of hours the boom experienced some flotation problems. After third system also experienced flotation problems, a field decision was made to discontinue use of this fireboom. This generation of Oil Stop boom did not prove to be a viable way to contain oil for burning.

As shown in Figure 4.32, Oil Stop systems used included 200 ft of guide boom on the leading edges connected to 300 ft of fireboom. Their guide boom is a standard containment boom in 100 ft sections with 12 ft long segmented chambers (8 chambers/section). During the DWH response, it was found that it was generally best practice to use fireboom for the entire 500 ft in order to burn larger volumes of contained oil. A full configuration of fire-resistant boom is also desirable in order to allow for 'full-boom' burns, and to handle shifts of burning oil within the U-configuration due to wind and/or back-and forth movements of towing vessels.

Regarding oil retention and wave performance, the PyroBoom oil containment capability was compromised under certain wind and wave conditions. As shown in Figure 4.33, PyroBoom tended to suffer from this fatigue stress during towing as the fabric tore more readily. This was observed when new boom was deployed before burning and on boom sections after burns.

As mentioned earlier, for fireboom to be effective, it has to contain oil floating on water before, during and after exposure to ISB of the oil. The more rigid construction

FIGURE 4.32 The Oil Stop fireboom deployed.

FIGURE 4.33   The PyroBoom deployed.

booms did not have as good wave response. This is mostly due to the boom construction and lower buoyancy to weight ratios, as listed below.

Buoyancy to Weight Ratios

Hydro-Fire Boom	6.3:1
PyroBoom	3.3:1
American Marine (3 m)	3.8:1

Two 500′ systems of (older generation) Kepner fireboom were deployed during the response effort. Both systems failed after approximately 5 min due to intense heat of the fires. It appeared that the outer, fireproof cover did not protect the underlying foam flotation.

4.12.8.3   Burn Fatigue/Durability/Repair

The Hydro-Fire Boom maintained its integrity and had good fatigue resistance for extended periods. Field observers reported as many as 10–14 burns (often, large burns) with the Hydro-Fire Boom. After a fireboom's extended use, localised degradation can take place at the hottest downwind portion (or apex) of the boom. The boom manufacturer of the water-cooled boom has already made modifications to improve its thermal protection. This has been accomplished by increasing the sea water flow rate to the boom and by enhancing the water distribution system within it. The boom was relatively quick to deploy and took from 30 to 40 min. The Hydro-Fire Boom retrieval was assisted greatly by the powered boom reels. Having a water-cooled flexible cover, this boom is easily handled, recovered or repaired while in the water. Typically the boom was left in the water overnight and towed by fishing vessels. The Hydro-Fire Boom did not show any signs of wear due to towing.

The repair ability of the Hydro-Fire Boom meant that 100-ft sections could be reused and the inflatable portion of the boom under the water-cooled cover was salvaged and re-blanketed.

Operators could extend the life of the Hydro-Fire Boom by adding foam flotation to any deflated areas or change deflated bladders. The Hydro-Fire Boom seemed to have the longest life, even during the most intense burns. It exhibited good seakeeping abilities, which extended the operating window when sea conditions deteriorated. The two longest continuous burns recorded with the Hydro-Fire Boom were 11 h 48 min and 10 h 20 min.

Another good performer was the American Marine fireboom. Although perhaps not as durable as the water-cooled boom, it was available in quantity and contributed significantly to the burn operation. The Poly Vinyl Chloride (PVC) cover protected the boom during handling and deployment. Fabric failures were only seen after extended high-temperature exposures. No tears were witnessed in the newer American Marine boom, which showed good thermal integrity. (There were, however, fewer burns per system than the Hydro-Fire Boom.) This boom is built like a traditional boom with fabric encasing the floats. Stainless steel connexes were typically undamaged and were able to be changed with the boom in the water. This boom also has a mid-tension stainless steel cable. The longest continuous burn recorded with the American Marine/3M was 11 h 21 min.

The PyroBoom is a fence-type boom constructed with stainless steel hemispheres on each side and high-temperature silicone-coated refractory fabric. The wind and wave conditions experienced during the DWH response occasionally affected the PyroBoom's stability, allowing oil to splash over. The structural integrity was subject to compromise after repeated burns but could often be controlled by alternating the most intense portions of a burn to different sections of a U-configuration.

The PyroBoom aluminium end connectors were a problem as they would melt and weld together. This prevented the operators from easily taking out bad sections or rotating the boom's leading ends into the apex. Completing such repairs while deployed in the water was nearly impossible, and recovering the boom on deck for such repairs often led to additional damage of the fabric.

Small wire rope was sometimes retrofitted between spheres and connectors and sphere to sphere to extend boom life. The smooth spheres made the boom easy to deploy on smooth decks but difficult to deploy over railings on some of the ships. The tensile strength of the upper fabric after several burns appeared weaker as evidenced by some fabric failures. The longest recorded continuous burn with the PyroBoom systems was 3 h 13 min.

4.12.8.4 Handling and Operational Observations

Regarding storage volume, the inflatable booms took up significantly less space on deck. The boom system's storage volume has a significant impact on logistics, especially when considering air lift transport. Shipping and delivery to a port of call usually involves connexes and crate packaging. Figure 4.3 portrays the three

FIGURE 4.34 American Marine/3M on the left, 2 reels of Hydro-Fire Boom in the centre and PyroBoom on the right.

types of fireboom (as labelled) placed on the back deck of a supply vessel for offshore transport from Venice Dock to the burn region during the DWH response. Figure 4.34 shows 1,000 ft of Hydro-Fire Boom, 1,000 ft of American Marine fireboom and 400 ft of PyroBoom. The photo was taken at Venice Dock, after loading operations.

Hydro-Fire Boom, provided on reels, offers speed, simplicity and stress reduction during deployment and recovery. PyroBoom, which is non-water-cooled and non-inflated, provides simplicity of use and a range of options for storage and transport. The fireboom shipping volumes shown in Table 4.3 represent critical information when performing planning and logistics. The DWH response required the shipment of fireboom systems from all over the world.

Over the last 35 years many manufacturers have tried to produce fire-resistant booms. Using ASTM guidelines, along with years of research by public and private sectors, such efforts have paid off and were big factors in the success of the Gulf of Mexico, in-situ controlled burn operations. Hydro-Fire Boom systems collected the most oil and were responsible for the highest volume of oil burns per system. Other systems also contributed significantly. Dry type booms, while successful, lost their oil retention capabilities more quickly than the water-cooled booms.

TABLE 4.3
Fireboom Shipping Volumes

Boom Type	Size Description	Shipping Volume
Hydro-Fire Boom	*Reel size*: 122 in. × 89 in. × 103 in./500 ft	0.05 cu.ft/ft
PyroBoom	Overall size 30 in., freeboard 11 in.	1.135 cu.ft/ft
American Marine/3M	Overall size 31 in., freeboard 12 in.	1.01 cu.ft/ft

The DWH response experience made it clear that the success of a fireboom is not only determined by its capability to contain oil and maintain a large fire; it must also sustain its oil containment capability and endure the constant fatigue stresses imposed by the varying wind and wave action. Effective firebooms must also retain their structural and thermal integrity while deployed for burning, and while on the water, waiting for the next burn. Along with the massive scale of the DWH response came the opportunity to try a wide range of available fireboom designs.

After the DWH experience most manufacturers have been improving their designs based on lessons learned during this incident. The overall collective fireboom performance during this unprecedented response effort expanded our understanding of controlled burn fireboom strategies and tactics.

4.13 CONCLUSION

The DWH response case was used in this chapter to illustrate the operational approaches and tactics that proved to be very successful in rapidly and safely removing large volumes of oil from the Gulf of Mexico marine environment. Preferred response options are highly situational and depend on different factors. Controlled ISB can be initiated on a pre-approved or case-by-case basis, and there is generally a short operational time window during which it can be effectively utilised; therefore, quick, informed decision making is imperative.

Controlled ISB has 'come of age', and after the DWH success, the response option has become more globally recognised. It is a response option that has been around for a little more than five decades. It has been proven through research and testing. It has proven itself in the field during this recent unprecedented response.

The new protocols described in this chapter were designed and developed for CISB operations from what was learned during this response. There will always be environmental and resource trade-offs regarding the selection of spill response methods, but the fundamental goal is to prevent spilled oil from reaching the shoreline. A combination of spill response options must be considered to decide how best to manage floating slicks in specific spill incident conditions. CISB has proven to be a very safe and effective way to remove large amounts of spilled oil from an offshore spill scenario like the recent DWH incident. Understanding the volume of oil that has been removed is crucial for measuring the success of CISB operations and building confidence with regulatory agencies and the public. This knowledge of the estimated burn volume also helps to address the necessary environmental considerations such as air quality and wildlife safety. The ability to evaluate field operations accurately in real time helps operations decision-makers to better direct resources, choose the most appropriate/effective tactics and measure their progress, especially when there is competition for the same resources, which is often the case. The measurement of the volume burned has increased confidence in the approach and, under the right conditions, has helped to establish CISB as a viable and measurable response method in the event of an offshore oil spill.

REFERENCES

Allen, A.A., Mabile, N.J., Jaeger, D., and Costanzo, D. 2011. The use of controlled burning during the Gulf of Mexico Deepwater Horizon MC-252 oil spill response, *International Oil Spill Conference Proceedings*, American Petroleum Institute, Washington, DC, abs194.

ASTM F1788, 2017. *Standard Guide for In-Situ Burning of Oil Spills on Water: Environmental and Operational Considerations*, ASTM, Conshohocken, PA: ASTM International.

ASTM F3195. 2016. *Standard Guide for Estimating the Volume of Oil Consumed in an In-Situ Burn*, ASTM, Conshohocken, PA: ASTM International.

Fingas, M.F., and Punt, M. 2000. *In-Situ Burning: A Cleanup Technique for Oil Spills on Water*, Environment Canada Special Publication, Ottawa, ON.

Mabile, N. 2010. Fire boom performance evaluation: Controlled burning during the deepwater horizon spill operational period, Internal Report, BP, Houston, TX.

Mabile, N. 2012a. Controlled in-situ burning: Transition from alternative technology to conventional spill response option, *Proceedings of the 35th AMOP Technical Seminar on Environmental. Contamination and Response*, New York, pp. 584–605.

Mabile, N.J. 2012b. Considerations for the application of controlled in-situ burning, *2012 Society of Petroleum Engineers: SPE/APPEA International Conference on Health, Safety and Environment in Oil and Gas Exploration and Production,* Perth, Australia, Vol. 3, pp. 2556–2575.

NOAA, Office of Response and Restoration, Emergency Response Division. https://response.restoration.noaa.gov/erd, accessed 2010.

USCG. 2003. In-situ burn operations manual: Oil spill response offshore, United States Coast Guard Report CG-D-06-03, *Springfield, VA*.

5 The Newfoundland Offshore Burn Experiment

Merv Fingas and Patrick Lambert

CONTENTS

5.1 OBJECTIVES

The offshore experiment was designed to meet four primary objectives:

1. To obtain measurements of critical burn parameters and to collect and analyse chemical emissions needed for comparison with data sets and models that are often based on laboratory and medium-scale tests.
2. To obtain samples for analysis of the smoke plume, water and gaseous emissions needed to determine whether the environmental impact of burning is acceptable.
3. To conduct a large-scale oil burning experiment in realistic open ocean conditions to demonstrate contained burning as a spill response technique.
4. To develop a response protocol that will establish operational strategies for burning and safety procedures under a variety of environmental and operational conditions.

5.2 OPERATIONS

5.2.1 OPERATIONAL DETAILS

The Newfoundland Offshore Burn Experiment (NOBE) took place on the Grand Banks in a 34 km^2 (ca. 10 nmi^2) area, coordinates 47° 40′ N, 52° W. The location is about 42 km (25 nm) east of the port of St. John's, Newfoundland. The experiment was conducted on 12 August 1993. The time and place were chosen to minimise potential ecological damage, should there be an unplanned release during the experiment and interference with the fishery. Two replicate experiments were planned during which 50 m^3 (13,200 gal) of oil were to be discharged in a controlled manner into a fireproof boom and ignited. The actual amount discharged was slightly less than this. Table 5.1 shows the overall results of the experiment; Table 5.2 shows

---

TABLE 5.1
Burn Summary

Burn 1
Oil volume discharged: 48.3 m^3
Burn and Pump time: 1.5 h
Residue in fireproof boom: 0.2 m^3 (max.)
Residue in backup boom: 0.2 m^3 (max.)
Efficiency: >99%

Burn 2
Oil volume discharged: 28.9 m^3
Burn and pump time: 1.3 h
Residue in fireproof boom: 0.1 m^3 (max.)
Residue in backup boom: 0.3 m^3 (max.)
Efficiency: >99%

TABLE 5.2
Weather and Sea Condition Averages

Burn	Time	Air Temperature (°C)	Relative Humidity (%)	Wave Height (m)	Water Temperature (°C)	Wind Velocity (km/h)
1 - Start	10:30	12.1	88.5	0.63	9	5–8
1 - End	11:40					
2 - Start	14:06	13.9	84.4	0.8	8.4	8–11
2 - End	15:10					

the relevant weather conditions. Papers on the experiment and results have been published (Blenkinsopp et al. 1996, 1997; Environment Canada 1997; Fingas et al. 1994a, 1994b, 1995).

A sophisticated array of state-of-the-art sensing, sampling and data-gathering equipment was deployed from a variety of platforms. Sampling near the fire and in the smoke plume was conducted using remote-controlled boats, helicopters and a remote-operated vehicle (ROV) (submersible) that was deployed beneath the slick (Bissonnette, et al. 1994). At more distant locations, a tethered blimp, conventional helicopters, fixed-wing aircraft and a variety of vessels were used. As a contingency measure, a secondary oil containment boom and recovery system capable of picking up all the oil that was discharged was towed behind the fireboom.

The experiment involved the measurement of (1) emissions into the air, (2) levels of oil and related compounds in the water and (3) operational parameters relevant to in-situ burning. Data was collected and analysed to generate information on over 2,000 parameters.

The layout of each of the vessels involved in the experiment is shown in Figures 5.1 and 5.2. The procession was led by the 68-m Canadian Coast Guard (CCG) vessel *Sir Wilfred Grenfell* (hereafter referred to as the *Grenfell*) that served as the oil discharge vessel. The fireboom was towed directly behind the *Grenfell* and by two Boston whalers with 45-m tow lines. Two 4-m remote-controlled boats and an 11-m sea truck serving as a platform for the tethered blimp were approximately 50, 100 and 150 m, respectively, behind the apex of the fireboom. Five hundred metres behind the burn, the secondary containment boom was towed by two 14-m vessels (i.e. 250 m behind the fireboom).

A number of other vessels were stationed farther from the main procession. These included several Boston whalers from which routine sampling was conducted and other vessels that served as platforms from which the remote-controlled boats, remote-controlled helicopters and the ROV were operated. The command vessel was the 83-m CCG vessel *Ann Harvey*. Two 30-m vessels were chartered to accommodate scientific observers and visitors.

The oil was Alberta Sweet Mixed Blend, which had been brought by truck to Newfoundland and loaded aboard the *Grenfell* for discharge into the boom (Figure 5.3). The oil was released into a fire-resistant boom and burned within it.

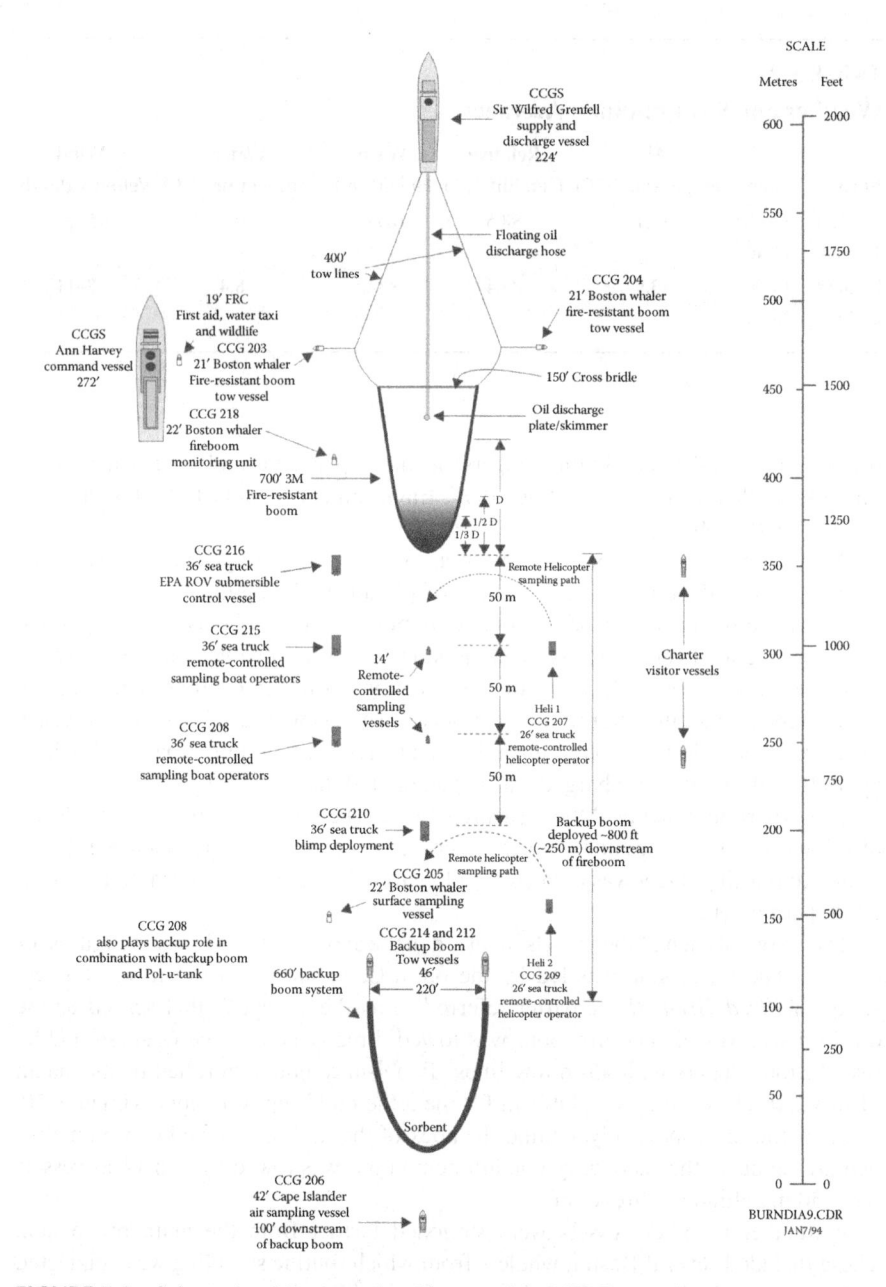

FIGURE 5.1 Schematic of the operational layout of the NOBE experiments.

Air emissions were monitored downwind using two remote-controlled boats and a research vessel, and from an airplane. The plume itself was sampled using an array of instruments on two remote-controlled helicopters and a blimp. Water samples were collected from the remote-controlled sampling boats, and air and water temperatures were measured from the same vessels. The fire-resistant boom was equipped with

FIGURE 5.2 Photograph of the layout of the experiment during burn 2.

FIGURE 5.3 Loading oil from four tank trucks into the *Grenfell* for the experimental discharge.

thermocouples to monitor temperatures directly affecting it and those in the water directly underneath the fire. A submersible was deployed under the burning slick to monitor temperatures and take video footage. A small boat was used to monitor and measure surface material that escaped and to take samples of the burn residue after the burn.

The oil was released from a supply-type ship through a skimmer so that if there were some problem, the flow could be reversed and the oil recovered. This is illustrated in Figures 5.4 through 5.6. A 700-ft section of boom was used. The amount of oil released in each spill was 50 m^3 or about 10,000 Imperial gallons. This is about the lower limit of a typical boom capacity. Once sufficient oil was in the boom to sustain combustion, it was ignited using a Helitorch.

The fire-resistant boom used was a commercial version along with some experimental sections. The middle sections near the burn were equipped with a number of thermocouples to measure the temperature on the boom. The boom was backed up by another boom, an offshore type, about one kilometre down current (see Figure 5.7). This second boom was loaded with sorbent to ensure that any escaping sheen was recovered. The fire-resistant boom was towed by a major vessel and the opening was maintained by two vessels towing outward at an angle of approximately 45°. Tow vessels were equipped with current meters to ensure that they were able to maintain a forward speed of 0.5 knots. Command and control operations took place from a major vessel of the CCG, the *Ann Harvey*. One helicopter was used both to ignite the slick and put out flares to guide the procession into the wind. Another helicopter was used to provide still and video footage for documentation. Two charter ships were engaged to bring out observers. They were also used as platforms for some of the documentation and air measurement.

FIGURE 5.4   The type of fire-resistant boom deployed at NOBE.

FIGURE 5.5 Deploying the fire-resistant boom from the rear of the *Grenfell*.

FIGURE 5.6 A view from the rear of the discharge vessel, the *Grenfell*, showing the discharge hose and the burn behind the hose.

FIGURE 5.7   A view of the experimental procession. The backup boom tow is seen at the rear.

Several smaller boats were used for other sampling purposes and for controlling the remote sampling boats and a remote-underwater vessel.

Burn 1 started after a second Helitorch run (Figure 5.8). Reports from the helicopters and both airplanes indicated that the smoke plume bifurcated after about 2 km downwind (Figure 5.9). A small part remained with the inversion layer at about 0.5 km, and the main portion split, with one portion turning southeast and one turning east after rising about 2 km. The pumping during burn 1 had to be stopped several times because the fire often spread back to the discharge point. The average discharge and burn rate for burn 1 were 915 L/min. The fire-resistant boom was inspected after the first burn. Some signs of fatigue in the stainless steel core were observed at a point about 10 cm from the stiffeners. Some of the Nextel fire-resistant fabric was missing from these areas (Figures 5.10 through 5.12). The boom was still fit for another burn, however.

The crews re-fitted the burn equipment and the sampling array for the second burn, which began in mid-afternoon. The first run of the Helitorch ignited the oil. Some oil was again splashed over; however, unlike the first burn no sheening whatsoever was observed. The oil outside the boom burned completely, leaving only small patches of residue which drifted back into the secondary recovery boom. Figure 5.13 shows some oil burning behind the boom. The wind was 8–11 km/h and this resulted in an approximate 45° angle for the plume. This burn was characterised by its 'classical', regular plume behaviour. The plume did bifurcate, however, somewhat about 2 km

FIGURE 5.8 Discharge of remaining fuel after successful ignite of burn 1.

FIGURE 5.9 Bifurcation of the smoke plume resulting from burn 1.

FIGURE 5.10 A side view of oil remaining in the fire-resistant boom after burn 1.

FIGURE 5.11 An aerial view of the oil remaining in the fire-resistant boom after burn 2.

FIGURE 5.12 View of the fire-resistant boom showing some sagging in nextel fabric.

FIGURE 5.13 Small amounts of oil burning behind the boom. As the water was entrained along behind the boom, so was this burning oil. Only small amounts of residue were released into the backup boom.

downwind, similar to the previous plume. Figure 5.14 shows oil in the backup boom; Figure 5.15 shows some oil burning behind the fire-resistant boom.

The pump rate for this burn averaged 610 L/min. Pumping was stopped after 1.25 h of burn time when some small pieces of the fire-resistant boom were released (Figures 5.16 through 5.19). The burn duration had already exceeded planned sampling times and most samplers had already been stopped.

FIGURE 5.14   A view of the backup boom after two burns showing the minor amount of residue that resulted from oil burning behind the boom.

FIGURE 5.15   Another small fire burning behind the main fire-resistant boom. This small fire also resulted in some residue which went into the backup boom.

FIGURE 5.16 A boom-monitoring crew whose responsibility was to monitor all aspects of the boom condition before, during and after fires.

FIGURE 5.17 A view of the end of burn 2 during which one of the boom floats released.

FIGURE 5.18   A close-up of the burning near the end of burn 2 showing one of the floats loose behind the main boom.

FIGURE 5.19   An aerial view of the fire-resistant boom at the end of burn 2. This shows a bit of smouldering still going on near the boom as well as small amounts of residue moving toward the boom apex.

FIGURE 5.20   A submersible unit which was manoeuvred under the boom during both burns to detect if oil was being lost under the fire-resistant boom. No losses were detected.

The fire-resistant boom was again inspected for damage and it was found that a prototype section with a middle tension member had lost three of its float logs. Inspection of this section at the factory showed that the section had not been properly constructed. The apex of the boom was still holding oil. The boom was in generally good condition, but one would not have used the apex for another burn.

Boom losses were carefully monitored by four methods: a dedicated team equipped with surface sampling equipment, a submersible under the boom end (Figure 5.20), the boom inspection team as noted earlier and the teams controlling the backup boom. These efforts showed that little oil was lost and that most of the small amount of oil entering the backup boom was residue from burns behind the boom.

5.3   OPERATIONAL RESULTS

5.3.1   OIL PROPERTIES

Table 5.3 shows the properties of both the oil and the residue (from burn 2). These data show that the residue was still floating and appeared like a heavy oil (Jokuty and Fingas, 1994).

TABLE 5.3
Properties of the Oil and Residue

Parameter	Starting Crude Oil	Residue
Weathering percentage	0.04	40%–48%
Density	0.8437 g/mL (15°C)	0.9365 g/mL (15°C)
Viscosity	11 mPa.s (15°C) (shear rate 500s⁻¹) Newtonian visc.	130500 mPa.s (15°C) (shear rate 1s⁻¹) Non-Newtonian visc.
Pour point	$-21°C$	34°C
Interfacial tension	21.4 dynes/cm (15°C) (air/oil) 13.3 dynes/cm (15°C) (oil/sea)	Not measurable at 15°C Not measurable at 15°C
Emulsion formation (f_i) and stability (f_f)	$f_i = 0$ (15°C) $f_f = 0$ (15°C)	$f_i = 0$ (15°C) $f_f = 0$ (15°C)
Asphaltene content	0.7 wt%	2.3 wt%
Wax content	10.1 wt%	13.8 wt%
Flash point	$-13°C$	>90°C
Water content	0.54 wt%	14.01 wt%
Sulphur content	0.15 wt%	0.40%

5.3.2 PUMP RATE AND SLICK AREA

The oil pump rates were monitored and the results are shown in Figures 5.21 and 5.22. The pump rates were controlled in steps to ensure that the fire did not progress past the bridle on the boom, as shown in Figure 5.23. Therefore, the area of the fire changed throughout the burn.

The burn area was variable with the pump rate and burn rate. Both burns varied from about 50 to 200 m² with about the mid-range of this being an average.

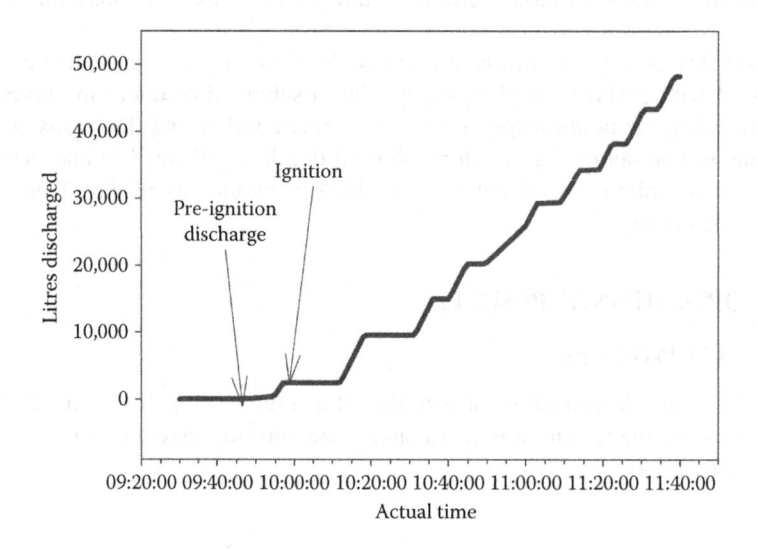

FIGURE 5.21 The oil pump rate with time for burn 1.

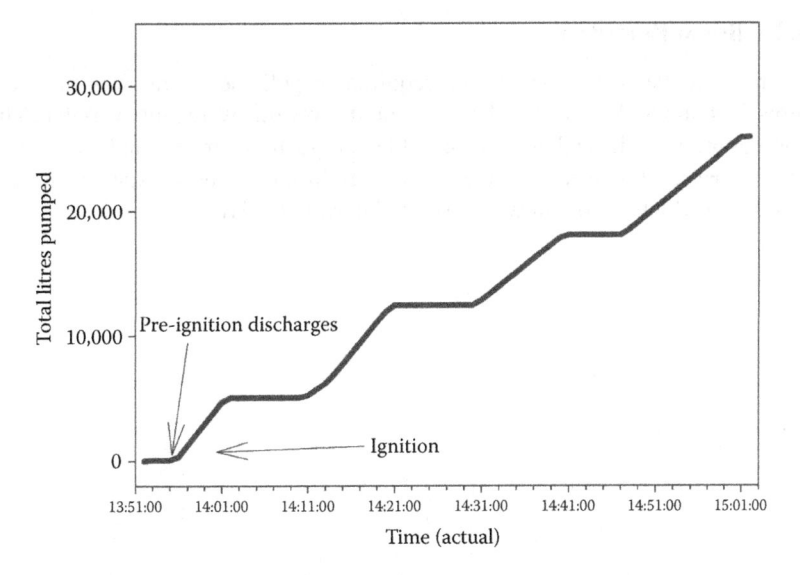

FIGURE 5.22   The oil pump rate with time for burn 2.

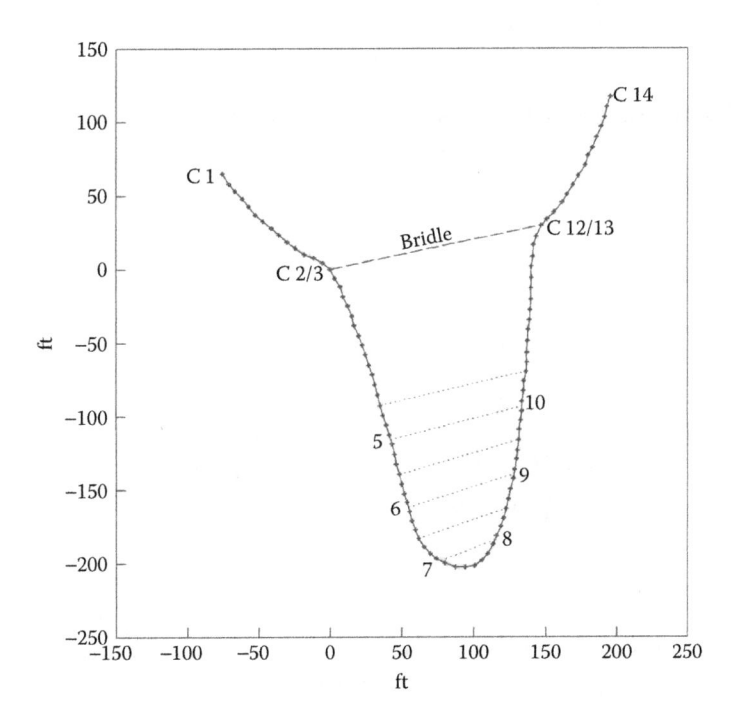

FIGURE 5.23   The boom shape during one portion of burn 1. This was determined using photogrammetric analysis of areas after the burns.

5.3.3 BOOM PARAMETERS

The strain on the boom tows (force required to pull the boom) was monitored. Figures 5.24 and 5.25 show that the forces on the tow line were quite variable. These tow forces are very dependent on external factors such as currents and winds. It was observed that the tow forces on the booms fell during the burn experiment, as the currents into which the boom was towed fell during the day.

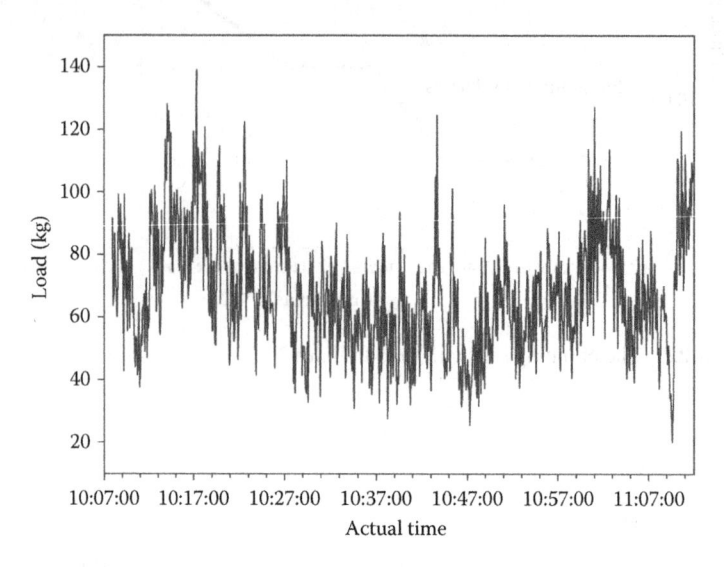

FIGURE 5.24   The load cell output on the boom tow line for burn 1.

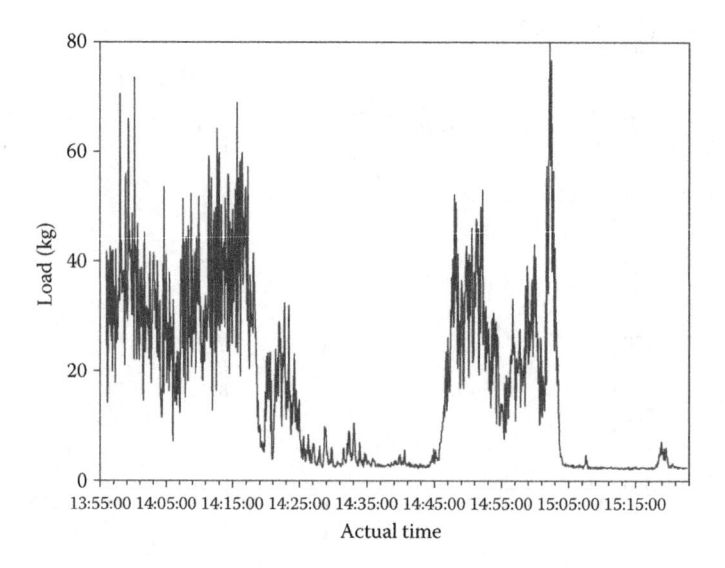

FIGURE 5.25   The load cell output on the boom tow line for burn 2.

Temperatures were recorded at several points on the fireboom. Three sections were monitored with thermocouples at four locations in the vertical plane. The very apex of the boom was also monitored; however, the electronics failed somewhere during burn 1. Figure 5.26 shows the temperatures on the boom during burn 1 and Figure 5.27 shows the temperatures during burn 2. These figures show that the temperatures at the top edge of the fireboom often reached 1,000°C and the temperatures below were substantially lower. Thermocouple probes known to be in the water show no increase in water temperatures. The high variability in temperatures depended on the amount of boom fill. As can be seen, the forward sections of the boom showed low temperatures at times, this being due to lack of burning at that point in time.

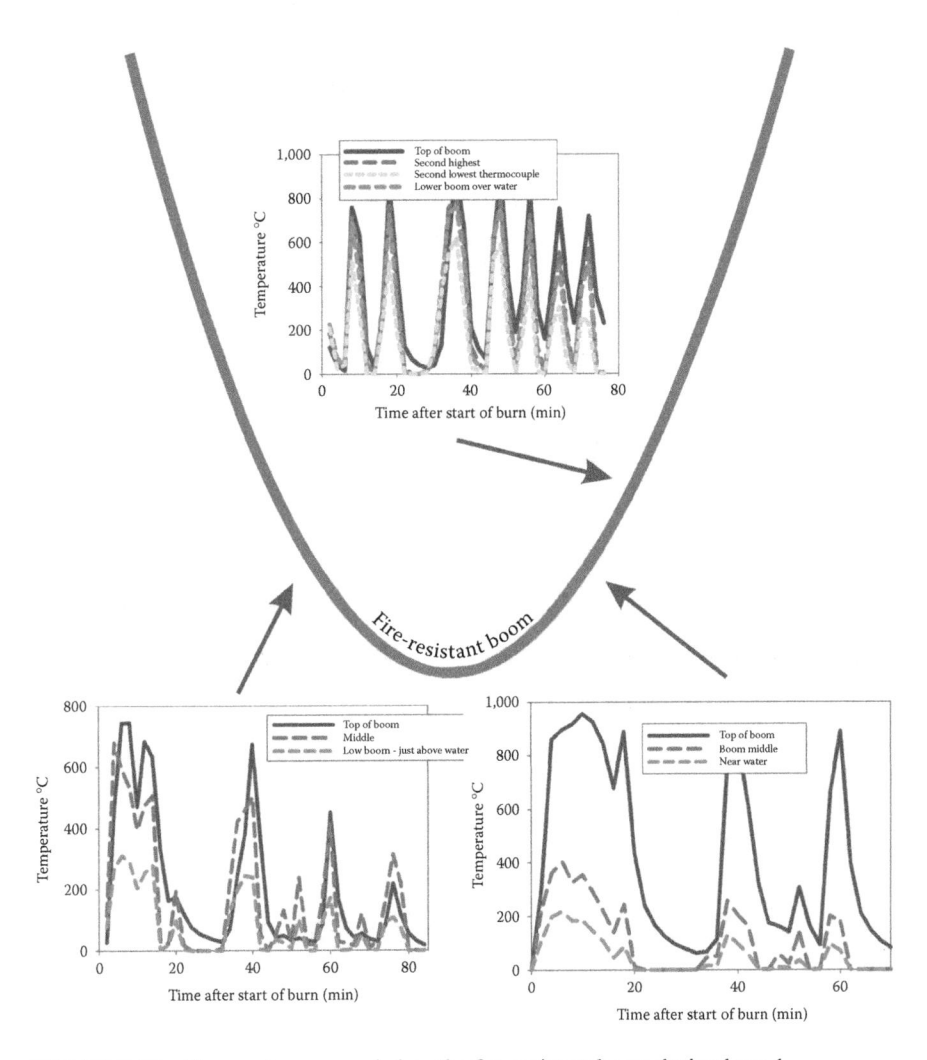

FIGURE 5.26   Temperatures recorded on the fire-resistant boom during burn 1.

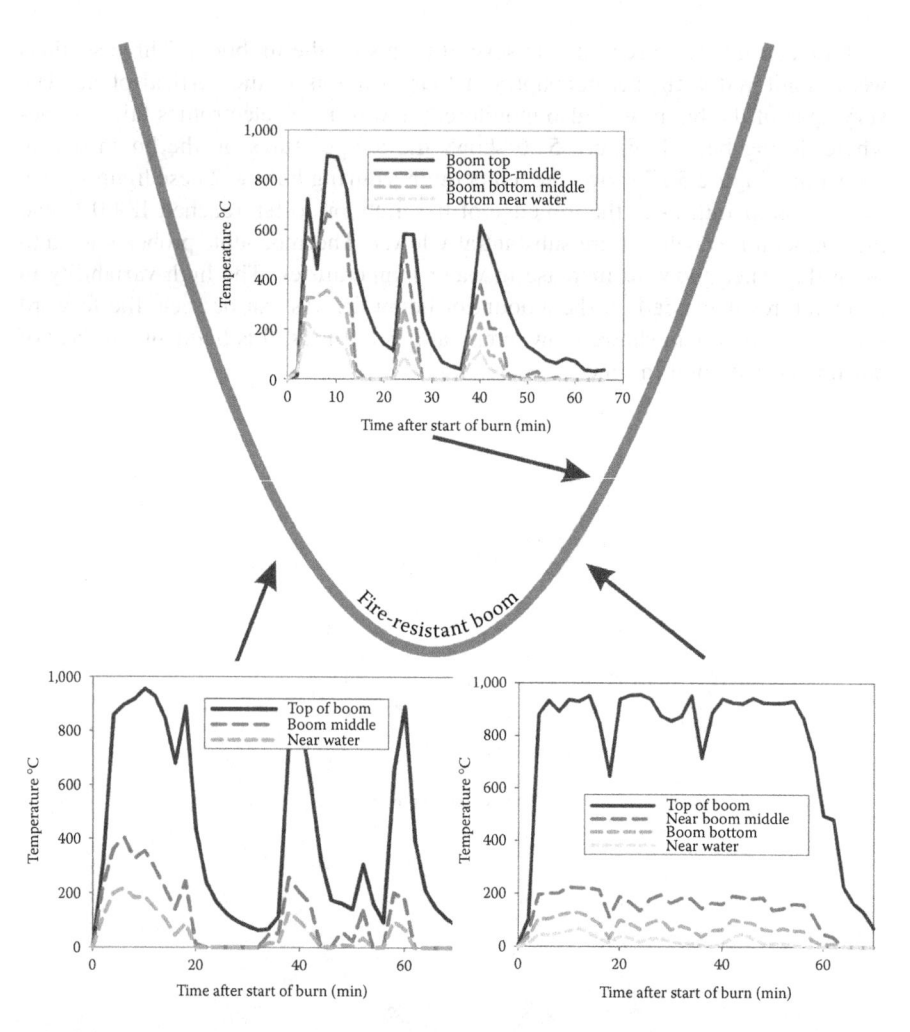

FIGURE 5.27 Temperatures recorded on the fire-resistant boom during burn 2.

TABLE 5.4
Summary of Analytical Methods

Sample	Sampler	Measurement Taken	Secondary Measurement	Additional Parameters
Soot at sea level	High volume sampler	Dioxins and Dibenzofurans	Particulates	PAHs
	Sampling pump medium volume	PAHs	Particulates	
	RAM	Particulates		
	Cascade sampler	Particle size	PAHs	
Soot in smoke	Sampling pump low volume	PAHs	Particulates	Metals
	Blimp, remote-controlled helicopter, research aircraft			
Gases	Summa canister	Volatile organic compounds	CO_2	
	Sampling pump low volume	Volatile organic compounds		
	CO_2 meter	Carbon dioxide		
	SO_2 meter	Sulphur dioxide		
	NO_2 meter	Nitrogen dioxide		
	CO meter	Carbon monoxide		
Oil		PAHs	Metals	Full analysis
Burn residue		PAHs	Metals	Full analysis
Water under burn		PAHs	Organics	Toxicity

5.3.4 EMISSIONS MEASUREMENT

Sampling methodologies and target emissions are summarised in Table 5.4. Sample locations and platforms are illustrated in Figure 5.28. Summary methods are given below. Detailed methods are described in the literature.

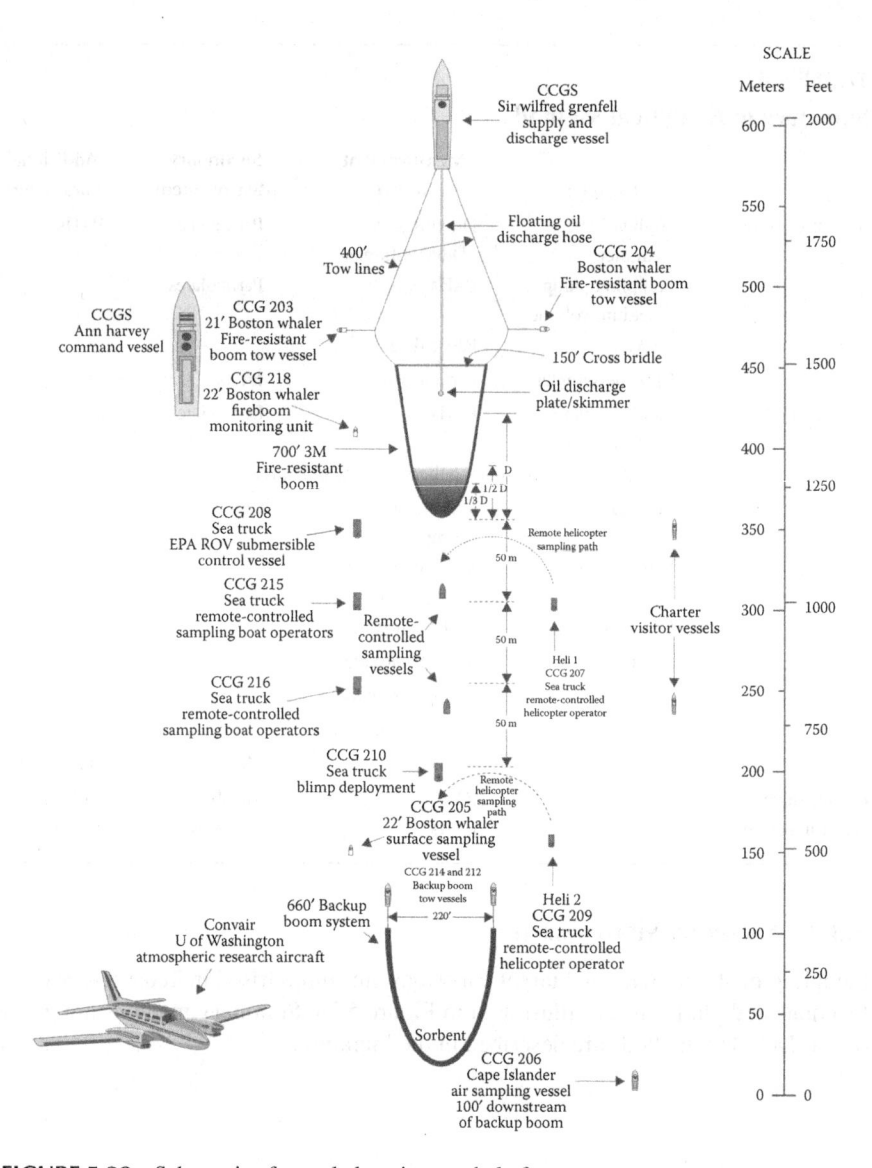

FIGURE 5.28 Schematic of sample locations and platforms.

5.4 DESCRIPTION OF SAMPLING AND ANALYTICAL PLATFORMS

Surface sampling was largely carried out using four remote-controlled boats approximately 4 m in length (Figures 5.29 and 5.30). Most of the samplers or instruments listed in Table 5.4 were contained in each boat. These remote-controlled boats operated near the fire and up to about 150 m behind the apex of the fire-resistant boom. A larger manned boat was deployed about 400 m behind the burn to take samples and measurement in the far field. This was operated by EPA personnel.

FIGURE 5.29 A remote-controlled sample boat.

FIGURE 5.30 A powered barge was used as a platform to control 2 remote-controlled boats.

Samples inside the smoke plume were taken by 3 platforms, a remote-controlled helicopter (Figure 5.31), a blimp and a research aircraft. The latter was operated by the University of Washington and is illustrated in Figure 5.32. The research aircraft took various samples and was equipped with a laser particle measuring instrument. The aircraft had the capability to provide cross-section data on the plumes. This aircraft took longitudinal runs as well as cross-sections through the plume at

FIGURE 5.31   A remote-controlled helicopter exiting the smoke plume after taking samples.

FIGURE 5.32   The University of Washington Convair aircraft which had particle-size measuring and sampling equipment aboard.

various points. The small remote-controlled helicopter operated at distances generally 50 m behind the apex of the fireboom. The blimp, operated by the National Institute of Standards and Technology (NIST), was situated about 200 m behind the apex of the fireboom.

Two small boats took water and oil samples as required both near and in the fire-resistant boom and along the path as well as in the backup boom.

5.5 WATER SAMPLES

Sigma 800SL samplers were used to collect water samples for both toxicity and organic compound analyses. One sampler contained 4 × 1-gallon sterile glass bottles with Teflon-lined caps (for toxicity testing) and the other contained 24 × 350 mL water (for organic measurements). The samplers were located on the deck of a sampling boat, and Teflon tubing was attached to a pole which was lowered into the water to approximately 1 m depth. All samples were collected and placed in refrigerated coolers and shipped to analytical laboratories within 24 hours of collection. The smaller bottles were sent to the Environment Canada's laboratory for chemistry analysis; the larger bottles were sent to EVS consultants in British Colombia for toxicity analysis (Blenkinsopp et al. 1996, 1997). The water was analysed for Polycyclic aromatic Hydrocarbons (PAHs) and water-soluble compounds. Most water samples revealed no PAHs above detection limits; however, 3 of the samples showed very low levels of naphthalenes. The water showed no toxicity to the aquatic species tested.

5.6 PARTICULATES

All burns produce an abundance of particulate matter. Smaller particulates are of greater health concern than larger ones. At the time of NOBE, portable instruments to measure small particulates were not common and were not used extensively. In other experiments, TPM-10 concentrations were sometimes about 0.7 of the total particulate concentration (total suspend particles [TSP]), as would be expected, but sometimes were the same as the TSP. The PM-2.5 concentrations were sometimes 0.5 of the TSP, as would be expected, but sometimes were closer to the PM-10 values.

5.6.1 TOTAL PARTICULATES BY HIGH VOLUME SAMPLER

High-volume air sampling was performed using a General Metal Works model PS-1 instrument modified to be fitted onto the deck of the sample boats. The PS-1 sampling heads were rinsed with hexane prior to loading the media. The sampling media consisted of 3 in. diameter glass fibre filter and a polyurethane foam (PUF) filter that was 3 in. thick, density 0.022 g/cm^3. After collection, the filters were wrapped in aluminium foil and placed in the glass jar, which originally contained the new PUF filter, wrapped in foil and refrigerated.

The flow rate varied between 128 and 279 L/min, the lower flow rates corresponding to older instruments. The volume of air going through the samplers during the experiment varied between 9,200 and 13,600 L. All glass fibre filters were pre-weighed and then weighed after the experiment. The PUF filter and fibre filter were then combined for extraction and subsequent GC–MSD analysis for PAHs. The results yield a single value of total particulates for each burn; however, the samples were also used to measure PAHs on the soot.

5.6.2 PARTICULATES BY CASCADE IMPACTOR

An 8-stage non-variable Anderson cascade impactor was used to collect various fractions of particulates at a constant sampling rate of 28.3 L/min. The instrument was mounted on the mast of the remote-controlled sampling boat approximately 1 m above deck level. The filters consisted of quartz fibre. The inside of the sampling unit was rinsed with hexane prior to the experiment. After sampling was complete the quartz fibre filters were wrapped in aluminium foil before being placed into an envelope and refrigerated. All filters were pre-weighed and then weighed after the experiment. The variations in weights were so small that all eight filters for each stack were combined for extraction and GC–MSD analysis for PAHs. As with the high-volume samples, the samples taken by the cascade impactor were used for PAH analysis.

5.6.3 Real-Time Particulate Measurement

A Measurement Instruments East (MIE) Ram-1 instrument was used to perform real-time total aerosol monitoring and to measure relative concentrations of airborne particulates. This instrument responds to a physical particle size of 0.1–10 μ. The flow rate of this instrument is 2 L/min and Tygon tubing was connected from the instrument to the mast to allow sampling at 1 m above deck level. The instrument was connected to a data logger which recorded the data every minute. The values obtained are shown in Figures 5.33 and 5.34.

5.6.4 PAHs on Particulates by Cyclone Sampler

A Cyclone sampler equipped with a Gilian 513A pump was used to collect particulates smaller than 5 μ on a PVC (37 mm) filter placed inside a cassette (Tygon tubing connections). Tygon tubing was connected from the instrument to the mast to allow sampling at 1 m above deck level. Flow rates varied between 1.7 and 2 L/min and collected air volumes between 57 and 122 L. The cassette was capped and refrigerated.

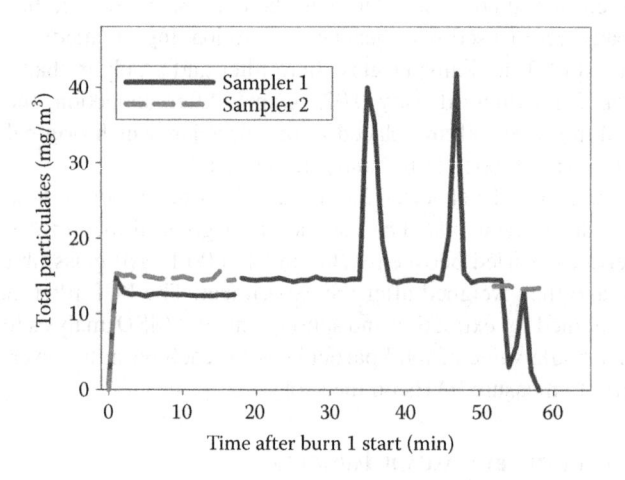

FIGURE 5.33 The total particulates obtained for burn 1 using a Ram instrument.

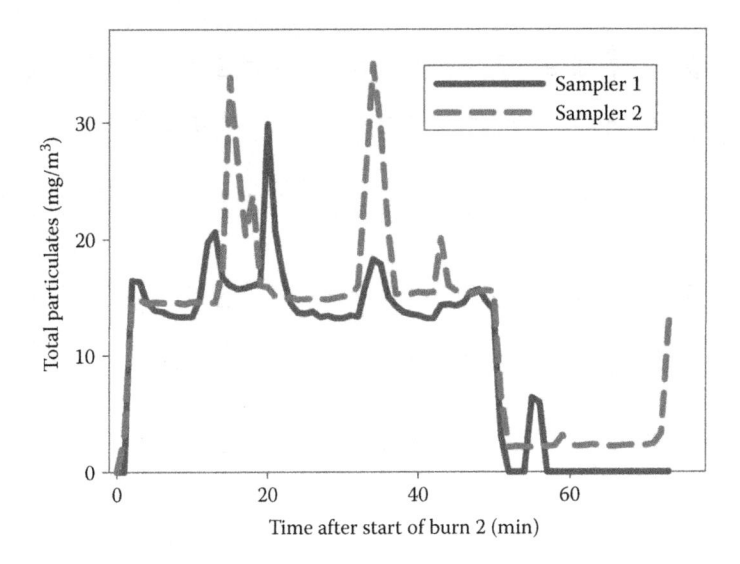

FIGURE 5.34 The total particulates obtained for burn 2 using a Ram instrument.

All filters were pre-weighed and then weighed after the experiment under the same conditions. The filters were extracted and analysed for PAH content using GC–MSD.

5.7 VOLATILE ORGANIC COMPOUNDS

Volatile organic compounds (VOCs) are hydrocarbons having a significant concentration in the vapour phase. One hundred and forty-eight VOCs were measured from samples taken in Summa canisters and some on carbon absorption tubes. The concentrations of VOCs were about the same in the two burns. Concentrations were under human health limits even at the closest monitoring station. VOC concentrations were about three times higher when the oil was not burning and was just evaporating. Unfortunately, this is difficult to measure at all burns, because measurements are usually not conducted during the evaporation only phase. At long distances of about 500 m, the concentrations of VOCs were low whether the oil was evaporating or not.

5.7.1 Volatile Organic Compounds by Summa Canister

Multiple 6 L SUMMA canisters pre-evacuated to 0.05 Torr were used to collect air for analysis for VOCs, CO and CO_2. The flow controller (restricted orifice) was adjusted to 500 cc/min for the evaporation period and to 100 cc/min for the burn. Table 5.5 presents tabular and summary results. More than 130 separate SUMMA canisters were sampled and analysed. Figures 5.35 through 5.43 illustrate the VOC results.

TABLE 5.5
VOC Summary Concentrations μg/m³ (Averages)

Type	Background	Evaporation	Burn	Burn	Background	Evaporation	Burn	Background	Burn	Evaporation	Burn	Burn	Evaporation	Burn	Burn
Burn #	1	1	1	1	1	1	1	1	1	2	2	2	2	2	2
Platform	All	Boat 2	Boat 2	Boat 1	Far	Far	Far	Airplane	Airplane	Boat 1	Boat 1	Boat 2	Far	Far	Airplane
Propene	17.5	70.3	3.1	1.9	2.2	7.8	2.3	5.9	9.8	5.6	47.1	4.4	4.5	6.2	9.6
1-Butene/ 2-Methylpropene	25.6	31.2	6.0	3.9	4.1	1.4	2.1	0.5	1.0	10.3	7.3	20.5	7.4	6.7	0.8
1,3-Butadiene	1.4	1.7	0.4	0.9	0.2	1.7	0.4	0.0	0.3	1.1	2.1	1.2	0.0	1.2	0.2
t-2-Butene	11.3	1.8	1.6	1.5	0.3	9.8	0.4	0.0	0.2	1.0	0.3	11.6	0.6	1.5	0.0
2,2-Dimethylpropane	0.2	0.0	0.0	0.4	0.0	0.0	0.3	0.0	0.0	16.6	8.9	2.4	0.0	0.0	0.0
1-Butyne	0.0	0.0	0.0	0.0	0.0	0.7	0.0	0.0	0.0	0.0	0.0	0.0	0.0	0.0	0.0
c-2-Butene	11.7	1.8	1.5	1.3	0.4	0.0	0.4	0.0	0.1	1.7	1.1	9.7	0.2	1.2	0.0
1-Pentene	4.0	7.4	1.0	0.5	0.7	0.6	0.3	0.0	0.1	1.4	1.3	3.7	1.1	1.2	0.0
Isoprene (2-Methyl-1, 3-Butadiene)	0.8	0.8	0.1	0.3	0.2	4.5	0.0	0.1	0.1	0.2	0.6	0.7	0.0	0.7	0.1
t-2-Pentene	4.5	0.7	1.0	0.9	0.1	2.8	0.6	0.0	0.2	1.7	1.1	6.5	0.2	2.1	0.0
c-2-Pentene	6.0	0.7	0.7	0.6	0.0	0.0	0.3	0.0	0.2	0.8	0.7	4.9	0.0	1.2	0.0
2,2-Dimethylbutane	2.0	0.6	0.6	2.8	0.0	0.5	2.4	0.0	0.3	129.2	59.9	19.6	0.0	3.2	0.1
t-4-Methyl-2-Pentene	0.3	0.0	0.0	0.2	0.0	0.0	0.0	0.0	0.0	0.0	0.0	0.2	0.0	0.0	0.0
c-4-Methyl-2-Pentene	1.3	0.0	0.0	0.1	0.0	0.0	0.0	0.0	0.0	0.0	0.0	0.8	0.0	0.7	0.0
1-Hexene/ 2-Methyl-1-Pentene	2.6	6.7	1.0	0.5	0.6	0.0	0.0	0.0	0.2	1.5	1.3	2.5	1.0	1.2	0.0
t-2-Hexene	1.0	0.5	0.2	0.2	0.0	0.0	0.2	0.0	0.1	0.8	0.5	1.3	0.0	0.9	0.0
t-3-Methyl-2-Pentene	1.0	0.0	0.1	0.1	0.0	13.9	0.0	0.0	0.1	0.1	0.3	1.2	0.0	0.7	0.0
c-2-Hexene	0.7	0.4	0.1	0.1	0.0	0.0	0.0	0.0	0.0	0.5	0.2	1.0	0.0	0.7	0.0
c-3-Methyl-2-Pentene	1.4	0.3	0.3	0.2	0.0	1.1	0.2	0.0	0.1	0.4	0.3	1.6	0.0	0.9	0.0
2,2-Dimethylpentane	0.4	0.5	0.2	2.1	0.0	2.0	1.1	0.0	0.1	113.9	48.4	15.4	0.0	0.5	0.1

(*Continued*)

TABLE 5.5 (Continued)

VOC Summary Concentrations µg/m³ (Averages)

Type	Background	Evaporation	Burn	Burn	Background	Evaporation	Burn	Background	Burn	Evaporation	Burn	Burn	Evaporation	Burn	Burn
Burn #	1	1	1	1	1	1	1	1	1	2	2	2	2	2	2
Platform	All	Boat 2	Boat 2	Boat 1	Far	Far	Far	Airplane	Airplane	Boat 1	Boat 1	Boat 2	Far	Far	Airplane
2,4-Dimethylpentane	1.3	1.2	0.6	4.8	0.0	4.1	2.5	0.9	0.9	245.6	105.5	35.4	0.0	1.6	0.5
2,2,3-Trimethylbutane	0.5	1.1	0.4	0.5	0.1	0.0	0.3	0.0	0.0	24.9	11.1	3.8	0.2	0.4	0.0
Benzene	6.5	2.7	1.6	4.0	0.2	0.2	3.2	0.1	3.3	72.1	39.6	16.5	0.8	8.3	1.3
2-Methylhexane	2.2	3.7	1.8	12.5	0.0	0.0	6.2	0.0	0.8	13.1	186.0	91.4	0.0	5.9	0.5
2,3-Dimethylpentane	1.8	2.1	1.4	8.7	0.0	0.0	3.8	0.5	0.8	379.5	173.3	67.1	0.0	2.6	0.5
2,2,4-Trimethylpentane	2.6	2.5	0.7	0.0	0.2	0.0	0.8	11.6	7.5	43.7	6.2	9.5	0.4	1.8	5.1
t-2-Heptene	0.6	4.9	0.3	0.0	0.7	1.0	0.0	0.0	0.0	0.5	0.7	0.8	0.0	0.7	0.0
c-2-Heptene	1.2	6.8	0.3	0.0	0.6	0.0	0.0	0.0	0.0	0.3	0.4	0.7	0.0	0.6	0.0
2,2-Dimethylhexane	0.1	0.0	0.1	0.0	0.0	1.1	0.8	0.0	0.1	90.1	12.7	0.0	0.0	0.2	0.0
2,5-Dimethylhexane	0.7	1.3	0.6	2.5	0.0	0.7	0.9	1.5	1.1	92.0	42.0	15.5	0.0	0.8	0.7
2,4-Dimethylhexane	1.0	1.8	0.7	3.1	0.0	0.7	1.2	1.9	1.4	133.6	65.9	21.7	0.0	1.0	1.0
2,3,4-Trimethylpentane	0.8	0.5	0.2	0.8	0.0	0.7	0.4	4.1	2.6	22.7	12.6	4.3	0.0	0.7	1.8
Toluene	20.2	14.9	5.6	8.7	0.4	0.0	9.8	9.4	8.5	194.5	95.7	45.4	1.7	25.0	3.3
2-Methylheptane	2.3	11.4	3.8	16.6	0.0	0.0	5.0	0.0	0.7	502.6	249.7	93.1	0.0	2.6	0.4
4-Methylheptane	0.9	2.7	0.8	2.8	0.0	0.0	1.5	0.0	0.1	112.0	67.3	18.4	0.0	0.8	0.1
1-Methylcyclohexene	8.5	0.0	0.0	0.0	0.0	0.5	0.0	0.0	0.0	0.0	0.0	0.0	0.0	39.4	0.0
3-Methylheptane	2.7	8.5	3.1	13.0	0.0	1.4	4.4	0.1	0.6	417.9	217.7	78.1	0.0	3.7	0.3
c-1,3-Dimethylcyclohexane	0.8	8.2	3.1	12.9	0.0	1.3	3.8	0.0	0.5	442.3	191.2	85.7	0.0	0.6	0.4
t-1,4-Dimethylcyclohexane	0.4	3.9	1.6	6.3	0.0	0.0	1.9	0.0	0.2	201.1	90.2	46.3	0.0	0.3	0.2
2,2,5-Trimethylhexane	0.3	0.1	0.0	0.2	0.0	1.0	0.1	1.7	1.1	4.6	2.1	1.0	0.0	0.2	0.8

(Continued)

TABLE 5.5 (Continued)
VOC Summary Concentrations µg/m³ (Averages)

Type	Background	Evaporation	Burn	Burn	Background	Background	Evaporation	Burn	Background	Burn	Evaporation	Burn	Burn	Evaporation	Burn	Burn
Burn #	1	1	1	1	1	1	1	1	1	1	2	2	2	2	2	2
Platform	All	Boat 2	Boat 2	Boat 1	Boat 1	Far	Far	Far	Airplane	Airplane	Boat 1	Boat 1	Boat 2	Far	Far	Airplane
c-1,4/t-1,3-Dimethylcyclohexane	0.5	2.9	0.9	3.2	0.0	0.0	0.0	0.9	0.0	0.1	114.8	66.7	22.0	0.0	0.4	0.1
c-2-Octene	0.7	3.7	0.1	0.0	0.1	0.0	0.0	0.0	0.0	0.0	0.1	0.0	0.1	0.4	0.1	0.0
Ethylbenzene	10.1	5.2	2.0	2.8	0.1	0.3	0.3	2.5	0.4	0.6	45.2	29.2	14.8	0.4	6.7	0.2
m/p-Xylene	35.1	22.3	7.5	9.3	0.3	0.0	0.0	7.7	0.3	1.2	148.3	99.2	49.6	1.4	20.8	0.3
Styrene	13.3	0.9	0.3	0.2	0.0	0.0	0.0	0.2	18.9	18.7	0.5	0.6	1.3	0.0	0.3	4.6
o-Xylene	11.2	7.5	2.5	2.9	0.1	0.0	0.0	2.5	0.1	0.4	40.8	29.2	15.2	0.5	7.6	0.1
Nonane	3.1	37.9	10.8	17.2	0.1	3.6	0.1	3.2	0.1	0.5	248.5	196.9	78.4	0.4	1.6	0.2
iso-Propylbenzene	0.5	1.6	0.4	0.5	0.0	0.1	0.1	0.2	0.0	0.0	7.2	6.0	2.5	0.0	0.5	0.0
3,6-Dimethyloctane	0.2	3.3	0.7	1.4	1.3	0.0	0.0	0.3	0.0	0.0	18.4	12.7	4.1	0.0	0.1	0.0
n-Propylbenzene	1.7	5.8	1.2	1.0	0.1	0.0	0.0	0.5	0.0	0.1	12.1	8.1	4.7	0.2	1.8	0.1
3-Ethyltoluene	5.0	15.5	3.2	2.9	0.1	0.8	0.8	1.5	0.1	0.2	22.7	20.0	12.5	0.3	5.6	0.1
4-Ethyltoluene	2.5	8.1	1.6	1.3	0.0	0.0	0.0	0.6	0.0	0.1	9.2	8.5	5.7	0.2	2.7	0.1
c-1,3-Dimethylcyclohexane	2.9	11.4	2.4	1.8	0.1	0.0	0.0	0.6	0.1	0.1	15.1	15.0	8.2	0.2	2.8	0.1
t-1,4-Dimethylcyclohexane	1.9	6.4	1.5	1.2	0.0	0.0	0.0	0.6	0.0	0.1	9.0	8.4	5.0	0.2	2.0	0.0
2,2,5-Trimethylhexane	10.2	38.0	8.9	5.4	0.4	0.0	0.0	2.0	0.2	0.5	33.1	33.4	24.4	0.7	9.1	0.3
Decane	5.8	53.0	15.4	13.1	0.2	0.4	0.4	1.3	0.1	0.4	63.6	81.5	41.9	0.4	1.2	0.0
iso-Butylbenzene	0.2	0.9	0.2	0.2	0.0	1.3	0.0	0.0	0.0	0.0	0.1	1.0	0.7	0.0	0.1	0.0
sec-Butylbenzene	0.2	1.3	0.4	0.3	0.0	0.0	0.1	0.1	0.0	0.0	2.1	1.9	1.2	0.0	0.1	0.0
1,2,3-Trimethylbenzene	2.5	10.4	3.0	1.8	0.0	0.4	0.4	0.5	0.1	0.1	10.3	11.4	7.7	0.2	1.8	0.1

(Continued)

TABLE 5.5 (Continued)
VOC Summary Concentrations µg/m³ (Averages)

Type	Background	Evaporation	Burn	Burn	Background	Evaporation	Burn	Background	Burn	Evaporation	Burn	Burn	Evaporation	Burn	Burn
Burn #	1	1	1	1	1	1	1	1	1	2	2	2	2	2	2
Platform	All	Boat 2	Boat 2	Boat 1	Far	Far	Far	Airplane	Airplane	Boat 1	Boat 1	Boat 2	Far	Far	Airplane
p-Cymene	0.8	2.3	0.9	0.4	0.1	0.8	0.0	0.0	0.2	0.5	3.2	2.1	0.0	0.0	0.0
1,2-Dichlorobenzene	0.1	0.2	0.1	0.0	0.0	0.1	0.0	0.0	0.0	0.0	0.0	0.1	0.0	0.0	0.0
Indane	0.8	1.7	0.6	0.4	0.0	0.0	0.2	0.0	0.0	2.1	1.9	1.8	0.0	0.9	0.0
1,3-Diethylbenzene	0.6	1.9	0.7	0.3	0.0	0.4	0.1	0.0	0.0	0.3	1.3	1.4	0.0	0.4	0.0
1,4-Diethylbenzene	2.2	5.7	2.5	0.0	0.0	0.3	0.1	0.0	0.1	1.8	6.6	2.9	0.0	1.5	0.0
n-Butylbenzene	0.6	2.2	0.9	0.5	0.0	0.2	0.1	0.0	0.0	0.5	2.0	2.0	0.0	0.4	0.0
1,2-Diethylbenzene	0.2	0.6	0.3	0.1	0.0	0.3	0.0	0.0	0.0	0.1	0.5	0.5	0.0	0.1	0.0
Undecane	10.2	63.6	20.6	14.2	0.2	0.2	1.2	0.2	0.7	31.3	53.9	36.5	0.4	1.1	0.3
Naphthalene	3.0	6.3	2.9	1.4	0.4	0.7	0.3	0.3	0.6	2.8	4.8	5.1	0.5	1.6	0.4
Dodecane	10.7	60.7	18.9	14.0	0.3	0.7	1.3	0.4	1.1	24.9	36.5	31.7	0.6	1.2	0.7
Hexylbenzene	3.9	11.7	1.1	0.3	0.9	0.0	2.1	0.0	0.0	0.6	1.1	0.9	0.0	1.0	0.0
TOTAL VOC	289.7	597.0	157.2	214.1	15.7	0.3	88.5	59.7	69.0	4146.7	2492.8	1128.8	24.6	201.7	35.9

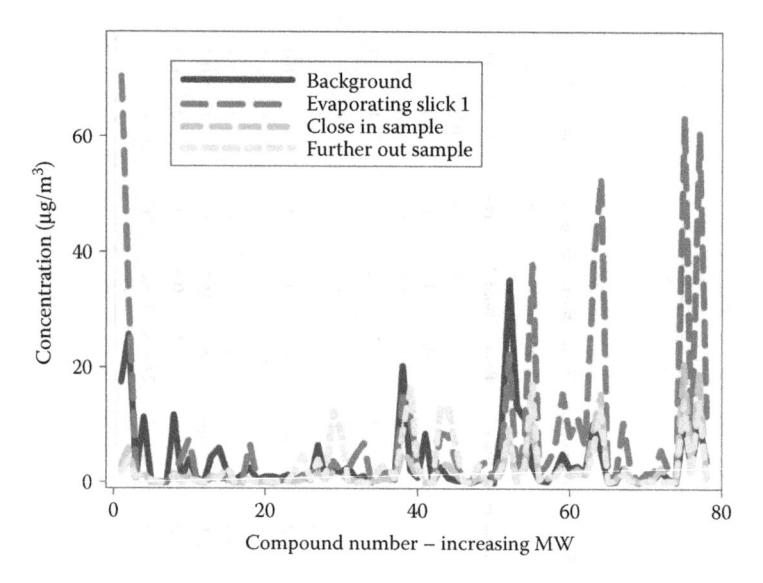

FIGURE 5.35 The summary VOC concentrations from slick 1 and burn 1 taken from a close sample boat. This shows that the evaporating slick yields higher concentrations than burning or background. The background concentrations and the concentrations during burning are similar.

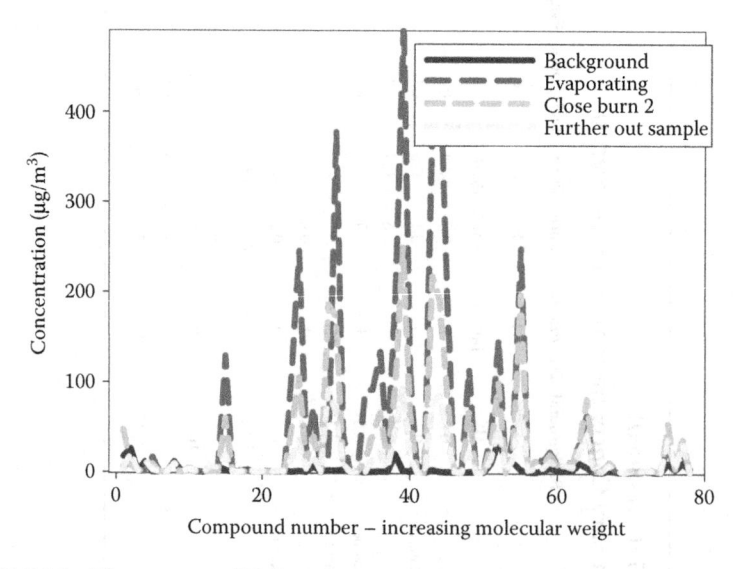

FIGURE 5.36 The summary VOC concentrations from slick 2 and burn 2 taken from a close-in sample boat. This shows that the evaporating slick yields higher concentrations than burning or background. The background concentrations and the concentrations during burning are similar but in this case the burn concentrations are somewhat higher.

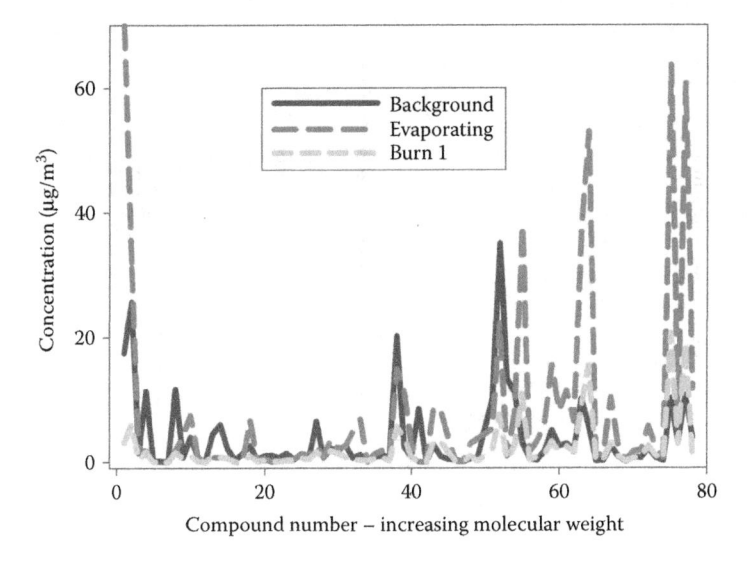

FIGURE 5.37 The summary VOC concentrations from slick 1 and burn 1 taken from a slightly further out sample boat. This shows that the evaporating slick yields higher concentrations than burning or background. The background concentrations and the concentrations during burning are similar.

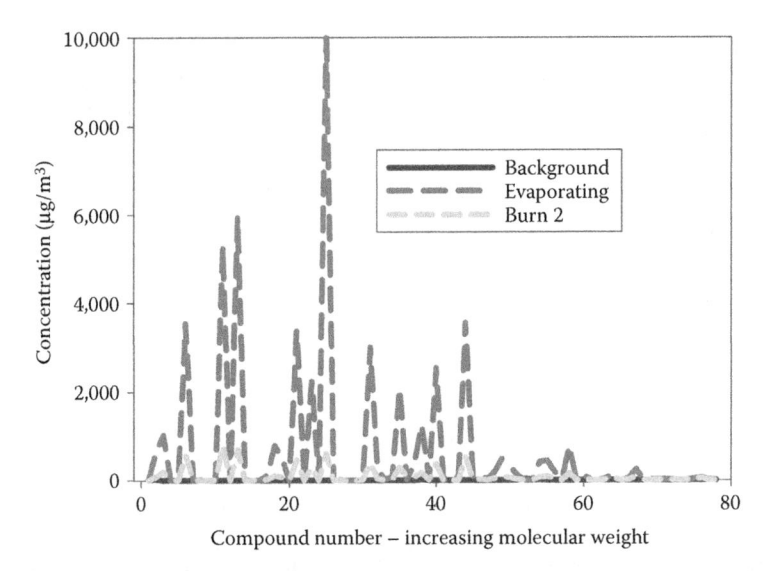

FIGURE 5.38 The summary VOC concentrations from slick 2 and burn 2 taken from a slightly further out sample boat. This shows that the evaporating slick yields higher concentrations than burning or background. The background concentrations and the concentrations during burning are similar.

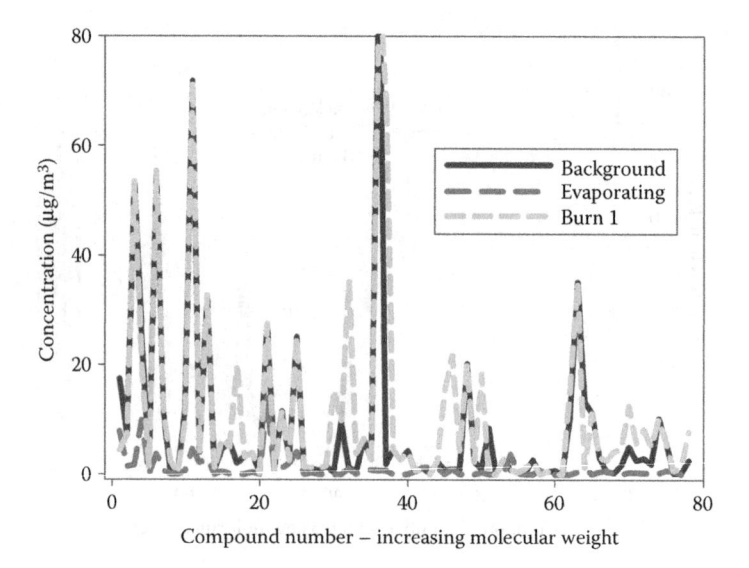

FIGURE 5.39 The summary VOC concentrations from slick 1 and burn 1 taken from a distant sample boat, more than 0.5 km downwind. In these cases, the background concentrations and the concentrations during burning are similar. This distant station received large amounts of emissions from the more than 15 vessels operating in front of it. At this distance, the evaporating oil did not yield a significant signal.

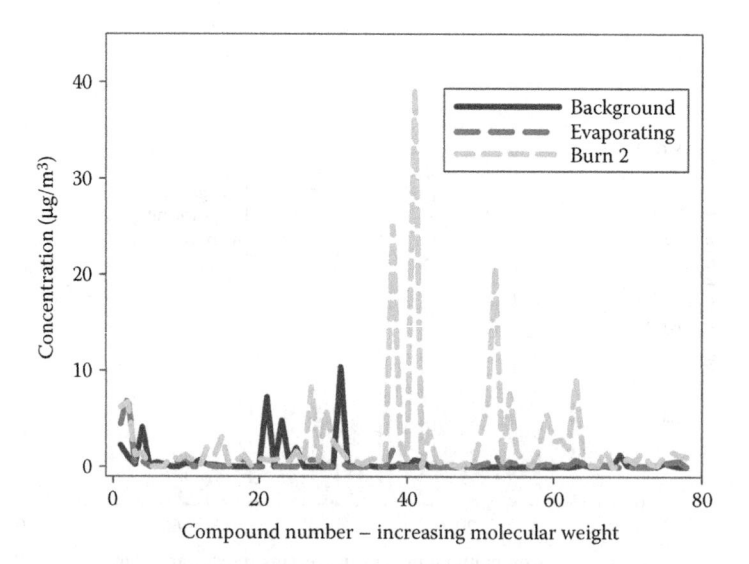

FIGURE 5.40 The summary VOC concentrations from slick 2 and burn 2 taken from a distant sample boat, more than 0.5 km downwind. In these cases, the background concentrations and the concentrations during burning are similar, with the latter being greater. This distant station received large amounts of emissions from the more than 15 vessels operating in front of it. At this distance, the evaporating oil did not yield a significant signal.

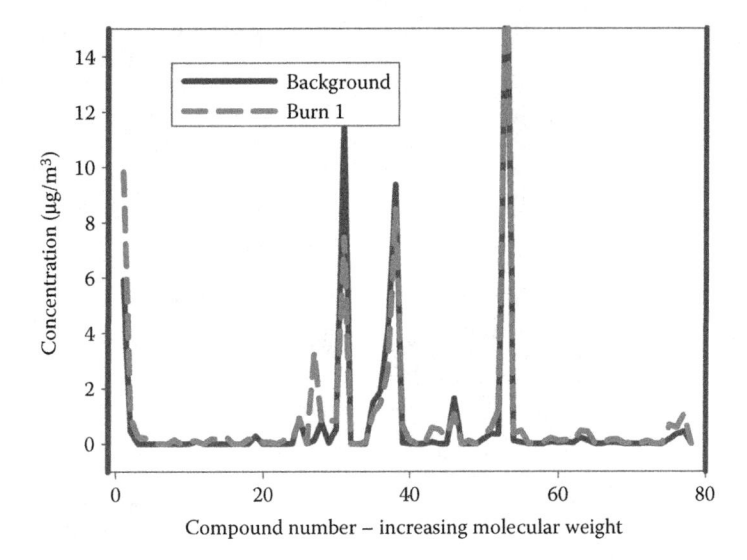

FIGURE 5.41 The summary VOC concentrations from burn 1 taken from the scientific sampling aircraft flown at between 0.5 km and 10 km downwind. In these cases, the background concentrations and the concentrations during burning are similar.

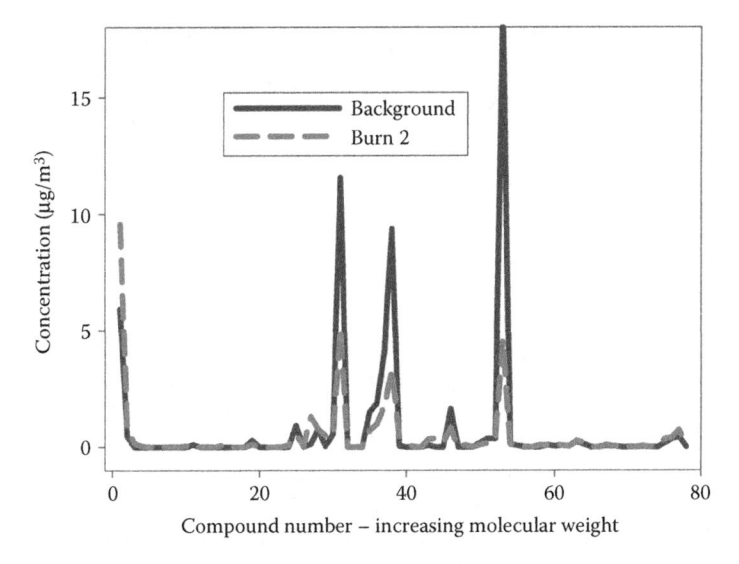

FIGURE 5.42 The summary VOC concentrations from burn 12 taken from the scientific sampling aircraft flown at between 0.5 km and 10 km downwind. In these cases, the background concentrations and the concentrations during burning are similar, with the background being generally somewhat higher.

FIGURE 5.43 Summa canisters which are used to sample air for later analysis for volatile organic compounds.

5.7.2 METALS

Heavy metals in the starting oil were found both on the soot and in the residue. Heavy metals on soot were collected using personal sampling pumps and filters. Analysis was by Inductively-Coupled Plasma (ICP), using standard techniques. A Gilian Aircon 2 was used to pump air through a tube on which a 37-mm canister containing a 0.8 μm pre-weighed cellulose ester filter was attached. Tygon tubing was connected from the instrument to the mast to allow sampling at 1 m above deck level. The elements to be measured included Mo, Zn, Pb, Ni, Fe, Cr, Mg, V, Cu, Ti and Ba. The flow rate of the pump was set at approximately 2 L/min yielding total volumes of 62–144 L. The cassette was capped and refrigerated after the experiment. The results in Table 5.6 show that the three metals of interest most associated with oil, nickel (Ni), vanadium (V) and chromium (Cr), are low in all three of the starting oil, soot and residue.

TABLE 5.6
Heavy Metals in Oil, Residue and Soot Samples (ppm)

Oil or Residue Source	Sample #	Mo	Zn	Pb	Ni	Fe	Cr	V	Cu	Ti	Ba
Fresh crude, Hughenden, truck #3	#1	<2.4	<0.6	<5.1	<2.3	<0.9	<0.9	<0.5	<0.3	<0.5	<0.05
Fresh crude, Hughenden, truck #2	#2a	<2.4	<0.6	<5.1	<2.3	<0.9	<0.9	<0.5	<0.3	<0.5	<0.05
Fresh crude, Hughenden, truck #2	#2b	<2.4	<0.6	<5.1	<2.3	<0.9	<0.9	<0.5	<0.3	<0.5	<0.05
Fresh crude, Hughenden, truck #1	#3	<2.4	6.3	<5.1	5	14.2	<0.9	<0.5	<0.3	<0.5	1.3
Weathered oil, Hughenden, truck #1	#4	<2.4	3	<5.1	<2.3	9.4	<0.9	<0.5	<0.3	3.1	11.3
Weathered oil, St. John's, truck #3 + 1	#5a	<2.4	8.7	<5.1	<2.3	16.2	<0.9	<0.5	1	<0.5	<0.05
Weathered oil, St. John's, truck #3 + 2	#5b	<2.4	1.1	<5.1	<2.3	<0.9	<0.9	<0.5	<0.3	<0.5	5.7
Oil, from apex of fireboom before burn 1	#7	<2.4	7.1	<5.1	<2.3	<0.9	<0.9	<0.5	<0.3	<0.5	<0.05
Residue, between fireboom and backup boom during burn 1	#11	<2.4	<0.6	<5.1	<2.3	<0.9	<0.9	0.6	<0.3	<0.5	0.4
Residue, from apex of fireboom after burn 1	#12	<2.4	<0.6	<5.1	<2.3	4.4	<0.9	3.9	1.2	<0.5	<0.05
Residue, between fireboom and backup boom during burn 2	#14	<2.4	<0.6	<5.1	<2.3	20.5	<0.9	4.7	0.6	3.5	1.6
Residue, from apex of rowboom after burn 2	#15	<2.4	<0.6	<5.1	<2.3	3.1	<0.9	2.8	<0.3	<0.5	<0.05
Residue, from side of R/C boats collected Aug 14, 93	#16	<2.4	<0.6	<5.1	<2.3	162.5	<0.9	6.9	<0.3	<0.5	<0.05
Soot Sample Location	**Soot Samples**										
Burn 2, Boat 1	A1	<5.5	8.7	<11.6	<5.2	25	<2.0	<1.1	<0.7	15.1	17.3
Burn 2, Boat 1	A2	<5.5	30.1	<11.6	<5.2	35.7	<2.0	<1.1	<0.7	35.5	46.9
Burn 1, Boat 4	A3	<5.5	<1.3	<11.6	<5.2	20.6	<2.0	<1.1	<0.7	<1.2	<0.1
Burn 1, Boat 4	A4	<5.5	<1.3	<11.6	<5.2	11.1	<2.0	<1.1	<0.7	12.5	17.8
Burn 2, Boat 2	A5	<5.5	<1.3	<11.6	<5.2	18.4	4.1	<1.1	<0.7	9.4	10.2
Burn 2, Boat 2	A6	<5.5	7.7	<11.6	<5.2	60.2	13.7	<1.1	<0.7	33.4	37.2
Burn 1, Boat 2	A7	<5.5	16.7	<11.6	<5.2	<2.0	<2.0	<1.1	<0.7	5	4.7
Burn 1, Boat 2	A8	<5.5	3.9	<11.6	<5.2	28.3	<2.0	<1.1	<0.7	35.6	60.3
Burn 1A, far downwind	A9	<5.5	12.9	<11.6	<5.2	10.3	<2.0	<1.1	<0.7	<1.2	<0.1

(Continued)

TABLE 5.6 (Continued)
Heavy Metals in Oil, Residue and Soot Samples (ppm)

Soot Sample Location	Soot Samples	Mo	Zn	Pb	Ni	Fe	Cr	V	Cu	Ti	Ba
Burn 1B, far downwind	A10	<5.5	9.9	<11.6	<5.2	24.6	<2.0	<1.1	<0.7	49.1	61
Burn 2A, far downwind	A11	<5.5	22.9	<11.6	<5.2	13.9	<2.0	<1.1	<0.7	19.8	23.4
Burn 2B, far downwind	A12	<5.5	8.7	<11.6	<5.2	9.7	<2.0	<1.1	<0.7	21.2	20.7
Trip blank	A13	<5.5	<1.3	<11.6	<5.2	<2.0	<2.0	<1.1	<0.7	<1.2	0.3
Trip blank	A14	<5.5	<1.3	<11.6	<5.2	2.2	<2.0	<1.1	<0.7	1.2	2.1
Burn 1, Boat 1	A15	<5.5	<1.3	<11.6	<5.2	3.9	<2.0	<1.1	<0.7	<1.2	<0.1
Burn 1, Boat 1	A16	<5.5	4.2	<11.6	<5.2	<2.0	<2.0	<1.1	<0.7	<1.2	<0.1
Background, Boat 1	A17	<5.5	<1.3	<11.6	<5.2	<2.0	<2.0	<1.1	<0.7	<1.2	<0.1
Background, Boat 1	A18	<5.5	<1.3	<11.6	<5.2	7.3	<2.0	<1.1	0.8	4	<0.1
Background, CCG 206	A19	<5.5	<1.3	<11.6	<5.2	<2.0	2.3	1.1	<0.7	<1.2	4.8

5.7.3 CARBONYLS

Oil burns produce low amounts of the small aldehydes (formaldehyde, acetaldehyde etc) and ketones (acetone etc), collectively known as carbonyls. Carbonyls from crude oil fires are found at very low concentrations. These would not be a health concern.

A Gilian 513A was used to pump air through a DNPH (2,4-dinitrophenylhydrazine)-silica cartridge attached via a Tygon tube. Tygon tubing was connected from the instrument to the mast to allow sampling at 1 m above deck level. The cartridge contains 350 mg of silica coated with 1.0 mg of DNPH. The flow rate was set between 185–250 cc/min and the pumped air volumes were between 1.1 and 18 L. The sample was wrapped in aluminium, placed in a small amber vial and refrigerated.

Results are shown in Table 5.7 and graphically in Figure 5.44. These show that carbonyl concentrations are much higher during the evaporation period than during the burn.

5.7.4 CARBON DIOXIDE

Carbon dioxide is the result of combustion and is found in increased concentrations around a burn. Normal atmospheric levels are about 350 ppm, and levels near a burn can be around 500–800 ppm. There is no human danger at this level. The three-dimensional distributions of carbon dioxide around a burn have been measured. Concentrations of carbon dioxide are highest at ground level and fall to near background levels at higher than about 50 m. Concentrations at ground level are as high as 10 times that of the plume. Distribution along the ground is broader than for particulates. Figures 5.45 and 5.46 summarise the average carbon dioxide concentrations around the burns. This shows that the carbon dioxide concentrations are the highest directly behind the burn and on the ground. Levels in the plume are just above background levels. Data are summarised in Tables 5.8 and 5.9.

Carbon dioxide was measured in one of 5 ways. Samples were captured by SUMMA canisters or Tedlar bags and analysed later by standard chromatographic methods. Direct-reading instruments were also used and values recorded. These included the CD-1 Å carbon dioxide meter and the Cannonball and the Metrosonics aq-501 meters. Each of these meters was calibrated with standard carbon dioxide concentrations before each burn.

5.7.5 CARBON MONOXIDE, SULPHUR DIOXIDE AND NITRIC OXIDE

The Exotox 75 was used to analyse all three gases. Its flow rate was 300 mL/min. The Cannonball and Metrosonics instruments were used for carbon monoxide and sulphur dioxide only at a flow rate of 1 L/min. Tubing was connected from the instrument to the mast to allow sampling at 1 m above deck level. The data was logged every 30 s.

Carbon monoxide levels are usually at or below the lowest detection levels of the instruments and thus did not pose any hazard to humans. Carbon monoxide has only been measured at burns where the burn is inefficient, such as when water is sprayed

TABLE 5.7
Carbonyl Concentrations (All Concentrations in µg/m³)

Burn 1

Description	Evaporation 1A	Evaporation 1B	Evaporation	Evaporation	Evaporation	Evaporation	Burn	Burn	Burn	Burn	Burn	Burn	Evaporation	Evaporation	Burn	Burn	Blank
Location	Far	Far	Boat 2	Boat 2	Boat 4	Boat 4	Boat 4	Boat 4	Boat 2	Boat 2	Far	Far	Boat 1	Boat 1	Boat 1	Boat 1	
Compounds																	
Formaldehyde	0.66	0.72	0.94	0.84	0.86	1.46	0.14	0.10	0.14	0.12	0.08	0.08	0.001	0.001	0.09	0.12	0.001
Acetaldehyde	0.68	0.80	1.07	1.88	0.91	2.12	0.15	0.15	0.23	0.19	0.10	0.12	0.002	0.002	0.24	0.27	0.000
Acetone	0.46	0.35	1.40	0.60	2.27	0.78	0.09	0.12	0.07	0.08	0.09	0.05	0.001	0.002	0.11	0.09	0.000
Acrolein	0.00	0.00	0.00	0.00	0.00	0.00	0.00	0.00	0.00	0.00	0.00	0.00	0.000	0.000	0.00	0.00	0.000
Propinaldehyde	0.08	0.11	0.15	0.35	0.13	0.37	0.03	0.02	0.04	0.04	0.01	0.02	0.000	0.000	0.04	0.04	0.000
Crotonaldehyde	0.00	0.00	0.00	0.00	0.00	0.00	0.00	0.00	0.00	0.00	0.00	0.00	0.000	0.000	0.00	0.00	0.000
2-Butanone	0.00	0.00	0.00	0.00	0.00	0.00	0.00	0.00	0.00	0.00	0.00	0.00	0.000	0.000	0.00	0.00	0.000
iso/- Butylaldehyde	0.00	0.00	0.00	0.00	0.00	0.31	0.00	0.00	0.00	0.00	0.00	0.00	0.000	0.000	0.04	0.04	0.000
Benzaldehyde	0.00	0.00	0.00	0.00	0.00	0.00	0.00	0.00	0.00	0.00	0.00	0.00	0.000	0.000	0.00	0.00	0.000
2-Pentanone	0.00	0.00	0.00	0.00	0.00	0.00	0.00	0.00	0.00	0.00	0.00	0.00	0.000	0.000	0.00	0.00	0.000
Isovaleraldehyde	0.00	0.00	0.00	0.00	0.00	0.00	0.00	0.00	0.00	0.00	0.00	0.00	0.000	0.000	0.00	0.00	0.000
Valeraldehyde	0.24	0.10	0.34	0.13	0.21	0.15	0.01	0.02	0.01	0.01	0.01	0.00	0.000	0.000	0.02	0.02	0.000
o/m/ pToluldehyde	0.00	0.00	0.00	0.00	0.00	0.00	0.00	0.00	0.00	0.00	0.00	0.00	0.000	0.000	0.00	0.00	0.000
MIBK	0.00	0.10	0.12	0.09	0.11	0.12	0.01	0.02	0.00	0.00	0.01	0.00	0.000	0.000	0.01	0.01	0.000
Hexanal	0.44	0.06	0.08	0.05	0.06	0.06	0.00	0.00	0.01	0.00	0.01	0.00	0.000	0.000	0.00	0.00	0.000
2,5-Dimethyl	0.00	0.00	0.00	0.00	0.00	0.00	0.00	0.00	0.00	0.00	0.00	0.00	0.000	0.000	0.00	0.00	0.000
TOTAL	2.56	2.24	4.11	3.93	4.55	5.37	0.44	0.43	0.49	0.45	0.31	0.27	0.01	0.01	0.55	0.59	0.001

(Continued)

TABLE 5.7 (*Continued*)
Carbonyl Concentrations (All Concentrations in µg/m³)

Burn 2

Description	Burn 2A	Burn 2B	Blank	Evaporation	Evaporation	Burn	Burn	Blank	Background	Background	Background	Background	Evaporation	Evaporation	Evaporation	Burn	Burn
Location	Far	Far	Far	Boat 1	Boat 1	Boat 1	Boat 1		Boat 1	Boat 1	Far	Far	Far	Far	Boat 2	Boat 2	Boat 2
Compounds	D-19	D-20	D-21	D-22	D-23	D-24	D-25	D-1	D-30	D-31	D-33	D-34	D-4	D-5	D-14	D-15	D-16
Formaldehyde	0.10	0.07	0.00	0.59	0.59	0.07	0.08	0.001	0.10	0.09	0.42	0.44	1.00	0.80	0.85	0.11	0.11
Acetaldehyde	0.15	0.10	0.00	1.12	1.12	0.11	0.08	0.000	0.20	0.21	0.30	0.29	1.20	1.04	0.96	0.13	0.15
Acetone	0.06	0.04	0.00	1.70	1.77	0.04	0.07	0.000	0.06	0.10	0.78	0.73	1.38	0.76	0.30	0.04	0.07
Acrolein	0.00	0.00	0.00	0.00	0.00	0.00	0.00	0.000	0.00	0.00	0.00	0.00	0.00	0.00	0.00	0.00	0.00
Propinaldehyde	0.02	0.01	0.00	0.15	0.19	0.02	0.02	0.000	0.05	0.04	0.08	0.07	0.16	0.19	0.14	0.02	0.02
Crotonaldehyde	0.00	0.00	0.00	0.00	0.00	0.00	0.00	0.000	0.00	0.00	0.00	0.00	0.00	0.00	0.00	0.00	0.00
2-Butanone	0.00	0.00	0.00	0.00	0.00	0.00	0.00	0.000	0.00	0.00	0.00	0.00	0.00	0.00	0.00	0.00	0.00
iso/- Butylaldehyde	0.00	0.00	0.00	0.00	0.00	0.00	0.00	0.000	0.00	0.00	0.00	0.00	0.00	0.00	0.00	0.00	0.00
Benzaldehyde	0.00	0.00	0.00	0.00	0.00	0.00	0.00	0.000	0.00	0.00	0.00	0.00	0.00	0.00	0.00	0.00	0.00
2-Pentanone	0.00	0.00	0.00	0.00	0.00	0.00	0.00	0.000	0.00	0.00	0.00	0.00	0.00	0.00	0.00	0.00	0.00
Isovaleraldehyde	0.00	0.00	0.00	0.00	0.00	0.00	0.00	0.000	0.00	0.00	0.00	0.00	0.00	0.00	0.00	0.00	0.00
Valeraldehyde	0.01	0.00	0.00	0.07	0.07	0.01	0.01	0.000	0.01	0.01	0.03	0.04	0.14	0.09	0.07	0.01	0.01
o/m/ p- Tolualdehyde	0.00	0.00	0.00	0.00	0.00	0.00	0.00	0.000	0.00	0.00	0.00	0.00	0.00	0.00	0.00	0.00	0.00
MIBK	0.00	0.00	0.00	0.03	0.03	0.01	0.00	0.000	0.01	0.01	0.01	0.01	0.07	0.07	0.24	0.00	0.00
Hexanal	0.00	0.00	0.00	0.04	0.05	0.00	0.00	0.000	0.01	0.00	0.01	0.02	0.06	0.03	0.04	0.01	0.00
2,5-Dimethyl	0.00	0.00	0.00	0.00	0.00	0.00	0.00	0.000	0.00	0.00	0.00	0.00	0.00	0.00	0.00	0.00	0.00
TOTAL	0.34	0.23	0.00	3.70	3.82	0.25	0.26	0.001	0.44	0.47	1.63	1.60	4.02	2.98	2.60	0.32	0.36

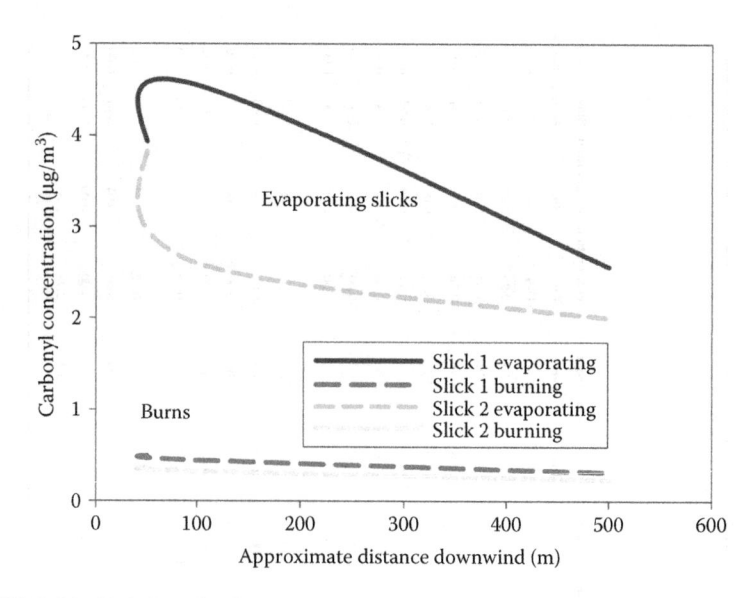

FIGURE 5.44 Relative distributions of carbonyls downwind during the evaporation and burn periods.

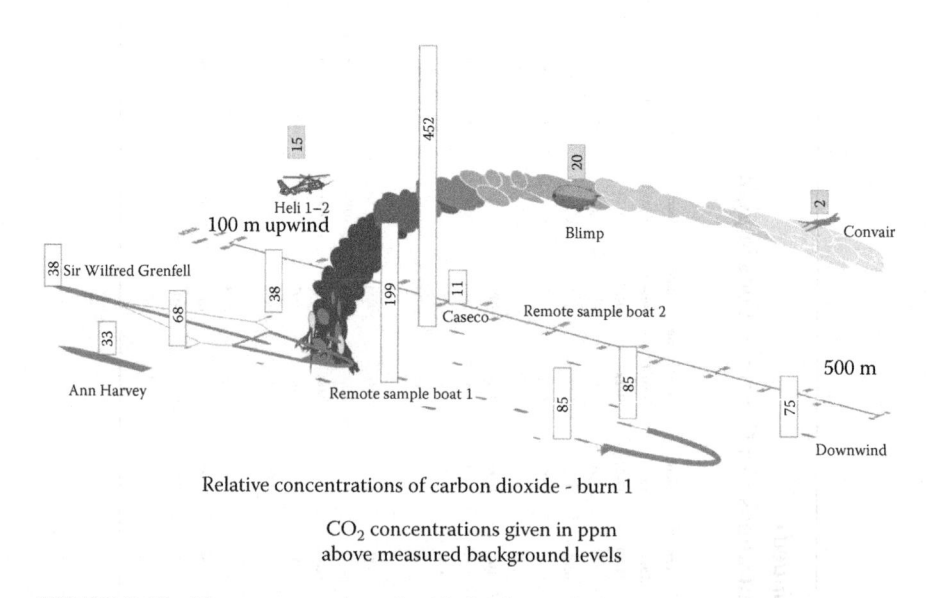

Relative concentrations of carbon dioxide - burn 1

CO_2 concentrations given in ppm above measured background levels

FIGURE 5.45 The average carbon dioxide levels on platforms measured around burn 1. This shows that the concentrations of CO_2 are highest just behind the burn at ground level.

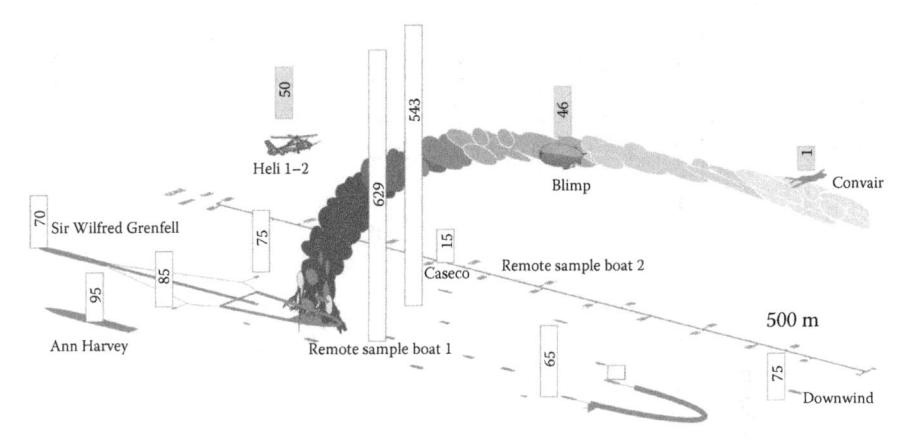

Relative concentrations of carbon dioxide - burn 2

CO_2 concentrations given in ppm
above measured background levels

FIGURE 5.46   The average carbon dioxide levels on platforms measured around burn 2. This shows that the concentrations of CO_2 are highest just behind the burn at ground level.

into the fire. Carbon monoxide appears to be distributed in the same way as carbon dioxide. Table 5.10 summarises carbon monoxide measurements.

The levels of nitrogen dioxide measured at the burns are shown in Table 5.11. These levels were low and would not constitute a health threat.

Sulphur dioxide per se is usually not detected at significant levels or sometimes not even at measurable levels. Sulphuric acid, or sulphur dioxide that has reacted with water, is detected at fires. Sulphur dioxide must be measured using impingers (injection into water solution) rather than conventional gas detectors.

A known volume of air was drawn through a Tygon tube attached to a filter (37 mm methyl cellulose ester, 0.8 μm) contained in a cassette and through a midget bubbler (25 mL) containing 15 mL of 0.3 N hydrogen peroxide. Again, the end of the tubing was connected to the mast to allow sampling at 1 m above deck level. The flow rate of the Gilian 513A pump was set between 90 and 200 cc/min and air volumes between 3 and 14 L passed through the impinger. Samples were transferred in amber vials and refrigerated. Subsequent analysis was performed using National Institute for Occupational Safety and Health (NIOSH) S308 method.

TABLE 5.8
Summary of Carbon Dioxide Analysis of Air Samples (ppm above Background) – Burn 1

							Corrected for Ambient Concentrations Except for 'Background'							
	RS-1	RS-2	Downwind	R/C Heli 1	R/C Heli 2	BLIMP	Convair	Wilfred	Ann Harvey	Casaco	CCG 203	CCG 204	CCG 212	CCG 214
			Station					Grenfell						
Approximate distance downwind m	50	50	500	50	50	100	100–10,000	Upwind	Beside	Beside	Various, Close	Various, Close	Various, Close	Various, Close
SUMMA Canister														
Background Aug 07, actual reading	317	317	317				387 (1)							
Background, actual reading				396	346		365							
Pre-ignition – background		380	388	368				370	387	345	523	367	365	375
Pre-ignition – actual reading		63	71	51				53	70	28	206	50	48	58
Burn	516	769	374					355	350	328	385	355	399	402
Burn	199	452	57					38	33	11	68	38	82	85
Burn, front of plume				310										
Burn, under plume				347										
Burn, under plume				30										
Burn, under plume, 20 m high					380									
Burn, under plume, 20 m high					63									
Burn, under plume, 40 m high					308									
Burn, under plume, 40 m high					0									
Burn, above cloud, 11:08							406	Actual reading						
Burn, above cloud, 11:08							41							

(Continued)

TABLE 5.8 (*Continued*)

Summary of Carbon Dioxide Analysis of Air Samples (ppm above Background) – Burn 1

	RS-1	RS-2	Downwind Station	R/C Heli 1	R/C Heli 2	BLIMP	Convair	Wilfred Grenfell	Ann Harvey	Casaco	CCG 203	CCG 204	CCG 212	CCG 214
								Corrected for Ambient Concentrations Except for 'Background'						
Burn, smoke 1mi DW at 1,000 ft, 10:39							376	Actual reading						
Burn, smoke 1mi DW, 1,000 ft, 10:39							11							
Burn, smoke 3mi DW, 11:27							380	Actual reading						
Burn, smoke 3mi DW, 11:27							15							
Burn, smoke 20mi DW, 11:43							374	Actual reading						
Burn, smoke 20mi DW, 11:43							9							
Tedlar Bag														
Background, actual reading						390								
Burn						416								
Burn						26								
Post-burn						369								
Post-burn						0								
Metrosonics aq-501														
Background, actual reading			419<427<471											
Pre-ignition, actual reading			399<415<455											
Burn, actual reading			401<417<464											

TABLE 5.9

Summary of Carbon Dioxide Analysis of Air Samples (ppm) – Burn 2

	RS-1	RS-2	Downwind Station	R/C Heli 1	R/C Heli 2	BLIMP	Convair	Wilfred Grenfell	Ann Harvey	Casaco	CCG 203	CCG 204	CCG 212	CCG 214
									Not Corrected for Ambient Concentrations					
Approximate distance downwind m	50	50	500	50	50	100	100–10,000	Upwind	Beside	Beside	~, Close	~, Close	~, Close	~, Close
SUMMA Canister														
Background, actual reading	317	317	317				378							
Pre-ignition – background	354		378					325	363	344	410	407	355	455
Burn	946	860	392					387	412	332	402	392	316	382
Burn, front of plume														
Burn, under plume														
Burn, 2mi DW at 1,000 ft, plume, 14:13							379							
Burn, 2mi DW at 1,000 ft, plume, 14:24							376							
Burn, 2mi DW at 1,000 ft, plume, 14:32							386							
Burn, smoke 9mi DW, 14:58							374							

(Continued)

TABLE 5.9 (*Continued*)

Summary of Carbon Dioxide Analysis of Air Samples (ppm) - Burn 2

	RS-1	RS-2	Downwind Station	R/C Heli 1	R/C Heli 2	BLIMP	Convair	Ann Wilfred Grenfell	Harvey	Casaco	CCG 203	CCG 204	CCG 212	CCG 214
						Not Corrected for Ambient Concentrations								
Burn, cross-section 6mi DW at 1,400 ft, 15:32							384							
Burn, cross-section 6mi DW at 1,700 ft, 15:38							376							
Post-burn, 16:23							362							
Tedlar Bag														
Pre-ignition						385								
Burn						436								
Post-burn						380								
Metrosonics aq-501														
Background, actual reading		419<428<449												
Pre-ignition, actual reading		413<421<456												
Burn, actual reading		415<435<453												
Post-burn, actual reading		424<431<443												

TABLE 5.10
Summary of Carbon Monoxide Values (ppm)

Burn 1– Cannonball Data

	Remote Station 1	Remote Station 2
Minimum	0.0	0.0
Average	0.5	0.1
Maximum	3.7	1.9
	Burn 1, R/C boat 4, RS-1	*Burn 1, R/C boat 2, RS-2*
Minimum	0.0	0.0
Average	0.2	0.3
Maximum	3.7	1.9
	Burn 1, R/C boat 2, RS-1	*Burn 1, R/C boat 1, RS-2*
Minimum	0.1	0.0
Average	0.5	0.0
Maximum	1.8	0.0

Metrosonics Data

	Burn 1 A	Burn 1 B
Minimum	0.0	0.0
Average	0.6	0.7
Maximum	16.0	3.0

Burn 2 – Cannonball Data

	Remote Station 1	Remote Station 2
Minimum	0.00	0.10
Average	0.11	0.67
Maximum	1.50	2.10

Metrosonics Data

	Burn 2 A	Burn 2 B
Minimum	1.0	0.0
Average	2.1	2.3
Maximum	39.0	5.0

The measured concentrations of sulphuric acid were below concern levels and appear to correspond to the sulphur content of the oil. The sulphur dioxide (sulphuric acid by impinger) values measured are summarised in Table 5.12.

5.7.6   WATER SAMPLING AND ANALYSIS

Sigma 800SL samplers were used to collect water samples for both toxicity and organic compound analyses. One sampler contained 4 × 1-gallon sterile glass bottles with Teflon-lined caps (for toxicity testing) and the other contained 24 × 350 mL

TABLE 5.11
Nitrogen Dioxide Levels (ppm)

By Exotox Burn 1

	Remote Station 1	Remote Station 2
Minimum	0.00	0.00
Average	0.34	0.11
Maximum	0.50	0.50

By Exotox Burn 2

	Remote Station 1	Remote Station 2
Minimum		0.40
Average		0.42
Maximum		0.50

TABLE 5.12
Sulphur Dioxide Levels Measured (ppm)

Burn 1

	RS-1	RS-2	Downwind Station
Exotox			
Minimum	0.0	0.0	
Average	0.5	0.0	
Maximum	2.5	0.4	
Cannonball			
Minimum	0.0	0.0	0.0
Average	0.0	0.0	0.4
Maximum	0.1	0.0	1.0
Impinger			
Average	10.3	10.6	6.3

Burn 2

	RS-1	RS-2	Downwind Station
Exotox			
Minimum		0.0	
Average		0.1	
Maximum		0.6	
Cannonball			
Minimum	0.0	0.0	0.0
Average	0.1	0.0	0.0
Maximum	0.7	0.2	0.0
Impinger			
Average	13.4	12.9	5.4

water (for organic measurements). The samplers were located on the deck of a sampling boat and Teflon tubing was attached to a pole which was lowered into the water to approximately 1 m.

All samples were collected and placed in refrigerated coolers and shipped to the laboratory within 24 h of collection. The smaller bottles were sent to the Environmental Technology Centre for chemistry analysis and the larger bottles (although not full) were sent to EVS consultants for toxicity analysis. The PAH analysis results are summarised in Table 5.13. Only columns where compounds were found are included. Most results were below detection limit.

5.7.7 Dioxins and Dibenzofurans

The high-volume samples taken on the remote-controlled boats and on the downwind station were also analysed for dioxins and dibenzofurans. The summary data are presented in Table 5.14. The values are at background levels. This confirms previous studies which show that dioxins and dibenzofurans are not produced by oil fires.

5.7.8 Polyaromatic Hydrocarbons on Soot, in Air and on Residue

PAHs are aromatic compounds found in crude oil and are often produced as a result of combustion. Many PAHs, particularly the larger PAHs, are toxic to humans and the environment. Crude oil burns result in PAH downwind of the fire, but the concentration on the particulate matter is often an order of magnitude less than the concentration in the starting oil and sometimes several orders of magnitude less. Overall, more PAHs are destroyed by the fires than are created. Table 5.15 summarises this information and shows that, despite increases in concentrations in some of the higher molecular weight PAHs, overall, they are destroyed in fires.

PAHs were sampled in air using filters and sorbent tubes initially, but later from particulates collected on high volume samplers. Analysis of PAHs was also conducted from various particulate sampling including fractionation samplers, PM-10, PM-2.5 or cascade samplers, and filters from low- and medium-volume pumps. Analysis was by standard methods using GC–MS.

Analysis of the PAHs in various starting oil samples shows variances, as seen in Table 5.16. Sample 7 was taken in front of the boom just before burn 1 and shows a bit of weathering. The PAHs on most particulates collected were too low to measure except for those taken by high-volume samplers. The PAHs on back filters and resin traps of high-volume samplers are summarised in Table 5.17. These PAHs would be a mixture of gaseous and PAHs absorbed on particulates.

TABLE 5.13

Water Sample Analysis

	Water Samples during Burn	Pre-Burn Water Samples	First Half Burn Water Samples			Second Half Burn Samples					Post-Burn Samples		
Sample Number	All Platforms	BR1-B4-PI1	BR1-B4-DB1-2	BR2-B1-DB1-1	BR2-B1-DB1-2	BRI-B2-DB2-1	BR1-B2-DB2-2	BR1-B4-DB2-1	BR1-B4-DB2-2	BR2-B1-DB2-1	BRI-B2-PB1	BR1-B2-PB2	BR2-B1-PB2
Compound	(μg/L H_2O)	(μg/L H_2O)	(μg/L H_2O)	(μg/L H_2O)	(μg/L H_2O)	(μg/L H_2O)	(μg/L H_2O)	(μg/L H_2O)	(μg/L H_2O)	(μg/L H_2O)	(μg/L H_2O)	(μg/L H_2O)	(μg/L H_2O)
Naphthalene	0	0.15	0.12	0.15	0.15	0.1	0.1	0.1	0.1	0.15	0.11	0.12	0.15
2-methyl-naphthalene	0	0.15	0	0	0	0	0	0	0	0	0	0	0
1-methyl-naphthalene	0	0.15	0	0	0	0	0	0	0	0	0	0	0
Biphenyl	0	0	0	0	0	0	0	0	0	0	0	0	0
2,6-dimethyl-naphthalene	0	0	0	0	0	0	0	0	0	0	0	0	0
Acenaphthylene	0	0	0	0	0	0	0	0	0	0	0	0	0
Acenaphthene	0	0	0	0	0	0	0	0	0	0	0	0	0
2,3,5-trimethyl-naphthalene	0	0	0	0	0	0	0	0	0	0	0	0	0
Fluorene	0	0	0	0	0	0	0	0	0	0	0	0	0
Dibenzothiophene	0	0	0	0	0	0	0	0	0	0	0	0	0
Phenanthrene	0	0	0	0	0	0	0	0	0	0	0	0	0
1-methyl-phenanthrene	0	0	0	0	0	0	0	0	0	0	0	0	0
Fluoranthene	0	0	0	0	0	0	0	0	0	0	0	0	0

(Continued)

TABLE 5.13 (*Continued*)
Water Sample Analysis

Pyrene	0	0	0	0	0	0	0	0	0	0	0	0	0
Benz(a)anthracene/ chrysene	0	0	0	0	0	0	0	0	0	0	0	0	0
Benz(b)fluoranthene/ b(k)fluor.	0	0	0	0	0	0	0	0	0	0	0	0	0
Benzo(e)pyrene/ benzo(a)pyrene	0	0	0	0	0	0	0	0	0	0	0	0	0
Perylene	0	0	0	0	0	0	0	0	0	0	0	0	0
Indeno(1,2,3-cd) pyrene	0	0	0	0	0	0	0	0	0	0	0	0	0
Dibenz(a,h) anthracene	0	0	0	0	0	0	0	0	0	0	0	0	0
Benzo(ghi)perylene	0	0	0	0	0	0	0	0	0	0	0	0	0
Total PAH	0	0.45	0.12	0.15	0.15	0.1	0.1	0.1	0.1	0.15	0.11	0.12	0.15

Recovery Check[a]	CS-M96	BR1-B4-PI1	BR1-B4-DB1-2	BR2-B1-DB1-1	BR2-B1-DB1-2	BRI-B2-DB2-2	BR1-B2-DB2-1	BR1-B2-DB2-2	BR1-B4-DB2-1	BR2-B4-DB2-2	B1-DB2-1	BRI-B2-PB1	BR1-B2-PB2	BR2-B1-PB2
Compound	% Recovery	% Recovery	% Recovery	% Recovery	% Recovery	% Recovery	% Recovery	% Recovery	% Recovery	% Recovery	% Recovery	% Recovery	% Recovery
d10-acenaphthalene	50	40	50	40	50	60	40	50	50	50	50	30	40
d10-phenanthrene	60	50	70	50	70	70	50	60	60	70	70	40	50
d12-benz(a) anthracene	70	60	80	60	80	80	50	70	70	80	80	50	60
d12-perylene	60	50	80	60	80	70	50	60	60	70	80	40	50

[a] Legend: BR1, Burn #1; B1,Boat #1; CS, Control seawater; M96, 96 h Menidia test endpoint sample.
BR2, Burn #2; B2, Boat #2; BG, Background; EB, Early burn; LB, Late burn B4; Boat 4.

TABLE 5.14

Results of Dioxin and Dibenzofuran Analyses

Sample	Concentrations in pg								
Congener	Burn 1	Burn 2	Blank	Homologue	Burn 1	Burn 2	Blank	DL	NP
	pg				pg				
2378-TCDD	ND	ND	ND	TCDD	ND	ND	ND	4–5	0
12378-P5CDD[a]	ND	ND	ND	P5CDD	ND	ND	ND	4–5	0
123478-H6CDD[a]	ND	ND	ND	H6CDD	ND	ND	ND	8–10	0
123678-H6CDD[a]	ND	ND	ND	H7CDD	ND	ND	ND	6–10	0
123789-H6CDD[a]	ND	ND	ND	OCDD	23	23	65	10	1
1234678-H7CDD	ND	ND	ND						
OCDD	23	187	65	Total PCDD	23	23	65		
				Concentration – pg/m^3	1.9	15.5	na		
2378-TCDF[a]	52	11	ND	TCDF	336	53	ND	4	0–13
12378-P5CDF[a]	ND	ND	ND	P5CDF	64	ND	ND	4	0–3
23478-P5CDF[a]	ND	ND	ND	H6CDF	ND	ND	ND	6	0
123478-H6CDF[a]	ND	ND	ND	H7CDF	ND	ND	ND	6–8	0
123678-H6CDF[a]	ND	ND	ND	OCDF	ND	ND	ND	10	0
234678-H6CDF[a]	ND	ND	ND						
123789-H6CDF[a]	ND	ND	ND	Total PCDF	400	53	ND		
1234678-H7CDF	ND	ND	ND	Concentration – pg/m^3	32.3	4.4	na		
1234789-H7CDF	ND	ND	ND						
OCDF	ND	ND	ND						

(Continued)

TABLE 5.14 (*Continued*)

Results of Dioxin and Dibenzofuran Analyses

Surrogate	Amount Added, ng					
	Recovery %		Recovery %		Recovery %	
13C12-TCDD	1.00	89	1.00	95	1.00	94
13C12-TCDF	1.00	107	1.00	114	1.00	116
13C12-P5CDD	1.00	93	1.00	89	1.00	105
13C12-P5CDF	1.00	109	1.00	118	1.00	120
13C12-H6CDD	1.00	95	1.00	106	1.00	110
13C12-H6CDF	1.00	104	1.00	127	1.00	128
13C12-H7CDD	1.00	106	1.00	95	1.00	125
13C12-H7CDF	1.00	99	1.00	104	1.00	117
13C12-OCDD	2.00	86	2.00	79	2.00	115

Note: (1) Results are corrected for surrogate recovery.

(2) DL = detection limit (ng/analyte peak); NP = number of analyte peaks.

(3) [a] Value represents maximum possible amount. This isomer could co-elute with other isomer(s).

(4) ND = not detected; NDR = not detected due to incorrect ratio, the target analyte concentration, if present, is given in brackets.

TABLE 5.15
Summary of PAHs and Their Fate in NOBE

	Crude Oil (Alberta Sweet Mixed Blend)				
	Concentrations (µg/g)			Burn Destruction %	
	Starting Oil	Residue	Soot	As Residue	As Soot
Naphthalenes					
C0-N	206	61	9	99.998	100.000
C1-N	1049	220	8	99.999	100.000
C2-N	2246	578	11	99.998	100.000
C3-N	1749	589	3	99.998	100.000
C4-N	859	333	2	99.997	100.000
Sum	6109	1780	104	99.998	100.000
Phenanthrenes					
C0-P	109	106	5	99.993	100.000
C1-P	407	301	5	99.995	100.000
C2-P	450	361	12	99.995	100.000
C3-P	342	299	17	99.994	100.000
C4-P	173	145	12	99.994	100.000
Sum	1481	1211	30	99.994	100.000
Dibenzothiophenes					
C0-D	14	8	0	99.997	100.000
C1-D	28	21	0	99.995	100.000
C2-D	46	37	0	99.995	100.000
C3-D	26	26	1	99.993	100.000
Sum	114	91	2	99.994	100.000
Fluorenes					
C0-F	82	50	1	99.996	100.000
C1-F	176	92	0	99.996	100.000
C2-F	176	93	0	99.996	100.000
C3-F	144	95	1	99.995	100.000
Sum	578	329	15	99.996	100.000
Chrysenes					
C0-C	35	45	79	99.991	100.000
C1-C	45	58	29	99.991	100.000
C2-C	55	75	30	99.991	100.000
C3-C	50	58	55	99.992	100.000
Sum	185	235	54	99.991	100.000
Total alkylated	8467	3645	76	99.997	100.000
Priority PAHs					
Biphenyl	70.8	14	1	99.999	100.000
Acenaphthylene	7.61	33	1	99.971	100.000

(Continued)

TABLE 5.15 (*Continued*)
Summary of PAHs and Their Fate in NOBE

	Crude Oil (Alberta Sweet Mixed Blend)				
	Concentrations (µg/g)			Burn Destruction %	
	Starting Oil	Residue	Soot	As Residue	As Soot
Acenaphthene	15.95	6	0	99.997	100.000
Anthracene	2.09	11	1	99.967	100.000
Fluoranthene	2.45	42	20	99.881	99.999
Pyrene	18.28	173	61	99.934	100.000
Benz(a)anthracene	2.94	44	32	99.901	99.999
Benzo(b)fluoranthene	2.94	137	426	99.687	99.990
Benzo(k)fluoranthene	0.49	59	360	99.159	99.950
Benzo(e)pyrene	8.71	285	740	99.774	99.994
Benzo(a)pyrene	0.86	147	306	98.985	99.979
Perylene	1.72	40	77	99.844	99.997
Indeno(1,2,3cd)pyrene	0.74	14.5	54.6	99.868	99.995
Dibenz(a,h)anthracene	1.35	1.3	6.9	100.000	100.000
Benzo(ghi)perylene	2.94	19.4	76	99.959	99.998
Total target PAHs	140	1026.2	2162.5	99.950	99.999
Overall	8607	4671.2	2238.5	100.000	100.000

TABLE 5.16A
Alkylated PAH Homologue Distribution of Oil Samples (µg/g oil)

		Oil Samples before Burn						Mixed	Oil Residue Samples				
NOBE	Fresh	2A	2B	3	4	5A	5B	7	11	12	14	15	16
Naphthalenes													
C0-N	307	313	313	308	307	297	278	206	79	42	66	67	18
C1-N	1662	1496	1569	1592	1519	1540	1529	1049	222	217	128	190	82
C2-N	3073	2783	2855	2915	2755	2864	2842	2246	586	570	323	494	312
C3-N	3013	2902	2972	3013	2850	2915	2902	1749	678	499	271	523	411
C4-N	1144	1295	1275	1283	1255	1236	1212	859	387	279	237	326	278
Sum	9199	8789	8984	9111	8686	8852	8763	6109	1952	1607	1025	1600	1101
Phenanthrenes													
C0-P	175	155	153	156	146	152	153	109	142	69	214	150	193
C1-P	481	535	542	540	513	541	535	407	378	224	303	298	296
C2-P	497	489	479	475	463	452	463	450	452	270	319	360	337
C3-P	369	345	335	350	328	344	339	342	385	213	280	276	270
C4-P	200	182	181	197	186	191	191	173	183	106	123	132	125
Sum	1722	1706	1690	1718	1636	1680	1681	1481	1540	882	1239	1216	1221

(*Continued*)

TABLE 5.16A (*Continued*)
Alkylated PAH Homologue Distribution of Oil Samples (µg/g oil)

NOBE	Fresh	Oil Samples before Burn						Mixed	Oil Residue Samples				
		2A	2B	3	4	5A	5B	7	11	12	14	15	16
Dibenzothiophenes													
C0-D	16	17	17	17	16	15	16	14	10	6	12	10	9
C1-D	37	35	36	36	34	36	34	28	26	15	22	22	21
C2-D	60	50	49	51	47	47	48	46	45	28	34	40	33
C3-D	43	37	35	41	39	38	40	26	31	20	20	25	23
Sum	156	139	137	145	136	136	138	114	112	69	88	97	86
Fluorenes													
C0-F	123	108	119	110	116	123	116	82	66	33	78	58	58
C1-F	250	218	216	217	218	210	210	176	115	69	91	83	74
C2-F	269	237	255	260	240	252	253	176	109	76	87	91	92
C3-F	174	187	172	181	175	169	180	144	122	67	77	69	78
Sum	816	750	762	768	749	754	759	578	412	245	333	301	302
Chrysenes													
C0-C	33	30	31	34	30	32	32	35	56	33	8	57	78
C1-C	55	51	48	49	48	48	50	45	72	44	73	56	68
C2-C	70	66	64	72	62	66	66	55	91	58	82	76	76
C3-C	34	32	36	36	36	35	36	50	69	47	69	65	59
Sum	192	179	179	191	176	181	184	185	288	182	232	254	281
TOTAL	12085	11563	11752	11933	11383	11603	11525	8467	4304	2985	2917	3468	2991

TABLE 5.16B
Alkylated PAH Homologue Distribution of Oil Samples (µg/g oil)

Residue Sample Description

7	Weathered crude	8/12/93	• Collected from apex of fireboom
			• Collected before burn 1
11	Residue	8/12/93	• Collected from water surface between fireboom and backup boom
			• Collected during burn 1
12	Residue	8/12/93	• Collected from apex of fireboom
			• Collected after burn 1
14	Residue	8/12/93	• Collected from water surface between fireboom and backup boom
			• Collecting during burn 2
15	Residue	8/12/93	• Collected from apex of backup boom
			• Collected before burn 2
16	Residue	8/14/93	• Collected from sides of remote-controlled sample boats

TABLE 5.17
PAHs on High-Volume Sampling Equipment Polyurethane Filters and Resins

Type	Burn 1 Burn 1, Boat 2	Background Background	Background Background, Boat 1	Burn 1 Burn 1, Boat 1	Burn 1 Burn 1, Boat 4	Burn 1 Burn 1, Boat 1	Burn 2 Burn 2, Boat 1	Burn 1 Burn 1, Boat 4	Burn 2 Burn 2, Boat 1	Burn 2 Burn 2, Boat 2	Burn 2 Burn 2, Boat 2	Burn 1 Burn 1, Boat 2	Burn 2 Burn 2, Far Downwind	Burn 1 Burn 1, Far Downwind	Burn 2 Burn 2, Far Downwind	Burn 1 Burn 1, Far Downwind
Description / Compound								Loading (µg/m³)								
Naphthalene	1.11	0.50	0.94	2.07	1.25	2.26	1.74	1.22	1.91	2.55	2.77	1.41	2.42	2.12	1.04	1.31
1-Methylnaphthalene	2.00	0.58	1.02	2.10	1.73	2.09	1.88	1.75	1.39	3.07	2.88	1.64	0.88	0.55	0.67	0.39
2-Methylnaphthalene	2.75	0.30	0.50	1.33	1.01	1.19	1.86	0.99	1.97	1.61	1.56	2.27	0.46	0.28	0.32	0.20
Biphenyl	0.14	0.12	0.15	0.30	0.31	0.30	0.18	0.13	0.15	0.17	0.20	0.17	0.15	0.03	0.04	0.03
2,6-Dimethylnaphthalene	1.62	0.49	0.64	1.00	0.67	0.87	1.41	0.63	0.94	0.93	0.88	0.94	0.21	0.16	0.16	0.12
Other Dimethylnaphthalenes	0.77	1.30	1.68	2.95	1.91	2.52	1.47	1.80	2.38	1.88	1.74	2.25	0.59	0.48	0.46	0.34
Acenaphthalene	0.71	0.02	0.03	0.29	0.02	0.24	1.24	0.02	0.71	0.14	0.14	0.14	0.05	0.03	0.04	0.02
Acenaphthene	0.08	0.03	0.03	0.05	0.03	0.05	0.07	0.04	0.01	0.03	0.03	0.01	0.01	0.01	0.01	0.01
2,3,5-Trimethylnaphthalene	0.36	0.22	0.24	0.26	0.12	0.15	0.32	0.01	0.24	0.14	0.12	0.21	0.04	0.03	0.05	0.03
Other Trimethylnaphthalenes	1.84	1.00	1.04	1.14	0.52	0.63	1.00	0.06	0.61	0.60	0.32	0.73	0.19	0.14	0.22	0.15
Fluorene	0.21	0.12	0.10	0.12	0.06	0.14	0.28	0.05	0.20	0.08	0.07	0.14	0.02	0.02	0.02	0.02
Phenanthrene	0.06	0.32	0.33	0.12	0.14	0.19	0.06	0.10	0.00	0.08	0.10	<0.008	0.06	0.03	0.07	0.02
Anthracene	0.04	0.02	0.35	0.12	0.15	0.02	0.11	0.10	0.08	0.01	0.11	0.02	<0.011	0.03	<0.008	0.02
1-Methylphenanthrene	0.02	0.05	0.05	0.01	0.01	0.01	0.03	<0.008	0.02	0.01	0.01	0.02	<0.011	<0.013	0.01	<0.011
Other Methylphenanthrenes	0.04	0.10	0.11	0.02	0.02	0.03	0.06	<0.008	0.03	0.02	0.02	0.01	<0.011	<0.013	0.02	0.01
Fluoranthene	0.05	0.02	0.01	0.02	0.03	0.04	0.31	0.02	0.23	0.03	0.04	0.04	<0.011	<0.013	0.01	<0.011

(Continued)

TABLE 5.17 (Continued)

PAHs on High-Volume Sampling Equipment Polyurethane Filters and Resins

Type	Burn 1	Background	Background	Burn 1	Burn 1	Burn 1	Burn 2	Burn 1	Burn 2	Burn 2	Burn 2	Burn 1	Burn 2	Burn 1	Burn 2	Burn 1
Description	Burn 1, Boat 2	Background	Background, Boat 1	Burn 1, Boat 1	Burn 1, Boat 4	Burn 1, Boat 1	Burn 2, Boat 1	Burn 1, Boat 4	Burn 2, Boat 1	Burn 2, Boat 2	Burn 2, Boat 2	Burn 1, Boat 2	Burn 2, Far Downwind	Burn 1, Far Downwind	Burn 2, Far Downwind	Burn 1, Far Downwind
Pyrene	0.05	0.05	0.06	0.02	0.01	0.03	0.31	0.01	0.23	0.03	0.03	0.04	<0.011	<0.013	0.01	<0.011
Benz(a)anthracene	0.01	<0.012	<0.012	<0.009	<0.009	<0.008	0.06	<0.008	0.05	<0.007	<0.008	0.01	<0.011	<0.013	<0.008	<0.011
Chrysene	0.06	<0.012	<0.012	<0.009	<0.009	<0.008	0.06	<0.008	0.00	<0.007	<0.008	<0.008	<0.011	<0.013	<0.008	<0.011
Benzo(b)fluoranthene	<0.008	<0.012	<0.012	<0.009	<0.009	<0.008	0.20	<0.008	0.15	0.01	0.01	<0.008	<0.011	<0.013	<0.008	<0.011
Benzo(k)fluoranthene	0.02	<0.012	<0.012	<0.009	<0.009	<0.008	0.01	<0.008	0.01	0.01	<0.008	<0.008	<0.011	<0.013	<0.008	<0.011
Benzo(e)pyrene	0.01	<0.012	<0.012	<0.009	<0.009	<0.008	0.06	<0.008	0.04	<0.007	<0.008	0.01	<0.011	<0.013	<0.008	<0.011
Benzo(a)pyrene	0.01	<0.012	<0.012	<0.009	<0.009	<0.008	0.08	<0.008	0.07	<0.007	<0.008	0.01	<0.011	<0.013	<0.008	<0.011
Perylene	0.05	<0.012	<0.012	<0.009	<0.009	<0.008	0.05	<0.008	0.01	<0.007	<0.008	<0.008	<0.011	0.02	<0.008	<0.011
Indeno(1,2,3-c,d)Pyrene	0.02	<0.012	<0.012	<0.009	<0.009	<0.008	0.15	<0.008	0.10	<0.007	<0.008	<0.008	<0.011	<0.013	<0.008	<0.011
Dibenz(a,h)anthracene	<0.008	<0.012	<0.012	<0.009	<0.009	<0.008	0.01	<0.008	0.01	<0.007	<0.008	<0.008	<0.011	<0.013	<0.008	<0.011
Benzo(g,h,i)Perylene	0.01	<0.012	<0.012	<0.009	<0.009	<0.008	0.10	<0.008	0.06	<0.007	<0.008	0.01	<0.011	<0.013	<0.008	<0.011
Total:	**12.04**	**5.25**	**7.28**	**11.92**	**8.00**	**10.77**	**13.12**	**6.91**	**11.59**	**11.39**	**11.05**	**10.07**	**5.07**	**3.93**	**3.15**	**2.67**
Surrogates (Percent Recovery)																
d8-Naphthalene	N/A	78	89	92	105	121	N/A	94	N/A	100	111	N/A	124	117	105	106
d10-Acenaphthene	105	88	88	109	96	93	93	92	4	103	98	N/A	91	85	83	76
d10-Phenanthrene	82	95	94	90	89	80	80	87	0	100	92	N/A	99	87	87	80
d12-Benz(a)Anthracene	78	100	99	99	101	99	72	100	0	97	94	N/A	94	94	103	97
d12-Perylene	61	82	82	77	73	84	61	70	3	84	85	N/A	70	74	89	80

5.7.9 Alkanes and Weathering in the Oil

Alkanes are a strong contributor to burning. The smallest and most volatile of these are rapidly burned off. This is shown in Figure 5.47. Table 5.18 shows the alkane distribution in various oil and residue samples. Table 5.19 shows the weathering summary. A most interesting result is that the residue appeared to be an oil with an evaporative loss of about 45%–50% by weight. The residue had a density of about 0.95 g/cc and a viscosity of about 100,000 mPa.s.

5.7.10 Overall Findings

The Newfoundland burns have revealed several facts about the fate, behaviour and quantity of the basic emissions from burning:

Gases: Combustion gases are very diffuse and do not have a spatial relationship with the plume. A good model is to view gas dispersal as following a doughnut-like pattern around the burn. This pattern is deformed by increasing wind velocities. Generally, gas concentrations downwind are very low.

Particulate matter/soot: Particulate matter at ground level is a matter of concern very close to the fire and under the plume. The concentration of particulates in the smoke plume may not be a concern past about 500 m. The level of respirable particulates, those which have a size less than 10 μ, was researched further in later studies.

Water emissions: No compounds, other than small amounts of naphthalene, have yet been detected in the water resulting from any burns. The aquatic toxicity of the water under a burn is either not measurable or not extant.

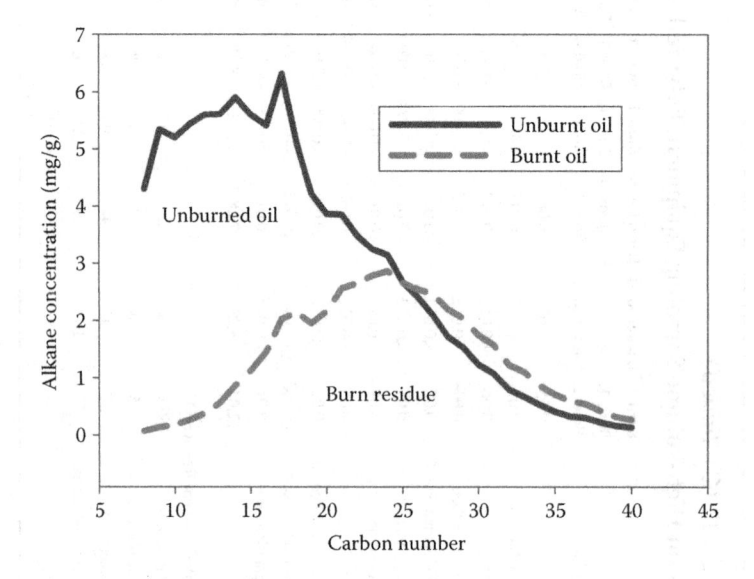

FIGURE 5.47 The fate of alkanes in the two burns as shown by their concentrations in the unburnt oil and the burn residue.

TABLE 5.18

Alkane Analysis of Oils and Residues (mg/g)

NOBE Oil	Starting Oils							Residues	(see Table 5.16 for residue description)			
	2A	2B	3	4	5A	5B	7	11	12	14	15	16
Alkane												
C8	4.75	4.85	4.11	4.09	4.13	4.36	3.83	0.03	0.19	0.02	0.05	–
C9	5.59	5.87	5.32	4.95	5.27	5.37	5.01	0.05	0.34	0.04	0.11	–
C10	5.49	5.55	5.25	4.88	5.11	5.17	4.97	0.11	0.48	0.09	0.19	0.02
C11	5.60	5.61	5.52	5.10	5.38	5.46	5.41	0.18	0.66	0.14	0.32	0.04
C12	5.67	5.58	5.63	5.32	5.81	5.70	5.52	0.29	0.82	0.21	0.47	0.12
C13	5.96	5.77	5.64	5.26	5.74	5.50	5.42	0.49	0.98	0.31	0.71	0.33
C14	6.29	6.04	5.90	5.59	6.06	5.75	5.72	0.79	1.34	0.52	1.03	0.67
C15	5.96	5.70	5.54	5.34	5.72	5.53	5.40	1.09	1.51	0.74	1.29	1.01
C16	5.75	5.54	5.35	5.15	5.51	5.31	5.27	1.45	1.77	1.05	1.62	1.39
C17	6.72	6.42	6.33	6.13	6.36	6.16	6.10	2.35	2.22	1.34	2.28	1.95
Pristane	2.98	2.96	2.80	2.68	2.86	2.76	2.68	1.04	1.27	0.71	1.08	1.08
C18	5.34	5.19	5.04	5.03	5.16	5.01	4.93	2.38	2.23	1.52	2.38	2.16
Phytane	2.60	2.58	2.43	2.34	2.50	2.38	2.34	1.07	1.10	0.73	1.10	1.09
C19	4.49	4.36	4.17	4.09	4.32	4.01	4.07	2.06	1.93	1.58	2.21	1.97
C20	4.07	3.93	3.94	3.79	3.92	3.71	3.71	2.33	2.08	1.83	2.31	2.27
C21	4.05	3.93	3.90	3.69	3.93	3.71	3.76	2.78	2.45	2.12	2.77	2.68
C22	3.54	3.67	3.41	3.47	3.43	3.39	3.50	2.87	2.39	2.26	2.90	2.84
C23	3.36	3.28	3.27	3.12	3.33	3.22	3.16	3.15	2.59	2.37	2.95	2.86
C24	3.47	3.21	3.06	3.01	3.15	3.12	2.99	3.14	2.58	2.49	3.06	3.06
C25	2.64	2.66	2.69	2.75	2.72	2.67	2.62	2.89	2.45	2.33	2.82	2.81
C26	2.40	2.42	2.40	2.44	2.44	2.40	2.35	2.79	2.35	2.23	2.69	2.65
C27	2.17	2.13	2.15	2.00	2.13	2.05	2.02	2.75	2.29	2.13	2.60	2.54
C28	1.88	1.76	1.78	1.67	1.76	1.57	1.60	2.24	2.14	1.95	2.37	2.26
C29	1.63	1.59	1.57	1.50	1.49	1.40	1.55	2.00	2.02	1.86	2.21	2.11
C30	1.22	1.24	1.28	1.28	1.25	1.16	1.21	1.71	1.64	1.61	1.86	1.91
C31	1.08	1.13	1.16	1.04	1.08	1.01	1.04	1.63	1.52	1.50	1.54	1.68
C32	0.81	0.80	0.86	0.79	0.78	0.70	0.81	1.32	1.16	1.13	1.26	1.27
C33	0.72	0.67	0.76	0.67	0.69	0.56	0.59	1.25	1.07	0.95	1.12	1.13
C34	0.54	0.53	0.58	0.56	0.54	0.45	0.52	0.91	0.87	0.77	0.90	0.86
C35	0.42	0.39	0.42	0.44	0.43	0.33	0.42	0.66	0.71	0.69	0.73	0.73
C36	0.32	0.30	0.34	0.36	0.35	0.28	0.32	0.59	0.57	0.56	0.62	0.65
C37	0.31	0.27	0.31	0.34	0.33	0.25	0.29	0.53	0.52	0.50	0.56	0.59
C38	0.21	0.17	0.25	0.25	0.26	0.17	0.24	0.37	0.39	0.40	0.43	0.50
C39	0.15	0.13	0.17	0.20	0.18	0.12	0.18	0.30	0.32	0.32	0.30	0.34
C40	0.12	0.10	0.15	0.18	0.15	0.10	0.15	0.26	0.28	0.27	0.24	0.29
TOTAL n-alkanes	108	106	103	100	104	101	100	50	49	39	51	48

TABLE 5.19

Alkane Weathering in Oil and Residues

Sample No.	Total n-Alkanes (mg/g oil)	GCRTPH (mg/g oil)	TPH (mg/g oil)	GCRTPH/TPH	(C8+C10+C12+C14)/ (C22+C24+C26+C28)	Weathered Percent (%)	C17/C18	Pris./Phy.
2A	108	145	590	0.25	2.02	5.6	1.26	1.12
2B	106	152	582	0.26	2.00	6.0	1.24	1.15
3	103	181	666	0.27	1.98	6.5	1.25	1.16
4	100	158	614	0.26	1.86	9.2	1.22	1.15
5A	104	173	655	0.26	2.00	6.0	1.23	1.15
5B	101	166	645	0.26	2.00	6.0	1.23	1.16
7	99	161	609	0.26	1.91	8.1	1.24	1.14
11[a]	50	74	513	0.14	0.11	48.6	0.99	0.97
12	49	62	431	0.14	0.30	44.3	0.99	1.15
14	39	51	408	0.13	0.10	48.9	0.88	0.97
15	51	65	479	0.14	0.16	47.5	0.96	0.99
16	48	57	451	0.13	0.07	49.5	0.90	1.00
Ref. oil	108	172	632	0.27	2.05		1.27	1.09

TPH: Total Petroleum Hydrocarbons; GCR TPH: GC Resolvable TPH; Pris./Phy.: Ratio of pristane/phytane.

[a] Shaded area a indicates residues.

Organic compounds: No exotic or highly toxic compounds were generated as a result of the combustion process. Organic macro-molecules were in lesser concentration in the smoke and downwind than they were in the oil itself. VOCs were released in large concentrations by the fires, but in lesser concentrations than from the evaporating slick if not burning.

Residue: The residue from burns of this crude was lighter than water. Density appears to relate to efficiency. If a burn is highly efficient (>99.9%), then the residue may be neutrally buoyant. The residue resembled high-weathered oil and measurements showed this to be about 40 to 50% weathered (% weight loss). The residue contained a lower amount of PAHs than the starting oil, although proportionately high amounts of multi-ringed PAHs were present.

PAHs: It can be said that additional PAHs are not produced by in-situ oil fires. Oils contain significant quantities of PAHs. These are largely destroyed in combustion. The PAH concentrations in the smoke, both in the plume and the particulate precipitation at ground level, were much less than the starting oil. This also includes the concentration of multi-ringed PAHs that are often created in other combustion processes such as low-temperature incinerators and diesel engines. This finding is very different from that noted in earlier laboratory experiments. It is suspected that re-precipitation of large soot particles occurs in large-scale tests which does not occur in laboratory tests. These large soot particles are conducive to the accumulation of large multi-ringed PAHs. The burn residue does, however, show a slight increase in the concentration of multi-ringed PAHs. However, when considering the mass balance of the burn, most of the five- and six-ringed PAHs were destroyed by the fire.

ACKNOWLEDGEMENTS

The NOBE experiment was funded by more than 20 agencies, listed here in order of funding: Environment Canada, U.S. Minerals Management Service, CCG, Marine Spill Response Corporation, United States Coast Guard, American Petroleum Institute, U.S. Environmental Protection Agency, Canadian Association of Petroleum Producers, 3M Ceramics Division, Canadian Petroleum Products Institute, Alaska Clean Seas, Amoco Production, Program for Energy Research and Development (PERD), Imperial Oil Limited, Hibernia Development, Exxon Biomedical Services, Canmar/AMOCO Canada, East Coast Response Incorporated, and Beaufort Sea Co-op.

The authors of this chapter thank the many people who contributed to this study. Some are listed as authors in the referenced papers.

REFERENCES

Bissonnette, M.C., Fingas, M.F., Nelson, R.D., Beaudry, P., and Paré, J.R.P. 1994. Crude oil combustion at sea: The sampling of released products using remote-controlled boats, *Proceedings of the Seventeenth Arctic and Marine Oil Spill Program Technical Seminar*. Environment Canada, Ottawa, ON, pp. 1065–1081.

Blenkinsopp, S.A., Sergy, G., Doe, K., Wohlgeshaffen, G., Li, K., and Fingas, M.F. 1996. Toxicity of the weathered crude oil used at the newfoundland offshore burn experiment (NOBE) and the resultant burn residue. *Spill Science and Technology Bulletin*. 3(4). 277–280.

Blenkinsopp, S.A., Sergy, G., Doe, K., Wohlgeshaffen, G., Li, K., and Fingas, M.F. 1997. Evaluation of the toxicity of the weathered crude oil used at the newfoundland offshore burn experiment (NOBE) and the resultant burn residue, *Proceedings of the Twentieth AMOP Technical Seminar*. Environment Canada, Ottawa, ON, pp. 677–684.

Environment Canada, 1997, Data compilation, Newfoundland Offshore Burn Experiment (NOBE) Report, Emergencies Science Division, Environmental Technology Centre, Environment Canada, Ottawa, ON.

Fingas, M.F., Ackerman, F., Lambert, P., Li, K., Wang, Z., Mullin, J., Hannon, L. et al. 1995. The newfoundland offshore burn experiment: Further results of emissions measurement, *Proceedings of the Eighteenth Arctic and Marine Oil Spill Technical Seminar*, Edmonton, Alberta, pp. 915–995.

Fingas, M.F., Ackerman, F., Li, K., Lambert, P., Wang, Z., Bissonnette, M.C., Campagna, P.R. et al. 1994a. The newfoundland offshore burn experiment—NOBE—Preliminary results of emissions measurement, *Proceedings Arctic and Marine Oilspill Program (AMOP) Technical Seminar. No. 17b.* Environment Canada, Ottawa, ON, pp. 1099–1164.

Fingas, M.F., Halley, G., Ackerman, F., Vanderkooy, N., Nelson, R., Bissonnette, M.C., Laroche, N. et al. 1994b. The newfoundland offshore burn experiment—NOBE experimental design and overview, *Proceedings of the 17th Arctic and Marine Oilspill Program Technical Seminar*. Environment Canada, Ottawa, ON, pp. 1053–1061.

Jokuty, P., and Fingas, M.F. 1994. Oil analytical techniques for environmental purposes, *Proceedings of the Seventeenth AMOP Technical Seminar*, Environment. Canada, Ottawa, ON, pp. 245–260.

6 The Fate of Polycyclic Aromatic Hydrocarbons Resulting from In-Situ Oil Burns

Merv Fingas

CONTENTS

6.1 INTRODUCTION

Polycyclic aromatic hydrocarbons (PAHs) are ubiquitous in the environment and are considered to be of concern for both humans and the environment (Lima et al. 2005). The sources of PAHs are many, but they come primarily from combustion sources. Natural sources such as forest fires are a contributor. Crude oil contains PAHs; however, oil spills themselves are not a significant contributor of PAHs to the environment. Burning of petroleum could constitute an input of PAHs into the environment. It has been pointed out that the two most significant contributors are diesel engines and home heating using a type of fuel similar to diesel fuel. Heating with wood also contributes some PAHs. Sootier flames appear to be associated with more PAHs. The amount of PAH in the starting fuel also influences the amount of PAHs emitted by a burn. Generally low-temperature sources such as diesel engines are larger emitters of PAHs than high-temperature sources such as fires.

PAHs are compounds consisting of at least two benzene rings. Common PAHs and their substituted counterparts found in oils are shown in Table 6.1 (Fingas 2012). As these are easily separated, there are much data about their presence in oils. These compounds have also been used somewhat as indicators of the presence of certain types of oils. There exists a set of compounds designated by the U.S. EPA as priority PAHs (Wise et al. 2015). The list of 16 EPA priority chemicals is shown in Table 6.2 (Fingas 2012). The concern with these compounds is that some of them are known to be relatively toxic and some to be carcinogenic.

TABLE 6.1

PAHs and Alkylated PAHS Found in Oils and Petroleum

Carbon Number	Number of Rings	Name	Formula	Abbrev.	Molecular Weight	CAS Number	Solubility in Water[a] g/L	SIMS	Levels in ASMB µg/g Oil	Levels in ANS µg/g Oil	Levels in Diesel Fuel µg/g Oil	Levels in Heavy Fuel Oil µg/g Oil
10	2	Naphthalene	$C_{10}H_8$	C_0N	128.717	91-20-3	$30\text{--}30\ E^{-3}$	128	680	260	820	140
11	2	methyl-naphthalene	$C_{11}H_{10}$	C_1N	142.197	1	$26\text{--}28\ E^{-3}$	142	1,180	1,000	3,700	1,300
12	2	di-methyl-naphthalene	$C_{12}H_{12}$	C_2N	156.233	2	$\sim2.4\ E^{-3}$	156	1,600	1,800	7,000	2,900
12	2	ethyl-naphthalene	$C_{12}H_{12}$	C_2N	156.233	3	$\sim1\ E^{-2}$	156				
13	2	C3-naphthalenes	$C_{13}H_{14}$	C_3N	170.25	4	$\sim1.6\ E^{-3}$	170	1,560	1,700	6,600	2,900
14	2	C4-naphthalenes	$C_{14}H_{16}$	C_4N				184	450	820	2,800	1,400
14	3	Phenanthrene	$C_{14}H_{10}$	C_0P	178.229	85-01-8	$\sim1\ E^{-3}$	178	170	210	440	420
15	3	C1-phenanthrenes	$C_{15}H_{12}$	C_1P	192.256	5	$\sim7\ E^{-5}$	192	400	670	1,000	1,900
16	3	C2-phenanthrenes	$C_{16}H_{14}$	C_2P				206	400	710	620	2,900
17	3	C3-phenanthrenes	$C_{17}H_{16}$	C_1P				220	160	490	190	3,100
18	3	C4-phenanthrenes	$C_{18}H_{16}$	C_1P				234	60	300	50	2,200
12	3	Dibenzothiophene	$C_{12}H_8S$	C_0D	184.257	132-65-0	$5\ E^{-4}$	184	200	120	70	100
13	3	C1-dibenzothiophenes	$C_{13}H_{10}S$	C_1D				198	370	230	110	320
14	3	C2-dibenzothiophenes	$C_{14}H_{12}S$	C_2D				212	450	320	100	620
15	3	C3-dibenzothiophenes	$C_{15}H_{14}S$	C_3D				226	220	270	40	700
13	3	Fluorene	$C_{13}H_{10}$	C_0F	166.218	86-73-7	$\sim8\ E^{-5}$	166	50	140	570	220
14	3	C1-fluorenes	$C_{14}H_{12}$	C_0F	180.245	6		180	110	330	800	570

(Continued)

TABLE 6.1 (*Continued*)
PAHs and Alkylated PAHS Found in Oils and Petroleum

Carbon Number	Number of Rings	Name	Formula	Abbrev.	Molecular Weight	CAS Number	Solubility in Water[a] g/L	SIMS	Levels in ASMB µg/g Oil	Levels in ANS µg/g Oil	Levels in Diesel Fuel µg/g Oil	Levels in Heavy Fuel µg/g Oil
15	3	C2-fluorenes	$C_{15}H_{14}$	C_0F				194	150	450	760	1,000
16	3	C3-fluorenes	$C_{16}H_{16}$	C_0F				208	85	380	360	940
18	4	Chrysene	$C_{18}H_{12}$	C_0C	228.288	218-01-9	~1.5 E^{-6}	228	25	50	<1	380
19	4	C1-chrysenes	$C_{19}H_{14}$	C_0C	242.314	7		242		70	<1	1,200
20	4	C2-chrysenes	$C_{20}H_{16}$	C_0C				256		100	<1	1,800
21	4	C3-chrysenes	$C_{21}H_{18}$	C_0C				270		80		1,400
20	5	Perylene	$C_{20}H_{12}$		252.309	198-55-0	~4 E^{-7}	252				

[a] Approximate sign (~) indicates solubility values are variable; a range indicates the range of values.

[1] CAS numbers for methyl naphthalenes - *1*-90-12-0; *2*-91-57-6

[2] CAS numbers for di-methyl naphthalenes - *1,2*-573-98-8; *1,3*-575-41-7; *1,4*-571-58-4; *1,5*-571-61-9; *1,6*-575-43-9; *1,7*-575-37-1; *1,8*-569-41-5; *2,3*-581-40-8; *2,6*-581-42-0; *2,7*-582-16-1

[3] CAS numbers for ethyl naphthalenes - *1*-1127-76-0; *2*-939-27-5

[4] CAS numbers for 1,4,5-trimethyl naphthalene - 2131-41-1

[5] CAS numbers for methyl phenanthrenes - *1*-832-69-9; *3*-832-71-3; *4*-832-64-4

[6] CAS numbers for methyl fluorenes - *1*-1730-37-6; *9*-2523-37-7

[7] CAS numbers for methyl chrysenes - *3*-3351-31-3; *5*-3697-24-3; *6*-1705-85-7
ASMB is Alberta Sweet Mixed Crude oil; ANS is Alaska North Slope Oil.

TABLE 6.2
EPA Top 16 Priority PAHs

Compound	Molecular Formula	Structure	CAS No.	MW	Human[a] Carcino-genicity	Solubility in Water[b] µg/L	SIMS	Levels in ASMB µg/g Oil	Levels in ANS µg/g Oil	Levels in Diesel Fuel µg/g Oil	Levels in Heavy Fuel µg/g Oil
Naphthalene	$C_{10}H_8$		91-20-3	128.171		~3.5 E^{-2}	128	250	260	820	140
Acenaphthene	$C_{12}H_{10}$		83-32-9	154.207		3.4 E^{-3}	153	16	13	150	90
Acenaphthylene	$C_{12}H_8$		208-96-8	152.192		~4 E^{-3}	152	8	12	35	20
Fluorene	$C_{13}H_{10}$		86-73-7	166.218		~2 E^{-3}	166	80	140	560	220
Phenanthrene	$C_{14}H_{10}$		85-01-8	178.229		~1 E^{-3}	178	140	210	440	420
Anthracene	$C_{14}H_{10}$		120-12-7	178.229		4–7 E^{-5}	178	2	3	13	95
Fluoranthene	$C_{16}H_{10}$		206-44-0	202.25		2.7 E^{-4}	202	2	3	7	40
Pyrene	$C_{16}H_{10}$		129-00-0	202.25		0.9–1.4 E^{-4}	202	18	8	31	230
Benz(a) anthracene	$C_{18}H_{12}$		56-55-3	228.288	2	0.9–1.3 E^{-5}	228	3	5	0.3	200

(Continued)

TABLE 6.2 (Continued)
EPA Top 16 Priority PAHs

Compound	Molecular Formula	Structure	CAS No.	MW	Human[a] Carcino-genicity	Solubility in Water[b] µg/L	SIMS	Levels in ASMB µg/g Oil	Levels in ANS µg/g Oil	Levels in Diesel Fuel µg/g Oil	Levels in Heavy Fuel µg/g Oil
Chrysene	$C_{18}H_{12}$		218-01-9	228.288	1	$2\text{--}1.6\ E^{-6}$	228	30	50	<1	380
Benzo(b) fluoranthene	$C_{20}H_{12}$		205-99-2	252.309	3	$\sim 1.5\ E^{-6}$	252	3	5	0	50
Benzo(k) fluoranthene	$C_{20}H_{12}$		207-08-9	252.309		$\sim 1\ E^{-6}$	252	3	1	0	10
Benzo(a)pyrene	$C_{20}H_{12}$		50-32-8	252.309	4	$1.4\text{--}3.8\ E^{-6}$	252	1	2	0	150
Dibenz(a,h) anthracene	$C_{22}H_{14}$		53-70-3	278.346	4	$\sim 1\ E^{-6}$	278	1	1	0	20
Benzo(g,h,i) perylene	$C_{22}H_{12}$		191-24-2	276.33		$1.8\ \text{to}\ 2.6\ E^{-7}$	276	3	3	0	30
Indeno(1,2,3-cd) pyrene	$C_{22}H_{12}$		193-39-5	276.33	2	$\sim 2\ E^{-7}$	276	1	0.1	0	10

Source: Fingas, M., Introduction to oil chemistry and properties. *Proceedings of the Thirty-Fifth. Arctic and Marine Oil Spill Program Technical Seminar*, Environment Canada, Ottawa, ON, pp. 951–980, 2012.

[a] Indicates carcinogenic potency, 1 is a suspected carcinogen, 4 is a known potent carcinogen.

[b] Approximate sign (~) indicates solubility values are variable, a range indicates the range of values.

PAH compounds have the general formula C_nH_{2n-6r}, where r is the number of rings. The amount of PAH in a typical crude oil varies but ranges from 0% to 5%. In crude oils, the alkylated compounds occur more frequently than the parent un-alkylated rings. This can be of use in identifying the source of contamination, as many PAH pollution sources have more abundant parent compounds than alkylated ones.

Crude oil burns result in PAH downwind of the fire, but the concentration on the particulate matter is often an order of magnitude less than the concentration in the starting oil and sometimes several orders of magnitude less. Diesel contains significant levels of PAHs of smaller molecular size, the 2- to 3-ring PAHs predominating. Burning diesel results in more pyrogenic PAHs of larger molecular sizes. Larger PAHs are either created or concentrated by the fire. Larger PAHs, some of which are not even detectable in the diesel fuel, are found both in the soot and in the residue. The concentrations of these larger PAHs are low and often just above detection limits. The question is, are more PAHs are destroyed by the fires than are created?

The PAHs in oil may have any or several of the following fates: (1) burn to CO_2 and water, (2) conversion to other chemicals such as other PAHs or to oxygenated PAHs, (3) transmission through to the gaseous emissions of a fire, (4) absorption on the particulate emissions (soot) from the fire or (5) accumulation on the residue or unburned portion. This is illustrated in Figure 6.1.

Carried with soot particulates

Transformed to other PAHs or other compounds

Burned to CO_2 and water

Emitted as gaseous PAHs

Remain in residue

FIGURE 6.1   The possible fate pathways for PAHs in oil during a burn.

An equation might be written for the PAH mass balance as:

$$\text{PAHs Mass Balance} = \text{Starting mass of PAHs} - \text{burned} - \text{on residue}$$
$$- \text{on soot} - \text{converted} - \text{vaporised}$$

$$(6.1)$$

6.2 STUDIES ON THE FATE OF POLYCYCLIC AROMATIC HYDROCARBONS IN FIRES

Several early studies noted that the concentrations of PAHs were about the same in the residue as in the starting oil (Benner et al. 1990; Fingas 1991; Fingas et al. 1993, 1994b, 1995; Li et al. 1992). The observations at the burn experiments reported at these trials included the following information: (1) generally the PAH concentrations in the residue and the starting oil were similar, (2) for diesel burns there appeared to be more larger PAHs present in the residue than the concentrations of these same compounds in the starting oil and (3) test of the volatile emissions showed low concentrations of PAHs. Several interpretations of these results ensued; however, no firm conclusions could be made because the scientists were unsure of the mass balance of the burns between the soot and the residue compared to the starting oil. It is important to recognise that the concentration of PAHs in the starting oil and the residue is only one facet. An important facet of the puzzle is the efficiency of the burn, or how much of the oil (and PAHs contained therein) were burned. To illustrate this, presume that 100 kg of oil were burned with a starting concentration of 1% PAHs – this would mean that there was 1 kg of PAHs. If the burn efficiency was 95% and the concentration of the PAHs was still 1% in the residue, then 95% of the PAHs were destroyed by the fire. In addition, one should consider the PAHs lofted in the soot and emitted as vapour. The distribution of PAHs should also be studied.

A basic study on the fate of PAHs in diesel burns was carried out by Wang et al. (1998, 1999a, 1999b). This study used several mesoscale burns conducted in Mobile Bay, Alabama, to study various aspects of diesel fuel burning. The target PAHs in the diesel, residue and the soot samples collected during each burn were quantitatively characterised by gas chromatography/mass spectrometry (GC/MS) as shown in Table 6.3. A simple model based on mass balance of individual petroleum PAHs pre- and post-burn was proposed to estimate the destruction efficiencies of the PAHs. This study demonstrated that distributions of PAHs in the original diesel and soot were actually different. The average destruction efficiencies for the total target diesel PAHs including five alkylated PAH series and other EPA priority unsubstituted PAHs were greater than 99%. Using the model, 27.3 kg of the diesel PAHs were destroyed for each 1,000 kg of diesel burned. These were mostly

TABLE 6.3

Results of the PAH Destruction Studies in Diesel Fires

	Aromatic Rings	PAHs (µg/g)			PAH Destruction Percentage			
		In Starting Diesel Oil	In Residue	In Soot	Scenario 1	Scenario 2	Scenario 3	This Study
					Soot = 5% Residue = 0.1%	Soot = 10% Residue = 0.2%	Soot = 15% Residue = 0.3%	Soot = 1% Residue = 2%
C0-Naphthalene	2	232.1	42.1	10.4	99.8	99.1	98.8	100.00
C1-N	2	1025.2	111.2	8.7	99.9	99.9	99.8	100.00
C2-N	2	4599.2	1109.8	20.7	99.9	99.9	99.8	100.00
C3-N	2	6305.4	2745.4	12.5	99.9	99.9	99.8	99.99
C4-N	2	2984.1	1920.4	6.5	99.9	99.9	99.8	99.99
Sum		15146	5,929	59	99.9	99.9	99.8	99.99
C0-Phenanthrene	3	253.6	339.9	11.5	99.6	99.3	98.9	99.97
C1-P	3	909.6	1133.5	12.0	99.8	99.5	99.3	99.97
C2-P	3	1066.8	1741.2	28.8	99.7	99.4	99.1	99.97
C3-P	3	570.5	1196.5	28.6	99.5	99.1	98.6	99.96
C4-P	3	199.3	602.8	13.3	99.4	98.7	98.0	99.94
Sum		3,000	5,014	94	99.7	99.3	99.0	99.97
C0-Dibenzothiophene	3	511.4	460.8	1.4	99.9	99.8	99.7	99.98
C1-D	3	1507.1	1691.7	3.4	99.9	99.7	99.5	99.98
C2-D	3	2019.6	2934.6	18.1	99.8	99.6	99.4	99.97
C3-D	3	982.2	1783.2	19.9	99.7	99.4	99.1	99.96
Sum		5,020	6,870	43	99.8	99.6	99.4	99.97
C0-Fluorene	3	241.4	144.0	1.9	99.9	99.8	99.7	99.99

(Continued)

TABLE 6.3 (Continued)
Results of the PAH Destruction Studies in Diesel Fires

	Aromatic Rings	PAHs (µg/g)			PAH Destruction Percentage			
		In Starting Diesel Oil	In Residue	In Soot	Scenario 1	Scenario 2	Scenario 3	This Study
					Soot = 5% Residue = 0.1%	Soot = 10% Residue = 0.2%	Soot = 15% Residue = 0.3%	Soot = 1% Residue = 2%
C1-F	3	1002.3	802.1	1.7	99.9	99.8	99.7	99.98
C2-F	3	1421.7	1374.3	2.7	99.9	99.8	99.7	99.98
C3-F	3	1236.9	1498.3	6.0	99.8	99.7	99.6	99.98
Sum		3,902	3,819	12	99.9	99.8	99.7	99.98
C0-Chrysene	4	6.8	48.4	15.3	88.0	76.0	64.0	99.84
C1-C	4	9.1	90.4	5.8	96.0	90.0	85.0	99.79
C2-C	4	7.5	122.4	4.1	96.0	91.0	87.0	99.67
C3-C	4	1.9	57.1	2.1	91.0	83.0	74.0	99.39
Sum		25	318	27	93.0	86.0	80.0	99.73
Total		27,094	21,950	235	99.8	99.7	99.6	99.73
Other PAHs								
Biphenyl (Bph)	2	309.2	62.9	4	99.9	99.8	99.7	100.00
Acenaphthylene (Acl)	3	14.3	62.7	2.8	98.9	97.9	96.8	99.96
Acenaphthene (Ace)	3	74.1	30	35	99.9	99.8	99.7	99.99
Anthracene (An)	3	6.9	35.2	2.2	97.9	95.8	93.7	99.87
Fluoranthene (Fl)	4	2.6	44.2	21.1	57	15		99.39
Pyrene (Py)	4	7.3	69	24.3	92	65	47	99.93

(Continued)

TABLE 6.3 (*Continued*)

Results of the PAH Destruction Studies in Diesel Fires

		PAHs (µg/g)			PAH Destruction Percentage			
					Scenario 1	Scenario 2	Scenario 3	This Study
	Aromatic Rings	In Starting Diesel Oil	In Residue	In Soot	Soot = 5% Residue = 0.1%	Soot = 10% Residue = 0.2%	Soot = 15% Residue = 0.3%	Soot = 1% Residue = 2%
Benz[*a*]anthracene (BaA)	4	1	14.8	11	43			99.70
Benzo[*b*]fluoranthene (BbF)	5	0.2	9.3	29				97.34
Benzo[*k*]fluoranthene (BkF)	5	0.1	12.1	73.4				90.04
Benzo[*e*]pyrene (BeP)	5	0.4	13.1	34				98.30
Benzo[*a*]pyrene (BaP)	5	0.1	17.1	35.6				95.50
Perylene (Pe)	5	0.2	4.7	9				98.10
Benz[*a*]anthracene (BaA)	4	1	14.8	11	43			99.70
Benzo[*b*]fluoranthene (BbF)	5	0.2	9.3	29				97.34
Benzo[*k*]fluoranthene (BkF)	5	0.1	12.1	73.4				90.04
Total of three- to six-ring PAHs		107	347	380				99.95
Total of five- to six-ring PAHs		1	93	319				
Total of target PAHs		27,510	22,360	620	99.8	99.6	99.4	100.00

Source: Wang et al. (1999a).

Legend: ND = not detected, NG = newly generated, blank indicates that value cannot be calculated.

two- and three-ring PAHs and their alkylated homologues. Combustion also generated trace amounts of high-molecular-weight 5- and 6-ring PAHs as well as the 4-ring benz[a]anthracene. But the total mass of these pyrogenic PAHs was found to be extremely low: only 0.016, 0.032 and 0.048 kg of the 5- and 6-ring PAHs were generated by combustion in the three different scenarios for each 1,000 kg of diesel burned. It was concluded that in-situ burning is an effective measure for minimising the impact of an oil spill on the environment, greatly reducing exposure of ecosystems to the PAHs of spilled oils.

The worksheet for the Wang et al. (1999a) study is shown as Table 6.3. This shows the concentrations of PAHs in the starting diesel fuel, the soot and the residue. It should be noted that for this test, the gaseous emissions were also measured and did not contain very many PAHs, so few in fact that these could be discounted. At the time of this study (1999), the amount of soot produced was not known; however, the amount of residue could be estimated as less than 1% from simple estimates at the burn site. Wang et al. (1999a) created three scenarios, soot % = 5% and residue % was 0.1%; soot = 10% and residue % = 0.2% and soot = 15% and residue = 0.3%. With all of these estimates, it was shown that destruction of PAHs was greater than 99%. For some of the priority PAHs, however, the destruction efficiency was lower or could not be calculated because there was no compounds present.

Studies of the amount of soot produced show that the amount of soot produced by diesel fires was less than 5% and as low as 1% (Fingas et al. 1996; Fingas 2010). Thus, the last column in Table 6.3 shows a subsequent calculation using more modern values of soot of 1% and residue of 2%. The results of this are that, for each PAH, the destruction percentage is at least 98% and in most cases it is 99.99%. Diesel fires still result in some production of higher-molecular-weight PAHs – or at least an increase in their quantity – such that the destruction is less than the PAHs normally in the diesel fuel.

Table 6.3 shows two trends related to the destruction of PAHs in diesel burns:

1. It appears that the greater the number of rings, the poorer the PAH destruction percentage. A test of this in general is shown in Figure 6.2. Figure 6.2 appears to show the opposite trend; however, all PAH numbers are included in this figure. Thus, as a general rule, the ring size is not an indicator of PAH destruction percentage. However, when the priority PAH destruction data is plotted as shown in Figure 6.3, the expected trend appears. For the priority PAHs, the trend is generally that, as the ring numbers increases, the destruction percentage decreases.

2. It appears that for the alkylated PAHs, the destruction percentage decreases with increasing alkylation. A test of this is shown in Figure 6.4. This figure shows that the increasing alkylation is correlated with a slight decrease in destruction efficiency. A look at the vapour component shows that during

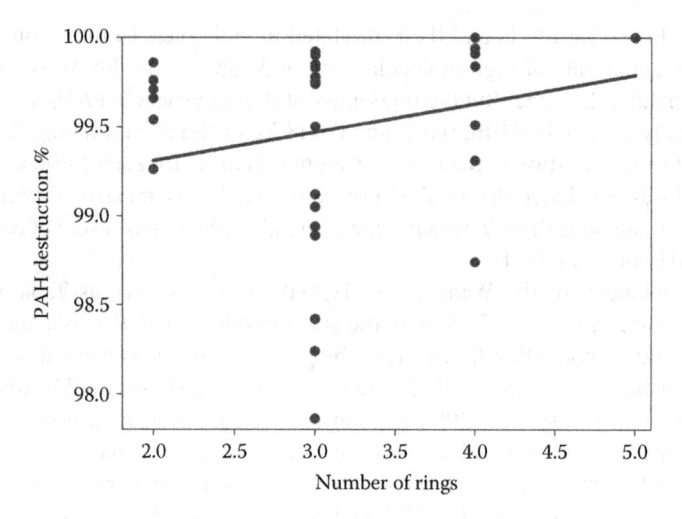

FIGURE 6.2   A plot of the PAH destruction efficiency in diesel burns versus the number of rings in the PAH. This plot is for all PAH data and appears to show that efficiency rises with the ring size; however, this may be skewed by the paucity of data on large PAHs.

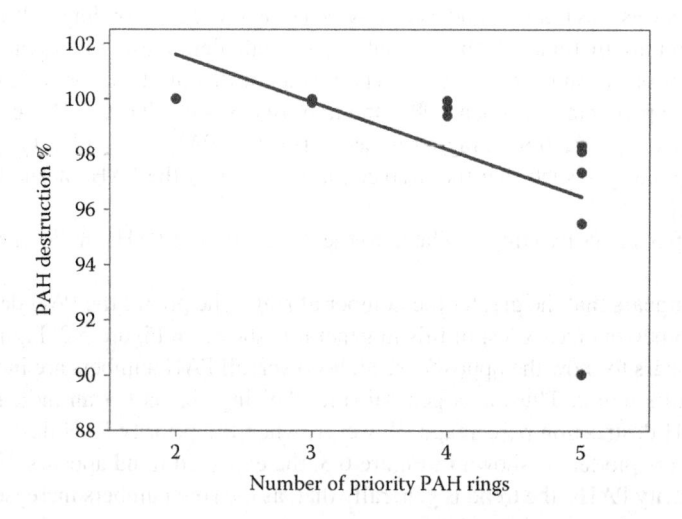

FIGURE 6.3   A plot of PAH destruction efficiency in diesel burns versus the number of rings in the PAH. This is for priority PAHs only and shows the expected trend: the larger the PAH, the less the destruction efficiency.

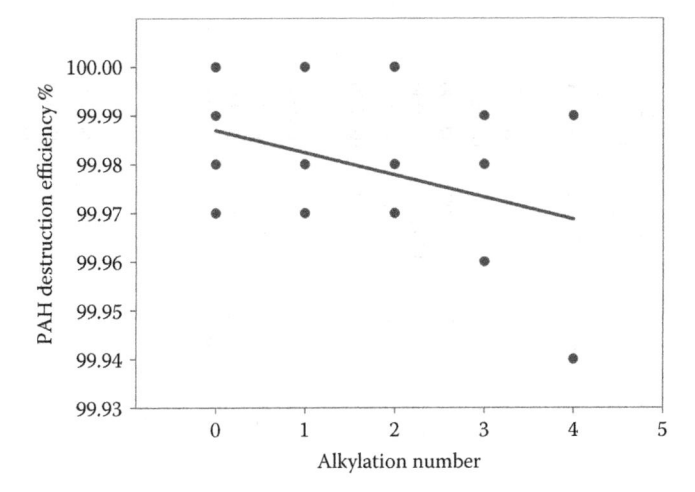

FIGURE 6.4 A plot of the PAH destruction efficiency versus the alkylation number of the PAH. This shows that the destruction efficiency decreases slightly with higher alkylation numbers.

3 diesel burns, the concentrations of the only measurable PAH, naphthalene, was 0.2 out of 169 μg/m³; 0.4 out of 153 and 0.7 out of 176 total μg/m³ of volatile organic compounds (VOCs). The background was 0.7 μg/m³; therefore, the PAHs in the vapour phase are negligible and often below background levels.

From Table 6.3 it can be seen that the PAHs in the soot are 620 μg/g, or 0.003% of that in the residue. With approximately the same percentages (1%) of residue or soot, it can be concluded that the PAHs on the soot are insignificant.

An extension of this is to examine the fate of PAHs in crude oil burns. The PAH dataset chosen for this study was that of the Newfoundland Offshore Burn Experiment (NOBE) which used a medium Alberta crude oil (Fingas et al. 1994a, 1994b, 1995). Table 6.4 shows the PAHs in the starting oil, soot and residue as measured from one burn at this study. Performing the mass calculation using the value of 1% for the soot and the measured oil burn efficiency of 99% (the maximum estimated at this burn), it was found that 99.99% of the PAHs are typically burned. This also shows that the PAH destruction is higher for crude oil than for diesel fuel. Data for diesel burns are also shown in Table 6.4. These data will be discussed later.

TABLE 6.4

Comparison of Crude Oil and Diesel Fuel PAH Fate in Burns

	Crude Oil (Alberta Sweet Mixed Blend)					Diesel Fuel				
	Concentrations (µg/g)			Burn destruction %		Concentrations (µg/g)			Burn destruction %	
	Starting Oil	Residue	Soot	As Residue	As Soot	Starting Oil	Residue	Soot	As Residue	As Soot
Naphthalenes										
C0-N	206	61	9	99.998	100.000	232.1	42.1	10.4	100.000	100.000
C1-N	1,049	220	8	99.999	100.000	1,025.2	111.2	8.7	100.000	100.000
C2-N	2,246	578	11	99.998	100.000	4,599.2	1,109.8	20.7	100.000	100.000
C3-N	1,749	589	3	99.998	100.000	6,305.4	2,745.4	12.5	100.000	100.000
C4-N	859	333	2	99.997	100.000	2,984.1	1,920.4	6.5	100.000	100.000
Sum	6,109	1,780	104	99.998	100.000	5,929	59	99.9	100.000	100.000
Phenanthrenes										
C0-P	109	106	5	99.993	100.000	253.6	339.9	11.5	99.999	100.000
C1-P	407	301	5	99.995	100.000	909.6	1,133.5	12	99.999	100.000
C2-P	450	361	12	99.995	100.000	1,066.8	1,741.2	28.8	99.999	100.000
C3-P	342	299	17	99.994	100.000	570.5	1,196.5	28.6	99.999	99.999
C4-P	173	145	12	99.994	100.000	199.3	602.8	13.3	99.998	99.999
Sum	1,481	1,211	30	99.994	100.000	5,014	94	99.7	100.000	100.000
Dibenzothiophenes										
C0-D	14	8	0	99.997	100.000	511.4	460.8	1.4	100.000	100.000
C1-D	28	21	0	99.995	100.000	1,507.1	1,691.7	3.4	99.999	100.000
C2-D	46	37	0	99.995	100.000	2,019.6	2,934.6	18.1	99.999	100.000
C3-D	26	26	1	99.993	100.000	982.2	1,783.2	19.9	99.999	100.000
Sum	114	91	2	99.994	100.000	6,870	43	99.8	100.000	100.000

(Continued)

TABLE 6.4 (*Continued*)

Comparison of Crude Oil and Diesel Fuel PAH Fate in Burns

	Crude Oil (Alberta Sweet Mixed Blend)					Diesel Fuel				
	Concentrations (µg/g)			Burn destruction %		Concentrations (µg/g)			Burn destruction %	
	Starting Oil	Residue	Soot	As Residue	As Soot	Starting Oil	Residue	Soot	As Residue	As Soot
Fluorenes										
C0-F	82	50	1	99.996	100.000	241.4	144	1.9	100.000	100.000
C1-F	176	92	0	99.996	100.000	1,002.3	802.1	1.7	100.000	100.000
C2-F	176	93	0	99.996	100.000	1,421.7	1,374.3	2.7	100.000	100.000
C3-F	144	95	1	99.995	100.000	1,236.9	1,498.3	6	99.999	100.000
Sum	578	329	15	99.996	100.000	3,819	12	99.9	100.000	100.000
Chrysenes										
C0-C	35	45	79	99.991	100.000	6.8	48.4	15.3	99.995	99.976
C1-C	45	58	29	99.991	100.000	9.1	90.4	5.8	99.996	99.993
C2-C	55	75	30	99.991	100.000	7.5	122.4	4.1	99.991	99.994
C3-C	50	58	55	99.992	100.000	1.9	57.1	2.1	99.983	99.988
Sum	185	235	54	99.991	100.000	318	27	93	100.000	99.997
Total alkylated	8,467	3,645	76	99.997	100.000	27,094	21,950	235	100.000	100.000
Priority PAHs										
Biphenyl	70.8	14	1	99.999	100.000	309.2	62.9	4	99.998	99.998
Acenaphthylene	7.61	33	1	99.971	100.000	14.3	62.7	2.8	100.000	100.000
Acenaphthene	15.95	6	0	99.997	100.000	74.1	30	0.5	99.995	99.997
Anthracene	2.09	11	1	99.967	100.000	6.9	35.2	2.2	99.988	99.915
Fluoranthene	2.45	42	20	99.881	99.999	2.6	44.2	21.1	99.996	99.965

(*Continued*)

TABLE 6.4 (*Continued*)

Comparison of Crude Oil and Diesel Fuel PAH Fate in Burns

| | Crude Oil (Alberta Sweet Mixed Blend) | | | | | Diesel Fuel | | | | |
| | Concentrations (µg/g) | | | Burn destruction % | | Concentrations (µg/g) | | | Burn destruction % | |
	Starting Oil	Residue	Soot	As Residue	As Soot	Starting Oil	Residue	Soot	As Residue	As Soot
Pyrene	18.28	173	61	99.934	100.000	7.3	69	24.3	100.000	99.825
Benz(a)anthracene	2.94	44	32	99.901	99.999	1	14.8	11		
Benzo(b)fluoranthene	2.94	137	426	99.687	99.990	0.2	9.3	29		
Benzo(k)fluoranthene	0.49	59	360	99.159	99.950	0.1	12.1	73.4	100.000	99.455
Benzo(e)pyrene	8.71	285	740	99.774	99.994	0.4	13.1	34		
Benzo(a)pyrene	0.86	147	306	98.985	99.979	0.1	17.1	35.6		
Perylene	1.72	40	77	99.844	99.997	0.2	4.7	9		
Indeno(1,2,3cd)pyrene	0.74	14.5	54.6	99.868	99.995		14.5	54.6		
Dibenz(a,h)anthracene	1.35	1.3	6.9	100.000	100.000		1.3	6.9		
Benzo(ghi)perylene	2.94	19.4	76	99.959	99.998		19.4	76	100.000	99.990
Total target PAHs	140	1,026.2	2,163	99.950	99.999	416.4	410.3	384	100.000	99.990
Overall	8,607	4,671.2	2,239	100.000	100.000	27,510.4	22,360	619	99.993	99.993

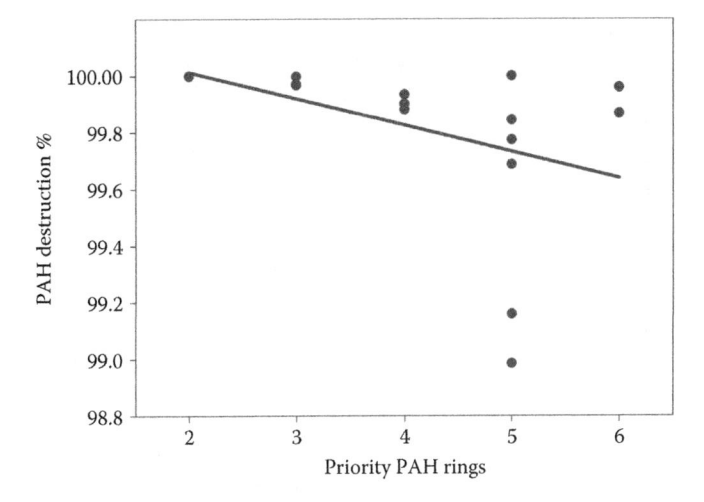

FIGURE 6.5 A plot of the PAH destruction efficiency for crude oil versus the number of rings for priority PAHs. This shows the expected decline in destruction with an increasing number of PAH rings.

An examination of the trends noted for diesel above are also examined for the crude oil burn shown in Table 6.4 (Alberta Medium crude). Figure 6.5 shows the comparison of the destruction of PAHs compared to the number of rings for the priority PAHs. This again shows the same trend as that for diesel burns, namely, that the more rings, the less efficient the PAH destruction is for the burn. It should be noted, however, that this effect is not very large and that the majority of the PAHs are still destroyed. The other effect of the change in burn efficiency with alkylation number is examined for a crude oil burn in Figure 6.6. This figure shows that the effect of

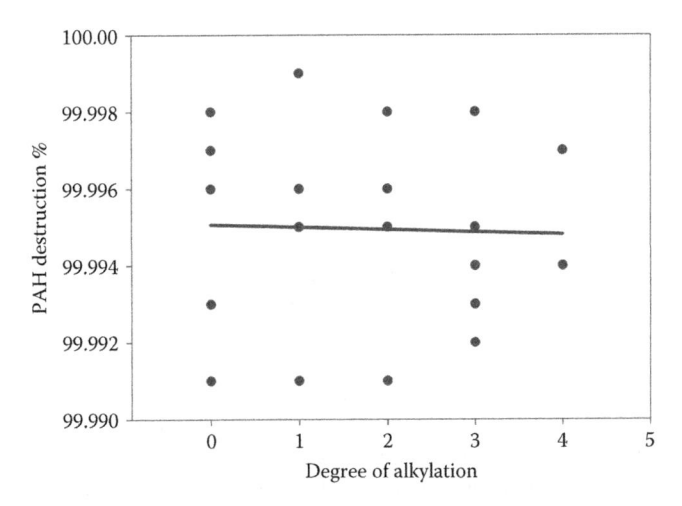

FIGURE 6.6 A plot of the PAH destruction efficiency for crude oil versus the degree of alkylation of the PAHs. This shows that destruction is about the same with the degree of alkylation in crude oil.

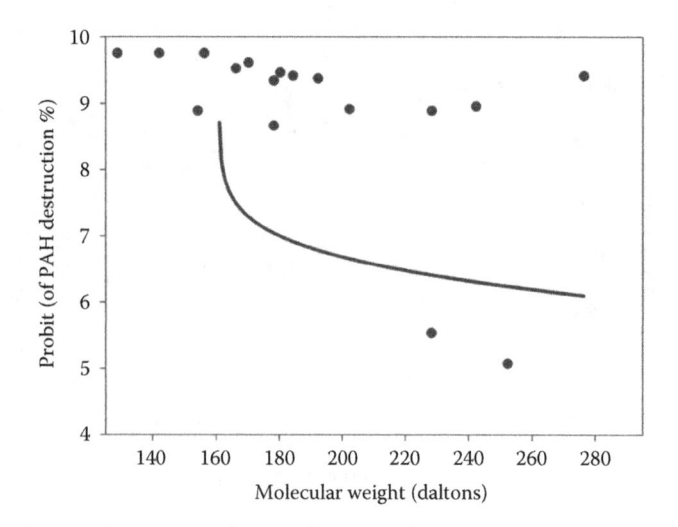

FIGURE 6.7 A probit plot of PAH destruction in diesel burns. This shows that there is a small fall-off with increasing molecular weight of the PAHs.

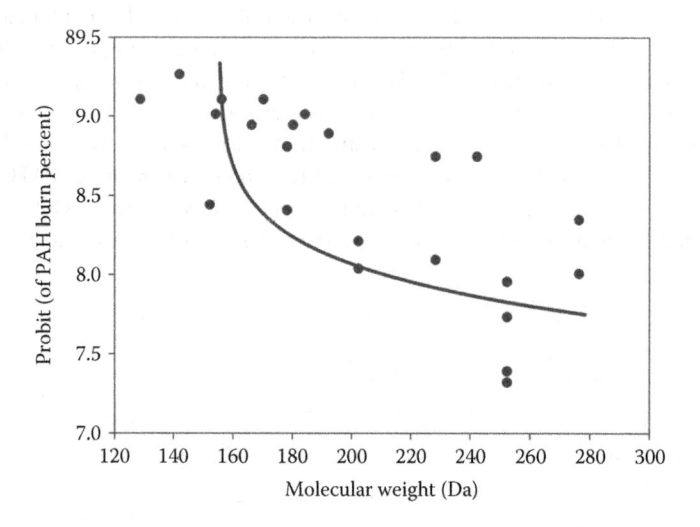

FIGURE 6.8 A probit plot of PAH destruction in crude oil burns. This shows that there is a fall-off with increasing molecular weight of the PAHs.

alkylation is not very prevalent or is absent for crude oil burns; that is, PAHs, whether alkylated or not and independent of degree of alkylation, burn about the same.

A further look at the destruction percentages compared to molecular weight was performed on the dataset shown in Table 6.4. Figures 6.7 and 6.8 show the comparison of the molecular weights to the PAH destruction for diesel fuel and for crude oil, in this case, Alberta medium sweet crude oil. Both figures show a slight decrease in destruction efficiency with increasing molecular weight. These graphs are shown as the probit of the destruction percentage, which is the probability as a normal distribution that the PAH destruction percentage would fall in a certain category. If this is shown as a linear function, the effect is only slight.

Another effect examined in Table 6.4 was that of the PAHs in the soot. It is noted that the destruction of the PAHs in the soot is usually 100%, indicating that this is not a source of PAH dissemination. This also indicates that PAHs on soot may be ignored in terms of the overall mass balance of PAHs in a burn.

Table 6.4 also includes data from a diesel fuel burn, different from the one analysed in Table 6.3. This table shows similar trends as does the data in Table 6.3; however, the trends as noted above are somewhat different. The alkylated PAHs with a greater degree of alkylation burn with about the same efficiency as those with a lower degree of alkylation.

A look at the vapour component shows that during the two NOBE burns, the concentrations of the only measurable PAH, naphthalene, was 0 to 7.3 $\mu g/m^3$ out of 6,000 to 24,000 total $\mu g/m^3$ of VOCs. The background was 0 to 6 $\mu g/m^3$; therefore, the PAHs in the vapour phase are negligible and sometimes below background levels.

Table 6.4 shows that the PAHs in the soot are 2,239 $\mu g/g$, or 2% of that in the residue if considering the alkylated compounds, more if considering the target PAHs. With approximately the same percentages ($<0.1\%$) of residue or soot, it can be concluded that the PAHs on the soot are not as significant fate for PAHs as in the residue. In the case of crude oil, this is more pronounced than for diesel fuel.

Test burns of diesel fuel were again conducted in 1997 at Mobil, Alabama (Fingas et al. 1998, 1999; Wang et al. 1998). The results of these are given in Table 6.5, which shows that similar results were obtained, namely that the destruction was nearly complete for all burns and all compounds. Further there does not appear to be differences in destruction percentages among the various alkylated PAHs nor among the priority PAHs. This appears to be contrary to the trends noted earlier. However, there are some higher concentrations in the residue of priority PAHs.

Similar burn tests were conducted in 1998 at Mobile, Alabama (Fingas et al. 2000, 2001). The results of these tests are given in Table 6.6. This table shows that similar results were obtained, namely that the destruction was nearly complete for all burns and all compounds. Again, there does not appear to be differences in destruction percentages among the various alkylated PAHs nor among the priority PAHs. However, there are some higher concentrations of priority PAHs in the residue.

Garrett et al. (2000) carried out burn tests on a small scale on Statfjord crude oil and found that the PAHs were reduced through burning. In addition to the 85% burned, an additional 40% of the PAHs were removed.

Burn tests on heavy and residual oils were conducted in 2003 at Ottawa, Ontario (Fingas et al. 2004, 2005). These burns were special in that they were carried out on heavy oils and bitumen, some mixtures and waste oils. The chosen heavy oils were poorly characterised before the burn. Bunker C was somewhat characterised, while the test oil was several years old and was used to test skimmers. Its origin was unknown; however, it was very viscous and low in PAHs. The Orimulsion and bitumen were of similar origin and were mixed samples from the supplier of the product in Venezuela. The results of these tests are given in Table 6.7, which shows that the PAH destruction was variable and low for the two Orimulsion burns and the test oil burns.

TABLE 6.5
Distribution of Alkylated PAHs and Other EPA Priority PAHs in Diesel and Residue Samples (Mobile 1997 Experiment)

Sample Type PAHs	Diesel Fresh (µg/g Oil)	Residue Boom 1 (µg/g Oil)	Residue Boom 2 (µg/g Oil)	Residue Boom 3 (µg/g Oil)	Residue Boom 4 (µg/g Oil)	PAH Destruction % Burn 1	PAH Destruction % Burn 2	PAH Destruction % Burn 3	PAH Destruction % Burn 4
Naphthalene									
C0-N	288	13	63	86	68	99.785	99.954	99.966	99.958
C1-N	1,981	57	380	408	556	99.65	99.948	99.951	99.964
C2-N	4,886	469	1,534	1,898	2,492	99.896	99.968	99.974	99.98
C3-N	4,669	1,010	1,933	2,688	3,431	99.954	99.976	99.983	99.986
C4-N	1,995	732	1,082	1,415	1,916	99.973	99.982	99.986	99.99
Sum	13,819	2,281	4,992	6,494	8,462	99.939	99.972	99.979	99.984
Phenanthrene									
C0-P	338	285	289	405	447	99.988	99.988	99.992	99.992
C1-P	1,193	1,331	1,218	1,618	1,858	99.991	99.99	99.993	99.994
C2-P	1,209	1,912	1,721	2,059	2,200	99.994	99.993	99.994	99.995
C3-P	709	1,489	1,217	1,360	1,470	99.995	99.994	99.995	99.995
C4-P	252	683	510	492	561	99.996	99.995	99.995	99.996
Sum	3,701	5,700	4,954	5,934	6,536	99.994	99.993	99.994	99.994
Dibenzothiophene									
C0-D	687	539	602	818	809	99.987	99.989	99.992	99.992
C1-D	1,667	1,963	1,939	2,602	2,554	99.992	99.991	99.994	99.993
C2-D	2,291	3,536	3,179	3,883	3,989	99.994	99.993	99.994	99.994
C3-D	1,069	2,219	1,753	2,017	2,037	99.995	99.994	99.995	99.995
Sum	5,714	8,256	7,472	9,321	9,388	99.993	99.992	99.994	99.994

(*Continued*)

TABLE 6.5 (*Continued*)

Distribution of Alkylated PAHs and Other EPA Priority PAHs in Diesel and Residue Samples (Mobile 1997 Experiment)

Sample Type	Diesel	Residue	Residue	Residue	Residue	PAH Destruction	PAH Destruction	PAH Destruction	PAH Destruction
	Fresh	Boom 1	Boom 2	Boom 3	Boom 4				
PAHs	(µg/g Oil)	(µg/g Oil)	(µg/g Oil)	(µg/g Oil)	(µg/g Oil)	% Burn 1	% Burn 2	% Burn 3	% Burn 4
Fluorene									
C0-F	221	80	125	168	187	99.972	99.982	99.987	99.988
C1-F	701	414	516	710	756	99.983	99.986	99.99	99.991
C2-F	1,152	1,186	1,102	1,539	1,756	99.99	99.99	99.993	99.993
C3-F	944	1,373	1,148	1,487	1,630	99.993	99.992	99.994	99.994
Sum	3,018	3,054	2,891	3,904	4,330	99.99	99.99	99.992	99.993
Chrysene									
C0-C	17	89	63	46	54	99.998	99.997	99.996	99.997
C1-C	24	110	80	49	59	99.998	99.997	99.995	99.996
C2-C	14	102	67	35	45	99.999	99.998	99.996	99.997
C3-C	6	48	32	15	14	99.999	99.998	99.996	99.996
Sum	61	349	242	145	173	99.998	99.997	99.996	99.996
Total	**26,313**	**19,639**	**20,551**	**25,799**	**28,890**	**99.987**	**99.987**	**99.99**	**99.991**
Other PAHs									
Biphenyl	285.4	15.1	100.9	91.7	92.3	99.811	99.972	99.969	99.969
Acenaphthalene	17.1	30.0	62.8	76.9	59.4	99.994	99.997	99.998	99.997
Acenaphthene	64.3	8.9	28.2	32.0	42.1	99.928	99.977	99.98	99.985
Anthracene	4.7	17.2	27.9	29.9	27.3	99.997	99.998	99.998	99.998
Fluoranthene	1.8	30.0	34.7	42.7	46.1	99.999	99.999	100	100

(*Continued*)

TABLE 6.5 (*Continued*)

Distribution of Alkylated PAHs and Other EPA Priority PAHs in Diesel and Residue Samples (Mobile 1997 Experiment)

Sample Type PAHs	Diesel Fresh (µg/g Oil)	Residue Boom 1 (µg/g Oil)	Residue Boom 2 (µg/g Oil)	Residue Boom 3 (µg/g Oil)	Residue Boom 4 (µg/g Oil)	PAH Destruction % Burn 1	PAH Destruction % Burn 2	PAH Destruction % Burn 3	PAH Destruction % Burn 4
Pyrene	22.4	64.5	73.5	82.5	99.6	99.997	99.997	99.997	99.998
Benz(a) anthracene	2.4	27.4	17.7	19.8	21.6	99.999	99.999	99.999	99.999
Benzo(b) fluoranthene	0.4	16.0	9.5	14.3	16.7	100	100	100	100
Benzo(k) fluoranthene	0.1	19.8	10.6	15.4	17.4	100	100	100	100
Benzo(e)pyrene	0.7	16.6	10.2	8.4	10.4	100	99.999	99.999	99.999
Benzo(a)pyrene	0.2	29.8	15.5	16.2	18.0	100	100	100	100
Perylene	0.2	9.2	4.7	3.7	4.2	100	100	100	100
Indeno(1,2,3cd) pyrene	0.0	1.6	7.0	7.5	7.5	100	100	100	100
Dibenz(a,h) anthracene	0.0	1.8	0.7	0.5	1.6	100	100	100	100
Benzo(ghi) perylene	0.0	22.0	10.4	11.6	12.0	100	100	100	100
Total	**400**	**310**	**414**	**453**	**476**	**99.987**	**99.99**	**99.991**	**99.992**
Total PAHs	**26,713**	**19,949**	**20,966**	**26,252**	**29,366**	**99.987**	**99.987**	**99.99**	**99.991**

TABLE 6.6

Distribution of Alkylated PAHs and Other EPA Priority PAHs in Diesel and Residue Samples (Mobile 1998 Experiment)

| Sample Type | Diesel | Residue | Residue | Residue | Residue | PAH | PAH | PAH | PAH |
| | Fresh | Boom 1 | Boom 2 | Boom 3 | Boom 4 | Destruction | Destruction | Destruction | Destruction |
PAHs	(µg/g Oil)	(µg/g Oil)	(µg/g Oil)	(µg/g Oil)	(µg/g Oil)	% Burn 1	% Burn 2	% Burn 3	% Burn 4
Naphthalene									
C0-N	288	13	63	86	68	99.785	99.954	99.966	99.958
C1-N	1,981	57	380	408	556	99.65	99.948	99.951	99.964
C2-N	4,886	469	1,534	1,898	2,492	99.896	99.968	99.974	99.98
C3-N	4,669	1,010	1,933	2,688	3,431	99.954	99.976	99.983	99.986
C4-N	1,995	732	1,082	1,415	1,916	99.973	99.982	99.986	99.99
Sum	13,819	2,281	4,992	6,494	8,462	99.939	99.972	99.979	99.984
Phenanthrene									
C0-P	338	285	289	405	447	99.988	99.988	99.992	99.992
C1-P	1,193	1,331	1,218	1,618	1,858	99.991	99.99	99.993	99.994
C2-P	1,209	1,912	1,721	2,059	2,200	99.994	99.993	99.994	99.995
C3-P	709	1,489	1,217	1,360	1,470	99.995	99.994	99.995	99.995
C4-P	252	683	510	492	561	99.996	99.995	99.995	99.996
Sum	3,701	5,700	4,954	5,934	6,536	99.994	99.993	99.994	99.994
Dibenzothiophene									
C0-D	687	539	602	818	809	99.987	99.989	99.992	99.992
C1-D	1,667	1,963	1,939	2,602	2,554	99.992	99.991	99.994	99.993
C2-D	2,291	3,536	3,179	3,883	3,989	99.994	99.993	99.994	99.994
C3-D	1,069	2,219	1,753	2,017	2,037	99.995	99.994	99.995	99.995
Sum	5,714	8,256	7,472	9,321	9,388	99.993	99.992	99.994	99.994

(Continued)

TABLE 6.6 (*Continued*)
Distribution of Alkylated PAHs and Other EPA Priority PAHs in Diesel and Residue Samples (Mobile 1998 Experiment)

Sample Type	Diesel Fresh	Residue Boom 1	Residue Boom 2	Residue Boom 3	Residue Boom 4	PAH Destruction	PAH Destruction	PAH Destruction	PAH Destruction
PAHs	(µg/g Oil)	(µg/g Oil)	(µg/g Oil)	(µg/g Oil)	(µg/g Oil)	% Burn 1	% Burn 2	% Burn 3	% Burn 4
Fluorene									
C0-F	221	80	125	168	187	99.972	99.982	99.987	99.988
C1-F	701	414	516	710	756	99.983	99.986	99.99	99.991
C2-F	1,152	1,186	1,102	1,539	1,756	99.99	99.99	99.993	99.993
C3-F	944	1,373	1,148	1,487	1,630	99.993	99.992	99.994	99.994
Sum	3,018	3,054	2,891	3,904	4,330	99.99	99.99	99.992	99.993
Chrysene									
C0-C	17	89	63	46	54	99.998	99.997	99.996	99.997
C1-C	24	110	80	49	59	99.998	99.997	99.995	99.996
C2-C	14	102	67	35	45	99.999	99.998	99.996	99.997
C3-C	6	48	32	15	14	99.999	99.998	99.996	99.996
Sum	61	349	242	145	173	99.998	99.997	99.996	99.996
Total	**26,313**	**19,639**	**20,551**	**25,799**	**28,890**	**99.987**	**99.987**	**99.99**	**99.991**
Other PAHs									
Biphenyl	285.4	15.1	100.9	91.7	92.3	99.811	99.972	99.969	99.969
Acenaphthalene	17.1	30.0	62.8	76.9	59.4	99.994	99.997	99.998	99.997
Acenaphthene	64.3	8.9	28.2	32.0	42.1	99.928	99.977	99.98	99.985
Anthracene	4.7	17.2	27.9	29.9	27.3	99.997	99.998	99.998	99.998
Fluoranthene	1.8	30.0	34.7	42.7	46.1	99.999	99.999	100	100
Pyrene	22.4	64.5	73.5	82.5	99.6	99.997	99.997	99.997	99.998

(*Continued*)

TABLE 6.6 (*Continued*)

Distribution of Alkylated PAHs and Other EPA Priority PAHs in Diesel and Residue Samples (Mobile 1998 Experiment)

Sample Type	Diesel Fresh	Residue Boom 1	Residue Boom 2	Residue Boom 3	Residue Boom 4	PAH Destruction	PAH Destruction	PAH Destruction	PAH Destruction
PAHs	(µg/g Oil)	(µg/g Oil)	(µg/g Oil)	(µg/g Oil)	(µg/g Oil)	% Burn 1	% Burn 2	% Burn 3	% Burn 4
Benz(a)anthracene	2.4	27.4	17.7	19.8	21.6	99.999	99.999	99.999	99.999
Benzo(b) fluoranthene	0.4	16.0	9.5	14.3	16.7	100	100	100	100
Benzo(k) fluoranthene	0.1	19.8	10.6	15.4	17.4	100	100	100	100
Benzo(e)pyrene	0.7	16.6	10.2	8.4	10.4	100	99.999	99.999	99.999
Benzo(a)pyrene	0.2	29.8	15.5	16.2	18.0	100	100	100	100
Perylene	0.2	9.2	4.7	3.7	4.2	100	100	100	100
Indeno(1,2,3cd) pyrene	0.0	1.6	7.0	7.5	7.5	100	100	100	100
Dibenz(a,h) anthracene	0.0	1.8	0.7	0.5	1.6	100	100	100	100
Benzo(ghi) perylene	0.0	22.0	10.4	11.6	12.0	100	100	100	100
Total	**400**	**310**	**414**	**453**	**476**	**99.987**	**99.99**	**99.991**	**99.992**
Total PAHs	**26,713**	**19,949**	**20,966**	**26,252**	**29,366**	**99.987**	**99.987**	**99.99**	**99.991**

TABLE 6.7
Summary of PAHs in Actual Deepwater Horizon Burns

	Concentration in µg/m³																					
	Starting Oil Samples					Residues Scrapped from Boom						Destruction Percentage							
	Ave. #1	Ave. #2	Ave. #3	Ave. #4	Ave. Oil	1	2	3	4	5	6	1	2	3	4	5	6	Ave.
C0-Naphthalene	25	31	11	2.34	31	2.99	3.4	4.11	3	8.54	16.2	99.76	99.78	99.25	97.44	99.45	98.96	99.11
C1-N	491	347	245	168	491	16.7	288	3.67	15.1	53.7	16.1	99.93	98.34	99.97	99.82	99.78	99.93	99.63
C2-N	1129	856	719	737	1129	120	1046	13.9	122	121.0	32.7	99.79	97.56	99.96	99.67	99.79	99.94	99.45
C3-N	794	658	591	631	794	194	812	18.4	155	104.0	35.1	99.51	97.53	99.94	99.51	99.74	99.91	99.36
C4-N	301	253	230	244	301	94.6	308	9.78	71.1	46.8	16.7	99.37	97.57	99.92	99.42	99.69	99.89	99.31
C0-Fluorene	111	94	83.4	91.8	111	23.5	117	4.05	22.1	17.7	8.0	99.58	97.51	99.9	99.52	99.68	99.86	99.34
C1-F	313	283	269	274	313	110	325	15.8	84.9	62.6	25.5	99.3	97.7	99.88	99.38	99.6	99.84	99.28
C2-F	258	234	223	231	258	109	270	18.4	87.3	66.8	27.7	99.16	97.69	99.84	99.24	99.48	99.79	99.2
C3-F	149	137	134	139	149	71.3	154	14.2	56.9	47.1	19.4	99.04	97.75	99.79	99.18	99.37	99.74	99.14
C0-Dibenzothiophene	37.3	34.3	32	34.3	37	13.6	40	2.34	11.5	8.8	4.1	99.27	97.67	99.85	99.33	99.52	99.78	99.24
C1-D	112	108	103	102	112	50.8	117	9.76	42.4	34.2	14.3	99.09	97.83	99.81	99.17	99.39	99.75	99.17
C2-D	120	113	113	111	120	59	124	14	49.8	41.7	19.1	99.01	97.8	99.75	99.1	99.31	99.68	99.11
C3-D	65.2	62.1	60	60.8	65	35.8	66.5	9.87	32	26.7	13.0	98.9	97.86	99.67	98.95	99.18	99.6	99.03
C0-Phenanthrene	252	232	222	227	252	96.6	239	22.7	90.5	68.3	38.4	99.23	97.94	99.8	99.2	99.46	99.7	99.22
C1-P	514	496	479	482	514	245	536	54.1	209	170.0	76.4	99.05	97.84	99.77	99.13	99.34	99.7	99.14
C2-P	389	373	356	355	389	213	384	50.9	172	148.0	69.1	98.9	97.94	99.71	99.03	99.24	99.65	99.08
C3-P	174	170	156	158	174	97.3	179	26.9	79.9	75.6	36.6	98.88	97.89	99.66	98.99	99.13	99.58	99.02
C4-P	66.7	59	53.9	53	67	36.4	65	10.5	29.5	28.5	14.9	98.91	97.8	99.61	98.89	99.15	99.56	98.98

(*Continued*)

TABLE 6.7 (*Continued*)

Summary of PAHs in Actual Deepwater Horizon Burns

	Concentration in µg/m³																	
	Starting Oil Samples					Residues Scrapped from Boom						Destruction Percentage						
	Ave. #1	Ave. #2	Ave. #3	Ave. #4	Ave. Oil	1	2	3	4	5	6	1	2	3	4	5	6	Ave.
Anthracene	4.17	3.58	2.5	2.49	4	2.36	4.34	2.48	2.16	2.6	4.4	98.87	97.57	98.01	98.26	98.71	97.83	98.21
Fluoranthene	8.34	8.61	7.89	8.42	9	8.8	7.79	12.8	8.83	8.5	8.5	97.89	98.19	96.75	97.9	98.12	98.12	97.83
C0-Pyrene	19.3	18.6	17.2	17.2	19	17.8	18.2	17.7	19.9	20.4	20.0	98.15	98.04	97.94	97.69	97.85	97.9	97.93
C1-P	98	91.9	82	84.2	98	61.8	100	29.5	58.8	62.5	33.7	98.74	97.82	99.28	98.6	98.72	99.31	98.75
C2-P	108	97.6	90.2	94.4	108	70.8	113	25.3	61.5	70.2	33.9	98.69	97.68	99.44	98.7	98.7	99.37	98.76
C3-P	93	83.1	76.8	79.4	93	62.9	94.8	24.5	53.1	64.9	33.2	98.65	97.72	99.36	98.66	98.6	99.29	98.71
C4-P	50.2	45.4	43.8	44.6	50	39	55.4	16.8	36.8	38.0	22.5	98.45	97.56	99.23	98.35	98.48	99.1	98.53
C0-Naphthobenzo-thiophene	14.5	15.3	13.6	14.8	15	11.5	15	5.3	11.4	11.3	6.2	98.41	98.03	99.22	98.45	98.49	99.17	98.63
C1-NBT	53.1	54.7	51.1	52.4	55	42.1	54.6	19.6	41.2	42.6	24.9	98.41	98	99.23	98.43	98.45	99.1	98.6
C2-NBT	42.9	44.1	41.1	42.3	44	36	45	17.4	31.9	36.4	25.2	98.32	97.96	99.15	98.49	98.35	98.86	98.52
C3-NBT	22	21.7	21.5	22.7	23	20.4	23.7	9.28	17.8	19.8	15.7	98.14	97.82	99.14	98.43	98.28	98.64	98.41
Benz(a)anthracene	6.64	5.92	5.3	5.8	7	6.83	7.27	6.31	6.93	7.6	8.1	97.94	97.54	97.62	97.61	97.84	97.68	97.71
C0-C	52.5	51.5	50.3	54.2	54	45	54.4	22.1	43.1	46.2	28.9	98.29	97.89	99.12	98.41	98.29	98.93	98.49
C1-C	109	106	98.5	103	109	86	115	40.8	81.4	87.1	54.5	98.42	97.82	99.17	98.41	98.4	99	98.54
C2-C	94	88.9	84.5	87.9	94	79.4	98.4	35	68.4	79.9	57.0	98.31	97.79	99.17	98.44	98.3	98.79	98.47
C3-C	46.3	45.3	40.4	42.1	46	41.4	49.5	16.8	31.3	39.8	32.5	98.21	97.82	99.17	98.51	98.27	98.59	98.43
C4-C	16.2	14.5	13.1	14.3	16	14.5	14.4	6.71	9.56	20.6	17.2	98.2	98.01	98.97	98.66	97.43	97.85	98.19
Benzo(b)fluoranthene	5.16	6.12	5.67	5.52	6	6.85	5.91	5.98	8.68	8.4	8.9	97.35	98.07	97.89	96.86	97.21	97.03	97.4

(*Continued*)

TABLE 6.7 (Continued)

Summary of PAHs in Actual Deepwater Horizon Burns

| | Starting Oil Samples | | | | | Concentration in μg/m³ | | | | | | | | | | | | | |
| | | | | | | Residues Scrapped from Boom | | | | | | Destruction Percentage | | | | | | |
	Ave. #1	Ave. #2	Ave. #3	Ave. #4	Ave. Oil	1	2	3	4	5	6	1	2	3	4	5	6	Ave.
Benzo(k)fluoranthene	0.89	0.8	0.8	1.07	1	1.36	1.04	3.92	2.72	1.9	3.5	96.95	97.39	90.2	94.94	96.28	92.92	94.78
Benzo(e)pyrene	11.1	11.2	10.7	10.7	11	11.2	11.6	8.17	11.6	12.6	10.9	97.97	97.92	98.47	97.82	97.71	98.02	97.99
Benzo(a)pyrene	2.22	1.98	1.8	1.71	2	3.46	2.17	6.5	4.7	5.2	7.4	96.88	97.8	92.76	94.5	94.83	92.59	94.89
Perylene	0.95	0.74	0.68	0.64	1	0.65	0.83	0.93	1.26	1.1	1.1	98.64	97.78	97.27	96.06	97.74	97.88	97.56
Indeno(1,2,3cd)pyrene	0	0	0	0	0	2.01	0	4.32	3.25	2.3	6.8							
Dibenz(a,h)anthracene	2.91	2.9	2.73	3.21	3	3.6	2.93	2.34	2.85	3.7	3.8	97.53	97.98	98.28	98.22	97.55	97.49	97.84
Benzo(ghi)perylene	2.5	2.63	2.46	2.74	3	5.93	2.36	6.95	5.6	5.1	10.0	95.26	98.2	94.35	95.91	96.59	93.36	95.61
Total PAHs	6166	5363	4841	4893	6166	2271	5972	651	1959	1830	962	99.3	97.8	99.7	99.2	99.4	99.7	99.2

Source: Shigenaka, G. et al., Physical and chemical characteristics of in-situ burn residue and other environmental oil samples collected during the Deepwater Horizon spill response, National Oceanic and Atmospheric Administration, Emergency Response Division Seattle, WA, 2015.

Ave. refers to the average of multiple determinations for the same burn.

Efficiency was low for the test oils, as low as 2%. The bitumen burns also had a lower burn efficiency. The PAH destruction percentages are higher compared to what one would expect, but that is because the boiling points of the PAHs are lower than the rest of the components in the oil and would be burned early in the combustion process. This series of burn experiments can be used to compare burn efficiencies with PAH removal percentage. This is shown in Table 6.9. This figure shows there is somewhat of a relationship, namely that the higher the burn efficiency, the higher the PAH removal rate. PAH on soot data were also available; it was very low and would not account for any amount of PAH in terms of mass balance.

Lin et al. (2005) performed a series of test burns on 30 cm pots containing marsh plants. Both diesel fuel and Louisiana crude oil were used. PAH removals of 99.4 and 98.9 for the diesel and crude, respectively, were observed. It was also noted that there were increases in the priority PAHs as had been observed by Wang et al. (1999a), especially for the diesel fuel. However, the PAH destruction overall was quantitative.

During the Deepwater Horizon (DWH) spill in the United States, 411 successful burns of weathered and sometimes emulsified oil were carried out. Shigenaka et al. (2014, 2015, 2017) carried out a series of test burns as well as took samples of oil burn residue at the site of the DWH burns. The PAHs in these and the starting oil were determined, as shown in Table 6.8. The table shows that the destruction percentage of the burns are typically well over 99%, with some of the priority PAHs being less – typically around 98%. This is consistent with previous data. It should be noted that some of the samples were scraped from the fire-resistant boom and thus they may have a slightly different composition from the residue. Shigenaka et al. (2014) indicated that these samples may not be burned as much as the residue; however, from this analysis it appears that they are.

Fritt-Rasmussen et al. (2013) conducted several small-scale burns on Troll B crude oil and noted that there was an increase in some of the 16 priority PAH compounds. A mass balance was not calculated. Fritt-Rasmussen et al. (2015) carried out a review of burn residue noting that there was an increase in some of the 16 priority PAH compounds

Shigenaka et al. (2014) also carried out a series of small test burns to assess the chemical changes that burns undergo. These are summarised in Table 6.9, which includes a very small burn of Macondo oil, pan burns of South Louisiana crude and burns of emulsified Macondo oil. The burn efficiencies were not measured, but they can be estimated from comments about the burnability. The Macondo oil burned with a high destruction percentage, as did the Louisiana crude oil. With the estimation of 80% burn efficiency for the Macondo emulsion, PAHs were still largely destroyed.

Gullett et al. (2017) studied the particulate matter emanating from the DWH burns. They sampled PAHs on particulate matter captured by volumetric samplers as well as on the sail of an aerostat flown in the smoke plumes. The mean sum of PAHs detected in the sail PAH extract and PM 2.5 Filters were 80.1 and 68.2 $\mu g/g$ of the particulate matter, respectively, accounting for less than 1% of the total particulate matter mass. At the PM collection rate from the fires of 0.088 g particulate matter/g C12, this resulted in an emission factor of 4.5 mg PAH/kg of oil burned. These results are much lower than would be expected from the PAH content of the oil, indicating an overall destruction of PAHs.

TABLE 6.8

Summary of PAHs in Actual Deepwater Horizon Burns

| | Starting Oil Samples | | | | | Concentration in µg/m³ | | | | | | Destruction Percentage | | | | | | |
| | | | | | | Residues Scrapped from Boom | | | | | | | | | | | | |
	Ave. #1	Ave. #2	Ave. #3	Ave. #4	Ave. Oil	1	2	3	4	5	6	1	2	3	4	5	6	Ave.
C0-Naphthalene	25	31	11	2.34	31	2.99	3.4	4.11	3	8.54	16.2	99.76	99.78	99.25	97.44	99.45	98.96	99.11
C1-N	491	347	245	168	491	16.7	288	3.67	15.1	53.7	16.1	99.93	98.34	99.97	99.82	99.78	99.93	99.63
C2-N	1129	856	719	737	1129	120	1046	13.9	122	121.0	32.7	99.79	97.56	99.96	99.67	99.79	99.94	99.45
C3-N	794	658	591	631	794	194	812	18.4	155	104.0	35.1	99.51	97.53	99.94	99.51	99.74	99.91	99.36
C4-N	301	253	230	244	301	94.6	308	9.78	71.1	46.8	16.7	99.37	97.57	99.92	99.42	99.69	99.89	99.31
C0-Fluorene	111	94	83.4	91.8	111	23.5	117	4.05	22.1	17.7	8.0	99.58	97.51	99.9	99.52	99.68	99.86	99.34
C1-F	313	283	269	274	313	110	325	15.8	84.9	62.6	25.5	99.3	97.7	99.88	99.38	99.6	99.84	99.28
C2-F	258	234	223	231	258	109	270	18.4	87.3	66.8	27.7	99.16	97.69	99.84	99.24	99.48	99.79	99.2
C3-F	149	137	134	139	149	71.3	154	14.2	56.9	47.1	19.4	99.04	97.75	99.79	99.18	99.37	99.74	99.14
C0-Dibenzothiophene	37.3	34.3	32	34.3	37	13.6	40	2.34	11.5	8.8	4.1	99.27	97.67	99.85	99.33	99.52	99.78	99.24
C1-D	112	108	103	102	112	50.8	117	9.76	42.4	34.2	14.3	99.09	97.83	99.81	99.17	99.39	99.75	99.17
C2-D	120	113	113	111	120	59	124	14	49.8	41.7	19.1	99.01	97.8	99.75	99.1	99.31	99.68	99.11
C3-D	65.2	62.1	60	60.8	65	35.8	66.5	9.87	32	26.7	13.0	98.9	97.86	99.67	98.95	99.18	99.6	99.03
C0-Phenanthrene	252	232	222	227	252	96.6	239	22.7	90.5	68.3	38.4	99.23	97.94	99.8	99.2	99.46	99.7	99.22
C1-P	514	496	479	482	514	245	536	54.1	209	170.0	76.4	99.05	97.84	99.77	99.13	99.34	99.7	99.14
C2-P	389	373	356	355	389	213	384	50.9	172	148.0	69.1	98.9	97.94	99.71	99.03	99.24	99.65	99.08
C3-P	174	170	156	158	174	97.3	179	26.9	79.9	75.6	36.6	98.88	97.89	99.66	98.99	99.13	99.58	99.02
C4-P	66.7	59	53.9	53	67	36.4	65	10.5	29.5	28.5	14.9	98.91	97.8	99.61	98.89	99.15	99.56	98.98
Anthracene	4.17	3.58	2.5	2.49	4	2.36	4.34	2.48	2.16	2.6	4.4	98.87	97.57	98.01	98.26	98.71	97.83	98.21
Fluoranthene	8.34	8.61	7.89	8.42	9	8.8	7.79	12.8	8.83	8.5	8.5	97.89	98.19	96.75	97.9	98.12	98.12	97.83
C0-Pyrene	19.3	18.6	17.2	17.2	19	17.8	18.2	17.7	19.9	20.4	20.0	98.15	98.04	97.94	97.69	97.85	97.9	97.93

(Continued)

TABLE 6.8 (*Continued*)
Summary of PAHs in Actual Deepwater Horizon Burns

	Starting Oil Samples					Residues Scrapped from Boom						Destruction Percentage						
												Concentration in µg/m³						
	Ave. #1	Ave. #2	Ave. #3	Ave. #4	Ave. Oil	1	2	3	4	5	6	1	2	3	4	5	6	Ave.
C1-P	98	91.9	82	84.2	98	61.8	100	29.5	58.8	62.5	33.7	98.74	97.82	99.28	98.6	98.72	99.31	98.75
C2-P	108	97.6	90.2	94.4	108	70.8	113	25.3	61.5	70.2	33.9	98.69	97.68	99.44	98.7	98.7	99.37	98.76
C3-P	93	83.1	76.8	79.4	93	62.9	94.8	24.5	53.1	64.9	33.2	98.65	97.72	99.36	98.66	98.6	99.29	98.71
C4-P	50.2	45.4	43.8	44.6	50	39	55.4	16.8	36.8	38.0	22.5	98.45	97.56	99.23	98.35	98.48	99.1	98.53
C0-Naphthoben-zothiophene	14.5	15.3	13.6	14.8	15	11.5	15	5.3	11.4	11.3	6.2	98.41	98.03	99.22	98.45	98.49	99.17	98.63
C1-NBT	53.1	54.7	51.1	52.4	55	42.1	54.6	19.6	41.2	42.6	24.9	98.41	98	99.23	98.43	98.45	99.1	98.6
C2-NBT	42.9	44.1	41.1	42.3	44	36	45	17.4	31.9	36.4	25.2	98.32	97.96	99.15	98.49	98.35	98.86	98.52
C3-NBT	22	21.7	21.5	22.7	23	20.4	23.7	9.28	17.8	19.8	15.7	98.14	97.82	99.14	98.43	98.28	98.64	98.41
Benz(a)anthracene	6.64	5.92	5.3	5.8	7	6.83	7.27	6.31	6.93	7.6	8.1	97.94	97.54	97.62	97.61	97.84	97.68	97.71
C0-C	52.5	51.5	50.3	54.2	54	45	54.4	22.1	43.1	46.2	28.9	98.29	97.89	99.12	98.41	98.29	98.93	98.49
C1-C	109	106	98.5	103	109	86	115	40.8	81.4	87.1	54.5	98.42	97.82	99.17	98.41	98.4	99	98.54
C2-C	94	88.9	84.5	87.9	94	79.4	98.4	35	68.4	79.9	57.0	98.31	97.79	99.17	98.44	98.3	98.79	98.47
C3-C	46.3	45.3	40.4	42.1	46	41.4	49.5	16.8	31.3	39.8	32.5	98.21	97.82	99.17	98.51	98.27	98.59	98.43
C4-C	16.2	14.5	13.1	14.3	16	14.5	14.4	6.71	9.56	20.6	17.2	98.2	98.01	98.97	98.66	97.43	97.85	98.19
Benzo(b)fluoranthene	5.16	6.12	5.67	5.52	6	6.85	5.91	5.98	8.68	8.4	8.9	97.35	98.07	97.89	96.86	97.21	97.03	97.4
Benzo(k)fluoranthene	0.89	0.8	0.8	1.07	1	1.36	1.04	3.92	2.72	1.9	3.5	96.95	97.39	90.2	94.94	96.28	92.92	94.78

(*Continued*)

TABLE 6.8 (*Continued*)

Summary of PAHs in Actual Deepwater Horizon Burns

	Concentration in µg/m³																	
	Starting Oil Samples					Residues Scrapped from Boom						Destruction Percentage						
	Ave. #1	Ave. #2	Ave. #3	Ave. #4	Ave. Oil	1	2	3	4	5	6	1	2	3	4	5	6	Ave.
Benzo(e)pyrene	11.1	11.2	10.7	10.7	11	11.2	11.6	8.17	11.6	12.6	10.9	97.97	97.92	98.47	97.82	97.71	98.02	97.99
Benzo(a)pyrene	2.22	1.98	1.8	1.71	2	3.46	2.17	6.5	4.7	5.2	7.4	96.88	97.8	92.76	94.5	94.83	92.59	94.89
Perylene	0.95	0.74	0.68	0.64	1	0.65	0.83	0.93	1.26	1.1	1.1	98.64	97.78	97.27	96.06	97.74	97.88	97.56
Indeno(1,2,3cd) pyrene	0	0	0	0	0	2.01	0	4.32	3.25	2.3	6.8							
Dibenz(a,h) anthracene	2.91	2.9	2.73	3.21	3	3.6	2.93	2.34	2.85	3.7	3.8	97.53	97.98	98.28	98.22	97.55	97.49	97.84
Benzo(ghi)perylene	2.5	2.63	2.46	2.74	3	5.93	2.36	6.95	5.6	5.1	10.0	95.26	98.2	94.35	95.91	96.59	93.36	95.61
Total PAHs	6166	5363	4841	4893	6166	2271	5972	651	1959	1830	962	99.3	97.8	99.7	99.2	99.4	99.7	99.2

Source: Shigenaka, G. et al., Physical and chemical characteristics of in-situ burn residue and other environmental oil samples collected during the Deepwater Horizon spill response, National Oceanic and Atmospheric Administration, Emergency Response Division Seattle, WA, 2015.

TABLE 6.9
Summary of PAHs in Laboratory Burns

Concentration in µg/m³

	PAH Concentrations in Starting/Residue Pairs										Destruction Percentage				
	Fresh	Residue	Weath.	Residue	Mac.	Residue	Emul.	Residue	Emul.	Residue	Fresh	Weath.	Mac.	Emul.	Emul.
C0-Naphthalene	534	18.8	5.68	33.7	840	6.55	5.68	33.73	0.114	2.34	99.93	88.13	99.98	88.12	58.95
C1-N	1278	74.93	1461	124.7	1700	9.163	1461	124.7	3.59	4.0	99.88	99.83	99.99	99.83	97.79
C2-N	1449	185.3	1699	257.7	1800	20	1699	257.7	105	94.1	99.74	99.7	99.98	99.7	98.21
C3-N	918	215	1072	250.7	1200	24.63	1072	250.7	231	216.7	99.53	99.53	99.96	99.53	98.12
C4-N	387	120	465	131	430	19.17	465	131	141	134.0	99.38	99.44	99.91	99.44	98.1
C0-Fluorene	114	38.87	135	45.95	160	3.43	135	45.95	27.7	25.4	99.32	99.32	99.96	99.32	98.17
C1-F	242	111	294	112.7	340	13.2	294	112.7	111	107.0	99.08	99.23	99.92	99.23	98.07
C2-F	287	169	325	153.7	410	26.07	325	153.7	150	147.3	98.82	99.05	99.87	99.05	98.04
C3-F	214	152.3	242	128.3	270	32.37	242	128.3	138	131.3	98.58	98.94	99.76	98.94	98.1
C0-Dibenzo-thiophene	25.8	14.43	31	14.15	67.6	2.3	31	14.15	16.9	16.7	98.88	99.09	99.93	99.09	98.02
C1-D	74.7	50.63	89.7	45.15	180	11.21	89.7	45.15	61.2	61.7	98.64	98.99	99.88	98.99	97.99
C2-D	98.5	77.1	115	64.15	250	25.23	115	64.15	90.5	92.9	98.44	98.88	99.8	98.88	97.95
C3-D	74.8	61.43	82.8	50.5	160	26.97	82.8	50.5	60.4	62.5	98.36	98.78	99.66	98.78	97.93
C0-Phenanthrene	218	149.3	259	144.7	400	23.53	259	144.7	115	115.3	98.63	98.88	99.88	98.88	97.99
C1-P	466	343.3	571	293.3	860	62.43	571	293.3	316	323.3	98.53	98.97	99.86	98.97	97.95
C2-P	490	403.7	584	334.3	800	97.23	584	334.3	326	328.3	98.35	98.86	99.76	98.86	97.99
C3-P	329	283	344	222.3	490	87.37	344	222.3	183	183.0	98.28	98.71	99.64	98.71	98

(*Continued*)

TABLE 6.9 (Continued)
Summary of PAHs in Laboratory Burns

| | Concentration in μg/m³ | | | | | | | | | | Destruction Percentage | | | | |
| | PAH Concentrations in Starting/Residue Pairs | | | | | | | | | | | | | | |
	Fresh	Residue	Weath.	Residue	Mac.	Residue	Emul.	Residue	Emul.	Residue	Fresh	Weath.	Mac.	Emul.	Emul.
C4-P	143	133.7	163	109.6	210	66.47	163	109.6	74.4	55.0	98.13	98.66	99.37	98.66	98.52
Anthracene	4.82	11.47	5.66	13.15	8.71	0	5.66	13.15	0.933	1.7	95.24	95.35	100	95.35	96.26
Fluoranthene	4.02	17.47	4.31	23.8	3.59	6.433	4.31	23.8	3.27	3.2	91.31	88.96	96.42	88.96	98.03
C0-Pyrene	10.5	28.4	12.2	29.45	8.57	10.92	12.2	29.45	8.13	10.1	94.59	95.17	97.45	95.17	97.52
C1-P	67.5	85.53	78	69.3	69.1	24.07	78	69.3	44.8	48.6	97.47	98.22	99.3	98.22	97.83
C2-P	84	106.9	97.8	86.5	87	32.8	97.8	86.5	57.5	58.7	97.45	98.23	99.25	98.23	97.96
C3-P	101	127.7	117	97.03	85.4	48.17	117	97.03	60.6	61.0	97.47	98.34	98.87	98.34	97.99
C4-P	65	82.03	73.8	67.75	56.1	42.37	73.8	67.75	45.1	47.6	97.48	98.16	98.49	98.16	97.89
C0-Naphtho-benzothiophene	6.28	6.53	7.41	5.927	17.9	5.12	7.41	5.927	6.73	7.1	97.92	98.4	99.43	98.4	97.89
C1-NBT	19.9	27.03	25.2	21.95	57.5	29.53	25.2	21.95	23.7	24.8	97.28	98.26	98.97	98.26	97.91
C2-NBT	22.9	30.2	27.6	25.7	54.2	42.5	27.6	25.7	23.9	26.8	97.36	98.14	98.43	98.14	97.76
C3-NBT	17.2	25.47	22	21.7	40.1	30.9	22	21.7	15.8	17.3	97.04	98.03	98.46	98.03	97.81
Benz(a)anthracene	4.91	10.57	5.45	10.3	4.56	11.1	5.45	10.3	1.85	2.8	95.7	96.22	95.13	96.22	97
C0-C	36.4	48.83	39.4	40.25	49.1	27.63	39.4	40.25	37.3	38.3	97.32	97.96	98.87	97.96	97.95
C1-C	84.3	108.7	98.7	84.5	110	63.9	98.7	84.5	65.2	67.5	97.42	98.29	98.84	98.29	97.93
C2-C	94.8	118.3	105	101	120	77.13	105	101	55.8	60.1	97.5	98.08	98.71	98.08	97.85

(Continued)

TABLE 6.9 (*Continued*)
Summary of PAHs in Laboratory Burns

	Concentration in $\mu g/m^3$														
	PAH Concentrations in Starting/Residue Pairs										Destruction Percentage				
	Fresh	Residue	Weath.	Residue	Mac.	Residue	Emul.	Residue	Emul.	Residue	Fresh	Weath.	Mac.	Emul.	Emul.
C3-C	64.5	87.33	77.1	72.3	67.2	55.33	77.1	72.3	34.1	36.5	97.29	98.13	98.35	98.13	97.86
C4-C	41.5	54.87	45.2	43.2	34.9	23.8	45.2	43.2	16.5	19.5	97.36	98.09	98.64	98.09	97.63
Benzo(b)-fluoranthene	1.79	5.57	1.77	5.153	3.31	8.463	1.77	5.153	1.15	1.2	93.78	94.18	94.89	94.18	97.97
Benzo(k)-fluoranthene	0.802	2.397	1.06	2.38	1.37	7.85	1.06	2.38	0.792	0.9	94.02	95.51	88.54	95.51	97.7
Benzo(e)pyrene	5.38	8.013	6.15	7.347	8.38	14.6	6.15	7.347	4.9	5.0	97.02	97.61	96.52	97.61	97.95
Benzo(a)pyrene	0.986	3.55	0.99	3.72	0.899	12.6	0.994	3.72	0.546	0.8	92.8	92.52	71.97	92.52	97.13
Perylene	7.57	11.1	9.55	9.493	0.741	3.443	9.55	9.493	0.659	0.7	97.07	98.01	90.71	98.01	97.78
Indeno(1,2,3cd)-pyrene	0	1.837	0	1.531	0.518	9.207	0	1.531	0	0.0					
Dibenz(a,h)-anthracene	0.75	1.56	0.94	0.953	1.31	2.993	0.935	0.953	0.596	0.6	95.84	97.96	95.43	97.96	97.87
Benzo(ghi)-perylene	0.807	3.123	0.83	3.16	1.74	14.77	0.832	3.16	0.596	0.7	92.26	92.4	83.03	92.4	97.55
Total PAHs	8091	3616	9362	3246	11459	1163	5972	651	1959	1830	99.1	99.3	99.8	99.8	98.1

Source: Shigenaka, G. et al., Comparison of physical and chemical characteristics of in-situ burn residue and other environmental oil samples collected during the Deepwater Horizon spill, Report for Bureau of Safety and Environmental Enforcement, Washington, DC, 2014.

Stout and Payne (2016) studied burn residues from the DWH using 2 samples collected from floating residue and 3 recovered from the sea floor. The PAH analytical results are shown in Table 6.10. The table shows results similar to the above; however, one of the samples from the sea floor showed higher PAH content than the starting weathered oil. This may be due to higher burning efficiency, mixing with other oils or the sample may not be consistent with DWH burning. Stout and Payne presumed that the burns were 89% efficient; however, if one presumes that the burns were 95% efficient, the PAH destruction efficiencies become similar to those above. It is suggested that 95% or more burning was more probable.

6.3   SUMMARY AND CONCLUSION

The data summarised in Table 6.11 show that, for the most part, PAHs are destroyed in fires. Further this analysis shows that there are slight differences between crude oil and diesel burns, in that crude oil burns appear to be slightly more efficient in terms of destroying PAHs. The analysis of the PAHs going to various compartments clearly shows that most PAHs are destroyed in the fire, with some remaining with the residue. Burn efficiency does not appear to change the PAH distribution to any extent.

The PAH fate equation might be rewritten as follows:

$$\text{PAHs Mass Balance} = \text{Starting PAHs} - \text{Burned} - \text{On residue} \qquad (6.2)$$

The remaining terms have been dropped as they have been found to be less significant, including the PAHs converted, vaporised and on the soot.

The main points about PAHs in oil burns can be summarised as follows:

1. Some diesel burns ended up with minor production of higher-molecular-weight PAHs. This was not found to be significant in any of the crude oil burns. The concentration of the EPA priority 16 compounds is higher in the burn residue than the starting concentrations, and the concentrations of the alkylated PAHs are lower in the residue than that of the starting oil.
2. Diesel burning is somewhat less efficient than crude oil burning in the destruction of PAHs and does result in more soot with its incumbent PAH load.
3. The amount of PAH emitted as vapour is negligible.
4. The amount of PAH on the soot is variable; however, in most cases it is negligible. The amount of priority PAHs on the soot maybe somewhat elevated.
5. Burn conditions vary and result in variable PAH destruction results.
6. PAH destruction depends on the total burn efficiency.

Overall the situation with PAHs and burning might be summarised in Table 6.10.

TABLE 6.10

Stout and Payne Analysis of Deepwater Oil Burn Residues

		Concentrations in µg/g Oil								
		Unburned Floating Oil				Floating Residue		Sunken Burn Residue		
Abbreviations	PAH Compound	Fresh Oil	Average	Minimal Weathering	Maximum Weathering	Sample A	Sample B	Sample C	Sample D	Sample E
N0	Naphthalene	964	31	346	0.18	2.3	4.1	1.7	4.4	408
N1	C1-naphthalenes	2,106	234	1,607	0.53	13	20	4	38	1,132
N2	C2-naphthalenes	2,259	587	2,371	1.3	76	108	24	231	1,864
N3	C3-naphthalenes	1,597	609	1,646	3.1	112	184	57	428	1,310
N4	C4-naphthalenes	721	363	752	7.5	84	142	59	369	1,310
B	Biphenyl	204	41	197	0.23	2.4	3.8	0.31	10	6.3
DF	Dibenzofuran	30	11	35	0.18	1.5	2.4	0.69	3.9	54
AY	Acenaphthylene	8.9	3	9.4	0.06	4.1	4.7	8.6	21	32
AE	Acenaphthene	21	8.1	28	0.08	1.4	1.8	0.75	5.9	64
F0	Fluorene	150	60	180	0.76	10	19	6.8	38	56
F1	C1-fluorenes	308	185	391	5.3	41	74	30	137	33
F2	C2-fluorenes	404	293	485	22	86	144	86	319	297
F3	C3-fluorenes	286	247	334	40	99	152	101	309	1,311
A0	Anthracene	2.3	3.9	nd	nd	2.8	4.4	7.3	37	1,499
P0	Phenanthrene	310	201	411	14	59	105	54	193	1,025
PA1	C1-phenanthrenes/anthracenes	676	569	862	101	197	309	184	513	110
PA2	C2-phenanthrenes/anthracenes	657	650	805	180	263	372	292	698	18

(Continued)

TABLE 6.10 (*Continued*)
Stout and Payne Analysis of Deepwater Oil Burn Residues

		Concentrations in µg/g Oil								
		Unburned Floating Oil				Floating Residue		Sunken Burn Residue		
Abbreviations	PAH Compound	Fresh Oil	Average	Minimal Weathering	Maximum Weathering	Sample A	Sample B	Sample C	Sample D	Sample E
PA3	C3-phenanthrenes/anthracenes	381	355	412	98	180	245	192	435	19
PA4	C4-phenanthrenes/anthracenes	148	150	176	40	84	113	84	193	39
DBT0	Dibenzothiophene	53	33	69	1.5	8.2	15	6.4	23	379
DBT1	C1-dibenzothiophenes	153	128	202	16	44	67	27	84	655
DBT2	C2-dibenzothiophenes	197	199	245	51	82	117	79	187	505
DBT3	C3-dibenzothiophenes	146	162	178	53	77	105	78	163	30
DBT4	C4-dibenzothiophenes	72	80	86	29	41	64	41	79	314
BF	Benzo(b)fluorene	11	7.5	17	nd	6	8.1	nd	31	730
FL0	Fluoranthene	4.1	3.6	5.8	0.66	8	11	20	100	762
PY0	Pyrene	16	15	21	2	18	24	37	187	437
FP1	C1-fluoranthenes/pyrenes	80	76	106	13	51	71	nd	194	181
FP2	C2-fluoranthenes/pyrenes	130	124	177	17	89	123	114	213	nd
FP3	C3-fluoranthenes/pyrenes	158	154	208	33	119	166	149	255	34
FP4	C4-fluoranthenes/pyrenes	125	139	177	43	112	146	127	200	132
NBT0	Naphthobenzothiophenes	18	27	29	13.6	17	22	21	38	187
NBT1	C1-naphthobenzothiophenes	56	79	91	38.8	50	68	60	93	152
NBT2	C2-naphthobenzothiophenes	80	100	114	39	77	100	84	126	84
NBT3	C3-naphthobenzothiophenes	58	70	90	19	59	75	63	90	12

(Continued)

TABLE 6.10 (*Continued*)

Stout and Payne Analysis of Deepwater Oil Burn Residues

		Concentrations in μg/g Oil								
		Unburned Floating Oil				Floating Residue		Sunken Burn Residue		
Abbreviations	PAH Compound	Fresh Oil	Average	Minimal Weathering	Maximum Weathering	Sample A	Sample B	Sample C	Sample D	Sample E
NBT4	C4-naphthobenzothiophenes	37	48	64	11.3	37	49	42	64	41
BA0	Benz[a]anthracene	7.3	5	7.6	nd	7.7	10	15	59	114
C0	Chrysene/triphenylene	56	68	64.1	40.1	47	60	64	125	185
BC1	C1-chrysenes	129	139	160	59	111	145	127	204	194
BC2	C2-chrysenes	158	153	194	43	137	180	148	268	150
BC3	C3-chrysenes	156	129	192	23	143	183	150	263	106
BC4	C4-chrysenes	90	77	113	15	88	111	90	155	21
BBF	Benzo[b]fluoranthene	6.1	7	6.8	4.4	8.8	11	16	46	50
BJKF	Benzo[jk]fluoranthene	0.5	0.04	nd	nd	3.7	3.8	11	36	65
BAF	Benzo[a]fluoranthene	0.7	nd	nd	nd	2.6	3	5.8	18	44
BEP	Benzo[e]pyrene	12	14	15	5.3	15	18	26	57	35
BAP	Benzo[a]pyrene	3.2	2.4	3.8	nd	7.8	8.9	20	84	14
PER	Perylene	1	0.78	1.4	nd	1.9	2.4	2.4	12	51
IND	Indeno[1,2,3-cd]pyrene	1.2	0.59	1.2	nd	4.2	4.6	11	43	98
DA	Dibenz[a,h]anthracene	2.5	1.9	2.6	nd	3.2	3.6	5.7	11	122
GHI	Benzo[g,h,i]perylene	2.3	2.4	2.6	0.72	6.7	7.7	16	57	113

(*Continued*)

TABLE 6.10 (Continued)
Stout and Payne Analysis of Deepwater Oil Burn Residues

		Concentrations in µg/g Oil								
		Unburned Floating Oil				Floating Residue		Sunken Burn Residue		
Abbreviations	PAH Compound	Fresh Oil	Average	Minimal Weathering	Maximum Weathering	Sample A	Sample B	Sample C	Sample D	Sample E
T19	Hopane	69	112	81	144	84	104	117	149	61
	Total Concentrations									
	TEM (C9-C44)	681,000	601,144	736,920	518,000	274,000	380,000	311,000	624,000	682,000
	TPAH51(ΣN0-GHI)	13,252	6,643	13,685	1,087	2,797	3,990	2,850	7,549	16,585
	HPAH (ΣBF-GHI)	1,399	1,443	1,861	421	1,228	1,615	1,424	3,030	4,116
	Priority Pollutants (Prio 1)	1,555	412	1,089	0	194	284	294	1,049	5,046
	Fraction PAHs Destroyed by burning					95	93	94	88	87
	Fraction Priority PAHs Destroyed burning					95	92	92	72	−35
	If 98% burn efficiency estimated									
	Fraction PAHs Destroyed by burning					99	99	99	98	98
	Fraction Priority PAHs Destroyed burning					99	99	99	95	76

TEM is total extractable material; TPAH is total polyaromatic aromatic hydrocarbons; HPAH is higher-molecular-weight PAHs.

TABLE 6.11

Summary of Burn Results and Assessment of PAH Destruction

Oil Type	Burn Type	Location	Year	References	% Burn Efficiency	Type (Measured or Estimated)	Measurements	Total PAHs Starting µg/g	Total PAHs Residue µg/g	Overall PAH Destruction %
Alberta medium	Field burn	Offshore	1993	Fingas et al. (1994a)	99.9	Meas.	Soot, gas, residue	8,607	4,671	99.95
Alberta medium	Field burn 1	Offshore	1993	Fingas et al. (1994a)	99.9	Meas.	Residue	8,607	5,330	99.94
Alberta medium	Field burn 1	Offshore	1993	Fingas et al. (1994a)	99.9	Meas.	Residue	8,607	4,011	99.95
Alberta medium	Field burn 2	Offshore	1993	Fingas et al. (1994a)	99.9	Meas.	Residue	8,607	3,943	99.95
Alberta medium	Field burn 2	Offshore	1993	Fingas et al. (1994a)	99.9	Meas.	Residue	8,607	4,494	99.95
Alberta medium	Field burn 2	Offshore	1993	Fingas et al. (1994a)	99.9	Meas.	Residue	8,607	4,017	99.95
Diesel fuel	Test burn 1	Meso tank	1992	Fingas et al. (1993)	98	Meas.	Residue	5,920	439	99.85
Diesel fuel	Test burn 2	Meso tank	1992	Fingas et al. (1993)	98	Meas.	Residue	5,825	551	99.81
Diesel fuel	Test burn 3	Meso tank	1992	Fingas et al. (1993)	98	Meas.	Residue	5,806	635	99.78
Diesel fuel	Test burn 4	Meso tank	1992	Fingas et al. (1993)	98	Meas.	Residue	5,697	519	99.82
Diesel fuel	Test burn 5	Meso tank	1992	Fingas et al. (1993)	98	Meas.	Residue	5,589	353	99.87

(Continued)

TABLE 6.11 (Continued)
Summary of Burn Results and Assessment of PAH Destruction

Oil Type	Burn Type	Location	Year	References	% Burn Efficiency	Type (Measured or Estimated)	Measurements	Total PAHs Starting µg/g	Total PAHs Residue µg/g	Overall PAH Destruction %
Diesel fuel	Tank burn	Meso tank	1994	Wang et al. (1999a)	99	Meas.	Soot, gas, residue	27,510	22,360	99.19
Diesel fuel	Test burn 1	Meso tank	1994	Fingas et al. (1996)	98	Meas.	Residue	5,060	1,994	99.21
Diesel fuel	Test burn 2	Meso tank	1994	Fingas et al. (1996)	98	Meas.	Residue	4,966	2,464	99.01
Diesel fuel	Test burn 3	Meso tank	1994	Fingas et al. (1996)	98	Meas.	Residue	5,243	2,991	98.86
Diesel fuel	Test burn 1	Meso tank	1997	Fingas et al. (1998)	99	Meas.	Soot, gas, residue	26,713	19,949	99.25
Diesel fuel	Test burn 2	Meso tank	1997	Fingas et al. (1998)	99	Meas.	Soot, gas, residue	26,713	20,966	99.22
Diesel fuel	Test burn 3	Meso tank	1997	Fingas et al. (1998)	99	Meas.	Soot, gas, residue	26,713	26,252	99.02
Diesel fuel	Test burn 4	Meso tank	1997	Fingas et al. (1998)	99	Meas.	Soot, gas, residue	26,713	29,366	98.9
Diesel fuel	Tank burn	Meso tank	1998	Fingas et al. (2000)	99	Meas.	Soot, gas, residue	25,906	25,810	99
Statfjord crude	Tank burn	Small tank	1999	Garrett et al. (2000)	85	Meas.	Residue			85% + 40%
Bunker C	Tank burn 2	Small tank	2003	Fingas et al. (2005)	63.8	Meas.	Soot, residue	29,419	31	99.96

(Continued)

TABLE 6.11 (*Continued*)
Summary of Burn Results and Assessment of PAH Destruction

Oil Type	Burn Type	Location	Year	References	% Burn Efficiency	Type (Measured or Estimated)	Measurements	Total PAHs Starting µg/g	Total PAHs Residue µg/g	Overall PAH Destruction %
Test oil –residual	Tank burn 3	Small tank	2003	Fingas et al. (2005)	2	Meas.	Soot, residue	849	305	64.81
Test oil –residual	Tank burn 4	Small tank	2003	Fingas et al. (2005)	2	Meas.	Soot, residue	849	424	51.1
Orimulsion	Tank burn 5	Small tank	2003	Fingas et al. (2005)	65.6	Meas.	Soot, residue	1,459	1,275	69.95
Orimulsion	Tank burn 6	Small tank	2003	Fingas et al. (2005)	61.3	Meas.	Soot, residue	1,459	1,322	64.94
Bunker C	Tank burn 7	Small tank	2003	Fingas et al. (2005)	65.9	Meas.	Soot, residue	29,419	1,034	98.8
Bunker C	Tank burn 8	Small tank	2003	Fingas et al. (2005)	70	Meas.	Soot, residue	29,419	39	99.96
Bitumen	Tank burn 9	Small tank	2003	Fingas et al. (2005)	12.3	Meas.	Soot, residue	1,459	1,471	87.60
Bitumen	Tank burn 10	Small tank	2003	Fingas et al. (2005)	12.5	Meas.	Soot, residue	1,459	1,020	91.27
Diesel fuel	Test burn	Small tank	2004	Lin et al. (2005)			Residue			99.40
Louisiana crude	Test burn	Small tank	2004	Lin et al. (2005)			Residue			98.90
Macondo	Field burn 1	Offshore	2010	Shigenaka et al. (2014)	98	Est.	Residue	6,166	2,271	99.26

(*Continued*)

TABLE 6.11 *(Continued)*

Summary of Burn Results and Assessment of PAH Destruction

Oil Type	Burn Type	Location	Year	References	% Burn Efficiency	Type (Measured or Estimated)	Measurements	Total PAHs Starting μg/g	Total PAHs Residue μg/g	Overall PAH Destruction %
Macondo	Field burn 2	Offshore	2010	Shigenaka et al. (2014)	98	Est.	Residue	5,363	5,972	97.77
Macondo	Field burn 3	Offshore	2010	Shigenaka et al. (2014)	98	Est.	Residue	4,841	651	99.73
Macondo	Field burn 4	Offshore	2010	Shigenaka et al. (2014)	98	Est.	Residue	4,893	1,959	99.2
Macondo	Field burn 5	Offshore	2010	Shigenaka et al. (2014)	98	Est.	Residue	4,893	1,830	99.25
Macondo	Field burn 6	Offshore	2010	Shigenaka et al. (2014)	98	Est.	Residue	4,893	962	99.61
Macondo	Small 5 g	Lab	2012	Shigenaka et al. 2014	98	Est.	Residue	11,459	1,163	99.8
South Louisiana	Test 500 mL	Lab	2012	Shigenaka et al. 2014	98	Est.	Residue	8,091	3,616	99.11
Weathered Lou	Test 500 mL	Lab	2012	Shigenaka et al. (2014)	98	Est.	Residue	9,362	3,246	99.31
Emulsified mac	Test 500 mL	Lab	2012	Shigenaka et al. (2014)	80	Est.	Residue	9,362	3,246	93.07
Emulsified mac	Test 500 mL	Lab	2012	Shigenaka et al. (2014)	80	Est.	Residue	2,660	2,643	80.13
Macondo	Actual burn	Offshore	2010	Garrett et al. (2016)	Large	Est.	Particulate matter			Large
Macondo	Actual burn	Offshore	2010	Stout and Payne (2016)	89	Est.	Residue	274,000	274,000	95

(Continued)

TABLE 6.11 (*Continued*)

Summary of Burn Results and Assessment of PAH Destruction

Oil Type	Burn Type	Location	Year	References	% Burn Efficiency	Type (Measured or Estimated)	Measurements	Total PAHs Starting µg/g	Total PAHs Residue µg/g	Overall PAH Destruction %
Macondo	Actual burn	Offshore	2010	Stout and Payne (2016)	89	Est.	Residue	380,000	380,000	93
Macondo	Actual burn	Offshore	2010	Stout and Payne (2016)	89	Est.	Residue	311,000	311,000	94
Macondo	Actual burn	Offshore	2010	Stout and Payne (2016)	89	Est.	Residue	624,000	624,000	89
Macondo	Actual burn	Offshore	2010	Stout and Payne (2016)	89	Est.	Residue	682,000	682,000	88
Macondo	Actual burn	Offshore	2010	Stout and Payne (2016)	98	Est. here	Residue	274,000	274,000	99
Macondo	Actual burn	Offshore	2010	Stout and Payne (2016)	98	Est. here	Residue	380,000	380,000	99
Macondo	Actual burn	Offshore	2010	Stout and Payne (2016)	98	Est. here	Residue	311,000	311,000	99
Macondo	Actual burn	Offshore	2010	Stout and Payne (2016)	98	Est. here	Residue	624,000	624,000	98
Macondo	Actual burn	Offshore	2010	Stout and Payne (2016)	98	Est. here	Residue	682,000	682,000	98

REFERENCES

Benner, A.B., Bryner, N.P., Wise, S.A., Mulholland, G.W., Lao, R.C., and Fingas, M.F. 1990. Polycyclic aromatic hydrocarbon emissions from the combustion of crude oil on water, *Environ. Sci. Techn.* 24(9), 1418–1427.

Fingas, M.F. 1991. In-situ burning of oil spills, in *World Catalog of Oil Spill Response Products*, R. Schulze and H.L. Hoffman (Eds.), Elkridge, MD: World Catalogue, pp. A1–A6.

Fingas, M.F. 2010. Soot production from in-situ oil fires: Review of the literature and calculation of values from experimental spills, *Proceedings of the Thirty-Third Arctic and Marine Oil Spill Programme (AMOP) Technical Seminar*, Environment Canada, Ottawa, ON, pp. 1017–1054.

Fingas, M. 2012. Introduction to oil chemistry and properties, *Proceedings of the Thirty-Fifth. Arctic and Marine Oil Spill Program (AMOP) Technical Seminar*, Environment Canada, Ottawa, ON, pp. 951–980.

Fingas, M.F., Ackerman, F., Lambert, P., Li, K., Wang, Z., Mullin, L. et al. 1995. The Newfoundland Offshore Burn Experiment: Further results of emissions measurement, *Proceedings of the Eighteenth Arctic and Marine Oil Spill Program (AMOP) Technical Seminar*, Edmonton, Alberta, pp. 915–995.

Fingas, M.F., Ackerman, F., Lambert, P., Li, K., Wang, Z., Nelson, R. et al. 1996b. Emissions from mesoscale in-situ oil (diesel) fires: The Mobile 1994 experiments, *Proceedings of the 1996 AMOP Technical Seminar*, Environment Canada, Ottawa, ON, pp. 907–978.

Fingas, M.F., Ackerman, F., Li, K., Lambert, P., Wang, Z., Bissonnette, M.C. et al. 1994b. The Newfoundland Offshore Burn Experiment – NOBE – Preliminary results of emissions measurement, *Proceedings of the Seventeenth Arctic and Marine Oil Spill Program (AMOP) Technical Seminar*, 2, 1099–1164.

Fingas, M.F., Fieldhouse, B., Brown, C.E., and Gamble, L. 2004. In-situ burning of heavy oils and orimulsion: Mid-scale burns, *Proceedings of the Twenty-Seventh Arctic and Marine Oil Spill Program (AMOP) Technical Seminar*, Environment Canada, Ottawa, ON, pp. 207–233.

Fingas, M.F., Halley, G., Ackerman, F., Vanderkooy, N., Nelson, R., Bissonnette, M.C. et al. 1994a. The Newfoundland Offshore Burn Experiment – NOBE experimental design and overview, *Proceedings of the Seventeenth Arctic and Marine Oil Spill Program (AMOP). Technical Seminar*, Environment Canada, Ottawa, ON, pp. 1053–1061.

Fingas, M., Lambert, P., Ackerman, F., Fieldhouse, B., Nelson, R., Goldthorp, M. et al., 1998. Particulate and carbon dioxide emissions from diesel fires: The mobile 1997 experiments (1998), *Proceedings of Twenty-First Arctic and Marine Oil Spill Program (AMOP) Technical Seminar*, Environment Canada, Ottawa, ON, pp. 569–598.

Fingas, M.F., Li, K., Ackerman, F., Campagna, P.R., Turpin, R.D., Getty, S.J. et al. 1993. Emissions from mesoscale in-situ oil fires: The mobile 1991 and 1992 tests, *Proceedings of the Sixteenth Arctic and Marine Oil Spill Program (AMOP) Technical Seminar*, Environment Canada, Ottawa, ON, pp. 749–821.

Fingas, M.F., Li, K., Ackerman, F., Wang, Z., Lambert, P., Gamble, R.L. et al. 1996a. Soot production from in-situ oil fires: Review of the literature, measurement and estimation techniques and calculation of values from experimental spills, *Proceedings of the Arctic and Marine Oil Spill Program (AMOP) Technical Seminar No. 19b*. Environment Canada, Ottawa, ON, pp. 999–1032.

Fingas, M.F., Wang, Z., Fieldhouse, B., Brown, C.E., Yang, C., Landriault, M., and Cooper, D. 2005. In-situ burning of heavy oils and Orimulsion: Analysis of soot and residue, *Proceedings of the Twenty-Eighth Arctic and Marine Oil Spill Program (AMOP) Technical Seminar*, Environment Canada, Ottawa, ON, pp. 333–348.

Fingas, M.F., Wang, Z., Lambert, P., Ackerman, F., Fieldhouse, B., Nelson, R. et al. 1999. Emissions from mesoscale in-situ (diesel) fires: Gases, PAHs and VOCs from the Mobile 1997 experiments, *Proceedings of the Twenty-Second Arctic and Marine Oil Spill Program (AMOP) Technical Seminar*, Environment Canada, Ottawa, ON, pp. 567–597.

Fingas, M.F., Wang, Z., Lambert, P., Ackerman, F., Li, K., Goldthorp, M. et al. 2000. Emissions from mesoscale in-situ (diesel) fires: Emissions from the Mobile 1998 experiments, *Proceedings of the Twenty-Third Arctic and Marine Oil Spill Program (AMOP)*. Environment Canada, Ottawa, ON, pp. 857–901.

Fingas, M.F., Wang, Z., Lambert, P., Ackerman, F., Li, K., Goldthorp, M. et al. 2001. Emissions from Mesoscale *in-situ* Oil (Diesel) Fires: Emissions from the Mobile 1998 Experiments, *Proceedings of the Twenty-Third Arctic and Marine Oil Spill Program (AMOP) Technical Seminar*, Environment Canada, Ottawa, ON, pp. 1471–1478.

Fritt-Rasmussen, J., Ascanius, B.E., Brandvik, P.J., Villumsen, A., and Stenby, E.H. 2013. Composition of in-situ burn residue as a function of weathering conditions, *Mar. Pollut. Bull.* 67, 75–86.

Fritt-Rasmussen, J., Wegeberg, S., and Gustavson K. 2015. Review on burn residues from in situ burning of oil spills in relation to Arctic waters, *Water, Air, Soil Poll.* 226, 329–339.

Garrett, R.M., Guénette, C.C., Haith, C.E., and Prince, R.C. 2000. Pyrogenic polycyclic aromatic hydrocarbons in oil burn residues, *Environ. Sci. Techn.* 34(10), 1934–1937.

Gullett, B.K., Aurell, J., Holder, A., Mitchell, W., Greenwell, D., Hays, M. et al. 2017. Characterization of emissions and residues from simulations of the Deepwater Horizon surface oil burns, *Mar. Poll. Bull.* 117, 392–405.

Li, K., Caron, T., Landriault, M., Paré, J.R.J., and Fingas, M. 1992. Measurement of volatiles, semi-volatiles and heavy metals in an oil burn test, *Proceedings of the Fifteenth Arctic and Marine Oil Spill Program (AMOP) Technical Seminar*, Environment Canada, Ottawa, ON, pp. 561–379.

Lima, A.L.C., Farrington, J.W., and Reddy, C.M. 2005. Combustion-derived polycyclic aromatic hydrocarbons in the environment—A review, *Environ. For.* 6(2), 109–131.

Lin, Q., Mendelssohn, I.A., Carney, K., Miles, S.M., Bryner, N.P., and Walton, W.D. 2005. In-situ burning of oil in coastal marshes. 2. Oil spill cleanup efficiency as a function of oil type, marsh type, and water depth, *Environ. Sci. Techn.*, 39(6), 1855–1860.

Shigenaka, G., Overton, E., Meyer, B., Gao, H., and Miles, S. 2014. Comparison of physical and chemical characteristics of in-situ burn residue and other environmental oil samples collected during the Deepwater Horizon spill, Report for Bureau of Safety and Environmental Enforcement, Washington, DC, 128 p.

Shigenaka, G., Overton, E., and Meyer, B. 2015. Physical and chemical characteristics of in-situ burn residue and other environmental oil samples collected during the Deepwater Horizon spill response, National Oceanic and Atmospheric Administration. Emergency Response Division Seattle, WA, 11 p.

Shigenaka, G., Meyer, B., Overton, E., and Miles, S. 2017. Physical and chemical characterization of in-situ burn residue encountered by a deep-water fishery in the Gulf of Mexico, *International Oil Spill Conference Proceedings*, 2017, 1020–1040.

Stout, S.A., and Payne, J.R. 2016. Chemical composition of floating and sunken in-situ burn residues from the Deepwater Horizon oil spill, *Mar. Poll. Bull.* 108, 186–202.

Wang, Z., Fingas, M.F., Sigouin, L., Landraiult, M., Li, K., Lambert, P. et al. 1998. Quantitative characterization of PAH in burn residue and soot samples and differentiation of pyrogenic and petrogenic PAHs from PAHs – The 1994 Mobile burn study, *Proceedings of the Twenty-First Arctic and Marine Oil Spill Program (AMOP) Technical Seminar*, Environment Canada, Ottawa, ON, pp. 673–703.

Wang, Z., Fingas, M.F., Shu, Y.Y., Sigouin, L., Landraiult, M., Lambert, P. et al. 1999a. Quantitative characterization of PAH in burn residue and soot samples and differentiation of pyrogenic and petrogenic PAHs from PAHs – The 1994 Mobile burn study, *Environ. Sci. Techn.* 33, 3100–3109.

Wang, Z., Fingas, M.F., Landraiult, M., Sigouin, L., and Lambert, P. 1999b. Distribution of PAHs in burn residue and soot samples and differentiation of pyrogenic and petrogenic PAHs from PAHs – The 1994 and 1997 Mobile burn study, in *Diesel Fuels*, C. Song, C. Hsu and I. Mochida (Eds.), Taylor and Francis, New York, pp. 237–253.

Wise, S.A., Sander, L.C., and Schantz, M.M. 2015. Analytical methods for determination of polycyclic aromatic hydrocarbons (PAHs) – A historical perspective on the 16 U.S. EPA priority pollutant PAHs, *Polycyclic Aromatic Compounds*, 35, 187–247.

Appendix: Volatile Organic Compounds

Merv Fingas

Volatile organic compounds (VOCs) are organic compounds that have high enough vapour pressures to be gaseous at normal temperatures. When oil is burned, these compounds evaporate and are released – if not burned. Most of the compounds are burned, and thus only a small proportion of those present are released as vapour. VOCs include many different classes of compounds including benzene, toluene, ethylbenzene and xylenes (BTEX), polycyclic aromatic hydrocarbons (PAHs), aliphatics, carbonyls and so on. What VOCs have in common, is that they have lower vapour pressures than their larger congeners and thus are present as vapours and are emitted from fires in measurable quantities.

This Appendix will provide the reader information on many common VOCs that have been measured at a variety of test fires. This section provides data on individual compounds emitted from fires of crude oils and diesel. The standard format followed is illustrated in Figure A1.

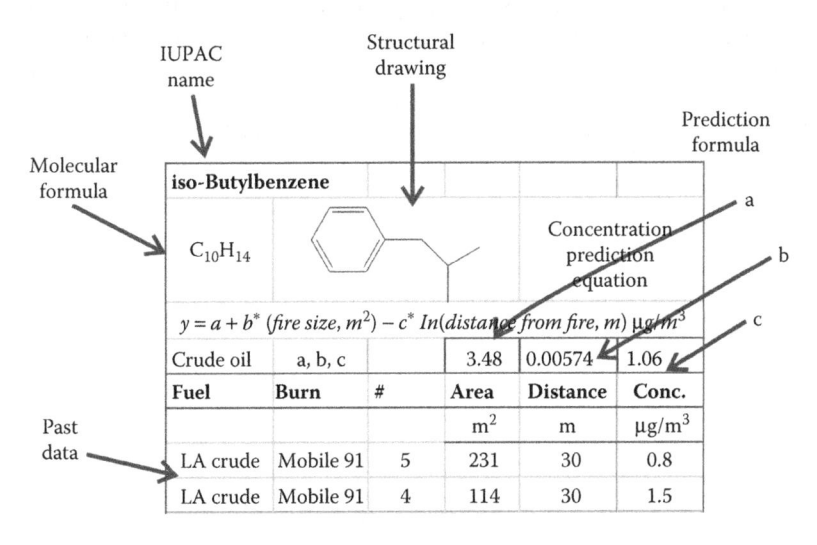

FIGURE A1 Illustration of the format that will be used in Appendix to provide information on VOC emissions.

It may be useful to comment on the information provided in this Appendix. First, this information has not been provided before, especially as complete as this. Emissions are provided separately for VOCs from crude oil fires and from diesel fires. It is important to recognise that these emissions differ, particularly because diesel may contain more VOCs of higher molecular weight and vapour pressure than many crude oils. This results in different VOC emission patterns.

The information contained in the Appendix, in the form as shown in Figure A1, includes the International Union of Pure and Applied Chemistry (IUPAC), or standard name, a structural drawing of the compound, the molecular formula, the calculated emission prediction equation of type, $y = a + b * (\text{fire size, m}^2) - c * \ln (\text{distance from fire, m})$. Concentrations are in $\mu g/m^3$. The correlation procedure to calculate the prediction equations involved collecting all valid emission data from the past studies, and then finding the best correlation procedure and the best equation that fit most data. A simple equation was found to universally fit the data using the software package, TableCurve 3D (San Raphael, CA). This software calculates up to 2,000 equations and sorts them in terms of fits (best regression coefficient, r^2). The regression coefficients of the equations given here ranged from typically 0.35 to 0.95. This varied because the one equation chosen may not have been optimal for the given compound correlation, or because of different environmental conditions.

The data from past trials is included in the table with two prime locations, Newfoundland Offshore Burn Experiment (NOBE) and Mobile, which is the test facility run by the United States Coast Guard in Mobile, Alabama. Two types of crude were burned, Alberta (AB) sweet light crude at NOBE and Louisiana (LA) crude at Mobile. The diesel burns took place at Mobile, Alabama. Burn test numbers are also given when someone wants to correlate different test burns.

VOC data on crude oil emissions are given in Tables A1.1 through A1.16, and diesel emissions are given in Tables A2.1 through A2.39. The Carbonyl data are separated from the other VOCs. Those Carbonyls emitting from crude oil burns are given in Table A3.1 and those emitted from diesel burns in Table A3.2.

The 95 compounds which are listed are summarised in Table A1.

TABLE A1
Compounds Included in Appendix

Compound	Crude oil	Diesel	Compound	Crude oil	Diesel
1,2,3-Trimethylbenzene	X	X	c-2-Butene	X	X
1,2,4-Trimethylbenzene	X	X	c-2-Heptene		X
1,2-Diethylbenzene		X	c-2-Hexene		X
1,3,5-Trimethylbenzene	X	X	c-2-Pentene		X
1,3-Butadiene		X	Cyclohexane	X	X
1,3-Diethylbenzene		X	Cyclohexene		X
1,4-Diethylbenzene	X	X	Cyclopentane	X	X
1-Butene/2-Methylpropene		X	Cyclopentene		X
1-Heptene		X	Decane	X	X
1-Hexene/2-Methyl-1-Pentene		X	Dodecane	X	X
1-Methylcyclohexene		X	Ethylbenzene	X	X
1-Methylcyclopentene		X	Formaldehyde	X	X
1-Nonene		X	Heptane	X	X
1-Octene		X	Hexylbenzene		X
1-Pentene		X	Indan (2,3-Dihydroindene)		X
2,2,3-Trimethylbutane	X	X	Indene		X
2,2,4-Trimethylpentane	X	X	Isobutane (2-Methylpropane)	X	X
2,2,5-Trimethylhexane	X	X	iso-Butylbenzene	X	
2,2-Dimethylbutane	X	X	Isoprene (2-Methyl-1,3-Butadiene)	X	
2,2-Dimethylpentane	X		iso-Propylbenzene	X	
2,2-Dimethylpropane	X	X	m,p-Xylene	X	X
2,3,4-Trimethylpentane	X	X	Methylcyclohexane	X	X
2,3-Dimethylbutane	X	X	Methylcyclopentane	X	X
2,3-Dimethylpentane	X	X	Naphthalene	X	X
2,4-Dimethylhexane	X	X	n-Butylbenzene	X	X
2,4-Dimethylpentane	X	X	Nonane	X	X
2,5-Dimethylhexane	X	X	n-Propylbenzene	X	X
2-Ethyltoluene	X	X	Octane	X	X
2-Methyl-1-Butene		X	o-Xylene	X	X
2-Methyl-2-Butene		X	p-Cymene (1-Methyl-4-iso-propylbenzene)	X	X
2-Methylbutane	X	X	Pentane	X	X
2-Methylheptane	X	X	Propane	X	X
2-Methylhexane		X	Propene	X	X
2-Methylpentane		X	Proprionaldehyde		X
3,6-Dimethyloctane		X	Propyne		X
3-Ethyltoluene		X	sec-Butylbenzene		X
3-Methylheptane		X	Styrene		X
3-Methylhexane	X	X	t-1,2-Dimethylcyclohexane		X
3-Methylpentane	X	X	t-2-Butene		X
4-Ethyltoluene	X	X	t-2-Heptene		X
4-Methylheptane	X	X	t-2-Hexene		X
Acetaldehyde	X	X	t-2-Octene		X
Acetone	X	X	t-2-Pentene		X
Benzene	X	X	t-3-Heptene		X
Butane	X	X	tert-Butylbenzene		X
Butyraldehydes		X	Toluene		X
c-1,3-Dimethylcyclohexane	X	X	Total	X	X
c-1,4/t-1,3-Dimethylcyclohexane	X	X	Undecane	X	X

TABLE A1.1
VOC Emissions from Crude Oil Burns

1,2,3-Trimethylbenzene

C_9H_{12}

Concentration Prediction Equation

$y = a + b*(\text{fire size}, m^2) - c*\ln(\text{distance from fire}, m) \ \mu g/m^3$

Crude oil	a, b, c		11.4	0.0106	2.53
Fuel	Burn	#	Area	Distance	Conc.
			m^2	m	$\mu g/m^3$
LA Crude	Mobile 91	4	114	30	10.3
AB Crude	NOBE	2	600	75	8.0
AB Crude	NOBE	2	600	125	4.0
AB Crude	NOBE	2	600	600	2.0
LA Crude	Mobile 92	23	181	30	0.9
LA Crude	Mobile 92	28	72.8	30	1.2
LA Crude	Mobile 92	30	37.2	30	0.8
LA Crude	Mobile 92	3	114	30	0.4
LA Crude	Mobile 92	4	114	30	10.4
LA Crude	Mobile 92	5	231	30	5.2

1,2,4-Trimethylbenzene

C_9H_{12}

Concentration Prediction Equation

$y = a + b*(\text{fire size}, m^2) - c*\ln(\text{distance from fire}, m) \ \mu g/m^3$

Crude oil	a, b, c		22.4	0.0239	4.58
Fuel	Burn	#	Area	Distance	Conc.
			m^2	m	$\mu g/m^3$
LA Crude	Mobile 92	4	114	30	27.0
AB Crude	NOBE 93	2	600	75	19.0
AB Crude	NOBE 93	2	600	125	10.0
AB Crude	NOBE 93	2	600	600	9.0
LA Crude	Mobile 92	23	181	30	3.4
LA Crude	Mobile 92	24	170	30	3.9
LA Crude	Mobile 92	28	72.8	30	3.0
LA Crude	Mobile 92	30	37.2	30	2.2
LA Crude	Mobile 92	3	114	30	3.3
LA Crude	Mobile 92	3	114	30	1.0
LA Crude	Mobile 92	4	114	30	27.3
LA Crude	Mobile 92	5	231	30	18.7

1,3,5-Trimethylbenzene

C_9H_{12}

Concentration Prediction Equation

$y = a + b*(\text{fire size}, m^2) - c*\ln(\text{distance from fire}, m) \ \mu g/m^3$

Crude oil	a, b, c		17.3	0.0191	4.28
Fuel	Burn	#	Area	Distance	Conc.
			m^2	m	$\mu g/m^3$
LA Crude	Mobile 91	4	114	30	12.1
AB Crude	Mobile 91	5	231	30	10.9
AB Crude	NOBE	2	600	75	11.0
AB Crude	NOBE	2	600	125	4.0
LA Crude	NOBE	2	600	600	3.0
LA Crude	Mobile 92	23	181	30	2.1
LA Crude	Mobile 92	24	170	30	2.8
LA Crude	Mobile 92	28	72.8	30	1.1
LA Crude	Mobile 92	30	37.2	30	0.8
LA Crude	Mobile 92	3	114	30	1.2
LA Crude	Mobile 92	3	114	30	0.8
LA Crude	Mobile 92	4	114	30	12.2
LA Crude	Mobile 92	5	231	30	10.9

1,4-Diethylbenzene

$C_{10}H_{14}$

Concentration Prediction Equation

$y = a + b*(\text{fire size}, m^2) - c*\ln(\text{distance from fire}, m) \ \mu g/m^3$

Crude oil	a, b, c		4.66	0.00529	0.947
Fuel	Burn	#	Area	Distance	Conc.
			m^2	m	$\mu g/m^3$
LA Crude	Mobile 91	4	114	30	4
AB Crude	NOBE	2	600	75	3
AB Crude	NOBE	2	600	600	2
AB Crude	Mobile 92	4	114	30	5.7
LA Crude	Mobile 92	28	72.8	30	0.6
LA Crude	Mobile 92	30	37.2	30	0.4
LA Crude	Mobile 92	3	114	30	0.6
LA Crude	Mobile 92	4	114	30	5.8
LA Crude	Mobile 92	5	231	30	2.1
LA Crude	Mobile 92	28	72.8	30	0.2

TABLE A1.2
VOC Emissions from Crude Oil Burns

2,2,3-Trimethylbutane

C_7H_{16} — Concentration Prediction Equation

$y = a + b*(fire\ size, m^2) - c*ln(distance\ from\ fire, m)\ \mu g/m^3$

Crude oil	a, b, c		25	0.0256	7.49
Fuel	Burn	#	Area	Distance	Conc.
			m^2	m	$\mu g/m^3$
LA Crude	Mobile 91	4	114	30	1.7
AB Crude	NOBE	2	600	75	10.0
AB Crude	NOBE	2	600	125	3.0
LA Crude	Mobile 92	23	181	30	2.7
LA Crude	Mobile 92	24	170	30	3.0
LA Crude	Mobile 92	28	72.8	30	3.0
LA Crude	Mobile 92	30	37.2	30	2.2
LA Crude	Mobile 92	3	114	30	2.7
LA Crude	Mobile 92	3	114	30	1.9
LA Crude	Mobile 92	4	114	30	1.9
LA Crude	Mobile 92	5	231	30	5.5

2,2,4-Trimethylpentane

C_8H_{18} — Concentration Prediction Equation

$y = a + b*(fire\ size, m^2) - c*ln(distance\ from\ fire, m)\ \mu g/m^3$

Crude oil	a, b, c		5.41	0.0131	1.66
Fuel	Burn	#	Area	Distance	Conc.
			m^2	m	$\mu g/m^3$
LA Crude	Mobile 91	4	114	30	0.9
LA Crude	Mobile 91	5	231	30	3.2
AB Crude	NOBE	2	600	75	4.0
AB Crude	NOBE	2	600	125	8.0
AB Crude	NOBE	2	600	600	2.0
LA Crude	Mobile 92	23	181	30	1.0
LA Crude	Mobile 92	28	72.8	30	1.4
LA Crude	Mobile 92	30	37.2	30	1.1
LA Crude	Mobile 92	3	114	30	2.0
LA Crude	Mobile 92	3	114	30	0.6
LA Crude	Mobile 92	4	114	30	1.1
LA Crude	Mobile 92	5	231	30	3.3
LA Crude	Mobile 92	24	170	30	1.4
LA Crude	Mobile 92	28	72.8	30	0.4
LA Crude	Mobile 92	3	114	30	0.6
LA Crude	Mobile 92	5	231	30	3.6

2,2,5-Trimethylhexane

C_9H_{20} — Concentration Prediction Equation

$y = a + b*(fire\ size, m^2) - c*ln(distance\ from\ fire, m)\ \mu g/m^3$

Crude oil	a, b, c		8.49	0.00806	2.58
Fuel	Burn	#	Area	Distance	Conc.
			m^2	m	$\mu g/m^3$
AB Crude	NOBE	2	600	75	2
AB Crude	NOBE	2	600	125	1
LA Crude	Mobile 92	3	114	30	0.3
LA Crude	Mobile 92	4	114	30	0.9
LA Crude	Mobile 92	5	231	30	2.2
LA Crude	Mobile 92	23	181	30	0.8
LA Crude	Mobile 92	24	170	30	1.0
LA Crude	Mobile 92	28	72.8	30	0.2
LA Crude	Mobile 92	30	37.2	30	0.2

2,2-Dimethylbutane

C_6H_{14} — Concentration Prediction Equation

$y = a + b*(fire\ size, m^2) - c*ln(distance\ from\ fire, m)\ \mu g/m^3$

Crude oil	a, b, c		61	0.105	19.3
Fuel	Burn	#	Area	Distance	Conc.
			m^2	m	$\mu g/m^3$
LA Crude	Mobile 91	4	114	30	5.6
LA Crude	Mobile 91	5	231	30	20.4
AB Crude	NOBE	2	600	75	56.0
AB Crude	NOBE	2	600	125	16.0
AB Crude	NOBE	2	600	600	3.0
LA Crude	Mobile 92	23	181	30	7.2
LA Crude	Mobile 92	24	170	30	9.7
LA Crude	Mobile 92	28	72.8	30	8.7
LA Crude	Mobile 92	30	37.2	30	7.8
LA Crude	Mobile 92	3	114	30	8.3
LA Crude	Mobile 92	3	114	30	1.6
LA Crude	Mobile 92	4	114	30	6.1
LA Crude	Mobile 92	5	231	30	20.5

TABLE A1.3
VOC Emissions from Crude Oil Burns

2,2-Dimethylpentane

C_7H_{16} — Concentration Prediction Equation

$y = a + b*(\text{fire size}, m^2) - c*\ln(\text{distance from fire}, m)\ \mu g/m^3$

Crude oil	a, b, c		52.3	0.0799	16.5

Fuel	Burn	#	Area (m^2)	Distance (m)	Conc. ($\mu g/m^3$)
AB Crude	NOBE	1	600	125	1
AB Crude	NOBE	2	600	75	48
AB Crude	NOBE	2	600	125	15
AB Crude	NOBE	2	600	600	1
LA Crude	Mobile 92	2	36	30	0.3
LA Crude	Mobile 92	2	231	60	1

2,2-Dimethylpropane

C_5H_{12} — Concentration Prediction Equation

$y = a + b*(\text{fire size}, m^2) - c*\ln(\text{distance from fire}, m)\ \mu g/m^3$

Crude oil	a, b, c		25.2	0.0271	7.93

Fuel	Burn	#	Area (m^2)	Distance (m)	Conc. ($\mu g/m^3$)
LA Crude	Mobile 91	4	114	30	1.6
LA Crude	Mobile 91	5	231	30	3.8
AB Crude	NOBE	2	600	75	9.0
AB Crude	NOBE	2	600	125	2.0
LA Crude	Mobile 92	23	181	30	2.4
LA Crude	Mobile 92	24	170	30	2.5
LA Crude	Mobile 92	28	72.8	30	0.3
LA Crude	Mobile 92	30	37.2	30	0.3
LA Crude	Mobile 92	3	114	30	1.0
LA Crude	Mobile 92	4	114	30	1.9
LA Crude	Mobile 92	5	231	30	3.8

2,3,4-Trimethylpentane

C_8H_{18} — Concentration Prediction Equation

$y = a + b*(\text{fire size}, m^2) - c*\ln(\text{distance from fire}, m)\ \mu g/m^3$

Crude oil	a, b, c		14	0.0249	4.53

Fuel	Burn	#	Area (m^2)	Distance (m)	Conc. ($\mu g/m^3$)
AB Crude	NOBE	2	600	75	12.0
AB Crude	NOBE	2	600	125	4.0
AB Crude	NOBE	2	600	600	1.0
LA Crude	Mobile 92	2	36	30	0.2
LA Crude	Mobile 92	2	231	60	0.2

2,3-Dimethylbutane

C_6H_{14} — Concentration Prediction Equation

$y = a + b*(\text{fire size}, m^2) - c*\ln(\text{distance from fire}, m)\ \mu g/m^3$

Crude oil	a, b, c		168	0.308	57

Fuel	Burn	#	Area (m^2)	Distance (m)	Conc. ($\mu g/m^3$)
LA Crude	Mobile 91	4	114	30	10.1
LA Crude	Mobile 91	5	231	30	36.9
AB Crude	NOBE	1	600	125	3.0
AB Crude	NOBE	2	600	75	195.0
AB Crude	NOBE	2	600	125	64.0
AB Crude	NOBE	2	600	600	4.0
LA Crude	Mobile 92	23	181	30	12.5
LA Crude	Mobile 92	24	170	30	5.0
LA Crude	Mobile 92	28	72.8	30	13.1
LA Crude	Mobile 92	30	37.2	30	11.7
LA Crude	Mobile 92	3	114	30	12.7

2,3-Dimethylpentane

C_7H_{16} — Concentration Prediction Equation

$y = a + b*(\text{fire size}, m^2) - c*\ln(\text{distance from fire}, m)\ \mu g/m^3$

Crude oil	a, b, c		173	0.294	56.8

Fuel	Burn	#	Area (m^2)	Distance (m)	Conc. ($\mu g/m^3$)
LA Crude	Mobile 91	4	114	30	14.7
AB Crude	NOBE 93	1	600	125	7.0
AB Crude	NOBE 93	2	600	75	170.0
AB Crude	NOBE 93	2	600	125	64.0
AB Crude	NOBE 93	2	600	600	3.0
LA Crude	Mobile 92	23	181	30	16.2
LA Crude	Mobile 92	24	170	30	22.2
LA Crude	Mobile 92	28	72.8	30	7.0
LA Crude	Mobile 92	30	37.2	30	6.6
LA Crude	Mobile 92	3	114	30	7.0
LA Crude	Mobile 92	3	114	30	7.8
LA Crude	Mobile 92	4	114	30	14.9
LA Crude	Mobile 92	5	231	30	53.5

TABLE A1.4
VOC Emissions from Crude Oil Burns

2,4-Dimethylhexane

C_8H_{18} Concentration Prediction Equation

$y = a + b*(\text{fire size}, m^2) - c*\ln(\text{distance from fire, m})\ \mu g/m^3$

Crude oil	a, b, c		72.2	0.109	22.7
Fuel	**Burn**	**#**	**Area**	**Distance**	**Conc.**
			m^2	m	$\mu g/m^3$
AB Crude	NOBE 93	1	600	125	2.0
AB Crude	NOBE 93	2	600	75	65.0
LA Crude	NOBE 93	2	600	125	21.0
LA Crude	NOBE 93	2	600	600	1.0
LA Crude	Mobile 92	2	36	30	1.0
LA Crude	Mobile 92	2	231	60	1.0

2,4-Dimethylpentane

C_7H_{16} Concentration Prediction Equation

$y = a + b*(\text{fire size}, m^2) - c*\ln(\text{distance from fire, m})\ \mu g/m^3$

Crude oil	a, b, c		99	0.164	32
Fuel	**Burn**	**#**	**Area**	**Distance**	**Conc.**
			m^2	m	$\mu g/m^3$
LA Crude	Mobile 91	4	114	30	7.1
AB Crude	NOBE 93	1	600	125	3.0
AB Crude	NOBE 93	2	600	75	104.0
AB Crude	NOBE 93	2	600	125	33.0
AB Crude	NOBE 93	2	600	600	2.0
LA Crude	Mobile 92	23	181	30	9.2
LA Crude	Mobile 92	24	170	30	10.5
LA Crude	Mobile 92	28	72.8	30	9.9
LA Crude	Mobile 92	30	37.2	30	8.5
LA Crude	Mobile 92	3	114	30	9.9
LA Crude	Mobile 92	3	114	30	5.0
LA Crude	Mobile 92	4	114	30	7.3
LA Crude	Mobile 92	5	231	30	25.5

2,5-Dimethylhexane

C_8H_{18} Concentration Prediction Equation

$y = a + b*(\text{fire size}, m^2) - c*\ln(\text{distance from fire, m})\ \mu g/m^3$

Crude oil	a, b, c		40.5	0.0787	14.3
Fuel	**Burn**	**#**	**Area**	**Distance**	**Conc.**
			m^2	m	$\mu g/m^3$
AB Crude	NOBE 93	1	600	125	2.0
AB Crude	NOBE 93	2	600	75	41.0
AB Crude	NOBE 93	2	600	125	15.0
AB Crude	NOBE 93	2	600	600	1.0
LA Crude	Mobile 92	23	181	30	3.3
LA Crude	Mobile 92	24	170	30	2.2
LA Crude	Mobile 92	4	114	30	3.4
LA Crude	Mobile 92	5	231	30	12.9

2-Ethyltoluene

C_9H_{12} Concentration Prediction Equation

$y = a + b*(\text{fire size}, m^2) - c*\ln(\text{distance from fire, m})\ \mu g/m^3$

Crude oil	a, b, c		5.98	0.00826	1.47
Fuel	**Burn**	**#**	**Area**	**Distance**	**Conc.**
			m^2	m	$\mu g/m^3$
LA Crude	Mobile 91	4	114	30	4.4
AB Crude	NOBE 93	2	600	75	6.0
AB Crude	NOBE 93	2	600	125	2.0
AB Crude	NOBE 93	2	600	600	2.0
LA Crude	Mobile 92	23	181	30	0.8
LA Crude	Mobile 92	24	170	30	1.0
LA Crude	Mobile 92	28	72.8	30	0.7
LA Crude	Mobile 92	30	37.2	30	0.6
LA Crude	Mobile 92	3	114	30	0.8
LA Crude	Mobile 92	4	114	30	4.5
LA Crude	Mobile 92	5	231	30	3.7

TABLE A1.5
VOC Emissions from Crude Oil Burns

2-Methylbutane

C_5H_{12}

Concentration Prediction Equation

$y = a + b*(fire\ size, m^2) - c*ln(distance\ from\ fire, m)\ \mu g/m^3$

Crude oil	a, b, c		2040	3.98	723
Fuel	Burn	#	Area	Distance	Conc.
			m²	m	µg/m³
LA Crude	Mobile 91	4	114	30	73.5
LA Crude	Mobile 91	5	231	30	328.1
AB Crude	NOBE 93	1	600	125	1.0
AB Crude	NOBE 93	2	600	75	2494.0
AB Crude	NOBE 93	2	600	125	668.0
AB Crude	NOBE 93	2	600	600	32.0
LA Crude	Mobile 92	23	181	30	119.3
LA Crude	Mobile 92	24	170	30	158.9
LA Crude	Mobile 92	28	72.8	30	11.8
LA Crude	Mobile 92	30	37.2	30	16.3
LA Crude	Mobile 92	3	114	30	11.4
LA Crude	Mobile 92	3	114	30	50.9
LA Crude	Mobile 92	4	114	30	85.7
LA Crude	Mobile 92	5	231	30	329.2

2-Methylheptane

C_8H_{18}

Concentration Prediction Equation

$y = a + b*(fire\ size, m^2) - c*ln(distance\ from\ fire, m)\ \mu g/m^3$

Crude oil	a, b, c		240	0.384	77.4
Fuel	Burn	#	Area	Distance	Conc.
			m²	m	µg/m³
AB Crude	NOBE 93	1	600	125	13
AB Crude	NOBE 93	2	600	75	246
AB Crude	NOBE 93	2	600	125	89
AB Crude	NOBE 93	2	600	600	3

2-Methylhexane

C_7H_{16}

Concentration Prediction Equation

$y = a + b*(fire\ size, m^2) - c*ln(distance\ from\ fire, m)\ \mu g/m^3$

Crude oil	a, b, c		225	0.33	68.9
Fuel	Burn	#	Area	Distance	Conc.
			m²	m	µg/m³
LA Crude	Mobile 91	4		30	33.39
LA Crude	Mobile 91	5	231	30	102.1
AB Crude	NOBE 93	1	600	125	9
AB Crude	NOBE 93	2	600	75	182
AB Crude	NOBE 93	2	600	125	88
AB Crude	NOBE 93	2	600	600	6
LA Crude	Mobile 92	23	181	30	37.58
LA Crude	Mobile 92	28	72.8	30	12.32
LA Crude	Mobile 92	30	37.2	30	11.11
LA Crude	Mobile 92	3	114	30	10.87
LA Crude	Mobile 92	3	114	30	15.57
LA Crude	Mobile 92	4	114	30	34.54

2-Methylpentane

C_6H_{14}

Concentration Prediction Equation

$y = a + b*(fire\ size, m^2) - c*ln(distance\ from\ fire, m)\ \mu g/m^3$

Crude oil	a, b, c		1200	2.21	411
Fuel	Burn	#	Area	Distance	Conc.
			m²	m	µg/m³
LA Crude	Mobile 91	4	114	30	61.3
LA Crude	Mobile 91	5	231	30	240.1
AB Crude	NOBE 93	1	600	125	24.0
AB Crude	NOBE 93	2	600	75	1373.0
AB Crude	NOBE 93	2	600	125	404.0
AB Crude	NOBE 93	2	600	600	16.0
LA Crude	Mobile 92	23	181	30	84.3
LA Crude	Mobile 92	24	170	30	110.1
LA Crude	Mobile 92	28	72.8	30	46.9
LA Crude	Mobile 92	30	37.2	30	41.3
LA Crude	Mobile 92	3	114	30	41.7
LA Crude	Mobile 92	3	114	30	37.3
LA Crude	Mobile 92	4	114	30	65.1
LA Crude	Mobile 92	5	231	30	241.6

TABLE A1.6
VOC Emissions from Crude Oil Burns

2-Methylpropane

C_4H_{10}

Concentration Prediction Equation

$y = a + b*(\text{fire size}, m^2) - c*\ln(\text{distance from fire}, m)\ \mu g/m^3$

Crude oil	a, b, c		205	0.296	63.5
Fuel	Burn	#	Area	Distance	Conc.
			m^2	m	$\mu g/m^3$
LA Crude	Mobile 91	3	114	30	11.9
LA Crude	Mobile 91	4	114	30	20.0
LA Crude	Mobile 91	5	231	30	88.0
LA Crude	Mobile 91	4	114	30	15.9
LA Crude	Mobile 91	5	231	30	87.5
AB Crude	Nobe	1	600	125	50.0
AB Crude	Nobe	2	600	500	1.0
LA Crude	Mobile 92	23	181	30	30.2
LA Crude	Mobile 92	24	170	30	41.8
LA Crude	Mobile 92	28	72.8	30	4.0
LA Crude	Mobile 92	30	37.2	30	6.6
LA Crude	Mobile 92	3	114	30	3.8

3,6-Dimethyloctane

$C_{10}H_{22}$

Concentration Prediction Equation

$y = a + b*(\text{fire size}, m^2) - c*\ln(\text{distance from fire}, m)\ \mu g/m^3$

Crude oil	a, b, c		78	0.023	18.6
Fuel	Burn	#	Area	Distance	Conc.
			m^2	m	$\mu g/m^3$
AB Crude	NOBE 93	1	600	125	1
AB Crude	NOBE 93	2	600	75	12
AB Crude	NOBE 93	2	600	125	4

3-Ethyltoluene

C_9H_{12}

Concentration Prediction Equation

$y = a + b*(\text{fire size}, m^2) - c*\ln(\text{distance from fire}, m)\ \mu g/m^3$

Crude oil	a, b, c		14	0.022	3.52
Fuel	Burn	#	Area	Distance	Conc.
			m^2	m	$\mu g/m^3$
LA Crude	Mobile 91	4	114	30	10.5
LA Crude	Mobile 91	5	231	30	10.6
AB Crude	NOBE 93	2	600	75	13.0
AB Crude	NOBE 93	2	600	125	6.0
AB Crude	NOBE 93	2	600	600	6.0

3-Ethyltoluene contd.

Fuel	Burn	#	Area	Distance	Conc.
LA Crude	Mobile 92	23	181	30	1.9
LA Crude	Mobile 92	24	170	30	2.4
LA Crude	Mobile 92	28	72.8	30	1.3
LA Crude	Mobile 92	30	37.2	30	1.1
LA Crude	Mobile 92	3	114	30	1.6
LA Crude	Mobile 92	3	114	30	0.7
LA Crude	Mobile 92	4	114	30	10.6
LA Crude	Mobile 92	5	231	30	10.6

2-Methylbutane

C_5H_{12}

Concentration Prediction Equation

$y = a + b*(\text{fire size}, m^2) - c*\ln(\text{distance from fire}, m)\ \mu g/m^3$

Crude oil	a, b, c		2221	4.58	821
Fuel	Burn	#	Area	Distance	Conc.
			m^2	m	$\mu g/m^3$
LA Crude	Mobile 91	4	114	30	73.5
LA Crude	Mobile 91	5	231	30	328.1
AB Crude	NOBE 93	1	600	125	1.0
AB Crude	NOBE 93	2	600	75	2,494.0
AB Crude	NOBE 93	2	600	125	668.0
AB Crude	NOBE 93	2	600	600	32.0

3-Methylheptane

C_8H_{18}

Concentration Prediction Equation

$y = a + b*(\text{fire size}, m^2) - c*\ln(\text{distance from fire}, m)\ \mu g/m^3$

Crude oil	a, b, c		240	0.384	77.4
Fuel	Burn	#	Area	Distance	Conc.
			m^2	m	$\mu g/m^3$
LA Crude	Mobile 91	4	114	30	28.6
AB Crude	NOBE 93	1	600	125	9.0
AB Crude	NOBE 93	2	600	75	213.0
AB Crude	NOBE 93	2	600	125	74.0
AB Crude	NOBE 93	2	600	600	4.0
LA Crude	Mobile 92	23	181	30	26.1
LA Crude	Mobile 92	24	170	30	17.3
LA Crude	Mobile 92	28	72.8	30	4.1
LA Crude	Mobile 92	30	37.2	30	3.5
LA Crude	Mobile 92	3	114	30	4.5
LA Crude	Mobile 92	3	114	30	9.3
LA Crude	Mobile 92	4	114	30	28.8
LA Crude	Mobile 92	5	231	30	114.5

TABLE A1.7
VOC Emissions from Crude Oil Burns

3-Methylhexane

C_7H_{16}

Concentration Prediction Equation

$y = a + b*(fire\ size, m^2) - c*ln(distance\ from\ fire, m)\ \mu g/m^3$

Crude oil	a, b, c		526	0.896	175
Fuel	**Burn**	**#**	**Area**	**Distance**	**Conc.**
			m^2	m	$\mu g/m^3$
LA Crude	Mobile 91	4	114	30	40.3
LA Crude	Mobile 91	5	231	30	159.2
AB Crude	NOBE 93	1	600	125	19.0
AB Crude	NOBE 93	2	600	75	512.0
AB Crude	NOBE 93	2	600	125	168.0
AB Crude	NOBE 93	2	600	600	6.0
LA Crude	Mobile 92	23	181	30	46.8
LA Crude	Mobile 92	24	170	30	64.0
LA Crude	Mobile 92	28	72.8	30	12.2
LA Crude	Mobile 92	30	37.2	30	9.6
LA Crude	Mobile 92	3	114	30	10.8
LA Crude	Mobile 92	3	114	30	4.8
LA Crude	Mobile 92	4	114	30	41.3
LA Crude	Mobile 92	5	231	30	159.5

3-Methylpentane

C_6H_{14}

Concentration Prediction Equation

$y = a + b*(fire\ size, m^2) - c*ln(distance\ from\ fire, m)\ \mu g/m^3$

Crude oil	a, b, c		822	1.41	272
Fuel	**Burn**	**#**	**Area**	**Distance**	**Conc.**
			m^2	m	$\mu g/m^3$
LA Crude	Mobile 91	4	114	30	50.5
LA Crude	Mobile 91	5	231	30	190.2
AB Crude	NOBE 93	1	600	125	28.0
AB Crude	NOBE 93	2	600	75	888.0
AB Crude	NOBE 93	2	600	125	285.0
AB Crude	NOBE 93	2	600	600	6.0
LA Crude	Mobile 92	23	181	30	64.0
LA Crude	Mobile 92	24	170	30	84.7
LA Crude	Mobile 92	28	72.8	30	72.2
LA Crude	Mobile 92	30	37.2	30	63.8
LA Crude	Mobile 92	3	114	30	71.5
LA Crude	Mobile 92	4	114	30	52.4
LA Crude	Mobile 92	5	231	30	191.1

4-Ethyltoluene

C_9H_{12}

Concentration Prediction Equation

$y = a + b*(fire\ size, m^2) - c*ln(distance\ from\ fire, m)\ \mu g/m^3$

Crude oil	a, b, c		4.79	0.0051	0.85
Fuel	**Burn**	**#**	**Area**	**Distance**	**Conc.**
			m^2	m	$\mu g/m^3$
LA Crude	Mobile 91	4	114	30	5.2
AB Crude	NOBE 93	2	600	75	5.0
AB Crude	NOBE 93	2	600	125	2.0
AB Crude	NOBE 93	2	600	600	3.0
LA Crude	Mobile 92	23	181	30	0.9
LA Crude	Mobile 92	24	170	30	1.3
LA Crude	Mobile 92	28	72.8	30	1.0
LA Crude	Mobile 92	30	37.2	30	0.7
LA Crude	Mobile 92	3	114	30	1.5
LA Crude	Mobile 92	4	114	30	5.3
LA Crude	Mobile 92	5	231	30	5.0

4-Methylheptane

C_8H_{18}

Concentration Prediction Equation

$y = a + b*(fire\ size, m^2) - c*ln(distance\ from\ fire, m)\ \mu g/m^3$

Crude oil	a, b, c		30.1	0.063	9.44
Fuel	**Burn**	**#**	**Area**	**Distance**	**Conc.**
			m^2	m	$\mu g/m^3$
AB Crude	NOBE 93	1	600	125	1.0
AB Crude	NOBE 93	2	600	75	66.0
AB Crude	NOBE 93	2	600	125	17.0
AB Crude	NOBE 93	2	600	600	1.0
LA Crude	Mobile 92	2	231	15	3.0
LA Crude	Mobile 92	1	36	85	1.0

TABLE A1.8
VOC Emissions from Crude Oil Burns

Benzene

C_6H_6

Concentration Prediction Equation

$y = a + b*(fire\ size, m^2) - c*ln(distance\ from\ fire, m)\ \mu g/m^3$

Crude oil	a, b, c		72	0.0242	14.1
Fuel	**Burn**	**#**	**Area**	**Distance**	**Conc.**
			m^2	m	$\mu g/m^3$
LA Crude	Mobile 91	4	114	30	25.1
LA Crude	Mobile 91	5	231	30	72.5
LA Crude	Mobile 91	1	36	23	55.6
LA Crude	Mobile 91	1	36	46	46.8
LA Crude	Mobile 91	2	231	23	68.9
LA Crude	Mobile 91	2	231	46	13.9
LA Crude	Mobile 91	2	231	76	0.0
LA Crude	Mobile 91	3	231	23	12.7
LA Crude	Mobile 91	4	231	23	13.7
LA Crude	Mobile 91	4	231	46	6.0
LA Crude	Mobile 91	6	231	23	11.2
AB Crude	NOBE 93	2	600	75	29.0
AB Crude	NOBE 93	2	600	125	6.0
AB Crude	NOBE 93	2	600	600	8.0
LA Crude	Mobile 92	23	181	30	42.4
LA Crude	Mobile 92	24	170	30	32.6
LA Crude	Mobile 92	28	72.8	30	2.8
LA Crude	Mobile 92	30	37.2	30	12.1
LA Crude	Mobile 92	3	114	30	5.0
LA Crude	Mobile 92	3	114	30	12.0
LA Crude	Mobile 92	4	114	30	26.7
LA Crude	Mobile 92	5	231	30	72.9

Butane

C_4H_{10}

Concentration Prediction Equation

$y = a + b*(fire\ size, m^2) - c*ln(distance\ from\ fire, m)\ \mu g/m^3$

Crude oil	a, b, c		1700	3.31	604
Fuel	**Burn**	**#**	**Area**	**Distance**	**Conc.**
			m^2	m	$\mu g/m^3$
LA Crude	Mobile 91	4	114	30	43.6
LA Crude	Mobile 91	5	231	30	245.8
AB Crude	NOBE 93	1	600	125	10.0
AB Crude	NOBE 93	2	600	75	2,102.0
AB Crude	NOBE 93	2	600	125	472.0
AB Crude	NOBE 93	2	600	600	11.0
LA Crude	Mobile 92	23	181	30	90.1
LA Crude	Mobile 92	24	170	30	121.4
LA Crude	Mobile 92	28	72.8	30	8.8
LA Crude	Mobile 92	30	37.2	30	9.6
LA Crude	Mobile 92	3	114	30	10.8
LA Crude	Mobile 92	3	114	30	34.9
LA Crude	Mobile 92	4	114	30	56.0
LA Crude	Mobile 92	5	231	30	246.8

c-1,3-Dimethylcyclohexane

C_8H_{18}

Concentration Prediction Equation

$y = a + b*(fire\ size, m^2) - c*ln(distance\ from\ fire, m)\ \mu g/m^3$

Crude oil	a, b, c		82.4	0.21	28
Fuel	**Burn**	**#**	**Area**	**Distance**	**Conc.**
			m^2	m	$\mu g/m^3$
AB Crude	NOBE 93	1	600	125	11.0
AB Crude	NOBE 93	2	600	75	190.0
AB Crude	NOBE 93	2	600	125	84.0
AB Crude	NOBE 93	2	600	600	1.0
LA Crude	Mobile 92	2	231	15	3.0
LA Crude	Mobile 92	1	36	85	1.0

TABLE A1.9
VOC Emissions from Crude Oil Burns

c-1,4/t-1,3-Dimethylcyclohexane

C_8H_{18}

Concentration Prediction Equation

$y = a + b*(fire\ size, m^2) - c*ln(distance\ from\ fire, m)\ \mu g/m^3$

Crude oil	a, b, c		22.4	0.0626	6.74
Fuel	Burn	#	Area	Distance	Conc.
			m^2	m	$\mu g/m^3$
AB Crude	NOBE 93	1	600	125	2.0
AB Crude	NOBE 93	2	600	75	66.0
AB Crude	NOBE 93	2	600	125	21.0
LA Crude	Mobile 92	2	231	15	9.0
LA Crude	Mobile 92	1	36	85	1.0

c-2-Butene

C_4H_8

Concentration Prediction Equation

$y = a + b*(fire\ size, m^2) - c*ln(distance\ from\ fire, m)\ \mu g/m^3$

Crude oil	a, b, c		4.73	0.0108	1.6
Fuel	Burn	#	Area	Distance	Conc.
			m^2	m	$\mu g/m^3$
LA Crude	Mobile 91	28	72.8	30	0.3
LA Crude	Mobile 91	3	114	30	0.4
AB Crude	NOBE 93	2	600	600	1.0
AB Crude	NOBE 93	2	600	600	1.0
AB Crude	NOBE 93	2	600	50	5.0
LA Crude	Mobile 92	3	114	30	0.4

Cyclohexane

C_6H_{14}

Concentration Prediction Equation

$y = a + b*(fire\ size, m^2) - c*ln(distance\ from\ fire, m)\ \mu g/m^3$

Crude oil	a, b, c		726	1.43	256
Fuel	Burn	#	Area	Distance	Conc.
			m^2	m	$\mu g/m^3$
LA Crude	Mobile 91	4	114	30	39.3
LA Crude	Mobile 91	5	231	30	112.9
AB Crude	NOBE 93	1	600	125	40.0
AB Crude	NOBE 93	2	600	75	849.0
AB Crude	NOBE 93	2	600	125	286.0
AB Crude	NOBE 93	2	600	600	3.0
LA Crude	Mobile 92	23	181	30	42.6
LA Crude	Mobile 92	24	170	30	60.5
LA Crude	Mobile 92	28	72.8	30	7.8
LA Crude	Mobile 92	30	37.2	30	7.8
LA Crude	Mobile 92	3	114	30	7.4
LA Crude	Mobile 92	3	114	30	16.5
LA Crude	Mobile 92	4	114	30	39.7
LA Crude	Mobile 92	5	231	30	113.0

Cyclopentane

C_5H_{12}

Concentration Prediction Equation

$y = a + b*(fire\ size, m^2) - c*ln(distance\ from\ fire, m)\ \mu g/m^3$

Crude oil	a, b, c		262	0.526	93.8
Fuel	Burn	#	Area	Distance	Conc.
			m^2	m	$\mu g/m^3$
LA Crude	Mobile 91	4	114	30	7.2
AB Crude	NOBE 93	1	600	125	10.0
AB Crude	NOBE 93	2	600	75	310.0
AB Crude	NOBE 93	2	600	125	102.0
AB Crude	NOBE 93	2	600	600	2.0
LA Crude	Mobile 92	23	181	30	9.8
LA Crude	Mobile 92	24	170	30	12.9
LA Crude	Mobile 92	28	72.8	30	0.9
LA Crude	Mobile 92	30	37.2	30	1.5
LA Crude	Mobile 92	3	114	30	0.7
LA Crude	Mobile 92	3	114	30	1.3
LA Crude	Mobile 92	4	114	30	7.5
LA Crude	Mobile 92	5	231	30	28.6

TABLE A1.10
VOC Emissions from Crude Oil Burns

Decane

$C_{10}H_{22}$ — Concentration Prediction Equation

$y = a + b*(\text{fire size}, m^2) - c*\ln(\text{distance from fire, m})\ \mu g/m^3$

Crude oil	a, b, c		97	0.0899	24.5
Fuel	**Burn**	**#**	**Area**	**Distance**	**Conc.**
			m^2	m	$\mu g/m^3$
LA Crude	Mobile 91	4	114	30	68.5
AB Crude	NOBE 93	1	600	125	3.0
AB Crude	NOBE 93	2	600	75	72.0
AB Crude	NOBE 93	2	600	125	32.0
AB Crude	NOBE 93	2	600	600	1.0
LA Crude	Mobile 92	23	181	30	5.9
LA Crude	Mobile 92	24	170	30	11.9
LA Crude	Mobile 92	28	72.8	30	11.4
LA Crude	Mobile 92	30	37.2	30	9.1
LA Crude	Mobile 92	3	114	30	9.9
LA Crude	Mobile 92	3	114	30	1.3
LA Crude	Mobile 92	4	114	30	68.8
LA Crude	Mobile 92	5	231	30	38.3

Ethylbenzene

C_8H_{10} — Concentration Prediction Equation

$y = a + b*(\text{fire size}, m^2) - c*\ln(\text{distance from fire, m})\ \mu g/m^3$

Crude oil	a, b, c		25	0.0391	6.69
Fuel	**Burn**	**#**	**Area**	**Distance**	**Conc.**
			m^2	m	$\mu g/m^3$
LA Crude	Mobile 91	4	114	30	8.1
LA Crude	Mobile 91	5	231	30	22.9
LA Crude	Mobile 91	2	231	23	9.1
AB Crude	NOBE 93	2	600	75	14.0
AB Crude	NOBE 93	2	600	600	7.0
LA Crude	Mobile 92	23	181	30	5.2
LA Crude	Mobile 92	24	170	30	6.5
LA Crude	Mobile 92	28	72.8	30	2.5
LA Crude	Mobile 92	30	37.2	30	2.6
LA Crude	Mobile 92	3	114	30	3.3
LA Crude	Mobile 92	3	114	30	1.9
LA Crude	Mobile 92	4	114	30	8.4
LA Crude	Mobile 92	5	231	30	22.9

Dodecane

$C_{12}H_{26}$ — Concentration Prediction Equation

$y = a + b*(\text{fire size}, m^2) - c*\ln(\text{distance from fire, m})\ \mu g/m^3$

Crude oil	a, b, c		27.1	0.0368	7.43
Fuel	**Burn**	**#**	**Area**	**Distance**	**Conc.**
			m^2	m	$\mu g/m^3$
LA Crude	Mobile 91	4	114	30	15.4
AB Crude	NOBE 93	2	600	75	18.0
AB Crude	NOBE 93	2	600	125	14.0
AB Crude	NOBE 93	2	600	600	1.0
LA Crude	Mobile 92	23	181	30	LDL
LA Crude	Mobile 92	24	170	30	LDL
LA Crude	Mobile 92	28	72.8	30	0.5
LA Crude	Mobile 92	30	37.2	30	0.3
LA Crude	Mobile 92	3	114	30	0.4
LA Crude	Mobile 92	3	114	30	LDL
LA Crude	Mobile 92	4	114	30	15.6
LA Crude	Mobile 92	5	231	30	2.8

Heptane

C_7H_{16} — Concentration Prediction Equation

$y = a + b*(\text{fire size}, m^2) - c*\ln(\text{distance from fire, m})\ \mu g/m^3$

Crude oil	a, b, c		1170	2.11	400
Fuel	**Burn**	**#**	**Area**	**Distance**	**Conc.**
			m^2	m	$\mu g/m^3$
LA Crude	Mobile 91	4	114	30	72.2
LA Crude	Mobile 91	5	231	30	297.0
AB Crude	NOBE 93	1	600	125	56.0
AB Crude	NOBE 93	2	600	75	1224.0
AB Crude	NOBE 93	2	600	125	361.0
AB Crude	NOBE 93	2	600	600	6.0
LA Crude	Mobile 92	23	181	30	80.8
LA Crude	Mobile 92	24	170	30	116.1
LA Crude	Mobile 92	28	72.8	30	1.9
LA Crude	Mobile 92	30	37.2	30	3.5
LA Crude	Mobile 92	3	114	30	2.6
LA Crude	Mobile 92	3	114	30	32.8
LA Crude	Mobile 92	4	114	30	73.0
LA Crude	Mobile 92	5	231	30	297.1

TABLE A1.11
VOC Emissions from Crude Oil Burns

Hexane

C_6H_{14}

Concentration Prediction Equation

$y = a + b*(fire\ size, m^2) - c*ln(distance\ from\ fire, m)\ \mu g/m^3$

Crude oil a, b, c			5080	6.26	1690
Fuel	**Burn**	**#**	**Area** (m^2)	**Distance** (m)	**Conc.** ($\mu g/m^3$)
LA Crude	Mobile 91	4	114	30	98.2
AB Crude	Mobile 91	5	231	30	379.9
AB Crude	NOBE 93	1	600	125	103.0
AB Crude	NOBE 93	2	600	75	2521.0
AB Crude	NOBE 93	2	600	125	568.0
LA Crude	Mobile 92	23	181	30	137.3
LA Crude	Mobile 92	24	170	30	147.2
LA Crude	Mobile 92	28	72.8	30	52.5
LA Crude	Mobile 92	30	37.2	30	47.7
LA Crude	Mobile 92	3	114	30	56.5
LA Crude	Mobile 92	3	114	30	61.4
LA Crude	Mobile 92	4	114	30	101.5
LA Crude	Mobile 92	5	231	30	380.7

Indan

C_9H_{10}

Concentration Prediction Equation

$y = a + b*(fire\ size, m^2) - c*ln(distance\ from\ fire, m)\ \mu g/m^3$

Crude oil a, b, c			2.64	0.00305	0.557
Fuel	**Burn**	**#**	**Area** (m^2)	**Distance** (m)	**Conc.** ($\mu g/m^3$)
LA Crude	Mobile 91	28	72.8	30	1.3
LA Crude	Mobile 91	30	37.2	30	0.9
LA Crude	Mobile 91	3	114	30	1.4
LA Crude	Mobile 91	5	231	30	0.6
AB Crude	NOBE 93	2	600	75	3.0
AB Crude	NOBE 93	2	600	125	1.0
AB Crude	NOBE 93	2	600	600	1.0

Isobutane (2-Methylpropane)

C_4H_{10}

Concentration Prediction Equation

$y = a + b*(fire\ size, m^2) - c*ln(distance\ from\ fire, m)\ \mu g/m^3$

Crude oil a, b, c			414	1.05	165
Fuel	**Burn**	**#**	**Area** (m^2)	**Distance** (m)	**Conc.** ($\mu g/m^3$)
AB Crude	NOBE 93	2	600	75	441.0
AB Crude	NOBE 93	2	600	125	98.0
AB Crude	NOBE 93	2	600	600	5.0
AB Crude	NOBE 93	2	600	600	5.0
LA Crude	Mobile 92	1	36	15	5.0

iso-Butylbenzene

$C_{10}H_{14}$

Concentration Prediction Equation

$y = a + b*(fire\ size, m^2) - c*ln(distance\ from\ fire, m)\ \mu g/m^3$

Crude oil a, b, c			3.48	0.00574	1.06
Fuel	**Burn**	**#**	**Area** (m^2)	**Distance** (m)	**Conc.** ($\mu g/m^3$)
LA Crude	Mobile 91	5	231	30	0.8
LA Crude	Mobile 91	4	114	30	1.5
LA Crude	Mobile 91	5	231	30	0.8
LA Crude	Mobile 91	5	231	30	0.8
AB Crude	NOBE 93	2	600	75	5.0
AB Crude	NOBE 93	2	600	75	0.1
AB Crude	NOBE 93	2	600	600	0.0

Isoprene (2-Methyl-1,3-Butadiene)

C_5H_8

Concentration Prediction Equation

$y = a + b*(fire\ size, m^2) - c*ln(distance\ from\ fire, m)\ \mu g/m^3$

Crude oil a, b, c			17.4	0.0314	5.51
Fuel	**Burn**	**#**	**Area** (m^2)	**Distance** (m)	**Conc.** ($\mu g/m^3$)
AB Crude	NOBE 93	2	600	600	1.0
LA Crude	Mobile 92	30	37.2	30	0.7
LA Crude	Mobile 92	3	114	30	0.8
LA Crude	Mobile 92	5	231	30	6.5

TABLE A1.12
VOC Emissions from Crude Oil Burns

Isopropylbenzene

C_9H_{12}

Concentration Prediction Equation

$y = a + b*(\text{fire size}, m^2) - c*\ln(\text{distance from fire}, m)\ \mu g/m^3$

Crude oil	a, b, c		21.4	0.0178	6.41
Fuel	**Burn**	**#**	**Area**	**Distance**	**Conc.**
			m^2	m	$\mu g/m^3$
LA Crude	Mobile 91	4	114	30	2.9
AB Crude	NOBE 93	1	600	125	0.0
AB Crude	NOBE 93	2	600	75	5.0
AB Crude	NOBE 93	2	600	125	2.0
LA Crude	Mobile 92	23	181	30	1.3
LA Crude	Mobile 92	24	170	30	1.6
LA Crude	Mobile 92	28	72.8	30	0.7
LA Crude	Mobile 92	30	37.2	30	0.7
LA Crude	Mobile 92	3	114	30	0.6
LA Crude	Mobile 92	4	114	30	3.0
LA Crude	Mobile 92	5	231	30	4.5

Methylcyclohexane

C_7H_{16}

Concentration Prediction Equation

$y = a + b*(\text{fire size}, m^2) - c*\ln(\text{distance from fire}, m)\ \mu g/m^3$

Crude oil	a, b, c		1660	3.03	571
Fuel	**Burn**	**#**	**Area**	**Distance**	**Conc.**
			m^2	m	$\mu g/m^3$
LA Crude	Mobile 91	4	114	30	99.8
AB Crude	NOBE 93	1	600	125	81.0
AB Crude	NOBE 93	2	600	75	1762.0
AB Crude	NOBE 93	2	600	125	522.0
AB Crude	NOBE 93	2	600	600	3.0
LA Crude	Mobile 92	23	181	30	107.3
LA Crude	Mobile 92	24	170	30	146.2
LA Crude	Mobile 92	28	72.8	30	0.9
LA Crude	Mobile 92	30	37.2	30	4.0
LA Crude	Mobile 92	3	114	30	1.1
LA Crude	Mobile 92	3	114	30	47.9
LA Crude	Mobile 92	4	114	30	100.3
LA Crude	Mobile 92	5	231	30	380.1

Methylcyclopentane

C_6H_{12}

Concentration Prediction Equation

$y = a + b*(\text{fire size}, m^2) - c*\ln(\text{distance from fire}, m)\ \mu g/m^3$

Crude oil	a, b, c		2090	2.9	713
Fuel	**Burn**	**#**	**Area**	**Distance**	**Conc.**
			m^2	m	$\mu g/m^3$
LA Crude	Mobile 91	4	114	30	30.1
LA Crude	Mobile 91	5	231	30	114.1
AB Crude	NOBE 93	1	600	125	49.0
AB Crude	NOBE 93	2	600	75	1307.0
AB Crude	NOBE 93	2	600	125	382.0
LA Crude	Mobile 92	23	181	30	36.5
LA Crude	Mobile 92	24	170	30	49.4
LA Crude	Mobile 92	28	72.8	30	38.9
LA Crude	Mobile 92	30	37.2	30	34.1
LA Crude	Mobile 92	3	114	30	38.6
LA Crude	Mobile 92	3	114	30	8.6
LA Crude	Mobile 92	4	114	30	30.6
LA Crude	Mobile 92	5	231	30	114.5

TABLE A1.13
VOC Emissions from Crude Oil Burns

m,p-Xylene

C_8H_{10}

Concentration Prediction Equation

$y = a + b*(fire\ size, m^2) - c*ln(distance\ from\ fire, m)\ \mu g/m^3$

Crude oil	a, b, c		88.6	0.109	20.8
Fuel	**Burn**	**#**	**Area** m²	**Distance** m	**Conc.** μg/m³
LA Crude	Mobile 91	4	114	30	49.7
LA Crude	Mobile 91	5	231	30	119.5
AB Crude	NOBE 93	2	600	75	45.0
AB Crude	NOBE 93	2	600	600	21.0
LA Crude	Mobile 92	1	36	23	10.4
LA Crude	Mobile 92	1	36	46	11.6
LA Crude	Mobile 92	1	36	76	14.0
LA Crude	Mobile 92	2	231	23	33.3
LA Crude	Mobile 92	2	231	46	7.1
LA Crude	Mobile 92	4	231	23	7.3
LA Crude	Mobile 92	23	181	30	27.7
LA Crude	Mobile 92	24	170	30	34.2
LA Crude	Mobile 92	28	72.8	30	9.2
LA Crude	Mobile 92	30	37.2	30	10.0
LA Crude	Mobile 92	3	114	30	10.9
LA Crude	Mobile 92	3	114	30	9.0
LA Crude	Mobile 92	4	114	30	50.4
LA Crude	Mobile 92	5	231	30	119.7

n-Butylbenzene

$C_{10}H_{14}$

Concentration Prediction Equation

$y = a + b*(fire\ size, m^2) - c*ln(distance\ from\ fire, m)\ \mu g/m^3$

Crude oil	a, b, c		3.28	0.003	0.806
Fuel	**Burn**	**#**	**Area** m²	**Distance** m	**Conc.** μg/m³
LA Crude	Mobile 91	4	114	30	2.5
AB Crude	NOBE 93	2	600	75	1.0
AB Crude	NOBE 93	2	600	125	1.0
AB Crude	NOBE 93	2	600	75	2.0
AB Crude	NOBE 93	2	600	125	1.0
AB Crude	NOBE 93	2	600	75	2.0
AB Crude	NOBE 93	2	600	125	1.0
AB Crude	NOBE 93	2	600	75	2.0
AB Crude	NOBE 93	2	600	600	0.0
LA Crude	Mobile 92	23	181	30	0.2
LA Crude	Mobile 92	24	170	30	0.4
LA Crude	Mobile 92	28	72.8	30	0.2
LA Crude	Mobile 92	30	37.2	30	0.2
LA Crude	Mobile 92	3	114	30	0.2
LA Crude	Mobile 92	4	114	30	2.6
LA Crude	Mobile 92	5	231	30	1.0

Naphtalene

$C_{10}H_8$

Concentration Prediction Equation

$y = a + b*(fire\ size, m^2) - c*ln(distance\ from\ fire, m)\ \mu g/m^3$

Crude oil	a, b, c		5.92	0.00991	1.7
Fuel	**Burn**	**#**	**Area** m²	**Distance** m	**Conc.** μg/m³
LA Crude	Mobile 91	28	72.8	30	0.5
LA Crude	Mobile 91	30	37.2	30	0.9
LA Crude	Mobile 91	3	114	30	0.5
LA Crude	Mobile 91	4	114	30	0.8
LA Crude	Mobile 91	5	231	30	5.2
LA Crude	Mobile 91	4	114	30	5.1
AB Crude	NOBE 93	2	600	600	1.0
LA Crude	Mobile 92	23	181	30	0.2
LA Crude	Mobile 92	24	170	30	0.4
LA Crude	Mobile 92	28	72.8	30	0.2
LA Crude	Mobile 92	30	37.2	30	0.2
LA Crude	Mobile 92	3	114	30	0.2
LA Crude	Mobile 92	4	114	30	2.6
LA Crude	Mobile 92	5	231	30	1.0

Nonane

C_9H_{20}

Concentration Prediction Equation

$y = a + b*(fire\ size, m^2) - c*ln(distance\ from\ fire, m)\ \mu g/m^3$

Crude oil	a, b, c		232	0.328	70.5
Fuel	**Burn**	**#**	**Area** m²	**Distance** m	**Conc.** μg/m³
LA Crude	Mobile 91	4	114	30	58.7
AB Crude	NOBE 93	1	600	125	12.0
AB Crude	NOBE 93	2	600	75	192.0
AB Crude	NOBE 93	2	600	125	73.0
AB Crude	NOBE 93	2	600	600	1.0
LA Crude	Mobile 92	23	181	30	21.7
LA Crude	Mobile 92	24	170	30	25.3
LA Crude	Mobile 92	28	72.8	30	11.7
LA Crude	Mobile 92	30	37.2	30	11.0
LA Crude	Mobile 92	3	114	30	10.9
LA Crude	Mobile 92	3	114	30	5.7
LA Crude	Mobile 92	4	114	30	58.9
LA Crude	Mobile 92	5	231	30	103.6

TABLE A1.14
VOC Emissions from Crude Oil Burns

n-Propylbenzene

C_9H_{12}

Concentration Prediction Equation

$y = a + b*(fire\ size, m^2) - c*ln(distance\ from\ fire, m)\ \mu g/m^3$

Crude oil	a, b, c	6.85	0.0073	1.52
Fuel	**Burn**	**#**	**Area** m^2	**Distance** m	**Conc.** $\mu g/m^3$
LA Crude	Mobile 91	4	114	30	3.9
AB Crude	NOBE 93	2	600	75	6.0
AB Crude	NOBE 93	2	600	125	2.0
AB Crude	NOBE 93	2	600	600	2.0
LA Crude	Mobile 92	23	181	30	1.0
LA Crude	Mobile 92	24	170	30	1.2
LA Crude	Mobile 92	28	72.8	30	1.4
LA Crude	Mobile 92	30	37.2	30	0.7
LA Crude	Mobile 92	3	114	30	4.3
LA Crude	Mobile 92	4	114	30	4.0
LA Crude	Mobile 92	5	231	30	4.5

Octane

C_8H_{18}

Concentration Prediction Equation

$y = a + b*(fire\ size, m^2) - c*ln(distance\ from\ fire, m)\ \mu g/m^3$

Crude oil	a, b, c	513	0.776	162
Fuel	**Burn**	**#**	**Area** m^2	**Distance** m	**Conc.** $\mu g/m^3$
LA Crude	Mobile 91	4	114	30	52.1
LA Crude	Mobile 91	5	231	30	198.8
AB Crude	NOBE 93	1	600	125	24.0
AB Crude	NOBE 93	2	600	75	418.0
AB Crude	NOBE 93	2	600	125	151.0
AB Crude	NOBE 93	2	600	600	2.0
LA Crude	Mobile 92	23	181	30	45.9
LA Crude	Mobile 92	28	72.8	30	15.2
LA Crude	Mobile 92	30	37.2	30	14.5
LA Crude	Mobile 92	3	114	30	17.3
LA Crude	Mobile 92	3	114	30	13.6
LA Crude	Mobile 92	4	114	30	52.4
LA Crude	Mobile 92	5	231	30	198.9

ortho-Xylene

C_8H_{10}

Concentration Prediction Equation

$y = a + b*(fire\ size, m^2) - c*ln(distance\ from\ fire, m)\ \mu g/m^3$

Crude oil	a, b, c	26	0.0186	5.38
Fuel	**Burn**	**#**	**Area** m^2	**Distance** m	**Conc.** $\mu g/m^3$
LA Crude	Mobile 91	1	36	76	10.3
LA Crude	Mobile 91	2	231	23	11.4
LA Crude	Mobile 91	4	114	30	15.2
AB Crude	NOBE 93	2	600	75	12.0
AB Crude	NOBE 93	2	600	600	1.0
LA Crude	Mobile 92	23	181	30	6.9
LA Crude	Mobile 92	24	170	30	8.3
LA Crude	Mobile 92	28	72.8	30	3.0
LA Crude	Mobile 92	30	37.2	30	3.1
LA Crude	Mobile 92	3	114	30	3.4
LA Crude	Mobile 92	3	114	30	2.4
LA Crude	Mobile 92	4	114	30	15.5
LA Crude	Mobile 92	5	231	30	31.9

p-Cymene

$C_{10}H_{14}$

Concentration Prediction Equation

$y = a + b*(fire\ size, m^2) - c*ln(distance\ from\ fire, m)\ \mu g/m^3$

Crude oil	a, b, c	2.52	0.0055	0.0125
Fuel	**Burn**	**#**	**Area** m^2	**Distance** m	**Conc.** $\mu g/m^3$
LA Crude	Mobile 91	4	114	30	3.0
AB Crude	NOBE 93	2	600	75	9.0
AB Crude	NOBE 93	2	600	125	0.1
AB Crude	NOBE 93	2	600	500	0.0
LA Crude	Mobile 92	24	170	30	0.6
LA Crude	Mobile 92	28	72.8	30	1.5
LA Crude	Mobile 92	30	37.2	30	1.5
LA Crude	Mobile 92	3	114	30	7.6
LA Crude	Mobile 92	4	114	30	3.1
LA Crude	Mobile 92	5	231	30	2.0

TABLE A1.15
VOC Emissions from Crude Oil Burns

Pentane

C_5H_{12} — Concentration Prediction Equation

$y = a + b*(\text{fire size}, m^2) - c*\ln(\text{distance from fire, m})\ \mu g/m^3$

Crude oil a, b, c	2590	5.05	920

Fuel	Burn	#	Area (m^2)	Distance (m)	Conc. ($\mu g/m^3$)
LA Crude	Mobile 91	4	114	30	79.5
LA Crude	Mobile 91	5	231	30	350.5
AB Crude	NOBE 93	1	600	125	74.0
AB Crude	NOBE 93	2	600	75	3,247.0
AB Crude	NOBE 93	2	600	125	635.0
AB Crude	NOBE 93	2	600	600	17.0
LA Crude	Mobile 92	23	181	30	125.8
LA Crude	Mobile 92	24	170	30	167.2
LA Crude	Mobile 92	28	72.8	30	13.1
LA Crude	Mobile 92	30	37.2	30	23.6
LA Crude	Mobile 92	3	114	30	10.4
LA Crude	Mobile 92	3	114	30	56.3
LA Crude	Mobile 92	4	114	30	87.8
LA Crude	Mobile 92	5	231	30	351.1

Propane

C_3H_8 — Concentration Prediction Equation

$y = a + b*(\text{fire size}, m^2) - c*\ln(\text{distance from fire, m})\ \mu g/m^3$

Crude oil a, b, c	733	0.789	236

Fuel	Burn	#	Area (m^2)	Distance (m)	Conc. ($\mu g/m^3$)
LA Crude	Mobile 91	4	114	30	25.3
LA Crude	Mobile 91	5	231	30	84.9
AB Crude	NOBE 93	1	600	125	11.0
AB Crude	NOBE 93	2	600	75	269.0
AB Crude	NOBE 93	2	600	125	71.0
LA Crude	Mobile 92	23	181	30	37.9
LA Crude	Mobile 92	24	170	30	50.4
LA Crude	Mobile 92	28	72.8	30	10.6
LA Crude	Mobile 92	30	37.2	30	9.3
LA Crude	Mobile 92	3	114	30	12.9
LA Crude	Mobile 92	3	114	30	15.9
LA Crude	Mobile 92	4	114	30	30.3
LA Crude	Mobile 92	5	231	30	87.4

Propene

C_3H_6 — Concentration Prediction Equation

$y = a + b*(\text{fire size}, m^2) - c*\ln(\text{distance from fire, m})\ \mu g/m^3$

Crude oil a, b, c	21.8	0.062	8.28

Fuel	Burn	#	Area (m^2)	Distance (m)	Conc. ($\mu g/m^3$)
AB Crude	NOBE 93	2	600	75	30.0
AB Crude	NOBE 93	2	600	600	4.0
LA Crude	Mobile 92	5	231	30	1.4
LA Crude	Mobile 92	23	181	30	2.3
LA Crude	Mobile 92	28	72.8	30	1.9
LA Crude	Mobile 92	30	37.2	30	1.6
LA Crude	Mobile 92	3	114	30	1.9
LA Crude	Mobile 92	5	231	30	1.6

t-1,4-Dimethylcyclohexane

C_8H_{18} — Concentration Prediction Equation

$y = a + b*(\text{fire size}, m^2) - c*\ln(\text{distance from fire, m})\ \mu g/m^3$

Crude oil a, b, c	562	0.114	125

Fuel	Burn	#	Area (m^2)	Distance (m)	Conc. ($\mu g/m^3$)
AB Crude	NOBE 93	1	600	125	6.0
AB Crude	NOBE 93	2	600	75	90.0
AB Crude	NOBE 93	2	600	125	46.0

TABLE A1.16
VOC Emissions from Crude Oil Burns

Toluene

C_7H_8

Concentration Prediction Equation

$y = a + b*(\text{fire size, m}^2) - c*\ln(\text{distance from fire, m})\ \mu g/m^3$

Crude oil	a, b, c		195	0.13	42.9
Fuel	**Burn**	**#**	**Area**	**Distance**	**Conc.**
			m^2	m	$\mu g/m^3$
LA Crude	Mobile 91	4	114	30	38.8
LA Crude	Mobile 91	5	231	30	148.9
AB Crude	NOBE 93	2	600	75	170.0
AB Crude	NOBE 93	2	600	125	20.0
AB Crude	NOBE 93	2	600	600	1.0
LA Crude	Mobile 92	1	36	23	7.8
LA Crude	Mobile 92	1	36	46	8.3
LA Crude	Mobile 92	2	231	23	30.7
LA Crude	Mobile 92	2	231	46	5.6
LA Crude	Mobile 92	2	231	76	0.0
LA Crude	Mobile 92	4	231	23	6.9
LA Crude	Mobile 92	23	181	30	39.4
LA Crude	Mobile 92	24	170	30	51.0
LA Crude	Mobile 92	28	72.8	30	159.9
LA Crude	Mobile 92	30	37.2	30	143.7
LA Crude	Mobile 92	3	114	30	158.6
LA Crude	Mobile 92	3	114	30	16.4
LA Crude	Mobile 92	4	114	30	41.0
LA Crude	Mobile 92	5	231	30	150.2

Total VOCs

all VOCs

Concentration Prediction Equation

$y = a + b*(\text{fire size, m}^2) - c*\ln(\text{distance from fire, m})\ \mu g/m^3$

Crude oil	a, b, c		13400	24	4430
Fuel	**Burn**	**#**	**Area**	**Distance**	**Conc.**
			m^2	m	$\mu g/m^3$
LA Crude	Mobile 91	4	114	30	1322.3
LA Crude	Mobile 91	5	231	30	4231.6
AB Crude	NOBE 93	1	600	75	644.0
AB Crude	NOBE 93	1	600	125	615.0
AB Crude	NOBE 93	1	600	600	0.0
AB Crude	NOBE 93	2	600	75	22170.0
AB Crude	NOBE 93	2	600	125	6042.0
AB Crude	NOBE 93	2	600	600	485.0
LA Crude	Mobile 92	23	181	30	1506.4
LA Crude	Mobile 92	24	170	30	1597.8
LA Crude	Mobile 92	28	72.8	30	622.9
LA Crude	Mobile 92	30	37.2	30	580.7
LA Crude	Mobile 92	3	114	30	655.9
LA Crude	Mobile 92	3	114	30	536.4
LA Crude	Mobile 92	4	114	30	1396.9
LA Crude	Mobile 92	5	231	30	4251.0

Undecane

$C_{11}H_{24}$

Concentration Prediction Equation

$y = a + b*(\text{fire size, m}^2) - c*\ln(\text{distance from fire, m})\ \mu g/m^3$

Crude oil	a, b, c		50	0.0525	12.4
Fuel	**Burn**	**#**	**Area**	**Distance**	**Conc.**
			m^2	m	$\mu g/m^3$
LA Crude	Mobile 91	4	114	30	45.0
AB Crude	NOBE 93	2	600	75	37.0
AB Crude	NOBE 93	2	600	125	19.0
AB Crude	NOBE 93	2	600	600	1.0
LA Crude	Mobile 92	23	181	30	1.8
LA Crude	Mobile 92	24	170	30	2.0
LA Crude	Mobile 92	28	72.8	30	4.0
LA Crude	Mobile 92	30	37.2	30	2.7
LA Crude	Mobile 92	3	114	30	3.1
LA Crude	Mobile 92	3	114	30	LDL
LA Crude	Mobile 92	4	114	30	45.2
LA Crude	Mobile 92	5	231	30	9.1

TABLE A2.1
VOC Emissions from Diesel Fuel Burns

1,4-Diethylbenzene

$C_{10}H_{14}$

Concentration Prediction Equation

$y = a + b*(\text{fire size}, m^2) - c*\ln(\text{distance from fire}, m)\,\mu g/m^3$

	a, b, c	3.57	0.00179	0.836
Burn	#	Area	Distance	Conc.
		m^2	m	$\mu g/m^3$
Mobile 94	1	199	30	0.2
Mobile 94	2	231	30	1.7
Mobile 94	3	231	30	1.2
Mobile 94	3	231	50	0.3
Mobile 94	3	231	85	0.2
Mobile 94	3	231	45	0.4
Mobile 94	3	231	75	1.5
Mobile 97	1-1	25	15	0.2
Mobile 97	1-1	25	45	0.1
Mobile 97	2-1	25	15	1.0
Mobile 97	2-1	25	30	1.6
Mobile 97	2-2	25	15	0.5
Mobile 97	2-2	25	30	0.2
Mobile 97	2-3	25	15	1.1
Mobile 97	2-3	25	30	0.5
Mobile 97	3-1	25	15	0.3
Mobile 97	3-1	25	30	0.3
Mobile 97	3a-1	25	30	0.5
Mobile 97	4-1	25	15	0.7
Mobile 97	4-1	25	30	0.4
Mobile 97	4-2	25	15	5.1
Mobile 97	4-2	25	30	2.1
Mobile 97	4-3	25	15	3.1
Mobile 97	4-3	25	30	0.4
Mobile 97	5-1	25	15	1.0

1-Butene/2-Methylpropene

C_4H_{10} or C_4H_8

Concentration Prediction Equation

$y = a + b*(\text{fire size}, m^2) - c*\ln(\text{distance from fire}, m)\,\mu g/m^3$

	a, b, c	7.5	0.0404	2.43
Burn	#	Area	Distance	Conc.
		m^2	m	$\mu g/m^3$
Mobile 94	1	199	30	5.3
Mobile 94	1	199	45	1.4
Mobile 94	1	199	75	1.7
Mobile 94	2	231	50	8.4
Mobile 94	2	231	85	7.5
Mobile 94	2	231	30	11.5
Mobile 94	2	231	45	7.7
Mobile 94	2	231	75	2.5
Mobile 94	3	231	30	14.3
Mobile 94	3	231	50	10.8
Mobile 94	3	231	85	3.7
Mobile 94	3	231	45	5.8
Mobile 94	3	231	75	8.3
Mobile 97	1-1	25	15	0.7
Mobile 97	1-1	25	30	0.6
Mobile 97	1-1	25	45	0.7
Mobile 97	2-1	25	15	1.5
Mobile 97	2-1	25	30	1.7
Mobile 97	2-2	25	15	1.7
Mobile 97	2-2	25	30	0.8
Mobile 97	2-3	25	15	0.8
Mobile 97	2-3	25	30	0.8
Mobile 97	3-1	25	15	1.0
Mobile 97	3-1	25	30	1.1
Mobile 97	3a-1	25	15	1.3
Mobile 97	3a-1	25	30	1.3
Mobile 97	4-1	25	15	0.9
Mobile 97	4-1	25	30	0.8
Mobile 97	4-2	25	15	1.0
Mobile 97	4-2	25	30	1.6
Mobile 97	4-3	25	15	1.0
Mobile 97	4-3	25	30	1.1
Mobile 97	5-1	25	15	1.2
Mobile 97	5-1	25	30	0.7

TABLE A2.2
VOC Emissions from Diesel Fuel Burns

1-Heptene					1-Hexene/2-Methyl-1-Pentene				
C_7H_{14}			Concentration Prediction Equation		C_6H_{12}			Concentration Prediction Equation	
$y = a + b*(fire\ size, m^2) - c*ln(distance\ from\ fire, m)\ \mu g/m^3$					$y = a + b*(fire\ size, m^2) - c*ln(distance\ from\ fire, m)\ \mu g/m^3$				
	a, b, c	2.14	0.0202	0.717		a, b, c	1.01	0.00241	0.228
Burn	#	Area	Distance	Conc.	Burn	#	Area	Distance	Conc.
		m^2	m	$\mu g/m^3$			m^2	m	$\mu g/m^3$
Mobile 97	1-1	25	15	0.3	Mobile 97	1-1	25	15	0.2
Mobile 97	1-1	25	45	0.3	Mobile 97	2-1	25	15	0.5
Mobile 97	2-1	25	15	0.5	Mobile 97	2-2	25	15	0.3
Mobile 97	2-1	25	30	1.1	Mobile 97	2-3	25	15	0.2
Mobile 97	2-2	25	30	0.2	Mobile 97	3-1	25	15	0.3
Mobile 97	2-3	25	15	0.2	Mobile 97	3a-1	25	15	0.6
Mobile 97	2-3	25	30	0.2	Mobile 97	4-1	25	15	0.3
Mobile 97	3-1	25	30	0.5	Mobile 97	4-2	25	15	0.5
Mobile 94	3-1	221	15	5.0	Mobile 97	4-3	25	15	0.4
Mobile 94	3-1	110	75	0.5	Mobile 97	5-1	25	15	0.3
					Mobile 97	1-1	25	30	0.3
					Mobile 97	2-1	25	30	0.2
					Mobile 97	2-2	25	30	0.2
					Mobile 97	2-3	25	30	0.3
					Mobile 97	3-1	25	30	0.4
					Mobile 97	3a-1	25	30	0.7
					Mobile 97	4-1	25	30	0.3
					Mobile 97	4-2	25	30	0.9
					Mobile 97	4-3	25	30	0.3
					Mobile 97	5-1	25	30	0.2
					Mobile 97	1-1	25	45	0.3
					Mobile 94	2	231	30	0.4
					Mobile 94	2	231	15	1.5
					Mobile 94	2	231	50	1.0
					Mobile 94	3	231	75	0.1

TABLE A2.3
VOC Emissions from Diesel Fuel Burns

1-Methylcyclohexene

C_7H_{14}

Concentration Prediction Equation

$y = a + b*(\text{fire size}, m^2) - c*\ln(\text{distance from fire}, m)\ \mu g/m^3$

a, b, c	1.13	0.00563	0.392

Burn	#	Area (m^2)	Distance (m)	Conc. ($\mu g/m^3$)
Mobile 94	1	199	30	0.2
Mobile 94	1	199	50	0.1
Mobile 94	1	199	30	0.5
Mobile 94	2	231	85	0.6
Mobile 94	2	231	30	0.9
Mobile 94	2	231	45	0.7
Mobile 94	3	231	30	2.9
Mobile 94	3	231	50	1.3
Mobile 94	3	231	45	0.8
Mobile 97	1-1	25	45	0.0
Mobile 97	2-1	25	30	0.2
Mobile 97	2-2	25	15	0.1
Mobile 97	4-2	25	30	0.1
Mobile 97	4-3	25	15	0.1
Mobile 97	4-3	25	30	0.0

1-Methylcyclopentene

C_6H_{12}

Concentration Prediction Equation

$y = a + b*(\text{fire size}, m^2) - c*\ln(\text{distance from fire}, m)\ \mu g/m^3$

a, b, c	0.238	0.00116	0.0442

Burn	#	Area (m^2)	Distance (m)	Conc. ($\mu g/m^3$)
Mobile 94	2	231	85	0.2
Mobile 94	2	231	30	0.7
Mobile 94	2	231	45	0.4
Mobile 94	2	231	75	0.2
Mobile 94	3	231	30	0.6
Mobile 94	3	231	50	0.2
Mobile 94	3	231	85	0.1
Mobile 94	3	231	45	0.2
Mobile 97	1-1	25	15	0.0
Mobile 97	2-1	25	15	0.1
Mobile 97	2-1	25	30	1.0
Mobile 97	2-2	25	15	0.0
Mobile 97	2-3	25	15	0.0
Mobile 97	2-3	25	30	0.0
Mobile 97	3-1	25	30	0.0
Mobile 97	3a-1	25	15	0.1
Mobile 97	3a-1	25	30	0.1
Mobile 97	4-1	25	15	0.1
Mobile 97	4-1	25	30	0.1
Mobile 97	4-2	25	15	0.1
Mobile 97	4-2	25	30	0.4
Mobile 97	4-3	25	15	0.1
Mobile 97	4-3	25	30	0.0
Mobile 97	5-1	25	15	0.1

TABLE A2.4
VOC Emissions from Diesel Fuel Burns

1-Nonene

C_9H_{18}

Concentration Prediction Equation

$y = a + b*(\text{fire size}, m^2) - c*\ln(\text{distance from fire}, m)\ \mu g/m^3$

a, b, c	4.09	0.0088	1.33

Burn	#	Area m^2	Distance m	Conc. $\mu g/m^3$
Mobile 97	1-1	25	15	0.3
Mobile 97	2-1	25	15	0.2
Mobile 97	2-1	25	30	0.3
Mobile 97	4-2	25	15	0.7
Mobile 94	1	110	15	2.0
Mobile 94	1	221	75	0.1

1-Octene

C_8H_{16}

Concentration Prediction Equation

$y = a + b*(\text{fire size}, m^2) - c*\ln(\text{distance from fire}, m)\ \mu g/m^3$

a, b, c	0.777	0.000651	0.164

Burn	#	Area m^2	Distance m	Conc. $\mu g/m^3$
Mobile 97	1-1	25	15	0.3
Mobile 97	1-1	25	30	0.0
Mobile 97	1-1	25	45	0.3
Mobile 97	2-1	25	15	0.4
Mobile 97	2-2	25	15	0.3
Mobile 97	2-2	25	30	0.2
Mobile 97	2-3	25	15	0.2
Mobile 97	2-3	25	30	0.1
Mobile 97	3-1	25	30	0.4
Mobile 97	3a-1	25	30	0.2
Mobile 97	4-1	25	15	0.2
Mobile 97	4-2	25	15	0.7
Mobile 97	4-2	25	30	0.4
Mobile 97	4-3	25	15	0.4
Mobile 97	4-3	25	30	0.2
Mobile 97	5-1	25	15	0.3
Mobile 94	1	199	15	1.0
Mobile 94	2	231	15	0.0

1-Pentene

C_5H_{10}

Concentration Prediction Equation

$y = a + b*(\text{fire size}, m^2) - c*\ln(\text{distance from fire}, m)\ \mu g/m^3$

a, b, c	1.55	0.0248	0.635

Burn	#	Area m^2	Distance m	Conc. $\mu g/m^3$
Mobile 94	1	199	30	1.3
Mobile 94	1	199	50	1.0
Mobile 94	1	199	30	2.5
Mobile 94	2	231	50	0.9
Mobile 94	2	231	85	3.4
Mobile 94	2	231	30	2.4
Mobile 94	2	231	45	1.3
Mobile 94	3	231	30	14.1
Mobile 94	3	231	50	13.7
Mobile 97	1-1	25	15	0.2
Mobile 97	1-1	25	30	0.1
Mobile 97	1-1	25	45	0.2
Mobile 97	2-1	25	15	0.4
Mobile 97	2-1	25	30	0.3
Mobile 97	2-2	25	15	0.2
Mobile 97	2-2	25	30	0.2
Mobile 97	2-3	25	15	0.2
Mobile 97	2-3	25	30	0.2
Mobile 97	3-1	25	15	0.2
Mobile 97	3-1	25	30	0.4
Mobile 97	3a-1	25	15	0.3
Mobile 97	3a-1	25	30	0.4
Mobile 97	4-1	25	15	0.2
Mobile 97	4-1	25	30	0.2
Mobile 97	4-2	25	15	0.4
Mobile 97	4-2	25	30	0.8
Mobile 97	4-3	25	15	0.2
Mobile 97	4-3	25	30	0.2
Mobile 97	5-1	25	15	0.2
Mobile 97	5-1	25	30	0.1

TABLE A2.5
VOC Emissions from Diesel Fuel Burns

2,2,3-Trimethylbutane

C_7H_{16} Concentration Prediction Equation

$y = a + b*(\text{fire size}, m^2) - c*\ln(\text{distance from fire}, m)\ \mu g/m^3$

	a, b, c	0.694	0.00125	0.208

Burn	#	Area m^2	Distance m	Conc. $\mu g/m^3$
Mobile 94	2	231	30	0.3
Mobile 97	2-1	25	15	0.1
Mobile 97	2-1	25	30	0.1
Mobile 97	2-2	25	15	0.1
Mobile 97	2-3	25	30	0.1
Mobile 97	3a-1	25	15	0.1
Mobile 97	3a-1	25	30	0.1
Mobile 97	4-2	25	15	0.1
Mobile 97	4-2	25	30	0.1
Mobile 94	1	199	1	1.0
Mobile 94	2	231	85	0.0

2,2,4-Trimethylpentane

C_8H_{18} Concentration Prediction Equation

$y = a + b*(\text{fire size}, m^2) - c*\ln(\text{distance from fire}, m)\ \mu g/m^3$

	a, b, c	3.23	0.00263	0.801

Burn	#	Area m^2	Distance m	Conc. $\mu g/m^3$
Mobile 97	1-1	25	15	0.2
Mobile 97	2-2	25	15	0.3
Mobile 97	2-3	25	15	0.3
Mobile 97	3-1	25	15	1.7
Mobile 97	3a-1	25	15	2.4
Mobile 97	4-1	25	15	1.5
Mobile 97	4-2	25	15	0.7
Mobile 97	4-3	25	15	0.3
Mobile 97	5-1	25	15	0.5

2,2,4-Trimethylpentane contd.

Burn	#	Area m^2	Distance m	Conc. $\mu g/m^3$
Mobile 97	1-1	25	30	0.3
Mobile 97	2-1	25	30	1.0
Mobile 97	2-2	25	30	0.2
Mobile 97	2-3	25	30	0.3
Mobile 97	3-1	25	30	1.8
Mobile 97	3a-1	25	15	1.8
Mobile 97	4-1	25	30	1.3
Mobile 97	4-2	25	15	1.8
Mobile 97	5-1	25	30	0.3
Mobile 97	1-1	25	45	0.2
Mobile 94	1	199	30	0.4
Mobile 94	1	199	45	0.3
Mobile 94	1	199	75	0.3
Mobile 94	2	231	30	2.1
Mobile 94	3	231	30	1.6
Mobile 94	2	231	45	1.3
Mobile 94	3	231	45	0.6
Mobile 94	3	231	50	0.2
Mobile 94	2	231	75	0.2
Mobile 94	3	231	75	0.2
Mobile 94	2	231	85	0.2
Mobile 94	3	231	85	0.2

TABLE A2.6
VOC Emissions from Diesel Fuel Burns

2,2,5-Trimethylhexane

C_9H_{20} — Concentration Prediction Equation

$y = a + b*(\text{fire size}, m^2) - c*\ln(\text{distance from fire}, m)$

	a, b, c	1.09	0.00323	0.314
Burn	**#**	**Area**	**Distance**	**Conc.**
		m^2	m	$\mu g/m^3$
Mobile 94	2	231	30	0.2
Mobile 97	1-1	25	15	0.1
Mobile 97	1-1	25	45	0.0
Mobile 97	2-1	25	15	0.1
Mobile 97	2-1	25	15	1.0
Mobile 97	2-2	25	15	0.1
Mobile 97	2-2	25	30	0.1
Mobile 97	2-3	25	15	0.1
Mobile 97	2-3	25	30	0.1
Mobile 97	3-1	25	15	0.3
Mobile 97	3-1	25	30	0.3
Mobile 97	3a-1	25	15	0.5
Mobile 97	3a-1	25	30	0.1
Mobile 97	4-1	25	15	0.3
Mobile 97	4-1	25	30	0.3
Mobile 97	4-2	25	15	0.3
Mobile 97	4-2	25	30	0.2
Mobile 97	4-3	25	30	0.1
Mobile 97	5-1	25	15	0.1
Mobile 97	5-1	25	30	0.1
Mobile 94	2	231	15	1.5

2,2-Dimethylbutane

C_6H_{14} — Concentration Prediction Equation

$y = a + b*(\text{fire size}, m^2) - c*\ln(\text{distance from fire}, m)$

	a, b, c	1.69	0.00274	0.475
Burn	**#**	**Area**	**Distance**	**Conc.**
		m^2	m	$\mu g/m^3$
Mobile 94	1	199	30	0.2
Mobile 94	1	199	45	0.2
Mobile 94	2	231	85	0.1
Mobile 94	2	231	30	1.2
Mobile 94	2	231	45	0.3
Mobile 94	3	231	30	0.7
Mobile 94	3	231	50	0.2
Mobile 94	3	231	85	0.2
Mobile 94	3	231	45	0.6
Mobile 97	1-1	25	15	0.2
Mobile 97	1-1	25	30	0.2
Mobile 97	1-1	25	45	0.2
Mobile 97	2-1	25	15	0.1
Mobile 97	2-1	25	30	0.1
Mobile 97	2-3	25	15	0.1
Mobile 97	2-3	25	30	0.1
Mobile 97	3-1	25	15	0.6
Mobile 97	3-1	25	30	0.5
Mobile 97	3a-1	25	15	0.7
Mobile 97	3a-1	25	30	0.8
Mobile 97	4-1	25	15	0.7
Mobile 97	4-1	25	30	0.7
Mobile 97	4-2	25	15	0.2
Mobile 97	4-2	25	15	2.2
Mobile 97	4-3	25	15	0.2
Mobile 97	4-3	25	30	0.2
Mobile 97	5-1	25	15	0.1
Mobile 97	5-1	25	30	0.1
Mobile 94	1	199	30	0.2
Mobile 94	2	231	30	0.9
Mobile 94	3	231	30	0.4
Mobile 97	2-3	25	15	0.0

TABLE A2.7
VOC Emissions from Diesel Fuel Burns

2,2-Dimethylpropane					2,3,4-Trimethylpentane				
C_5H_{12}			Concentration	Prediction Equation	C_8H_{18}			Concentration	Prediction Equation
$y= a + b*(fire\ size, m^2) - c*ln(distance\ from\ fire, m)$					$y= a + b*(fire\ size, m^2) - c*ln(distance\ from\ fire, m)$				
	a, b, c	0.335	0.00145	0.0886		a, b, c	1.92	0.00285	0.542
Burn	#	Area	Distance	Conc.	Burn	#	Area	Distance	Conc.
		m^2	m	$\mu g/m^3$			m^2	m	$\mu g/m^3$
Mobile 97	1-1	25	15	0.0	Mobile 94	1	199	30	0.2
Mobile 97	1-1	25	45	0.0	Mobile 94	1	199	45	0.1
Mobile 97	2-1	25	30	0.3	Mobile 94	2	231	85	0.1
Mobile 97	2-2	25	15	0.0	Mobile 94	2	231	30	0.4
Mobile 97	3-1	25	15	0.1	Mobile 94	3	231	30	0.4
Mobile 97	3-1	25	30	0.1	Mobile 94	3	231	85	0.1
Mobile 97	3a-1	25	15	0.1	Mobile 94	3	231	45	0.6
Mobile 97	3a-1	25	30	0.1	Mobile 97	2-1	231	30	2.1
Mobile 97	4-1	25	15	0.1	Mobile 97	2-3	25	30	0.1
Mobile 97	4-1	25	30	0.1	Mobile 97	3-1	25	15	0.5
Mobile 97	4-2	25	15	0.1	Mobile 97	3-1	25	30	0.6
Mobile 97	4-2	25	30	0.2	Mobile 97	3a-1	25	15	0.8
Mobile 97	4-3	25	15	0.0	Mobile 97	3a-1	25	15	1.1
Mobile 97	4-3	25	30	0.1	Mobile 97	4-1	25	15	0.6
Mobile 97	5-1	25	15	0.0	Mobile 97	4-1	25	30	0.5
Mobile 94	1	199	85	0.0	Mobile 97	4-2	25	15	0.3
Mobile 94	2	231	15	0.6	Mobile 97	4-2	25	15	1.3
					Mobile 97	4-3	25	15	0.2
					Mobile 97	4-3	25	30	0.1

TABLE A2.8
VOC Emissions from Diesel Fuel Burns

2,4-Dimethylhexane					2,4-Dimethylpentane				
C_8H_{18}			Concentration Prediction Equation		C_7H_{16}			Concentration Prediction Equation	
$y = a + b*(fire\ size, m^2) - c*ln(distance\ from\ fire, m)$					$y = a + b*(fire\ size, m^2) - c*ln(distance\ from\ fire, m)$				
	a, b, c	2.23	0.00445	0.646		a, b, c	3.26	0.0062	1.02
Burn	#	Area	Distance	Conc.	Burn	#	Area	Distance	Conc.
		m^2	m	$\mu g/m^3$			m^2	m	$\mu g/m^3$
Mobile 94	1	199	50	0.1	Mobile 97	5-1	25	15	0.1
Mobile 94	1	199	30	0.7	Mobile 97	1-1	25	15	0.1
Mobile 94	1	199	45	0.1	Mobile 97	2-3	25	15	0.2
Mobile 94	2	231	50	0.6	Mobile 97	4-3	25	15	0.2
Mobile 94	2	231	85	0.1	Mobile 97	4-2	25	15	0.2
Mobile 94	2	231	30	2.6	Mobile 97	4-1	25	15	0.5
Mobile 94	2	231	45	0.5	Mobile 97	3-1	25	15	0.6
Mobile 94	2	231	75	0.3	Mobile 97	3a-1	25	15	0.7
Mobile 94	3	231	30	1.9	Mobile 97	2-3	25	30	0.1
Mobile 94	3	231	85	0.2	Mobile 97	2-2	25	30	0.1
Mobile 94	3	231	45	0.4	Mobile 97	4-3	25	30	0.1
Mobile 97	1-1	25	15	0.3	Mobile 94	1	199	30	0.4
Mobile 97	2-1	25	15	0.1	Mobile 97	4-1	25	30	0.5
Mobile 97	2-1	25	30	0.1	Mobile 97	3-1	25	30	0.6
Mobile 97	2-3	25	15	0.1	Mobile 97	3a-1	25	30	0.8
Mobile 97	3-1	25	15	0.3	Mobile 94	3	231	30	0.9
Mobile 97	3-1	25	30	0.3	Mobile 97	4-2	25	30	1.6
Mobile 97	3a-1	25	15	0.6	Mobile 94	2	231	30	1.8
Mobile 97	3a-1	25	30	0.9	Mobile 97	2-1	25	15	3.3
Mobile 97	4-1	25	15	0.4	Mobile 94	1	199	45	0.1
Mobile 97	4-1	25	30	0.3	Mobile 94	3	231	45	0.3
Mobile 97	4-2	25	15	0.4	Mobile 94	2	231	45	0.4
Mobile 97	4-2	25	30	1.1	Mobile 94	2	231	50	0.4
Mobile 97	4-3	25	15	0.5					
Mobile 97	4-3	25	30	0.2					
Mobile 97	5-1	25	30	0.1					

TABLE A2.9
VOC Emissions from Diesel Fuel Burns

2,5-Dimethylhexane

C_8H_{18}

Concentration Prediction Equation

$y = a + b*(\text{fire size}, m^2) - c*\ln(\text{distance from fire}, m)$

a, b, c	1.12	0.00228	0.298

Burn	#	Area	Distance	Conc.
Mobile 94	1	199	30	0.3
Mobile 94	2	231	30	1.7
Mobile 94	2	231	45	0.3
Mobile 94	2	231	75	0.2
Mobile 94	3	231	30	0.8
Mobile 94	3	231	85	0.1
Mobile 94	3	231	45	0.2
Mobile 97	1-1	25	15	0.1
Mobile 97	2-1	25	15	0.2
Mobile 97	2-1	25	30	0.2
Mobile 97	2-3	25	15	0.1
Mobile 97	2-3	25	30	0.1
Mobile 97	3-1	25	15	0.2
Mobile 97	3-1	25	30	0.3
Mobile 97	3a-1	25	15	0.4
Mobile 97	3a-1	25	30	0.5
Mobile 97	4-1	25	15	0.3
Mobile 97	4-1	25	30	0.2
Mobile 97	4-2	25	15	0.4
Mobile 97	4-2	25	30	0.7
Mobile 97	4-3	25	15	0.3
Mobile 97	4-3	25	30	0.2
Mobile 97	5-1	25	30	0.1

2-Ethyltoluene

C_9H_{12}

Concentration Prediction Equation

$y = a + b*(\text{fire size}, m^2) - c*\ln(\text{distance from fire}, m)$

a, b, c	3.32	0.00295	0.857

Burn	#	Area	Distance	Conc.
		m^2	m	$\mu g/m^3$
Mobile 94	1	199	30	0.3
Mobile 94	1	199	45	0.1
Mobile 94	2	231	50	0.5
Mobile 94	2	231	30	2.4
Mobile 94	2	231	45	0.4
Mobile 94	2	231	75	0.2
Mobile 94	3	231	30	1.6
Mobile 94	3	231	85	0.3
Mobile 94	3	231	45	0.6
Mobile 94	3	231	75	0.2
Mobile 97	1-1	25	15	0.2
Mobile 97	1-1	25	30	0.1
Mobile 97	1-1	25	45	0.1
Mobile 97	2-1	25	15	0.7
Mobile 97	2-1	25	30	0.7
Mobile 97	2-2	25	15	0.4
Mobile 97	2-2	25	30	0.2
Mobile 97	2-3	25	15	0.8
Mobile 97	2-3	25	30	0.4
Mobile 97	3-1	25	15	0.4
Mobile 97	3-1	25	30	0.5
Mobile 97	3a-1	25	15	0.5
Mobile 97	3a-1	25	30	0.7
Mobile 97	4-1	25	15	0.8
Mobile 97	4-1	25	30	0.6
Mobile 97	4-2	25	15	3.3
Mobile 97	4-2	25	30	2.0
Mobile 97	4-3	25	15	2.1
Mobile 97	4-3	25	30	0.7
Mobile 97	5-1	25	15	0.7
Mobile 97	5-1	25	30	0.1

TABLE A2.10
VOC Emissions from Diesel Fuel Burns

2-Methyl-1-Butene					2-Methyl-2-Butene				
C_5H_{10}			Concentration Prediction Equation		C_5H_{10}			Concentration Prediction Equation	
$y = a + b*(fire\ size, m^2) - c*ln(distance\ from\ fire, m)$					$y = a + b*(fire\ size, m^2) - c*ln(distance\ from\ fire, m)$				
a, b, c		0.951	0.00207	0.275	a, b, c		1.67	0.00406	0.53
Burn	#	Area	Distance	Conc.	Burn	#	Area	Distance	Conc.
		m^2	m	$\mu g/m^3$			m^2	m	$\mu g/m^3$
Mobile 94	2	231	85	0.2	Mobile 94	1	199	30	0.0
Mobile 94	2	231	30	0.4	Mobile 94	2	231	85	0.5
Mobile 94	3	231	50	0.0	Mobile 94	2	231	30	0.8
Mobile 97	1-1	25	45	0.1	Mobile 94	2	231	45	0.6
Mobile 97	2-1	25	15	0.2	Mobile 94	2	231	75	0.2
Mobile 97	2-1	25	30	0.1	Mobile 94	3	231	30	0.4
Mobile 97	2-2	25	15	0.1	Mobile 94	3	231	85	0.1
Mobile 97	2-2	25	30	0.1	Mobile 94	3	231	45	0.3
Mobile 97	2-3	25	15	0.1	Mobile 94	3	231	75	0.1
Mobile 97	2-3	25	30	0.1	Mobile 97	1-1	25	15	0.1
Mobile 97	3-1	25	15	0.2	Mobile 97	1-1	25	45	0.1
Mobile 97	3-1	25	30	0.2	Mobile 97	2-1	25	15	0.1
Mobile 97	3a-1	25	15	0.3	Mobile 97	2-1	25	30	0.1
Mobile 97	3a-1	25	30	0.3	Mobile 97	2-2	25	15	0.1
Mobile 97	4-1	25	15	0.3	Mobile 97	2-2	25	30	0.1
Mobile 97	4-1	25	30	0.2	Mobile 97	2-3	25	15	0.1
Mobile 97	4-2	25	15	0.2	Mobile 97	2-3	25	30	0.1
Mobile 97	4-2	25	15	1.1	Mobile 97	3-1	25	15	0.2
Mobile 97	4-3	25	15	0.1	Mobile 97	3-1	25	30	0.2
Mobile 97	4-3	25	30	0.1	Mobile 97	3a-1	25	15	0.3
Mobile 97	5-1	25	15	0.1	Mobile 97	3a-1	25	30	0.3
Mobile 97	5-1	25	30	0.1	Mobile 97	4-1	25	15	0.3
					Mobile 97	4-1	25	30	0.3
					Mobile 97	4-2	25	15	0.2
					Mobile 97	4-2	25	15	2.6
					Mobile 97	4-3	25	15	0.1
					Mobile 97	4-3	25	30	0.1
					Mobile 97	5-1	25	15	0.1

TABLE A2.11
VOC Emissions from Diesel Fuel Burns

2-Methylbutane

C_5H_{12}

Concentration Prediction Equation

$y = a + b*(fire\ size, m^2) - c*\ln(distance\ from\ fire, m)$

	a, b, c	43.1	0.0762	13.2

Burn	#	Area m^2	Distance m	Conc. $\mu g/m^3$
Mobile 94	1	199	30	0.2
Mobile 94	1	199	30	3.3
Mobile 94	1	199	45	2.5
Mobile 94	1	199	75	2.3
Mobile 94	2	231	50	3.8
Mobile 94	2	231	85	1.5
Mobile 94	2	231	30	12.3
Mobile 94	2	231	45	3.2
Mobile 94	2	231	75	2.3
Mobile 94	3	231	30	9.6
Mobile 94	3	231	50	3.5
Mobile 94	3	231	85	3.7
Mobile 94	3	231	15	56.2
Mobile 94	3	231	75	1.9
Mobile 97	1-1	25	15	2.4
Mobile 97	1-1	25	30	2.5
Mobile 97	1-1	25	45	2.1
Mobile 97	2-1	25	15	1.3
Mobile 97	2-1	25	30	1.0
Mobile 97	2-2	25	15	0.9
Mobile 97	2-2	25	30	0.9
Mobile 97	2-3	25	15	1.6
Mobile 97	2-3	25	30	1.0
Mobile 97	3-1	25	15	9.9
Mobile 97	3-1	25	30	9.6
Mobile 97	3a-1	25	15	14.5
Mobile 97	3a-1	25	30	13.3
Mobile 97	4-1	25	15	11.7
Mobile 97	4-1	25	30	11.3
Mobile 97	4-2	25	15	3.7
Mobile 97	4-2	25	30	38.4
Mobile 97	4-3	25	15	1.7
Mobile 97	4-3	25	30	2.1
Mobile 97	5-1	25	15	1.1
Mobile 97	5-1	25	30	0.8

2-Methylheptane

C_8H_{18}

Concentration Prediction Equation

$y = a + b*(fire\ size, m^2) - c*\ln(distance\ from\ fire, m)$

	a, b, c	7.87	0.0205	2.45

Burn	#	Area m^2	Distance m	Conc. $\mu g/m^3$
Mobile 94	1	199	30	0.9
Mobile 94	1	199	45	0.2
Mobile 94	2	231	50	2.5
Mobile 94	2	231	85	0.1
Mobile 94	2	231	30	12.9
Mobile 94	2	231	45	2.3
Mobile 94	2	231	75	1.4
Mobile 94	3	231	30	6.8
Mobile 94	3	231	85	1.0
Mobile 94	3	231	45	0.8
Mobile 97	1-1	25	15	0.7
Mobile 97	1-1	25	30	0.5
Mobile 97	1-1	25	45	0.4
Mobile 97	2-1	25	15	0.8
Mobile 97	2-1	25	30	1.5
Mobile 97	2-2	25	15	0.5
Mobile 97	2-3	25	15	0.4
Mobile 97	2 3	25	30	0.4
Mobile 97	3-1	25	15	0.3
Mobile 97	3-1	25	30	0.5
Mobile 97	3a-1	25	15	0.7
Mobile 97	3a-1	25	30	0.7
Mobile 97	4-1	25	15	1.0
Mobile 97	4-1	25	30	0.6
Mobile 97	4-2	25	15	2.8
Mobile 97	4-2	25	30	1.8
Mobile 97	4-3	25	15	1.9
Mobile 97	4-3	25	30	0.8
Mobile 97	5-1	25	15	1.0
Mobile 97	5-1	25	30	0.3

TABLE A2.12
VOC Emissions from Diesel Fuel Burns

2-Methylhexane

C_7H_{16} — Concentration Prediction Equation

$y = a + b*(fire\ size, m^2) - c*ln(distance\ from\ fire, m)$

	a, b, c	13.4	0.0399	4.44
Burn	**#**	**Area**	**Distance**	**Conc.**
		m^2	m	$\mu g/m^3$
Mobile 94	1	199	30	0.3
Mobile 94	1	199	50	0.4
Mobile 94	1	199	85	0.1
Mobile 94	1	199	30	8.1
Mobile 94	1	199	45	0.7
Mobile 94	1	199	75	0.6
Mobile 94	2	231	50	4.4
Mobile 94	2	231	30	23.1
Mobile 94	2	231	45	3.6
Mobile 94	3	231	30	14.8
Mobile 94	3	231	50	1.2
Mobile 94	3	231	45	2.8
Mobile 97	1-1	25	15	0.5
Mobile 97	1-1	25	30	0.5
Mobile 97	1-1	25	45	0.3
Mobile 97	2-1	25	15	0.4
Mobile 97	2-1	25	30	0.3
Mobile 97	2-2	25	15	0.3
Mobile 97	2-3	25	15	0.3
Mobile 97	2-3	25	30	0.3
Mobile 97	3-1	25	15	0.9
Mobile 97	3-1	25	30	1.1
Mobile 97	3a-1	25	15	1.6
Mobile 97	3a-1	25	30	1.5
Mobile 97	4-1	25	15	1.2
Mobile 97	4-1	25	30	1.0
Mobile 97	4-2	25	15	1.2
Mobile 97	4-2	25	30	2.6
Mobile 97	4-3	25	15	1.0
Mobile 97	4-3	25	30	0.7
Mobile 97	5-1	25	30	0.2

2-Methylpentane

C_6H_{14} — Concentration Prediction Equation

$y = a + b*(fire\ size, m^2) - c*ln(distance\ from\ fire, m)$

	a, b, c	15.7	0.0366	4.17
Burn	**#**	**Area**	**Distance**	**Conc.**
		m^2	m	$\mu g/m^3$
Mobile 94	1	199	85	1.8
Mobile 94	1	199	30	10.5
Mobile 94	1	199	45	5.3
Mobile 94	1	199	75	6.5
Mobile 94	2	231	50	16.2
Mobile 94	2	231	85	5.8
Mobile 94	2	231	30	14.3
Mobile 94	2	231	45	6.7
Mobile 94	2	231	75	4.4
Mobile 94	3	231	30	7.8
Mobile 94	3	231	50	7.2
Mobile 94	3	231	85	2.1
Mobile 94	3	231	45	7.7
Mobile 94	3	231	75	4.0
Mobile 97	2-1	25	30	1.0
Mobile 97	3-1	25	30	2.3
Mobile 97	3a-1	25	15	3.2
Mobile 97	3a-1	25	30	3.8
Mobile 97	4-1	25	15	3.0
Mobile 97	4-1	25	30	2.7
Mobile 97	4-2	25	15	1.4
Mobile 97	4-2	25	30	10.6

TABLE A2.13
VOC Emissions from Diesel Fuel Burns

3,6-Dimethyloctane

$C_{10}H_{22}$

Concentration Prediction Equation

$y = a + b*(fire\ size, m^2) - c*ln(distance\ from\ fire, m)$

a, b, c	-0.034	0.0259	1.27

Burn	#	Area	Distance	Conc.
		m^2	m	$\mu g/m^3$
Mobile 94	1	199	30	0.3
Mobile 94	1	199	45	0.1
Mobile 94	2	231	50	0.5
Mobile 94	2	231	30	0.8
Mobile 94	2	231	75	0.3
Mobile 94	3	231	30	2.4
Mobile 94	3	231	50	0.2
Mobile 94	3	231	85	0.2
Mobile 94	3	231	45	0.4
Mobile 94	3	231	75	0.2

3-Ethyltoluene

C_9H_{12}

Concentration Prediction Equation

$y = a + b*(fire\ size, m^2) - c*ln(distance\ from\ fire, m)$

a, b, c	5.74	0.004	1.44

Burn	#	Area	Distance	Conc.
		m^2	m	$\mu g/m^3$
Mobile 94	1	199	30	0.2
Mobile 94	1	199	50	0.2
Mobile 94	1	199	30	0.6
Mobile 94	1	199	45	0.2
Mobile 94	1	199	75	0.2
Mobile 94	2	231	50	0.9
Mobile 94	2	231	85	0.1
Mobile 94	2	231	30	4.8
Mobile 94	2	231	45	0.8
Mobile 94	3	231	30	3.0
Mobile 94	3	231	50	0.4

3-Ethyltoluene contd.

C_9H_{12}

Concentration Prediction Equation

$y = a + b*(fire\ size, m^2) - c*ln(distance\ from\ fire, m)$

a, b, c	5.74	0.004	1.44

Burn	#	Area	Distance	Conc.
		m^2	m	$\mu g/m^3$
Mobile 94	3	231	85	0.3
Mobile 94	3	231	45	1.3
Mobile 94	3	231	75	0.5
Mobile 97	1-1	25	15	0.3
Mobile 97	1-1	25	30	0.3
Mobile 97	1-1	25	45	0.2
Mobile 97	2-1	25	15	1.2
Mobile 97	2-1	25	30	1.0
Mobile 97	2-2	25	15	0.7
Mobile 97	2-2	25	30	0.3
Mobile 97	2-3	25	15	1.3
Mobile 97	2-3	25	30	0.7
Mobile 97	3-1	25	15	1.1
Mobile 97	3-1	25	30	1.2
Mobile 97	3a-1	25	15	0.8
Mobile 97	3a-1	25	30	0.8
Mobile 97	4-1	25	15	1.9
Mobile 97	4-1	25	30	1.4
Mobile 97	4-2	25	15	5.3
Mobile 97	4-2	25	30	4.8
Mobile 97	4-3	25	15	3.5
Mobile 97	4-3	25	30	1.1
Mobile 97	5-1	25	15	1.2
Mobile 97	5-1	25	30	0.2

TABLE A2.14
VOC Emissions from Diesel Fuel Burns

3-Methylheptane					3-Methylhexane				
C_8H_{18}			Concentration Prediction Equation		C_7H_{16}			Concentration Prediction Equation	
$y = a + b*(fire\ size, m^2) - c*ln(distance\ from\ fire, m)$					$y = a + b*(fire\ size, m^2) - c*ln(distance\ from\ fire, m)$				
	a, b, c	4.9	0.0124	1.51		a, b, c	34.1	0.0889	12.2
Burn	#	Area	Distance	Conc.	Burn	#	Area	Distance	Conc.
		m^2	m	$\mu g/m^3$			m^2	m	$\mu g/m^3$
Mobile 94	1	199	30	0.6	Mobile 94	1	199	30	0.6
Mobile 94	1	199	45	0.2	Mobile 94	1	199	50	0.4
Mobile 94	2	231	50	1.5	Mobile 94	1	199	85	0.1
Mobile 94	2	231	85	0.1	Mobile 94	1	199	30	11.3
Mobile 94	2	231	30	8.9	Mobile 94	1	199	45	0.8
Mobile 94	2	231	45	1.7	Mobile 94	1	199	75	0.6
Mobile 94	2	231	75	0.8	Mobile 94	2	231	50	5.7
Mobile 94	3	231	30	4.2	Mobile 94	2	231	85	1.7
Mobile 94	3	231	50	0.3	Mobile 94	2	231	30	29.9
Mobile 94	3	231	85	0.7	Mobile 94	2	231	45	4.4
Mobile 94	3	231	45	0.7	Mobile 94	2	231	75	2.8
Mobile 97	1-1	25	15	0.4	Mobile 94	3	231	30	18.9
Mobile 97	1-1	25	30	0.3	Mobile 94	3	231	50	1.6
Mobile 97	1-1	25	45	0.2	Mobile 94	3	231	85	2.1
Mobile 97	2-1	25	15	0.5	Mobile 94	3	231	45	3.6
Mobile 97	2-1	25	30	1.5	Mobile 94	3	231	75	2.2
Mobile 97	2-2	25	15	0.2	Mobile 97	2-1	25	30	1.0
Mobile 97	2-3	25	15	0.3	Mobile 97	3-1	25	15	1.0
Mobile 97	2-3	25	30	0.2	Mobile 97	3a-1	25	15	1.6
Mobile 97	3-1	25	15	0.3					
Mobile 97	3-1	25	30	0.5					
Mobile 97	3a-1	25	15	0.5					
Mobile 97	3a-1	25	30	0.7					
Mobile 97	4-1	25	15	0.7					
Mobile 97	4-1	25	30	0.5					
Mobile 97	4-2	25	15	1.5					
Mobile 97	4-2	25	30	1.1					
Mobile 97	4-3	25	15	1.0					
Mobile 97	4-3	25	30	0.6					
Mobile 97	5-1	25	15	0.6					
Mobile 97	5-1	25	30	0.2					

TABLE A2.15
VOC Emissions from Diesel Fuel Burns

3-Methylpentane					4-Ethyltoluene				
C_6H_{14}			Concentration Prediction Equation		C_9H_{12}			Concentration Prediction Equation	
$y = a + b*(fire\ size, m^2) - c*ln(distance\ from\ fire, m)$					$y = a + b*(fire\ size, m^2) - c*ln(distance\ from\ fire, m)$				
	a, b, c	15.7	0.0366	4.17		a, b, c	2.84	0.00266	0.717
Burn	#	Area	Distance	Conc.	Burn	#	Area	Distance	Conc.
		m^2	m	$\mu g/m^3$			m^2	m	$\mu g/m^3$
Mobile 94	1	199	30	0.8	Mobile 94	1	199	30	0.3
Mobile 94	1	199	45	0.4	Mobile 94	1	199	45	0.1
Mobile 94	1	199	75	0.5	Mobile 94	1	199	75	0.2
Mobile 94	2	231	50	2.7	Mobile 94	2	231	50	0.5
Mobile 94	2	231	85	0.3	Mobile 94	2	231	30	2.6
Mobile 94	2	231	15	7.5	Mobile 94	2	231	45	0.4
Mobile 94	2	231	45	2.0	Mobile 94	3	231	30	1.7
Mobile 94	2	231	75	1.1	Mobile 94	3	231	50	0.2
Mobile 94	3	231	30	3.5	Mobile 94	3	231	85	0.2
Mobile 94	3	231	50	0.6	Mobile 94	3	231	45	0.6
Mobile 94	3	231	85	0.9	Mobile 94	3	231	75	0.2
Mobile 94	3	231	45	2.1	Mobile 97	1-1	25	15	0.2
Mobile 94	3	231	75	0.4	Mobile 97	1-1	25	30	0.1
Mobile 97	1-1	25	15	0.6	Mobile 97	1-1	25	45	0.1
Mobile 97	1-1	25	30	0.9	Mobile 97	2-1	25	15	0.6
Mobile 97	1-1	25	45	0.6	Mobile 97	2-1	25	30	0.6
Mobile 97	2-1	25	15	0.3	Mobile 97	2-2	25	15	0.3
Mobile 97	2-1	25	30	1.0	Mobile 97	2-2	25	30	0.2
Mobile 97	2-3	25	15	0.4	Mobile 97	2-3	25	15	0.6
Mobile 97	2-3	25	30	0.4	Mobile 97	2-3	25	30	0.3
Mobile 97	3-1	25	15	1.5	Mobile 97	3-1	25	15	0.5
Mobile 97	3-1	25	30	1.6	Mobile 97	3-1	25	30	0.5
Mobile 97	3a-1	25	15	2.2	Mobile 97	3a-1	25	15	0.2
Mobile 97	3a-1	25	30	2.3	Mobile 97	3a-1	25	30	0.9
Mobile 97	4-1	25	15	2.2	Mobile 97	4-1	25	15	0.9
Mobile 97	4-1	25	30	2.0	Mobile 97	4-1	25	30	0.7
Mobile 97	4-2	25	15	0.9					
Mobile 97	4-2	25	30	6.4					
Mobile 97	4-3	25	15	0.6					
Mobile 97	4-3	25	30	0.6					
Mobile 97	5-1	25	15	0.3					
Mobile 97	5-1	25	30	0.2					

TABLE A2.16
VOC Emissions from Diesel Fuel Burns

4-Methylheptane

C_8H_{18} — Concentration Prediction Equation

$y = a + b*(fire\ size, m^2) - c*ln(distance\ from\ fire, m)$

a, b, c	1.62	0.00668	0.49

Burn	#	Area m^2	Distance m	Conc. $\mu g/m^3$
Mobile 94	1	199	30	0.3
Mobile 94	2	231	50	0.8
Mobile 94	2	231	30	4.5
Mobile 94	2	231	45	0.7
Mobile 94	3	231	30	1.4
Mobile 94	3	231	45	0.3
Mobile 97	1-1	25	30	0.3
Mobile 97	2-1	25	15	0.2
Mobile 97	2-1	25	30	0.6
Mobile 97	2-2	25	15	0.2
Mobile 97	2-3	25	15	0.2
Mobile 97	2-3	25	30	0.2
Mobile 97	3-1	25	15	0.2
Mobile 97	3-1	25	30	0.2
Mobile 97	3a-1	25	15	0.3
Mobile 97	4-1	25	15	0.4
Mobile 97	4-1	25	30	0.2
Mobile 97	4-2	25	15	0.7
Mobile 97	4-2	25	30	0.6
Mobile 97	4-3	25	15	0.5
Mobile 97	4-3	25	30	0.3
Mobile 97	5-1	25	15	0.2
Mobile 97	5-1	25	30	0.1

Benzene

C_6H_6 — Concentration Prediction Equation

$y = a + b*(fire\ size, m^2) - c*ln(distance\ from\ fire, m)$

a, b, c	27.4	0.0649	8.15

Burn	#	Area m^2	Distance m	Conc. $\mu g/m^3$
Mobile 94	1	199	30	2.0
Mobile 94	1	199	50	1.2
Mobile 94	1	199	30	5.1
Mobile 94	1	199	45	2.4
Mobile 94	1	199	75	1.4
Mobile 94	2	231	50	11.0
Mobile 94	2	231	85	2.6
Mobile 94	2	231	30	37.4
Mobile 94	2	231	45	17.5
Mobile 94	2	231	75	8.8
Mobile 94	3	231	30	28.1
Mobile 94	3	231	50	10.7
Mobile 94	3	231	85	4.9
Mobile 94	3	231	45	6.5
Mobile 94	3	231	75	3.1
Mobile 97	1-1	25	15	1.6
Mobile 97	1-1	25	30	1.4
Mobile 97	1-1	25	45	1.2
Mobile 97	2-1	25	15	6.1
Mobile 97	2-1	25	30	1.0
Mobile 97	2-2	25	15	2.6
Mobile 97	2-2	25	30	1.1
Mobile 97	2-3	25	15	2.7
Mobile 97	2-3	25	30	2.1
Mobile 97	3-1	25	15	3.9
Mobile 97	3-1	25	30	4.0
Mobile 97	3a-1	25	15	5.7
Mobile 97	3a-1	25	30	5.0
Mobile 97	4-1	25	15	5.0
Mobile 97	4-1	25	30	4.8
Mobile 97	4-2	25	15	9.7
Mobile 97	4-2	25	30	12.7
Mobile 97	4-3	25	15	6.2
Mobile 97	4-3	25	30	3.2
Mobile 97	5-1	25	15	4.1

TABLE A2.17
VOC Emissions from Diesel Fuel Burns

Butane					c-1,3-Dimethylcyclohexane				
C_4H_{10}			Concentration Prediction Equation		C_8H_{18}			Concentration Prediction Equation	
$y = a + b*(fire\ size, m^2) - c*ln(distance\ from\ fire, m)$					$y = a + b*(fire\ size, m^2) - c*ln(distance\ from\ fire, m)$				
	a, b, c	19.6	0.0286	5.55		a, b, c	5.81	0.022	1.95
Burn	#	Area	Distance	Conc.	Burn	#	Area	Distance	Conc.
		m^2	m	$\mu g/m^3$			m^2	m	$\mu g/m^3$
Mobile 94	1	199	30	2.3	Mobile 94	1	199	30	1.0
Mobile 94	1	199	45	2.1	Mobile 94	2	231	50	2.3
Mobile 94	1	199	75	2.2	Mobile 94	2	231	85	0.3
Mobile 94	2	231	50	2.7	Mobile 94	2	231	30	12.2
Mobile 94	2	231	85	1.7	Mobile 94	2	231	45	2.3
Mobile 94	2	231	30	5.2	Mobile 94	3	231	30	5.1
Mobile 94	2	231	45	2.1	Mobile 94	3	231	45	0.8
Mobile 94	2	231	75	1.8	Mobile 97	1-1	25	15	0.2
Mobile 94	3	231	30	3.2	Mobile 97	1-1	25	30	0.1
Mobile 94	3	231	50	2.4	Mobile 97	1-1	25	45	0.1
Mobile 94	3	231	85	2.2	Mobile 97	2-1	25	15	0.4
Mobile 94	3	231	45	4.9	Mobile 97	2-1	25	30	0.3
Mobile 94	3	231	75	1.7	Mobile 97	2-2	25	15	0.2
Mobile 97	1-1	25	15	3.2	Mobile 97	2-3	25	15	0.2
Mobile 97	1-1	25	30	3.2	Mobile 97	2-3	25	30	0.2
Mobile 97	1-1	25	45	2.7	Mobile 97	3a-1	25	15	0.2
Mobile 97	2-1	25	15	1.3	Mobile 97	3a-1	25	30	0.1
Mobile 97	2-1	25	30	2.0	Mobile 97	4-1	25	15	0.5
Mobile 97	2-2	25	15	1.0	Mobile 97	4-1	25	30	0.2
Mobile 97	2-2	25	30	0.9	Mobile 97	4-2	25	15	1.5
Mobile 97	2-3	25	15	1.2	Mobile 97	4-2	25	30	0.6
Mobile 97	2-3	25	30	1.0	Mobile 97	4-3	25	15	1.0
Mobile 97	3-1	25	15	7.8	Mobile 97	4-3	25	30	0.4
Mobile 97	3-1	25	30	8.0	Mobile 97	5-1	25	15	0.5
Mobile 97	3a-1	25	15	11.8	Mobile 97	5-1	25	30	0.1
Mobile 97	3a-1	25	30	11.0					
Mobile 97	4-1	25	15	10.1					
Mobile 97	4-1	25	30	9.9					
Mobile 97	4-2	25	15	4.6					
Mobile 97	4-2	25	15	22.7					
Mobile 97	4-3	25	15	1.5					
Mobile 97	4-3	25	30	2.3					
Mobile 97	5-1	25	15	1.1					

TABLE A2.18
VOC Emissions from Diesel Fuel Burns

c-1,4-Dimethylcyclohexane					c-2-Butene contd.				
C_8H_{18}		Concentration Prediction Equation			C_4H_8		Concentration Prediction Equation		
$y = a + b*(fire\ size, m^2) - c*ln(distance\ from\ fire,\ m)$					$y = a + b*(fire\ size, m^2) - c*ln(distance\ from\ fire,\ m)$				
	a, b, c	2.46	0.00776	0.837		a, b, c	0.673	0.00265	0.205
Burn	**#**	**Area**	**Distance**	**Conc.**	**Burn**	**#**	**Area**	**Distance**	**Conc.**
		m^2	m	$\mu g/m^3$			m^2	m	$\mu g/m^3$
Mobile 94	1	199	30	0.3	Mobile 94	2	231	30	0.7
Mobile 94	2	231	50	0.8	Mobile 94	2	231	45	0.8
Mobile 94	2	231	30	3.9	Mobile 94	2	231	75	0.5
Mobile 94	2	231	45	0.7	Mobile 94	3	231	30	1.0
Mobile 94	2	231	75	0.3	Mobile 94	3	231	50	0.3
Mobile 94	3	231	30	1.6	Mobile 94	3	231	85	0.3
Mobile 94	3	231	85	0.3	Mobile 94	3	231	45	0.2
Mobile 94	3	231	45	0.3	Mobile 94	3	231	75	0.4
Mobile 97	1-1	25	15	0.04	Mobile 97	1-1	25	45	0.1
Mobile 97	1-1	25	45	0.04	Mobile 97	2-1	25	15	0.1
Mobile 97	2-2	25	15	0.04	Mobile 97	2-1	25	30	0.5
Mobile 97	2-2	25	30	0.04	Mobile 97	2-2	25	15	0.1
Mobile 97	2-3	25	30	0.04	Mobile 97	2-3	25	15	0.0
Mobile 97	3-1	25	15	0.04	Mobile 97	2-3	25	30	0.1
Mobile 97	3-1	25	30	0.1	Mobile 97	3-1	25	15	0.1
Mobile 97	4-1	25	15	0.1	Mobile 97	3-1	25	30	0.1
Mobile 97	5-1	25	30	0.04	Mobile 97	3a-1	25	15	0.1
c-2-Butene					Mobile 97	3a-1	25	30	0.1
					Mobile 97	4-1	25	15	0.1
C_4H_8		Concentration Prediction Equation			Mobile 97	4-1	25	30	0.1
					Mobile 97	4-2	25	15	0.1
$y = a + b*(fire\ size, m^2) - c*ln(distance\ from\ fire,\ m)$					Mobile 97	4-2	25	30	0.2
	a, b, c	0.673	0.00265	0.205	Mobile 97	4-3	25	15	0.1
Burn	**#**	**Area**	**Distance**	**Conc.**	Mobile 97	4-3	25	30	0.1
		m^2	m	$\mu g/m^3$	Mobile 97	5-1	25	15	0.1
Mobile 94	1	199	30	0.2					
Mobile 94	1	199	50	0.1					
Mobile 94	1	199	30	0.3					
Mobile 94	1	199	45	0.2					
Mobile 94	2	231	50	0.5					
Mobile 94	2	231	15	1.2					

TABLE A2.19
VOC Emissions from Diesel Fuel Burns

c-2-Heptene

C_7H_{14} Concentration Prediction Equation

$y = a + b*(fire\ size, m^2) - c*ln(distance\ from\ fire, m)$

	a, b, c	2.02	0.00134	0.53
Burn	**#**	**Area**	**Distance**	**Conc.**
		m^2	m	$\mu g/m^3$
Mobile 94	2	231	30	0.6
Mobile 94	2	231	45	0.2
Mobile 97	2-1	25	30	0.3
Mobile 97	2-2	25	30	0.2
Mobile 94	2	231	85	0.0

c-2-Hexene

C_6H_{12} Concentration Prediction Equation

$y = a + b*(fire\ size, m^2) - c*ln(distance\ from\ fire, m)$

	a, b, c	1.91	0.00492	0.697
Burn	**#**	**Area**	**Distance**	**Conc.**
		m^2	m	$\mu g/m^3$
Mobile 94	2	231	45	0.2
Mobile 94	3	231	30	0.8
Mobile 94	3	231	50	0.3
Mobile 94	3	231	85	0.0
Mobile 97	3	25	15	0.1

c-2-Pentene

C_5H_{10} Concentration Prediction Equation

$y = a + b*(fire\ size, m^2) - c*ln(distance\ from\ fire, m)$

	a, b, c	0.596	0.00233	0.178
Burn	**#**	**Area**	**Distance**	**Conc.**
		m^2	m	$\mu g/m^3$
Mobile 97	2-1	25	15	0.1
Mobile 97	2-1	25	30	0.1
Mobile 97	3-1	25	15	0.1
Mobile 97	3-1	25	30	0.1
Mobile 97	3a-1	25	15	0.2
Mobile 97	3a-1	25	30	0.2
Mobile 97	4-1	25	15	0.1
Mobile 97	4-1	25	30	0.2
Mobile 97	4-2	25	15	0.1
Mobile 97	4-2	25	15	0.9
Mobile 97	4-3	25	15	0.1
Mobile 97	4-3	25	30	0.1
Mobile 97	5-1	25	15	0.1
Mobile 97	5-1	25	85	0.0
Mobile 97	5-1	25	85	0.0

TABLE A2.20
VOC Emissions from Diesel Fuel Burns

Cyclohexane

C_6H_{14} Concentration Prediction Equation

$y = a + b*(fire\ size, m^2) - c*ln(distance\ from\ fire, m)$

a, b, c	8.27	0.024	2.74

Burn	#	Area	Distance	Conc.
		m^2	m	$\mu g/m^3$
Mobile 94	1	199	30	1.0
Mobile 94	1	199	45	0.1
Mobile 94	1	199	75	0.2
Mobile 94	2	231	50	3.5
Mobile 94	2	231	85	0.2
Mobile 94	2	231	30	15.3
Mobile 94	2	231	45	3.0
Mobile 94	2	231	75	1.6
Mobile 94	3	231	30	6.6
Mobile 94	3	231	50	0.2
Mobile 94	3	231	85	1.0
Mobile 94	3	231	45	2.0
Mobile 97	1-1	25	15	0.3
Mobile 97	1-1	25	30	0.3
Mobile 97	1-1	25	45	0.2
Mobile 97	2-1	25	15	0.3
Mobile 97	2-1	25	30	0.9
Mobile 97	2-2	25	15	0.2
Mobile 97	2-2	25	30	0.1
Mobile 97	2-3	25	30	0.1
Mobile 97	3-1	25	15	0.3
Mobile 97	3-1	25	30	0.3
Mobile 97	3a-1	25	15	0.5
Mobile 97	3a-1	25	30	0.4
Mobile 97	4-1	25	15	0.6
Mobile 97	4-1	25	30	0.5
Mobile 97	4-2	25	15	0.8
Mobile 97	4-2	25	30	0.8
Mobile 97	4-3	25	15	0.8
Mobile 97	4-3	25	30	0.3
Mobile 97	5-1	25	15	0.4
Mobile 97	5-1	25	30	0.1

Cyclohexene

C_6H_{12} Concentration Prediction Equation

$y = a + b*(fire\ size, m^2) - c*ln(distance\ from\ fire, m)$

a, b, c	1.55	0.00346	0.479

Burn	#	Area	Distance	Conc.
		m^2	m	$\mu g/m^3$
Mobile 94	1	199	30	0.3
Mobile 94	1	199	50	0.2
Mobile 94	1	199	30	0.6
Mobile 94	1	199	45	0.1
Mobile 94	2	231	85	0.4
Mobile 94	2	231	30	0.4
Mobile 94	2	231	45	0.3
Mobile 94	3	231	30	1.6
Mobile 94	3	231	50	0.7

Cyclopentane

C_5H_{12} Concentration Prediction Equation

$y = a + b*(fire\ size, m^2) - c*ln(distance\ from\ fire, m)$

a, b, c	2.6	0.00684	0.811

Burn	#	Area	Distance	Conc.
		m^2	m	$\mu g/m^3$
Mobile 94	1	199	85	0.0
Mobile 94	2	231	30	1.3
Mobile 94	2	231	45	0.3
Mobile 94	3	231	30	0.6
Mobile 94	3	231	45	1.4
Mobile 97	1-1	25	15	0.1
Mobile 97	1-1	25	30	0.2
Mobile 97	1-1	25	45	0.1
Mobile 97	2-1	25	15	0.1
Mobile 97	2-1	25	15	3.4
Mobile 97	2-3	25	15	0.1
Mobile 97	2-3	25	30	0.1
Mobile 97	3-1	25	15	0.4
Mobile 97	3-1	25	30	0.4
Mobile 97	3a-1	25	15	0.6
Mobile 97	3a-1	25	30	0.7
Mobile 97	4-1	25	15	0.6
Mobile 97	4-1	25	30	0.5
Mobile 97	4-2	25	15	0.2
Mobile 97	4-2	25	15	2.1
Mobile 97	4-3	25	15	0.1
Mobile 97	4-3	25	30	0.2

TABLE A2.21
VOC Emissions from Diesel Fuel Burns

Cyclopentene					Decane				
C_5H_{10}			Concentration Prediction Equation		$C_{10}H_{22}$			Concentration Prediction Equation	
$y = a + b*(\text{fire size, m}^2) - c*\ln(\text{distance from fire, m})$					$y = a + b*(\text{fire size, m}^2) - c*\ln(\text{distance from fire, m})$				
a, b, c	0.229	0.0016	0.0066		a, b, c	23.1	0.0124	6.05	
Burn	#	Area	Distance	Conc.	Burn	#	Area	Distance	Conc.
		m²	m	µg/m³			m²	m	µg/m³
Mobile 94	2	231	85	0.2	Mobile 94	1	199	30	0.5
Mobile 94	2	231	30	0.5	Mobile 94	1	199	50	0.1
Mobile 94	2	231	45	0.4	Mobile 94	1	199	30	2.1
Mobile 94	2	231	75	0.2	Mobile 94	1	199	45	0.6
Mobile 94	3	231	30	0.4	Mobile 94	1	199	75	0.8
Mobile 97	1-1	25	15	0.0	Mobile 94	2	231	50	2.3
Mobile 97	1-1	25	30	0.0	Mobile 94	2	231	85	0.3
Mobile 97	1-1	25	45	0.0	Mobile 94	2	231	30	9.3
Mobile 97	2-1	25	15	0.1	Mobile 94	2	231	45	2.1
Mobile 97	2-1	25	30	0.1	Mobile 94	2	231	75	1.1
Mobile 97	2-2	25	15	0.1	Mobile 94	3	231	30	7.7
Mobile 97	2-2	25	30	0.0	Mobile 94	3	231	50	0.7
Mobile 97	2-3	25	15	0.0	Mobile 94	3	231	85	1.1
Mobile 97	2-3	25	30	0.0	Mobile 94	3	231	45	2.2
Mobile 97	3-1	25	30	0.0	Mobile 94	3	231	75	0.7
Mobile 97	3a-1	25	15	0.1	Mobile 97	1-1	25	15	1.0
Mobile 97	3a-1	25	30	0.1	Mobile 97	1-1	25	30	0.7
Mobile 97	4-1	25	15	0.0	Mobile 97	1-1	25	45	0.5
Mobile 97	4-1	25	30	0.0	Mobile 97	2-1	25	15	6.0
Mobile 97	4-2	25	15	0.1	Mobile 97	2-1	25	30	2.8
Mobile 97	4-2	25	30	0.3	Mobile 97	2-2	25	15	2.8
Mobile 97	4-3	25	15	0.1	Mobile 97	2-2	25	30	0.9
Mobile 97	4-3	25	30	0.0	Mobile 97	2-3	25	15	7.1
Mobile 97	5-1	25	15	0.1	Mobile 97	2-3	25	30	3.4
Mobile 97	5-1	25	30	0.0	Mobile 97	3-1	25	15	0.9
					Mobile 97	3-1	25	30	1.4
					Mobile 97	3a-1	25	15	0.2
					Mobile 97	3a-1	25	30	0.6
					Mobile 97	4-1	25	15	2.6
					Mobile 97	4-1	25	30	1.1
					Mobile 97	4-2	25	15	31.0
					Mobile 97	4-2	25	30	9.6
					Mobile 97	4-3	25	15	18.1
					Mobile 97	4-3	25	30	3.1

TABLE A2.22
VOC Emissions from Diesel Fuel Burns

Dodecane					Ethylbenzene				
$C_{12}H_{26}$		Concentration Prediction Equation			C_8H_{10}		Concentration Prediction Equation		
$y = a + b*(fire\ size, m^2) - c*ln(distance\ from\ fire, m)$					$y = a + b*(fire\ size, m^2) - c*ln(distance\ from\ fire, m)$				
a, b, c		139	0.121	40.1	a, b, c		6.53	0.00714	1.69
Burn	#	Area	Distance	Conc.	Burn	#	Area	Distance	Conc.
		m^2	m	$\mu g/m^3$			m^2	m	$\mu g/m^3$
Mobile 94	1	199	30	2.3	Mobile 94	1	199	30	0.8
Mobile 94	1	199	45	1.6	Mobile 94	1	199	45	0.3
Mobile 94	1	199	75	0.3	Mobile 94	1	199	75	0.4
Mobile 94	2	231	50	0.6	Mobile 94	2	231	50	1.6
Mobile 94	2	231	85	0.1	Mobile 94	2	231	85	0.3
Mobile 94	2	231	30	2.2	Mobile 94	2	231	15	6.5
Mobile 94	2	231	45	0.8	Mobile 94	2	231	45	1.6
Mobile 94	2	231	75	0.2	Mobile 94	2	231	75	0.8
Mobile 94	3	231	30	6.0	Mobile 94	3	231	30	4.3
Mobile 94	3	231	50	1.5	Mobile 94	3	231	50	0.4
Mobile 94	3	231	85	0.9	Mobile 94	3	231	85	0.9
Mobile 94	3	231	45	1.9	Mobile 94	3	231	45	1.5
Mobile 94	3	231	75	0.9	Mobile 94	3	231	75	0.3
Mobile 97	1-1	25	15	2.9	Mobile 97	1-1	25	15	0.8
Mobile 97	1-1	25	30	1.8	Mobile 97	1-1	25	30	0.7
Mobile 97	1-1	25	45	0.8	Mobile 97	1-1	25	45	0.5
Mobile 97	2-1	25	15	45.9	Mobile 97	2-1	25	15	1.2
Mobile 97	2-1	25	30	12.6	Mobile 97	2-1	25	30	1.0
Mobile 97	2-2	25	15	13.8	Mobile 97	2-2	25	15	0.7
Mobile 97	2-2	25	30	4.5	Mobile 97	2-2	25	30	0.3
Mobile 97	2-3	25	15	50.5	Mobile 97	2-3	25	15	1.0
Mobile 97	2-3	25	30	19.5	Mobile 97	2-3	25	30	0.6
Mobile 97	3-1	25	15	0.7	Mobile 97	3-1	25	15	1.6
Mobile 97	3-1	25	30	1.6	Mobile 97	3-1	25	30	1.7
Mobile 97	3a-1	25	30	0.5	Mobile 97	3a-1	25	15	1.7
Mobile 97	4-1	25	15	6.2	Mobile 97	3a-1	25	30	2.3
Mobile 97	4-1	25	30	1.5	Mobile 97	4-1	25	15	2.2
Mobile 97	4-2	25	15	135.6	Mobile 97	4-1	25	30	1.9
Mobile 97	4-2	25	30	37.2	Mobile 97	4-2	25	15	3.1
Mobile 97	4-3	25	15	74.6	Mobile 97	4-2	25	30	5.9
Mobile 97	4-3	25	30	2.7	Mobile 97	4-3	25	15	2.2
Mobile 97	5-1	25	15	36.7	Mobile 97	4-3	25	30	0.9
Mobile 97	5-1	25	30	0.4	Mobile 97	5-1	25	15	1.0
					Mobile 97	5-1	25	30	0.3

TABLE A2.23
VOC Emissions from Diesel Fuel Burns

Heptane					Hexane				
C_7H_{16} ∕∖∕∖∕ Concentration Prediction Equation					C_6H_{14} ∕∖∕∖∕ Concentration Prediction Equation				
$y = a + b*(fire\ size, m^2) - c*ln(distance\ from\ fire, m)$					$y = a + b*(fire\ size, m^2) - c*ln(distance\ from\ fire, m)$				
a, b, c	32.2	0.096	10.9		a, b, c	-0.077	-0.04	1.6	
Burn	#	Area	Distance	Conc.	Burn	#	Area	Distance	Conc.
		m^2	m	$\mu g/m^3$			m^2	m	$\mu g/m^3$
Mobile 94	1	199	30	0.4	Mobile 97	1-1	25	15	2.9
Mobile 94	1	199	30	12.4	Mobile 97	1-1	25	30	7.7
Mobile 94	1	199	45	0.9	Mobile 97	1-1	25	45	4.2
Mobile 94	1	199	75	0.6	Mobile 97	2-1	25	15	1.4
Mobile 94	2	231	50	12.7	Mobile 97	2-1	25	30	1.0
Mobile 94	2	231	85	2.2	Mobile 97	2-2	25	15	2.4
Mobile 94	2	231	30	59.9	Mobile 97	2-2	25	30	4.3
Mobile 94	2	231	45	9.9	Mobile 97	2-3	25	15	2.5
Mobile 94	2	231	75	5.4	Mobile 97	2-3	25	30	2.7
Mobile 94	3	231	30	33.3	Mobile 97	3-1	25	15	1.8
Mobile 94	3	231	50	2.2	Mobile 97	3-1	25	30	2.3
Mobile 94	3	231	85	4.1	Mobile 97	3a-1	25	15	2.9
Mobile 94	3	231	45	5.0	Mobile 97	3a-1	25	30	2.9
Mobile 94	3	231	75	2.4	Mobile 97	4-1	25	15	4.6
Mobile 97	1-1	25	15	1.0	Mobile 97	4-1	25	30	3.5
Mobile 97	1-1	25	30	0.9	Mobile 97	4-2	25	15	3.0
Mobile 97	1-1	25	45	0.8	Mobile 97	4-2	25	30	9.0
Mobile 97	2-1	25	15	1.1	Mobile 97	4-3	25	15	1.4
Mobile 97	2-1	25	30	1.0	Mobile 97	4-3	25	30	1.2
Mobile 97	2-2	25	15	0.6	Mobile 97	5-1	25	15	0.8
Mobile 97	2-2	25	30	0.3	Mobile 97	5-1	25	30	0.4
Mobile 97	2-3	25	15	0.6					
Mobile 97	2-3	25	30	0.5					
Mobile 97	3-1	25	15	0.8					
Mobile 97	3-1	25	30	1.4					
Mobile 97	3a-1	25	15	1.1					
Mobile 97	3a-1	25	30	1.3					
Mobile 97	4-1	25	15	1.6					
Mobile 97	4-1	25	30	1.1					
Mobile 97	4-2	25	15	3.0					
Mobile 97	4-2	25	30	2.9					
Mobile 97	4-3	25	15	2.5					
Mobile 97	4-3	25	30	1.6					
Mobile 97	5-1	25	15	1.4					
Mobile 97	5-1	25	30	0.5					

TABLE A2.24
VOC Emissions from Diesel Fuel Burns

Hexylbenzene					Indane (2,3-Dihydroindene)				
$C_{12}H_{18}$			Concentration Prediction Equation		C_9H_{10}			Concentration Prediction Equation	
$y = a + b*(fire\ size, m^2) - c*ln(distance\ from\ fire,\ m)$					$y = a + b*(fire\ size, m^2) - c*ln(distance\ from\ fire,\ m)$				
a, b, c	4.55	0.00942	1.38		a, b, c	0.761	0.00181	0.191	
Burn	#	Area	Distance	Conc.	Burn	#	Area	Distance	Conc.
		m^2	m	$\mu g/m^3$			m^2	m	$\mu g/m^3$
Mobile 94	1	199	75	0.4	Mobile 97	1-1	25	15	0.0
Mobile 94	2	25	30	0.3	Mobile 97	1-1	25	45	0.0
Mobile 94	3	231	50	0.6	Mobile 97	2-1	25	15	0.1
Mobile 97	1-1	25	30	0.1	Mobile 97	2-1	25	30	0.7
Mobile 97	2-1	25	15	1.2	Mobile 97	2-2	25	15	0.1
Mobile 97	2-1	25	30	0.4	Mobile 97	2-2	25	30	0.0
Mobile 97	2-2	25	15	0.5	Mobile 97	2-3	25	15	0.1
Mobile 97	2-3	25	30	0.5	Mobile 97	2-3	25	30	0.1
Mobile 97	3-1	25	15	0.0	Mobile 97	3-1	25	15	0.1
Mobile 97	4-2	25	15	3.8	Mobile 97	3-1	25	30	0.2
Mobile 97	4-2	25	30	0.7	Mobile 97	3a-1	25	15	0.1
Mobile 97	4-3	25	15	1.4	Mobile 97	3a-1	25	30	0.2
Mobile 97	5-1	25	15	0.6	Mobile 97	4-1	25	15	0.2
					Mobile 97	4-1	25	30	0.2
					Mobile 97	4-2	25	15	0.6
					Mobile 97	4-2	25	30	0.5
					Mobile 97	4-3	25	15	0.4
					Mobile 97	4-3	25	30	0.1
					Mobile 97	5-1	25	15	0.1
					Mobile 97	5-1	25	30	0.3

TABLE A2.25
VOC Emissions from Diesel Fuel Burns

Indene

C_9H_8

Concentration Prediction Equation

$y= a + b*(fire\ size, m^2) - c*ln(distance\ from\ fire, m)$

a, b, c	0.309	0.00142	0.0972

Burn	#	Area	Distance	Conc.
		m^2	m	$\mu g/m^3$
Mobile 94	1	199	75	0.1
Mobile 94	2	231	30	0.4
Mobile 94	3	231	30	0.3
Mobile 94	3	231	45	0.3
Mobile 94	3	231	45	0.3
Mobile 94	3	231	45	0.3
Mobile 97	3	25	15	0.1
Mobile 97	3	25	45	0.00

Isobutane (2-Methylpropane)

C_4H_{10}

Concentration Prediction Equation

$y= a + b*(fire\ size, m^2) - c*ln(distance\ from\ fire, m)$

a, b, c	7.22	0.00282	1.58

Burn	#	Area	Distance	Conc.
		m^2	m	$\mu g/m^3$
Mobile 94	1	199	30	1.2
Mobile 94	1	199	45	1.0
Mobile 94	1	199	75	1.1
Mobile 94	2	231	50	0.9
Mobile 94	2	231	85	0.6
Mobile 94	2	231	30	1.3
Mobile 94	2	231	45	0.8

Isobutane (2-Methylpropane) contd.

Burn	#	Area	Distance	Conc.
		m^2	m	$\mu g/m^3$
Mobile 94	2	231	75	0.7
Mobile 94	3	231	30	3.4
Mobile 94	3	231	50	1.4
Mobile 94	3	231	85	0.6
Mobile 94	3	231	15	9.0
Mobile 94	3	231	75	0.6
Mobile 97	1-1	25	15	1.5
Mobile 97	1-1	25	30	1.6
Mobile 97	1-1	25	45	1.5
Mobile 97	2-1	25	15	0.5
Mobile 97	2-1	25	30	5.3
Mobile 97	2-2	25	15	0.4
Mobile 97	2-2	25	30	0.4
Mobile 97	2-3	25	15	0.5
Mobile 97	2-3	25	30	0.4
Mobile 97	3-1	25	15	3.9
Mobile 97	3-1	25	30	3.9
Mobile 97	3a-1	25	15	6.6
Mobile 97	3a-1	25	30	5.5
Mobile 97	4-1	25	15	4.6
Mobile 97	4-1	25	30	4.7
Mobile 97	4-2	25	15	2.4
Mobile 97	4-2	25	30	5.0
Mobile 97	4-3	25	15	0.6
Mobile 97	4-3	25	30	0.9
Mobile 97	5-1	25	15	0.3
Mobile 97	5-1	25	30	0.3

TABLE A2.26
VOC Emissions from Diesel Fuel Burns

iso-Butylbenzene					Isoprene (2-Methyl-1,3-Butadiene)				
$C_{10}H_{14}$			Concentration Prediction Equation		C_5H_8			Concentration Prediction Equation	
$y= a + b*(fire\ size, m^2) - c*ln(distance\ from\ fire, m)$					$y= a + b*(fire\ size, m^2) - c*ln(distance\ from\ fire, m)$				
	a, b, c	0.36	0.002	-0.09		a, b, c	0.4	-9.5	1.9
Burn	**#**	**Area**	**Distance**	**Conc.**	**Burn**	**#**	**Area**	**Distance**	**Conc.**
		m^2	m	$\mu g/m^3$			m^2	m	$\mu g/m^3$
Mobile 94	2	231	30	0.4	Mobile 94	1	199	30	0.6
Mobile 97	1-1	25	15	0.0	Mobile 94	1	199	50	0.8
Mobile 97	2-1	25	15	0.1	Mobile 94	1	199	30	2.9
Mobile 97	2-1	25	30	0.2	Mobile 94	1	199	45	0.3
Mobile 97	2-2	25	15	0.1	Mobile 94	1	199	75	1.0
Mobile 97	2-3	25	15	0.1	Mobile 94	2	231	85	1.3
Mobile 97	2-3	25	30	0.1	Mobile 94	2	231	30	1.7
Mobile 97	3-1	25	15	0.0	Mobile 94	2	231	45	0.8
Mobile 97	3-1	25	30	0.0	Mobile 94	3	231	30	0.6
Mobile 97	3a-1	25	15	0.0	Mobile 94	3	231	50	2.4
Mobile 97	3a-1	25	30	0.0	Mobile 94	3	231	45	2.1
Mobile 97	4-1	25	15	0.1	Mobile 97	1-1	25	15	0.9
Mobile 97	4-1	25	30	0.0	Mobile 97	1-1	25	30	1.0
Mobile 97	4-2	25	15	0.6	Mobile 97	1-1	25	45	0.8
Mobile 97	4-2	25	30	0.2	Mobile 97	2-1	25	15	0.8
Mobile 97	4-3	25	15	0.4	Mobile 97	2-1	25	30	0.8
Mobile 97	4-3	25	30	0.1	Mobile 97	2-2	25	15	0.6
Mobile 97	5-1	25	15	0.1	Mobile 97	2-2	25	30	0.5
Mobile 97	5-1	25	30	0.0	Mobile 97	2-3	25	15	1.0
					Mobile 97	2-3	25	30	1.0
					Mobile 97	3-1	25	15	2.2
					Mobile 97	3-1	25	30	2.2
					Mobile 97	3a-1	25	15	1.0
					Mobile 97	3a-1	25	30	0.5
					Mobile 97	4-1	25	15	0.7
					Mobile 97	4-1	25	30	0.6
					Mobile 97	4-2	25	15	2.2
					Mobile 97	4-2	25	30	1.9
					Mobile 97	4-3	25	15	1.4
					Mobile 97	4-3	25	30	0.5
					Mobile 97	5-1	25	15	0.5
					Mobile 97	5-1	25	30	0.2

TABLE A2.27
VOC Emissions from Diesel Fuel Burns

iso-Propylbenzene					Methylcyclohexane				
C_9H_{12}		Concentration Prediction Equation			C_7H_{16}		Concentration Prediction Equation		
$y= a + b*(fire\ size, m^2) - c*ln(distance\ from\ fire, m)$					$y= a + b*(fire\ size, m^2) - c*ln(distance\ from\ fire, m)$				
	a, b, c	0.31	-8.3	4.7		a, b, c	27.9	0.0806	9.44
Burn	#	Area	Distance	Conc.	Burn	#	Area	Distance	Conc.
		m^2	m	$\mu g/m^3$			m^2	m	$\mu g/m^3$
Mobile 94	2	231	30	1.0	Mobile 94	1	199	30	4.3
Mobile 94	2	231	75	0.1	Mobile 94	1	199	45	0.3
Mobile 94	3	231	30	0.6	Mobile 94	1	199	75	0.2
Mobile 94	3	231	85	0.1	Mobile 94	2	231	50	10.6
Mobile 94	3	231	45	0.2	Mobile 94	2	231	85	0.5
Mobile 94	3	231	75	0.3	Mobile 94	2	231	30	51.5
Mobile 97	1-1	25	15	0.1	Mobile 94	2	231	45	9.4
Mobile 97	1-1	25	30	0.1	Mobile 94	2	231	75	5.0
Mobile 97	1-1	25	45	0.0	Mobile 94	3	231	30	21.5
Mobile 97	2-1	25	15	0.2	Mobile 94	3	231	50	0.6
Mobile 97	2-1	25	30	0.2	Mobile 94	3	231	85	3.5
Mobile 97	2-2	25	15	0.1	Mobile 94	3	231	45	2.9
Mobile 97	2-2	25	30	0.0	Mobile 94	3	231	75	0.5
Mobile 97	2-3	25	15	0.2	Mobile 97	1-1	25	15	0.7
Mobile 97	2-3	25	30	0.1	Mobile 97	1-1	25	30	0.6
Mobile 97	3-1	25	15	0.2	Mobile 97	1-1	25	45	0.5
Mobile 97	3-1	25	30	0.2	Mobile 97	2-1	25	15	1.0
Mobile 97	3a-1	25	15	0.2	Mobile 97	2-1	25	30	1.2
Mobile 97	3a-1	25	30	0.2	Mobile 97	2-2	25	15	0.5
Mobile 97	4-1	25	15	0.3	Mobile 97	2-2	25	30	0.2
Mobile 97	4-1	25	30	0.2	Mobile 97	2-3	25	15	0.5
Mobile 97	4-2	25	15	0.7	Mobile 97	2-3	25	30	0.4
Mobile 97	4-2	25	30	0.5	Mobile 97	3-1	25	15	0.4
Mobile 97	4-3	25	15	0.5	Mobile 97	3-1	25	30	0.4
Mobile 97	4-3	25	30	0.2	Mobile 97	3a-1	25	15	0.7
Mobile 97	5-1	25	15	0.2	Mobile 97	3a-1	25	30	0.7
Mobile 97	5-1	25	30	0.1	Mobile 97	4-1	25	15	1.1
					Mobile 97	4-1	25	30	0.7
					Mobile 97	4-2	25	15	3.1
					Mobile 97	4-2	25	30	1.6
					Mobile 97	4-3	25	15	2.7
					Mobile 97	4-3	25	30	1.1
					Mobile 97	5-1	25	15	1.4
					Mobile 97	5-1	25	30	0.4

TABLE A2.28
VOC Emissions from Diesel Fuel Burns

Methylcyclopentane

C_6H_{14} Concentration Prediction Equation

$y = a + b*(\text{fire size, m}^2) - c*\ln(\text{distance from fire, m})$

		a, b, c	5.21	0.0131	1.55

Burn	#	Area	Distance	Conc.
		m^2	m	$\mu g/m^3$
Mobile 94	1	199	30	0.2
Mobile 94	1	199	50	0.2
Mobile 94	1	199	30	0.7
Mobile 94	1	199	45	0.2
Mobile 94	1	199	75	0.2
Mobile 94	2	231	50	4.4
Mobile 94	2	231	85	0.6
Mobile 94	2	231	30	9.3
Mobile 94	2	231	45	4.2
Mobile 94	2	231	75	2.3
Mobile 94	3	231	30	4.1
Mobile 94	3	231	50	0.3
Mobile 94	3	231	85	0.6
Mobile 94	3	231	45	1.8
Mobile 94	3	231	75	0.2
Mobile 97	1-1	25	15	0.6
Mobile 97	1-1	25	30	1.1
Mobile 97	1-1	25	45	0.6
Mobile 97	2-1	25	15	0.4
Mobile 97	2-1	25	30	0.3
Mobile 97	2-2	25	15	0.4
Mobile 97	2-2	25	30	0.6
Mobile 97	2-3	25	15	0.5
Mobile 97	2-3	25	30	0.4
Mobile 97	3-1	25	15	0.8
Mobile 97	3-1	25	30	0.8
Mobile 97	3a-1	25	15	1.2
Mobile 97	4-1	25	15	1.4
Mobile 97	4-1	25	30	1.2
Mobile 97	4-2	25	15	0.8
Mobile 97	4-2	25	30	2.7
Mobile 97	4-3	25	30	0.3
Mobile 97	5-1	25	30	0.2

m,p-Xylene

C_8H_{10} Concentration Prediction Equation

$y = a + b*(\text{fire size, m}^2) - c*\ln(\text{distance from fire, m})$

		a, b, c	29.7	0.0458	8.13

Burn	#	Area	Distance	Conc.
		m^2	m	$\mu g/m^3$
Mobile 97	1-1	25	15	2.8
Mobile 97	1-1	25	30	2.2
Mobile 97	1-1	25	45	1.8
Mobile 97	2-1	25	15	4.7
Mobile 97	2-1	25	30	3.0
Mobile 97	2-2	25	15	2.9
Mobile 97	2-2	25	30	1.1
Mobile 97	2-3	25	15	4.4
Mobile 97	2-3	25	30	2.4
Mobile 97	3-1	25	15	5.0
Mobile 97	3-1	25	30	5.2
Mobile 97	3a-1	25	15	4.8
Mobile 97	3a-1	25	30	4.0
Mobile 97	4-1	25	15	8.7
Mobile 97	4-1	25	30	6.0
Mobile 97	4-2	25	15	16.3
Mobile 97	4-2	25	30	26.1
Mobile 97	4-3	25	15	11.7
Mobile 97	4-3	25	30	4.0
Mobile 97	5-1	25	15	4.1
Mobile 97	5-1	25	30	1.1
Mobile 94	1	199	30	3.8
Mobile 94	1	199	45	1.0
Mobile 94	1	199	75	1.4
Mobile 94	2	231	50	7.9
Mobile 94	2	231	85	1.0
Mobile 94	2	231	30	36.7
Mobile 94	2	231	45	7.1
Mobile 94	2	231	75	3.4
Mobile 94	3	231	30	20.9
Mobile 94	3	231	50	1.8
Mobile 94	3	231	85	3.5
Mobile 94	3	231	45	5.9

TABLE A2.29
VOC Emissions from Diesel Fuel Burns

Naphthalene					n-Butylbenzene				
$C_{10}H_8$			Concentration Prediction Equation		$C_{10}H_{14}$			Concentration Prediction Equation	
$y = a + b*(fire\ size, m^2) - c*ln(distance\ from\ fire, m)$					$y = a + b*(fire\ size, m^2) - c*ln(distance\ from\ fire, m)$				
	a, b, c	10.8	0.0146	3.05		a, b, c	1.63	0.00128	0.433
Burn	#	Area	Distance	Conc.	Burn	#	Area	Distance	Conc.
		m^2	m	$\mu g/m^3$			m^2	m	$\mu g/m^3$
Mobile 94	1	199	30	0.2	Mobile 94	2	231	30	0.4
Mobile 94	1	199	45	0.3	Mobile 94	3	231	75	0.1
Mobile 94	1	199	75	1.0	Mobile 97	1-1	25	15	0.1
Mobile 94	2	231	50	3.1	Mobile 97	1-1	25	30	0.1
Mobile 94	2	231	85	0.2	Mobile 97	1-1	25	45	0.1
Mobile 94	2	231	30	6.4	Mobile 97	2-1	25	15	0.3
Mobile 94	2	231	45	0.9	Mobile 97	2-1	25	30	0.4
Mobile 94	3	231	30	7.1	Mobile 97	2-2	25	15	0.1
Mobile 94	3	231	50	2.1	Mobile 97	2-2	25	30	0.1
Mobile 94	3	231	85	0.9	Mobile 97	2-3	25	15	0.3
Mobile 94	3	231	45	1.4	Mobile 97	2-3	25	30	0.2
Mobile 94	3	231	75	1.4	Mobile 97	3-1	25	15	0.1
Mobile 97	1-1	25	15	0.2	Mobile 97	3-1	25	30	0.1
Mobile 97	1-1	25	30	0.2	Mobile 97	3a-1	25	15	0.0
Mobile 97	1-1	25	45	0.2	Mobile 97	3a-1	25	30	0.1
Mobile 97	2-1	25	15	3.9	Mobile 97	4-1	25	15	1.9
Mobile 97	2-1	25	30	2.2	Mobile 97	4-1	25	30	0.1
Mobile 97	2-2	25	15	1.3	Mobile 97	4-2	25	15	1.2
Mobile 97	2-2	25	30	0.4	Mobile 97	4-2	25	30	0.5
Mobile 97	2-3	25	15	1.9	Mobile 97	4-3	25	15	0.8
Mobile 97	2-3	25	30	0.7	Mobile 97	4-3	25	30	0.1
Mobile 97	3-1	25	15	0.2	Mobile 97	5-1	25	15	0.3
Mobile 97	3-1	25	30	0.7	Mobile 97	5-1	25	30	0.1
Mobile 97	3a-1	25	15	0.1					
Mobile 97	3a-1	25	30	0.8					
Mobile 97	4-1	25	15	1.3					
Mobile 97	4-1	25	30	0.8					
Mobile 97	4-2	25	15	11.5					
Mobile 97	4-2	25	30	2.8					
Mobile 97	4-3	25	15	5.3					
Mobile 97	4-3	25	30	0.3					
Mobile 97	5-1	25	15	2.3					
Mobile 97	5-1	25	30	0.1					

TABLE A2.30
VOC Emissions from Diesel Fuel Burns

Nonane

C_9H_{20} — Concentration Prediction Equation

$y = a + b*(fire\ size, m^2) - c*ln(distance\ from\ fire, m)$

a, b, c	19.6	0.0284	5.68

Burn	#	Area m^2	Distance m	Conc. $\mu g/m^3$
Mobile 94	1	199	30	1.7
Mobile 94	1	199	45	0.2
Mobile 94	1	199	75	0.3
Mobile 94	2	231	50	3.9
Mobile 94	2	231	85	0.2
Mobile 94	2	231	30	18.2
Mobile 94	2	231	45	3.6
Mobile 94	2	231	75	1.7
Mobile 94	3	231	30	11.1
Mobile 94	3	231	50	0.6
Mobile 94	3	231	85	1.6
Mobile 94	3	231	45	1.6
Mobile 94	3	231	75	0.3
Mobile 97	1-1	25	15	1.2
Mobile 97	1-1	25	30	0.8
Mobile 97	1-1	25	45	0.6
Mobile 97	2-1	25	15	3.5
Mobile 97	2-1	25	30	1.8
Mobile 97	2-2	25	15	1.8
Mobile 97	2-2	25	30	0.7
Mobile 97	2-3	25	15	3.0
Mobile 97	2-3	25	30	1.6
Mobile 97	3-1	25	15	0.7
Mobile 97	3-1	25	30	1.5
Mobile 97	3a-1	25	15	0.3
Mobile 97	3a-1	25	30	0.7
Mobile 97	4-1	25	15	2.1
Mobile 97	4-1	25	30	1.1
Mobile 97	4-2	25	15	15.3
Mobile 97	4-2	25	30	5.3
Mobile 97	4-3	25	15	10.0
Mobile 97	4-3	25	30	2.9
Mobile 97	5-1	25	15	3.3
Mobile 97	5-1	25	30	1.1

n-Propylbenzene

C_9H_{12} — Concentration Prediction Equation

$y = a + b*(fire\ size, m^2) - c*ln(distance\ from\ fire, m)$

a, b, c	1.77	0.00178	0.435

Burn	#	Area m^2	Distance m	Conc. $\mu g/m^3$
Mobile 94	1	199	30	0.3
Mobile 94	2	231	30	1.7
Mobile 94	2	231	45	0.3
Mobile 94	2	231	75	0.2
Mobile 94	3	231	30	1.0
Mobile 94	3	231	50	0.0
Mobile 94	3	231	85	0.2
Mobile 94	3	231	45	0.4
Mobile 94	3	231	75	0.3
Mobile 97	1-1	25	15	0.1
Mobile 97	1-1	25	30	0.1
Mobile 97	1-1	25	45	0.1
Mobile 97	2-1	25	15	0.4
Mobile 97	2-1	25	30	0.4
Mobile 97	2-2	25	15	0.2
Mobile 97	2-2	25	30	0.1
Mobile 97	2-3	25	15	0.4
Mobile 97	2-3	25	30	0.2
Mobile 97	3-1	25	15	0.4
Mobile 97	3-1	25	30	0.4
Mobile 97	3a-1	25	15	0.2
Mobile 97	3a-1	25	30	0.6
Mobile 97	4-1	25	15	0.7
Mobile 97	4-1	25	30	0.5
Mobile 97	4-2	25	15	1.6
Mobile 97	4-2	25	30	1.6
Mobile 97	4-3	25	15	1.1
Mobile 97	4-3	25	30	0.3
Mobile 97	5-1	25	15	0.4
Mobile 97	5-1	25	30	0.1

TABLE A2.31
VOC Emissions from Diesel Fuel Burns

Octane					o-Xylene				

C_8H_{18} — Concentration Prediction Equation

$y = a + b*(fire\ size, m^2) - c*ln(distance\ from\ fire, m)$

C_8H_{10} — Concentration Prediction Equation

$y = a + b*(fire\ size, m^2) - c*ln(distance\ from\ fire, m)$

Burn	#	Area	Distance	Conc.	Burn	#	Area	Distance	Conc.
a, b, c		13.9	0.041	4.31	**a, b, c**		20.9	0.0356	6.3
		m^2	m	$\mu g/m^3$			m^2	m	$\mu g/m^3$
Mobile 94	1	199	30	2.3	Mobile 94	1	199	30	1.7
Mobile 94	1	199	45	0.2	Mobile 94	1	199	45	0.4
Mobile 94	1	199	75	0.2	Mobile 94	1	199	75	0.7
Mobile 94	2	231	50	5.3	Mobile 94	2	231	50	3.4
Mobile 94	2	231	85	16.3	Mobile 94	2	231	85	0.3
Mobile 94	2	231	30	27.5	Mobile 94	2	231	30	15.1
Mobile 94	2	231	45	4.9	Mobile 94	2	231	45	2.9
Mobile 94	2	231	75	2.5	Mobile 94	2	231	75	1.6
Mobile 94	3	231	30	13.5	Mobile 94	3	231	30	8.8
Mobile 94	3	231	50	0.6	Mobile 94	3	231	50	0.6
Mobile 94	3	231	85	2.1	Mobile 94	3	231	85	1.4
Mobile 94	3	231	45	1.9	Mobile 94	3	231	45	2.2
Mobile 94	3	231	75	0.2	Mobile 94	3	231	15	21.9
Mobile 97	1-1	25	15	1.2	Mobile 97	1-1	25	15	0.9
Mobile 97	1-1	25	30	1.0	Mobile 97	1-1	25	30	0.7
Mobile 97	1-1	25	45	0.8	Mobile 97	1-1	25	45	0.6
Mobile 97	2-1	25	15	2.0	Mobile 97	2-1	25	15	2.0
Mobile 97	2-1	25	30	2.0	Mobile 97	2-1	25	30	1.0
Mobile 97	2-2	25	15	0.9	Mobile 97	2-2	25	15	1.3
Mobile 97	2-2	25	30	0.5	Mobile 97	2-2	25	30	0.5
Mobile 97	2-3	25	15	1.2	Mobile 97	2-3	25	15	2.2
Mobile 97	2-3	25	30	0.7	Mobile 97	2-3	25	30	1.2
Mobile 97	3-1	25	15	0.5	Mobile 97	3-1	25	15	1.7
Mobile 97	3-1	25	30	1.3	Mobile 97	3-1	25	30	1.8
Mobile 97	3a-1	25	15	0.5	Mobile 97	3a-1	25	15	2.2
Mobile 97	3a-1	25	30	0.9	Mobile 97	3a-1	25	30	2.7
Mobile 97	4-1	25	15	1.8	Mobile 97	4-1	25	15	3.3
Mobile 97	4-1	25	30	1.0	Mobile 97	4-1	25	30	2.0
Mobile 97	4-2	25	15	7.4	Mobile 97	4-2	25	15	7.7
Mobile 97	4-2	25	30	3.2	Mobile 97	4-2	25	30	8.0
Mobile 97	4-3	25	15	4.7	Mobile 97	4-3	25	15	5.3
Mobile 97	4-3	25	30	2.0	Mobile 97	4-3	25	30	2.1
Mobile 97	5-1	25	15	2.1	Mobile 97	5-1	25	15	2.1
Mobile 97	5-1	25	30	0.7	Mobile 97	5-1	25	30	0.6

TABLE A2.32
VOC Emissions from Diesel Fuel Burns

p-Cymene

$C_{10}H_{14}$

Concentration Prediction Equation

$y = a + b*(fire\ size, m^2) - c*ln(distance\ from\ fire, m)$

a, b, c	1.02	0.000282	0.275

Burn	#	Area m^2	Distance m	Conc. $\mu g/m^3$
Mobile 94	1	199	30	0.2
Mobile 94	1	199	45	0.1
Mobile 94	1	199	75	0.1
Mobile 94	2	231	50	1.4
Mobile 94	2	231	85	0.1
Mobile 94	2	231	30	1.0
Mobile 94	2	231	45	0.2
Mobile 94	3	231	50	0.3
Mobile 94	3	231	45	1.4
Mobile 94	3	231	75	0.6
Mobile 97	1-1	25	15	0.1
Mobile 97	1-1	25	30	0.1
Mobile 97	1-1	25	45	0.1
Mobile 97	2-1	25	15	0.2
Mobile 97	2-1	25	30	0.2
Mobile 97	2-2	25	15	0.1
Mobile 97	2-3	25	15	0.2
Mobile 97	2-3	25	30	0.1
Mobile 97	3-1	25	15	0.1
Mobile 97	3-1	25	30	0.1
Mobile 97	3a-1	25	15	0.2
Mobile 97	3a-1	25	30	0.1
Mobile 97	4-1	25	30	0.2
Mobile 97	4-2	25	15	1.1
Mobile 97	4-2	25	30	0.5
Mobile 97	4-3	25	15	0.7
Mobile 97	4-3	25	30	0.3
Mobile 97	5-1	25	15	0.2

Pentane

C_5H_{12}

Concentration Prediction Equation

$y = a + b*(fire\ size, m^2) - c*ln(distance\ from\ fire, m)$

a, b, c	29.2	0.0587	9.1

Burn	#	Area m^2	Distance m	Conc. $\mu g/m^3$
Mobile 94	1	199	30	2.1
Mobile 94	1	199	45	1.5
Mobile 94	1	199	75	1.4
Mobile 94	2	231	50	3.5
Mobile 94	2	231	85	1.0
Mobile 94	2	231	30	11.9
Mobile 94	2	231	45	2.8
Mobile 94	2	231	75	1.8
Mobile 94	3	231	30	6.3
Mobile 94	3	231	50	2.2
Mobile 94	3	231	85	2.6
Mobile 94	3	231	15	33.2
Mobile 94	3	231	75	1.3
Mobile 97	1-1	25	15	1.9
Mobile 97	1-1	25	30	2.2
Mobile 97	1-1	25	45	1.9
Mobile 97	2-1	25	15	1.1
Mobile 97	2-1	25	30	1.0
Mobile 97	2-2	25	15	0.6
Mobile 97	2-2	25	30	0.6
Mobile 97	2-3	25	15	1.2
Mobile 97	2-3	25	30	0.7
Mobile 97	3-1	25	15	4.9
Mobile 97	3-1	25	30	5.4
Mobile 97	3a-1	25	15	7.5
Mobile 97	3a-1	25	30	7.2
Mobile 97	4-1	25	15	7.0
Mobile 97	4-1	25	30	6.7
Mobile 97	4-2	25	15	3.0
Mobile 97	4-2	25	30	32.5
Mobile 97	4-3	25	15	1.1
Mobile 97	4-3	25	30	1.8
Mobile 97	5-1	25	15	0.9

TABLE A2.33
VOC Emissions from Diesel Fuel Burns

Propane

C_3H_8 — Concentration Prediction Equation

$y = a + b*(fire\ size, m^2) - c*ln(distance\ from\ fire, m)$

a, b, c: 19.5 | 0.002 | 4.5

Burn	#	Area (m²)	Distance (m)	Conc. (µg/m³)
Mobile 94	1	199	30	11.3
Mobile 94	1	199	30	13.2
Mobile 94	1	199	45	4.5
Mobile 94	1	199	75	2.5
Mobile 94	2	231	50	3.9
Mobile 94	2	231	85	3.0
Mobile 94	2	231	30	14.9
Mobile 94	2	231	45	3.5
Mobile 94	2	231	75	4.5
Mobile 94	3	231	30	2.9
Mobile 94	3	231	50	5.9
Mobile 94	3	231	85	2.8
Mobile 94	3	231	15	17.9
Mobile 94	3	231	75	2.8
Mobile 97	1-1	25	15	6.6
Mobile 97	1-1	25	30	6.4
Mobile 97	1-1	25	45	4.9
Mobile 97	2-1	25	15	3.3
Mobile 97	2-1	25	30	3.4
Mobile 97	2-2	25	15	2.2
Mobile 97	2-2	25	30	2.3
Mobile 97	2-3	25	15	2.7
Mobile 97	2-3	25	30	2.4
Mobile 97	3-1	25	15	7.3
Mobile 97	3-1	25	30	7.9
Mobile 97	3a-1	25	15	11.4
Mobile 97	3a-1	25	30	10.0
Mobile 97	4-1	25	15	12.2
Mobile 97	4-1	25	30	11.9
Mobile 97	4-2	25	15	7.6
Mobile 97	4-2	25	30	9.6
Mobile 97	4-3	25	15	4.2
Mobile 97	4-3	25	30	6.5
Mobile 97	5-1	25	15	2.3
Mobile 97	5-1	25	30	1.1

Propene

C_3H_6 — Concentration Prediction Equation

$y = a + b*(fire\ size, m^2) - c*ln(distance\ from\ fire, m)$

a, b, c: 10.2 | 0.0436 | 3.25

Burn	#	Area (m²)	Distance (m)	Conc. (µg/m³)
Mobile 94	1	199	30	2.9
Mobile 94	1	199	50	1.5
Mobile 94	1	199	30	4.3
Mobile 94	1	199	45	2.0
Mobile 94	1	199	75	1.0
Mobile 94	2	231	50	8.0
Mobile 94	2	231	85	9.3
Mobile 94	2	231	30	21.1
Mobile 94	2	231	45	16.3
Mobile 94	2	231	75	8.1
Mobile 94	3	231	30	14.0
Mobile 94	3	231	50	6.6
Mobile 94	3	231	85	3.0
Mobile 94	3	231	45	2.2
Mobile 94	3	231	75	4.0
Mobile 97	1-1	25	15	0.5
Mobile 97	1-1	25	30	0.5
Mobile 97	1-1	25	45	0.5
Mobile 97	2-1	25	15	2.9
Mobile 97	2-1	25	30	2.4
Mobile 97	2-2	25	15	1.3
Mobile 97	2-2	25	30	0.9
Mobile 97	2-3	25	15	1.0
Mobile 97	2-3	25	30	1.0
Mobile 97	3-1	25	15	1.4
Mobile 97	3-1	25	30	1.3
Mobile 97	3a-1	25	15	1.9
Mobile 97	3a-1	25	30	2.5
Mobile 97	4-1	25	15	2.8
Mobile 97	4-1	25	30	2.8
Mobile 97	4-2	25	15	1.9
Mobile 97	4-2	25	30	1.5
Mobile 97	4-3	25	15	1.1
Mobile 97	4-3	25	30	0.9
Mobile 97	5-1	25	15	1.5
Mobile 97	5-1	25	30	0.5

TABLE A2.34
VOC Emissions from Diesel Fuel Burns

Propyne					sec-Butylbenzene				
C_3H_4			Concentration Prediction Equation		$C_{10}H_{14}$			Concentration Prediction Equation	
$y = a + b*(fire\ size, m^2) - c*ln(distance\ from\ fire, m)$					$y = a + b*(fire\ size, m^2) - c*ln(distance\ from\ fire, m)$				
	a, b, c	0.874	0.00155	0.236		a, b, c	0.882	0.00158	0.247
Burn	**#**	**Area**	**Distance**	**Conc.**	**Burn**	**#**	**Area**	**Distance**	**Conc.**
		m^2	m	$\mu g/m^3$			m^2	m	$\mu g/m^3$
Mobile 97	2-1	25	15	0.3	Mobile 94	2	231	85	0.0
Mobile 97	2-1	25	30	0.2	Mobile 94	3	231	30	0.3
Mobile 97	2-2	25	15	0.2	Mobile 97	1-1	25	15	0.0
Mobile 97	2-2	25	30	0.1	Mobile 97	2-1	25	15	0.2
Mobile 97	2-3	25	15	0.2	Mobile 97	2-1	25	30	0.2
Mobile 97	2-3	25	30	0.1	Mobile 97	2-2	25	15	0.1
Mobile 97	3-1	25	15	0.2	Mobile 97	2-3	25	15	0.2
Mobile 97	3-1	25	30	0.2	Mobile 97	2-3	25	30	0.1
Mobile 97	3a-1	25	15	0.2	Mobile 97	3-1	25	15	0.1
Mobile 97	3a-1	25	30	0.3	Mobile 97	3-1	25	30	0.1
Mobile 97	4-1	25	15	0.2	Mobile 97	3a-1	25	15	0.1
Mobile 97	4-1	25	30	0.2	Mobile 97	3a-1	25	30	0.1
Mobile 97	4-2	25	15	0.4	Mobile 97	4-1	25	15	0.1
Mobile 97	4-2	25	30	0.2	Mobile 97	4-1	25	30	0.1
Mobile 97	4-3	25	15	0.2	Mobile 97	4-2	25	15	0.9
Mobile 97	4-3	25	30	0.1	Mobile 97	4-2	25	30	0.4
Mobile 97	5-1	25	15	0.2	Mobile 97	4-3	25	15	0.6
Mobile 97	5-1	25	30	0.1	Mobile 97	4-3	25	30	0.2
Mobile 97	5-1	25	15	0.8	Mobile 97	5-1	25	15	0.2
Mobile 97	5-1	25	30	0.4	Mobile 97	5-1	25	30	0.0
Mobile 97	5-1	25	85	0.0					

TABLE A2.35
VOC Emissions from Diesel Fuel Burns

Styrene C_8H_8				t (trans) -1,2-Dimethylcyclohexane C_8H_{18}			
		Concentration	Prediction Equation			Concentration	Prediction Equation
$y= a + b*(fire\ size, m^2) - c*ln(distance\ from\ fire, m)$				$y= a + b*(fire\ size, m^2) - c*ln(distance\ from\ fire, m)$			
a, b, c	3.96	0.021	1.37	a, b, c	2.86	0.0111	0.933
Burn	**#**	**Area**	**Distance**	**Conc.**	**Burn**	**#**	**Area**	**Distance**	**Conc.**

Burn	#	Area (m^2)	Distance (m)	Conc. ($\mu g/m^3$)	Burn	#	Area (m^2)	Distance (m)	Conc. ($\mu g/m^3$)
Mobile 94	1	199	45	0.6	Mobile 94	1	199	30	0.5
Mobile 94	1	199	75	0.3	Mobile 94	2	231	50	1.3
Mobile 94	2	231	30	5.2	Mobile 94	2	231	30	6.6
Mobile 94	2	231	45	0.3	Mobile 94	2	231	45	1.2
Mobile 94	3	231	30	7.4	Mobile 94	2	231	75	0.7
Mobile 94	3	231	50	2.1	Mobile 94	3	231	30	2.8
Mobile 94	3	231	45	7.6	Mobile 94	3	231	85	0.5
Mobile 97	1-1	25	15	0.1	Mobile 94	3	231	45	0.4
Mobile 97	1-1	25	30	0.1	Mobile 97	1-1	25	15	0.2
Mobile 97	1-1	25	45	0.1	Mobile 97	1-1	25	30	0.2
Mobile 97	2-1	25	15	0.7	Mobile 97	1-1	25	45	0.1
Mobile 97	2-1	25	30	0.4	Mobile 97	2-1	25	15	0.4
Mobile 97	2-2	25	15	0.2	Mobile 97	2-1	25	30	0.3
Mobile 97	2-2	25	30	0.1	Mobile 97	2-2	25	15	0.2
Mobile 97	2-3	25	15	0.2	Mobile 97	2-2	25	30	0.2
Mobile 97	2-3	25	30	0.2	Mobile 97	2-3	25	15	0.2
Mobile 97	3-1	25	15	0.1	Mobile 97	2-3	25	30	0.2
Mobile 97	3-1	25	30	0.3	Mobile 97	3-1	25	15	0.1
Mobile 97	3a-1	25	15	0.1	Mobile 97	3-1	25	30	0.1
Mobile 97	3a-1	25	30	0.5	Mobile 97	3a-1	25	15	0.1
Mobile 97	4-1	25	15	0.4	Mobile 97	3a-1	25	30	0.1
Mobile 97	4-1	25	30	0.3	Mobile 97	4-1	25	15	0.4
Mobile 97	4-2	25	15	1.2	Mobile 97	4-1	25	30	0.2
Mobile 97	4-2	25	30	0.4	Mobile 97	4-2	25	15	1.5
					Mobile 97	4-2	25	30	0.6
					Mobile 97	4-3	25	15	1.0
					Mobile 97	4-3	25	30	0.5
					Mobile 97	5-1	25	15	0.5
					Mobile 97	5-1	25	30	0.2

TABLE A2.36
VOC Emissions from Diesel Fuel Burns

t *(trans)*-1,4-Dimethylcyclohexane					t *(trans)*-2-Butene				
C_8H_{18}		Concentration Prediction Equation			C_4H_8		Concentration Prediction Equation		
$y = a + b*(fire\ size, m^2) - c*ln(distance\ from\ fire, m)$					$y = a + b*(fire\ size, m^2) - c*ln(distance\ from\ fire, m)$				
	a, b, c	1.96	0.011	-0.67		a, b, c	0.898	0.00256	0.281
Burn	**#**	**Area**	**Distance**	**Conc.**	**Burn**	**#**	**Area**	**Distance**	**Conc.**
		m^2	m	$\mu g/m^3$			m^2	m	$\mu g/m^3$
Mobile 94	1	199	30	0.6	Mobile 94	1	199	30	0.2
Mobile 94	2	231	50	1.3	Mobile 94	1	199	50	0.1
Mobile 94	2	231	30	6.1	Mobile 94	1	199	30	0.3
Mobile 94	2	231	45	1.3	Mobile 94	1	199	45	0.2
Mobile 94	3	231	30	2.6	Mobile 94	2	231	15	1.2
Mobile 94	3	231	45	0.5	Mobile 94	2	231	30	0.5
Mobile 97	1-1	25	15	0.1	Mobile 94	2	231	45	0.6
Mobile 97	1-1	25	30	0.1	Mobile 94	3	231	30	0.9
Mobile 97	1-1	25	45	0.1	Mobile 94	3	231	50	0.3
Mobile 97	2-1	25	15	0.2	Mobile 97	2-1	25	15	0.1
Mobile 97	2-2	25	15	0.1	Mobile 97	2-1	25	30	0.1
Mobile 97	2-2	25	30	0.0	Mobile 97	2-2	25	15	0.1
Mobile 97	2-3	25	15	0.1	Mobile 97	2-3	25	30	0.1
Mobile 97	2-3	25	30	0.1	Mobile 97	3-1	25	15	0.1
Mobile 97	3a-1	25	15	0.1	Mobile 97	3-1	25	30	0.1
Mobile 97	3a-1	25	30	0.1	Mobile 97	3a-1	25	15	0.1
Mobile 97	4-1	25	15	0.2	Mobile 97	3a-1	25	30	0.2
Mobile 97	4-1	25	30	0.1	Mobile 97	4-1	25	15	0.1
Mobile 97	4-2	25	15	0.7	Mobile 97	4-1	25	30	0.1
Mobile 97	4-2	25	30	0.3	Mobile 97	4-2	25	15	0.1
Mobile 97	4-3	25	15	0.5	Mobile 97	4-2	25	30	0.3
Mobile 97	4-3	25	30	0.2	Mobile 97	4-3	25	15	0.1
Mobile 97	5-1	25	15	0.3	Mobile 97	4-3	25	30	0.1
Mobile 97	5-1	25	30	0.1	Mobile 97	5-1	25	15	0.1

TABLE A2.37
VOC Emissions from Diesel Fuel Burns

t *(trans)* -2-Heptene

C_7H_{14} Concentration Prediction Equation

$y = a + b*(fire\ size, m^2) - c*ln(distance\ from\ fire, m)$

a, b, c	1.89	0.00392	0.553

Burn	#	Area	Distance	Conc.
		m^2	m	µg/m^3
Mobile 94	1	199	30	0.6
Mobile 94	1	199	50	0.3
Mobile 94	1	199	45	0.2
Mobile 94	2	231	30	0.2
Mobile 94	3	231	15	2.2
Mobile 94	3	231	45	1.1
Mobile 94	3	25	15	0.2
Mobile 94	3	25	85	0.001

t *(trans)* -2-Hexene

C_7H_{14} Concentration Prediction Equation

$y = a + b*(fire\ size, m^2) - c*ln(distance\ from\ fire, m)$

a, b, c	0.377	0.032	1.53

Burn	#	Area	Distance	Conc.
		m^2	m	µg/m^3
Mobile 94	1	199	30	0.6
Mobile 94	1	199	50	0.2
Mobile 94	1	199	30	0.6
Mobile 94	1	199	45	0.3
Mobile 94	2	231	85	0.6
Mobile 94	2	231	45	0.4
Mobile 94	3	231	30	2.8
Mobile 94	3	231	50	1.0
Mobile 94	3	231	45	0.5

t *(trans)* -2-Octene

C_8H_{16} Concentration Prediction Equation

$y = a + b*(fire\ size, m^2) - c*ln(distance\ from\ fire, m)$

a, b, c	4.58	0.112	4.67

Burn	#	Area	Distance	Conc.
		m^2	m	µg/m^3
Mobile 94	1	199	30	0.6
Mobile 94	1	199	50	0.2
Mobile 94	1	199	30	2.4
Mobile 94	1	199	45	0.2
Mobile 94	2	231	50	1.6
Mobile 94	2	231	30	8.8
Mobile 94	2	231	45	2.4
Mobile 94	2	231	75	2.6
Mobile 94	3	231	50	1.2
Mobile 94	3	231	85	1.5
Mobile 94	3	231	45	2.6

t *(trans)* -3-Heptene

C_7H_{14} Concentration Prediction Equation

$y = a + b*(fire\ size, m^2) - c*ln(distance\ from\ fire, m)$

a, b, c	85.4	0.0688	25.7

Burn	#	Area	Distance	Conc.
		m^2	m	µg/m^3
Mobile 94	1	199	30	9.2
Mobile 94	2	231	50	1.2
Mobile 94	2	231	30	5.2
Mobile 94	3	231	30	24.7
Mobile 94	3	231	50	0.4
Mobile 97	2-1	25	30	0.1

TABLE A2.38
VOC Emissions from Diesel Fuel Burns

t (trans)-2-Pentene

C_5H_{10}

Concentration Prediction Equation

$y = a + b*(fire\ size, m^2) - c*ln(distance\ from\ fire, m)$

	a, b, c	2.24	0.00797	0.677
Burn	**#**	**Area**	**Distance**	**Conc.**
		m^2	m	$\mu g/m^3$
Mobile 97	2-1	25	30	1.1
Mobile 97	2-3	25	15	0.1
Mobile 97	3-1	25	15	0.2
Mobile 97	3-1	25	30	0.1
Mobile 97	3a-1	25	15	0.3
Mobile 97	3a-1	25	30	0.3
Mobile 97	4-1	25	15	0.2
Mobile 97	4-1	25	30	0.2
Mobile 97	4-2	25	15	0.1
Mobile 97	4-2	25	30	1.7
Mobile 97	4-3	25	30	0.1
Mobile 97	5-1	25	15	0.1
Mobile 97	5-1	25	85	0.01
Mobile 97	5-1	25	15	3.3
Mobile 97	5-1	25	15	2.0

tert-Butylbenzene

$C_{10}H_{14}$

Concentration Prediction Equation

$y = a + b*(fire\ size, m^2) - c*ln(distance\ from\ fire, m)$

	a, b, c	1.37	0.0026	0.411
Burn	**#**	**Area**	**Distance**	**Conc.**
		m^2	m	$\mu g/m^3$
Mobile 97	2-1	25	15	0.1
Mobile 97	2-3	25	45	0.0
Mobile 97	4-2	25	15	0.5
Mobile 97	4-2	25	30	0.2
Mobile 97	4-3	25	15	0.3
Mobile 97	5-1	25	15	0.1
Mobile 97	5-1	25	15	1.0
Mobile 97	5-1	25	85	0.002

Toluene

C_7H_8

Concentration Prediction Equation

$y = a + b*(fire\ size, m^2) - c*ln(distance\ from\ fire, m)$

	a, b, c	34.6	0.0696	8.94
Burn	**#**	**Area**	**Distance**	**Conc.**
		m^2	m	$\mu g/m^3$
Mobile 94	1	199	30	5.7
Mobile 94	1	199	45	2.0
Mobile 94	1	199	75	2.8
Mobile 94	2	231	50	15.3
Mobile 94	2	231	85	3.0
Mobile 94	2	231	30	50.6
Mobile 94	2	231	45	13.4
Mobile 94	2	231	75	7.1
Mobile 94	3	231	30	27.1
Mobile 94	3	231	50	4.2
Mobile 94	3	231	85	5.5
Mobile 94	3	231	45	45.2
Mobile 94	3	231	75	7.3
Mobile 97	1-1	25	15	4.5
Mobile 97	1-1	25	30	4.6
Mobile 97	1-1	25	45	3.1
Mobile 97	2-1	25	15	4.5
Mobile 97	2-1	25	30	3.0
Mobile 97	2-2	25	15	3.0
Mobile 97	2-2	25	30	1.9
Mobile 97	2-3	25	15	5.0
Mobile 97	2-3	25	30	2.9
Mobile 97	3-1	25	15	9.6
Mobile 97	3-1	25	30	9.7
Mobile 97	3a-1	25	15	13.7
Mobile 97	3a-1	25	30	10.0
Mobile 97	4-1	25	15	14.4
Mobile 97	4-1	25	30	15.1
Mobile 97	4-2	25	15	12.3
Mobile 97	4-2	25	30	55.2
Mobile 97	4-3	25	15	7.2
Mobile 97	4-3	25	30	3.8
Mobile 97	5-1	25	15	3.9
Mobile 97	5-1	25	30	1.7

TABLE A2.39
VOC Emissions from Diesel Fuel Burns

Total VOCs

Total VOCs — Concentration Prediction Equation

$y = a + b*(fire\ size, m^2) - c*ln(distance\ from\ fire, m)$

a, b, c	570	1.06	163

Burn	#	Area (m^2)	Distance (m)	Conc. ($\mu g/m^3$)
Mobile 94	1	199	30	64.3
Mobile 94	1	199	45	38.1
Mobile 94	1	199	85	29.9
Mobile 94	1	199	30	132.2
Mobile 94	1	199	45	44.3
Mobile 94	1	199	75	34.6
Mobile 94	2	231	50	174.8
Mobile 94	2	231	85	73.2
Mobile 94	2	231	30	650.0
Mobile 94	2	231	45	172.5
Mobile 94	2	231	75	85.1
Mobile 94	3	231	30	433.0
Mobile 94	3	231	50	107.4
Mobile 94	3	231	85	67.3
Mobile 94	3	231	45	290.2
Mobile 94	3	231	75	98.4
Mobile 97	1-1	25	15	50.1
Mobile 97	1-1	25	30	49.8
Mobile 97	1-1	25	45	37.7
Mobile 97	2-1	25	15	133.0
Mobile 97	2-1	25	30	107.0
Mobile 97	2-2	25	15	60.9
Mobile 97	2-2	25	30	31.1
Mobile 97	2-3	25	15	130.1
Mobile 97	2-3	25	30	67.2
Mobile 97	3-1	25	15	86.2
Mobile 97	3-1	25	30	97.4
Mobile 97	3a-1	25	15	119.9
Mobile 97	3a-1	25	30	115.5
Mobile 97	4-1	25	15	143.0
Mobile 97	4-1	25	30	115.9
Mobile 97	4-2	25	15	140.3
Mobile 97	4-2	25	30	115.7
Mobile 97	4-3	25	15	254.4
Mobile 97	4-3	25	30	65.7
Mobile 97	5-1	25	15	112.8
Mobile 97	5-1	25	30	21.9

Undecane

$C_{11}H_{24}$ — Concentration Prediction Equation

$y = a + b*(fire\ size, m^2) - c*ln(distance\ from\ fire, m)$

a, b, c	48.8	0.0395	13.8

Burn	#	Area (m^2)	Distance (m)	Conc. ($\mu g/m^3$)
Mobile 94	1	199	30	1.0
Mobile 94	1	199	50	0.5
Mobile 94	1	199	30	1.8
Mobile 94	1	199	45	0.8
Mobile 94	1	199	75	0.4
Mobile 94	2	231	50	1.0
Mobile 94	2	231	85	0.2
Mobile 94	2	231	30	4.1
Mobile 94	2	231	45	1.1
Mobile 94	2	231	75	0.5
Mobile 94	3	231	30	5.8
Mobile 94	3	231	50	0.9
Mobile 94	3	231	85	0.8
Mobile 94	3	231	45	2.7
Mobile 94	3	231	75	1.6
Mobile 97	1-1	25	15	1.7
Mobile 97	1-1	25	30	1.2
Mobile 97	1-1	25	45	0.7
Mobile 97	2-1	25	15	15.7
Mobile 97	2-1	25	30	5.4
Mobile 97	2-2	25	15	6.1
Mobile 97	2-2	25	30	2.0
Mobile 97	2-3	25	15	19.6
Mobile 97	2-3	25	30	8.8
Mobile 97	3-1	25	15	0.9
Mobile 97	3-1	25	30	1.4
Mobile 97	3a-1	25	15	0.1
Mobile 97	3a-1	25	30	0.5
Mobile 97	4-1	25	15	4.0
Mobile 97	4-1	25	30	1.1
Mobile 97	4-2	25	15	54.6
Mobile 97	4-2	25	30	22.1
Mobile 97	4-3	25	15	41.7
Mobile 97	4-3	25	30	3.2
Mobile 97	5-1	25	15	15.7
Mobile 97	5-1	25	30	1.6

TABLE A3.1
Carbonyl Emissions from Crude Oil Burns

Acetaldehyde

C_2H_4O

Concentration Prediction Equation

$y = a + b*(fire\ size, m^2) - c*\ln(distance\ from\ fire, m)\ \mu g/m^3$

Crude oil	a, b, c		27.1	0.106	12.6

Fuel	Burn	#	Area	Distance	Conc.
			m^2	m	$\mu g/m^3$
AB Crude	Nobe 93	1	600	75	12.8
AB Crude	Nobe 93	1	600	125	60.3
AB Crude	Nobe 93	1	600	75	12.8
AB Crude	Nobe 93	1	600	125	60.3
AB Crude	Nobe 93	1	600	125	31.7
LA Crude	Mobile 92	1	110	15	1.7
LA Crude	Mobile 92	1	231	30	5
AB Crude	Nobe 93	1	600	600	0.001
AB Crude	Nobe 93	1	600	900	0.001
AB Crude	Nobe 93	2	25	15	0.3

Formaldehyde

CH_2O

Concentration Prediction Equation

$y = a + b*(fire\ size, m^2) - c*\ln(distance\ from\ fire, m)\ \mu g/m^3$

Crude oil	a, b, c		58.4	0.103	20.1

Fuel	Burn	#	Area	Distance	Conc.
			m^2	m	$\mu g/m^3$
LA Crude	Mobile 92	1	36	15	2
AB Crude	Nobe 93	1	600	75	40.8
AB Crude	Nobe 93	1	600	125	26.8
AB Crude	Nobe 93	2	427	75	32
AB Crude	Nobe 93	2	427	125	7.6
AB Crude	Nobe 93	1	600	125	0.01

Acetone

C_3H_6O

Concentration Prediction Equation

$y = a + b*(fire\ size, m^2) - c*\ln(distance\ from\ fire, m)\ \mu g/m^3$

Crude oil	a, b, c		20.9	0.024	5.14

Fuel	Burn	#	Area	Distance	Conc.
			m^2	m	$\mu g/m^3$
AB Crude	Nobe 93	1	400	75	31.7
AB Crude	Nobe 93	1	400	125	1.7
AB Crude	Nobe 93	1	900	75	31.7
AB Crude	Nobe 93	1	900	125	1.7
AB Crude	Mobile 92	1	110	15	1.7
LA Crude	Mobile 92	1	231	30	5
AB Crude	Nobe 93	1	600	600	0.001
AB Crude	Nobe 93	1	600	900	0.001

TABLE A3.2
Carbonyl Emissions from Crude Oil Burns

Acetaldehyde

C_2H_4O

Concentration Prediction Equation

$y = a + b*(fire\ size, m^2) - c*ln(distance\ from\ fire, m)$

	a, b, c	41.1	0.181	12.5
Burn	**#**	**Area**	**Distance**	**Conc.**
		m^2	m	$\mu g/m^3$
Mobile 94	1	231	30	53.4
Mobile 94	1	231	50	42.6
Mobile 94	1	231	85	36.2
Mobile 94	2	231	30	31.6
Mobile 94	2	231	50	24.0
Mobile 94	2	231	85	21.5
Mobile 94	3	231	30	39.4
Mobile 94	3	231	50	31.9
Mobile 94	3	231	85	25.2
Mobile 97	3	25	15	10.0
Mobile 97	3	25	30	5.0

Acetone

C_3H_6O

Concentration Prediction Equation

$y = a + b*(fire\ size, m^2) - c*ln(distance\ from\ fire, m)$

	a, b, c	14.7	0.057	3.84
Burn	**#**	**Area**	**Distance**	**Conc.**
		m^2	m	$\mu g/m^3$
Mobile 94	1	231	30	17.7
Mobile 94	1	231	50	15.9
Mobile 94	1	231	85	12.1
Mobile 94	2	231	30	11.7
Mobile 94	2	231	50	6
Mobile 94	2	231	85	5.4
Mobile 94	3	231	30	19.5
Mobile 94	3	231	50	15.1
Mobile 94	3	231	85	12.1
Mobile 97	3	25	15	4
Mobile 97	3	25	30	2
Mobile 97	3	25	75	1
Mobile 97	3	25	85	0.2

Butyraldehydes

C_4H_8O

Concentration Prediction Equation

$y = a + b*(fire\ size, m^2) - c*ln(distance\ from\ fire, m)$

	a, b, c	22.5	0.034	5.67
Burn	**#**	**Area**	**Distance**	**Conc.**
		m^2	m	$\mu g/m^3$
Mobile 94	1	231	30	14
Mobile 94	1	231	50	12
Mobile 94	1	231	85	4
Mobile 94	2	231	30	9.9
Mobile 94	2	231	50	8

Butyraldehydes contd.

Burn	**#**	**Area**	**Distance**	**Conc.**
		m^2	m	$\mu g/m^3$
Mobile 94	2	231	85	4
Mobile 94	3	231	30	13.3
Mobile 94	3	231	50	8.6
Mobile 94	3	231	85	0.01
Mobile 97	3	25	15	5
Mobile 97	3	25	30	3
Mobile 97	3	25	75	1
Mobile 97	3	25	85	0.01

Formaldehyde

CH_2O

Concentration Prediction Equation

$y = a + b*(fire\ size, m^2) - c*ln(distance\ from\ fire, m)$

	a, b, c	35.4	0.107	9.17
Burn	**#**	**Area**	**Distance**	**Conc.**
		m^2	m	$\mu g/m^3$
Mobile 94	1	231	30	29.9
Mobile 94	1	231	50	17.5
Mobile 94	1	231	85	24
Mobile 94	2	231	30	44
Mobile 94	2	231	50	18
Mobile 94	2	231	85	16.5
Mobile 94	3	231	30	31.6
Mobile 94	3	231	50	21.5
Mobile 94	3	231	85	15.6
Mobile 97	3	25	15	8
Mobile 97	3	25	30	5
Mobile 97	3	25	75	2
Mobile 97	3	25	85	1

Proprionaldehyde

C_3H_6O

Concentration Prediction Equation

$y = a + b*(fire\ size, m^2) - c*ln(distance\ from\ fire, m)$

	a, b, c	19.6	0.037	4.85
Burn	**#**	**Area**	**Distance**	**Conc.**
		m^2	m	$\mu g/m^3$
Mobile 94	1	231	30	15.3
Mobile 94	1	231	50	12.7
Mobile 94	1	231	85	9.5
Mobile 94	2	231	30	9
Mobile 94	2	231	50	5.7
Mobile 94	2	231	85	2
Mobile 94	3	231	30	12.8
Mobile 94	3	231	50	11.7
Mobile 94	3	231	85	4
Mobile 97	3	25	15	6
Mobile 97	3	25	30	3
Mobile 97	3	25	75	1
Mobile 97	3	25	85	0.1

Index